Medicinal Chemistry Handbook

Medicinal Chemistry Handbook

Edited by **Thomas Haldane**

NY RESEARCH PRESS

New York

Published by NY Research Press,
23 West, 55th Street, Suite 816,
New York, NY 10019, USA
www.nyresearchpress.com

Medicinal Chemistry Handbook
Edited by Thomas Haldane

International Standard Book Number: 978-1-63238-319-8 (Hardback)

Printed in China.

Contents

Preface

Medical chemistry is a multidisciplinary field of science which combines analytical chemistry, organic chemistry, pharmacology, biological science, and botanical sciences. This branch involves chemical aspects related to identification and alteration of new chemical compounds to create a product which is suitable for therapeutic use. This field contributes significantly to the development of pharmaceutical drugs for market use. Essentially, medicinal chemistry is an interdisciplinary science which requires a strong background in organic chemistry along with a broad understanding of biological concepts. It is an essential branch as the need to develop and create drugs and medicines is never-ending.

Most of the compounds used for medical applications are organic compounds, which are further categorized into sub-classes like organic molecules and biologics, and medicinal proteins. Inorganic and organic metallic compounds are used as drugs. Drug design is a process of discovering new medications based on biological treatments. A drug is mainly a small organic molecule which is responsible for activation of body functions. Drug design involves designs of smaller molecules which are complementary in shape and charge to bio-molecular targets, with which they have to ultimately interact.

I would like to acknowledge the authors for their contribution to the book. I thank all the authors for allocating much of their scarce time to this project. I also wish to acknowledge the team at the publishing house, who helped at every stage of the publication process. Each chapter is a sole-standing publication that reflects the author´s unique interpretation of the data. However, the unifying theme of all chapters is medicinal chemistry. I hope that the book will enhance the knowledge of scientists in the complexities of medicinal approaches. I also hope that this book will stimulate both professionals and students to dedicate a part of their future research to understand relevant mechanisms and applications. Lastly, I wish to thank my family. I would not have been able to complete this book without their support.

Editor

Syntheses and Biological Activity of Some Derivatives of C-9154 Antibiotic

Isaac Asusheyi Bello,[1] George Iloegbulam Ndukwe,[1] Joseph Olorunju Amupitan,[1] Rachael Gbekele Ayo,[2] and Francis Oluwole Shode[3]

[1] *Department of Chemistry, Ahmadu Bello University, Zaria 810001, Nigeria*
[2] *Division of Agricultural Colleges, College of Agriculture, Ahmadu Bello University, Zaria 810001, Nigeria*
[3] *School of Chemistry and Biochemistry, University of Zululand, Empangeni 3880, South Africa*

Correspondence should be addressed to Isaac Asusheyi Bello, lobell_ng@yahoo.com

Academic Editor: Feng Liang

This research was undertaken to design several new antibiotics, by structurally modifying the C-9154 antibiotic, simultaneously improving its activity and lowering toxicity. This was achieved by synthesizing an analogue to the C-9154 antibiotic and seven derivatives of this analogue. The approach was to significantly reduce the polarity of the synthesized analogue in the derivatives to achieve increased permeability across cell membranes by conversion of the highly polar carboxylic group to an ester functional group. The compounds were fully characterized using infrared, GC-MS, and 1D and 2D NMR experiments. The *in vitro* biological activity of the compounds showed that the derivatives were more active than the analogue as was anticipated and both were more active than the standard drugs used for comparison. Work is ongoing to establish applications for the compounds as antiplasmodials, antivirals, anticancers/tumours, antitrypanosomiasis, anthelminthic, and as general antibiotics for human, veterinary, and even agricultural use as they had marked effect on both Gram-positive and Gram-negative bacteria and some fungi.

1. Introduction

Vuillemin (1889), a French bacteriologist, was first to suggest using the word "antibiosis," (against life) to describe the group of compounds that exhibited biological action against microorganisms [1]. Selman Waksman, the discoverer of streptomycin, later changed this term to antibiotic in 1942 [2]. The term "antibiotic" according to Waksman is used to describe any substance produced by a microorganism that is detrimental to the growth of other microorganisms in high dilution or at low concentration [3]. With current advances in medicinal chemistry, antibiotics are mostly semisynthetic modifications of various natural compounds [4]. These include, for example, the beta-lactam antibiotics, like the penicillins, the cephalosporins, and the carbapenems. Some antibiotic compounds are still being isolated from natural sources, for example, the aminoglycosides, whereas other antibiotics like the sulfonamides, the quinolones, and the oxazolidinones are produced entirely by chemical synthesis [4]. This implies that synthesis of antibiotic compounds plays an important and vital role in the fight against disease-causing organisms.

The problem of resistance to antibiotics on the part of the microorganism, the adverse side effects associated with antibiotics in current use, and the difficulty in obtaining these antibiotics in large (commercial) quantities from their natural sources imply that newer antibiotics have to be constantly sought to address these problems to give man the needed advantage in the ongoing battle between microbes and men. Synthesis of previously characterized antibiotics with structural modifications to imbue desirable qualities or remove undesirable ones provides a way to assist man in this great battle.

During studies on screening for antibiotics that showed activity against bacteria resistant to various known antibiotics, a new antibiotic with a broad antibacterial spectrum was isolated from the whole agar culture of a *Streptomyces* strain, NR-7GGI. This *Streptomyces* species was called *Streptomyces kurssanovii* and the isolated antibiotic referred to as fumaramidmycin [5]. Another researcher working independently and slightly earlier than the previous researcher also found that a new species of *Streptomyces*, *Streptomyces ishigakiensis*, produced a novel antibiotic which was named C-9154 [6]. The two new antibiotics were found from structural studies to be the same compound [6, 7]. This new antibiotic was found to inhibit the growths of various microorganism at concentrations between 10–>100 μg/mL [6]. It was also shown to be active against certain strains that were resistant to ampicillin, cephalosporin, chloramphenicol, gentamicin, kanamycin, macrolides, neomycin, sulfonamides, streptomycin, and tetracyclines at concentrations between 3.12–>200 μg/mL [5]. Its intraperitoneal LD_{50} value in mice was found to be 75–100 mg/kg while its oral LD_{50} was found to be 1.25–2.5 g/kg [5].

The structure of C-9154 (Figure 1) was determined using elemental analysis procedures, IR and UV measurements and NMR and GC-MS experiments [6, 7].

Analysis of the structure of C-9154 antibiotic showed that it was made up of two fragments, namely, phenylacetic acid and fumaramide [6, 7]. Analogues of the C-9154 antibiotic have been previously synthesized [7–10]. These analogues were shown to have antibacterial activity against *Staphylococcus aureus* and *Escherichia coli* ranging from 15 μg/mL [10] to 4000 μg/mL [9]. We now report the synthesis of an analogue and seven derivatives of the C-9154 antibiotic and their biological activity.

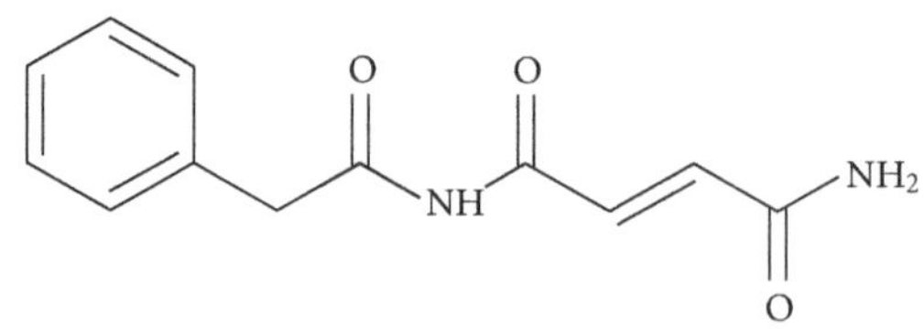

Figure 1: C-9154 antibiotic.

2. Materials and Method

Infrared spectra were determined using a PerkinElmer Spectrum 100 series Universal ATR. 1D and 2D NMR experiments were carried out using a Bruker av400 MHz NMR. The gc/ms spectra were taken using an Agilent Technologies 6890 series GC coupled with an Agilent 5973 Mass Selective detector. All weight measurements were taken using a Precisa 3100C top loading balance. Melting points were determined using a Micro Melting point apparatus (Reichart Austria), uncorrected. All chemicals and reagents unless otherwise stated were obtained from Merck Chemicals, Germany.

2.1. Synthesis of C-9154 Analogue. N-phenyl fumaramic acid was prepared according to reaction scheme 1 (Figure 2). Aniline (1.0 g, 11.0 mmol) in toluene (5 mL) was transferred to a round bottom flask containing maleic anhydride (1.3 g, 13.3 mmol) in toluene (5 mL). The mixture was refluxed at 100°C with continuous stirring for 2 hrs. The reaction was allowed to cool to room temperature and vacuum filtered using a Buchner funnel. The residue was washed using ethyl acetate and dried to afford a grey solid labeled IA/01/1 (1.9 g, 90.0%). TLC was used to determine that the reaction had gone to completion. IR, 1D and 2D NMR, and GC-MS were used to identify the compound as the desired N-phenyl fumaramic acid. Its melting point was determined to be 195-196°C.

2.2. Synthesis of Derivatives of C-9154 Analogue. This analogue was then converted to its ester derivatives using methanol, ethanol, n-propanol, isopropanol, n-butanol, 2-butanol, and t-butanol, respectively, and thionyl chloride ($SOCl_2$) as catalyst according to reaction scheme 2 (Figure 3).

Seven portions of IA/01/1 (0.5 g, 2.6 mmol) were individually transferred to seven round bottomed flasks each containing thionyl chloride (2 mL) in an ice bath. The excess thionyl chloride was removed using a rotary evaporator. Methanol (10 mL), ethanol (10 mL), n-propanol (10 mL), isopropanol (10 mL), n-butanol (10 mL), 2-butanol (10 mL), and t-butanol (10 mL) were respectively added to each flask and the mixtures refluxed for 10 minutes. At the end of the reactions as determined by TLC, saturated sodium carbonate (Na_2CO_3) solution was added to each flask until the solutions just turned alkaline as indicated by litmus paper. Water (20 mL) was added to each flask and the mixtures were transferred to individual separatory funnels. The mixtures were each extracted using dichloromethane (2 × 25 mL). The combined dichloromethane fractions were dried using anhydrous sodium sulphate (Na_2SO_4) and concentrated to give light yellow to clear oils. These were chromatographed on individual silica gel columns and eluted using ethyl acetate : hexane (3 : 7), to give the desired esters which crystallized on standing. All the esters except the 2-butyl esters were obtained as crystalline solids (Table 1). IR, 1D and 2D NMR, and GC-MS were used to identify the compounds as the desired esters.

2.3. Biological Screening. The analogue and its derivatives were successfully synthesized and characterized. These compounds were then subjected to biological screening *in vitro* to ascertain their activity and the concentration at which this activity was exhibited.

This was carried out using zones of inhibition measurements, minimum inhibitory concentration measurements, and minimum bactericidal/fungicidal concentration measurements. Fourteen (14) microorganisms were selected for the screening to include both gram-positive and gram-negative bacteria and some fungi. These microorganisms are: Methicillin-Resistant *Staphylococcus aureus, Staphylococcus aureus, Streptococcus pyogenes, Bacillus subtilis,* and *Corynebacterium ulcerans* for Gram-positive bacteria, *Escherichia coli, Proteus mirabilis, Pseudomonas aeruginosa, Salmonella typhii, Shigella dysenteriae,* and *Klebsiella pneumonia* for Gram-negative bacteria, and *Candida albicans, Aspergillus nigre*, and *Trichophyton rubrum* for fungi.

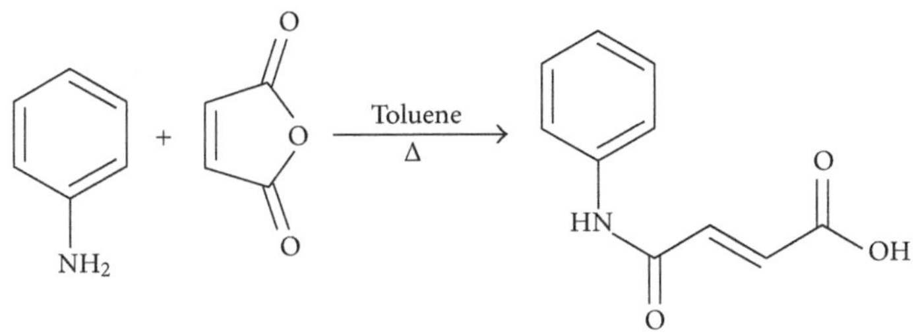

Figure 2: Reaction scheme 1.

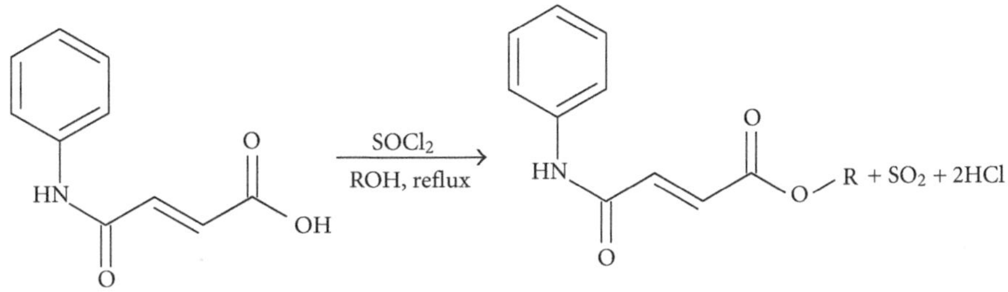

Where R = CH_3
CH_3CH_2
$CH_3CH_2CH_2$
$(CH_3)_2CH$
$CH_3CH_2CH_2CH_2$
$(CH_3CH_2)(CH_3)CH$

Figure 3: Reaction scheme 2.

2.4. Zones of Inhibition. The antimicrobial activity (Table 2) of the synthesized analogue and derivatives was determined using some pathogenic microorganisms obtained from the Department of Medical Microbiology, Ahmadu Bello University Teaching Hospital, Zaria, Nigeria. All isolates were checked for purity and maintained in slants of blood agar.

The analogue (0.1 mg) and the derivatives (0.05 mg) were each weighed and dissolved in DMSO (10 mL) to obtain a concentration of 10 μg/mL and 5 μg/mL, respectively. This was the initial concentration used to check the antimicrobial activities of the compounds.

Mueller Hinton, (Oxoid, England) and Sabouraud agar, (Oxoid, England) were the growth media used for the bacteria and fungi, respectively. The media were prepared according to the manufacturer's instructions, sterilized at 121°C for 15 minutes and were poured into sterile Petri dishes. The plates were allowed to cool and solidify. Diffusion method was the method used for screening the compounds. The sterilized media were seeded with a standard inoculum (0.1 mL) of the test microorganisms. This was spread evenly over the surface of the plate by using a sterile swab. The plates were dried at 37°C for 30 minutes. By the use of a standard cork-borer of 6 mm in diameter, a well was cut at the centre of each seeded plate. 0.1 mL of the compounds was then introduced into the well. The plates were then incubated at 37°C for 24 hrs for the bacteria and 30°C for 48 hrs for the fungi, after which the plates were observed for zones of inhibition of growth. The zones were measured using a pair of dividers and a ruler and the result recorded in millimeters.

The activity of the compounds was compared against two standard drugs; Sparfloxacin (antibacterial), and Fluconazole (antifungal) which were subjected to the same treatment as the synthesized compounds.

2.5. Minimum Inhibitory Concentration. The minimum inhibitory concentrations (Table 3) of the compounds were carried out using broth dilution method. Mueller Hinton, (Oxoid, England) and Sabouraud dextrose broth, (Oxoid, England) were prepared according to the manufacturer's instructions, and 10 mL was dispensed into test tubes and the broths were sterilized at 121°C for 15 minutes, the broths were allowed to cool.

McFarland's turbidity scale number 0.5 was prepared to give turbid solution. Normal saline was prepared, and the test microorganisms were inoculated and incubated at 37°C for 6 hrs. Dilution of the test microorganisms were done continuously in the normal saline until the turbidity matched that of the McFarland's scale by visual comparison. At that point the test microbe was at a concentration of 1.5×10^8 cfu/mL. Two-fold serial dilutions of the compounds in the broth were made to obtain the different concentrations of the compounds in the broth. Having obtained the different concentrations, 0.1 mL of the standard inoculum of the test microorganisms in the normal saline was then inoculated

Table 1: Synthesized C-9154 analogue and its derivatives.

Sample code	Type	Yield (mg)	Melting point (°C)	Physical state
IA/01/1	C-9154 analogue	90%	195-196	White powdery solid
IA/24/1/B	Methyl ester	160, (0.78 mmol)	74	Bright yellow crystalline solid
IA/21/1/B	Ethyl ester	100, (0.46 mmol)	63	Light brown crystalline solid
IA/22/1/B	n-propyl ester	170, (0.73 mmol)	55	Dark brown crystalline solid
IA/23/1/B	Isopropyl ester	105, (0.45 mmol)	70	Dark brown crystalline solid
IA/19/1/B	n-butyl ester	200, (0.81 mmol)	42	Yellow waxy solid
IA/20/1/B	2-butyl ester	150, (0.61 mmol)	40	Golden brown oil
IA/25/1/B	t-butyl ester	2, (~0.01 mmol)	Not determined	

The yield from the t-butyl derivative was too little and could not be optimized. This was probably due to steric hindrance from the methyl groups.

Table 2: Zones of inhibition (mm) of the analogue and derivatives.

	IA/01/1 (10 μg/mL)	IA/24/1/B (5 μg/mL)	IA/21/1/B (5 μg/mL)	IA/22/1/B (5 μg/mL)	IA/23/1/B (5 μg/mL)	IA/19/1/B (5 μg/mL)	IA/20/1/B (5 μg/mL)	DMSO	Sparfloxacin (20 μg/mL)	Fluconazole (50 μg/mL)
MRSA	20	32	29	35	32	34	31	0	22	0
S. aureus	28	30	32	30	32	31	32	0	27	0
S. pyogenes	24	0	30	37	36	0	29	0	24	0
B. subtilis	30	26	32	32	30	29	31	0	30	0
C. ulcerans	0	30	29	27	30	0	27	0	0	0
E. coli	27	27	29	22	25	29	27	0	27	0
P. mirabilis	22	25	27	27	24	31	26	0	22	0
P. aeruginosa	0	0	0	0	0	0	0	0	20	0
S. typhii	26	30	27	29	31	29	22	0	21	0
S. dysenteriae	24	27	27	35	32	31	24	0	27	0
K. pneumoniae	27	24	27	30	31	29	27	0	25	0
C. albicans	22	22	24	27	30	24	21	0	0	24
A. nigre	21	0	0	0	0	0	0	0	0	0
T. rubrum	0	0	0	0	0	0	0	0	0	20

into the different concentrations, and then incubated at 37°C for 24 hrs for the bacteria and 30°C for 48 hrs for the fungi, after which each test tube was observed for turbidity (growth). The MIC was the test tube with the lowest concentration of the compounds which showed no turbidity.

2.6. Minimum Bactericidal/Fungicidal Concentration. MBC (Table 3) was carried out to check whether test microorganisms were killed or only their growths were inhibited. Mueller Hinton, (Oxoid, England) and Sabouraud dextrose agar, (Oxoid, England) were prepared, sterilized, and poured into sterile Petri dishes. They were allowed to cool and solidify. The content of the MIC in the serial dilution were then subcultured onto the prepared media. They were then incubated at 37°C for 24 hrs for the bacteria and 30°C for 48 hrs for the fungi after which each plate was observed for colony growth. The MBC/MFC was the plate with lowest concentration of the compounds without colony growth.

3. Results and Discussion

A total of eight compounds were synthesized and fully characterized using 1D and 2D NMR experiments, infrared spectrophotometry, and gas chromatography-mass spectrometry.

The analogue was synthesized by reaction between the required N-phenyl amine (aniline) and maleic anhydride to get the desired fumaramic acid. This analogue was then converted to its methyl, ethyl, n-propyl, isopropyl, n-butyl, 2-butyl, and t-butyl esters using the Fischer-Speier esterification procedure. The results are presented below.

3.1. IA/01/1 (N-Phenylamino Fumaramic Acid). Grey powdery solid; melting point, 195-196°C.

^{13}C-NMR (400 MHz, $CDCl_3$ and DMSO-d_6). 165.1 (C10), 163.7 (C7), 136.6 (C1), 133.7 (C9), 132.7 (C8), 128.4 (C3 and C5), 125.0 (C4), 120.3 (C2 and C6).

^{1}H-NMR (400 MHz, $CDCl_3$ and DMSO-d_6): δ 6.23 (1H, d, J = 12.69 Hz, H-9), 6.55 (1H, d, J = 12.73 Hz, H-8), 7.09 (1H, t, J = 7.40 Hz, H-4), 7.27 (2H, *t*, J = 7.88 Hz, H-3 and H-5), 7.57 (2H, *d*, J = 7.84 Hz, H2 and H-6), 7.68 (1H, s, 1-NH), 10.79 (1H, s, 10-OH). EI-MS: m/z 173 {$[M\text{-}H]^+$, 100%}. IR_{vmax} (neat) cm^{-1}: 3271.36 (N-H), 3071.75 (O-H), 2883.57 (C-H), 1694.60, 1618.99 (C=O).

3.2. IA/24/1/B (Methyl N-Phenylamino Fumaramate). Bright yellow crystalline solid; melting point, 74°C.

^{13}C-NMR (400 MHz, $CDCl_3$). 167.4 (C7), 161.5 (C10), 137.8 (C1), 140.5 (C9), 124.9 (C8), 129.0 (C3 and C5), 124.6 (C4), 120.1 (C2 and C6), 52.8 (C11).

TABLE 3: Minimum inhibitory concentration (above) and minimum bactericidal/fungicidal concentration (below) of the analogue and derivatives (μg/mL).

	IA/01/1	IA/24/1/B	IA/21/1/B	IA/22/1/B	IA/23/1/B	IA/19/1/B	IA/20/1/B	DMSO	Sparfloxacin	Fluconazole
MRSA	2.5	0.625	1.25	0.625	0.625	0.625	0.625	ND	10	ND
	10	2.5	2.5	1.25	2.5	1.25	2.5			
S. aureus	2.5	0.625	0.625	0.625	0.625	0.625	0.625	ND	10	ND
	5	2.5	2.5	2.5	2.5	2.5	2.5			
S. pyogenes	2.5	ND	0.625	0.625	0.625	ND	1.25	ND	10	ND
	10	ND	2.5	1.25	1.25	ND	2.5			
B. subtilis	1.25	1.25	0.625	0.625	0.625	1.25	0.625	ND	5	ND
	5	5	1.25	2.5	2.5	2.5	1.25			
C. ulcerans	ND	0.625	1.25	1.25	0.625	ND	1.25	ND	ND	ND
	ND	2.5	2.5	2.5	2.5	ND	5			
E. coli	2.5	1.25	1.25	1.25	1.25	1.25	1.25	ND	10	ND
	5	2.5	2.5	5	5	2.5	5			
P. mirabilis	2.5	1.25	1.25	1.25	1.25	0.625	1.25	ND	10	ND
	10	5	2.5	2.5	5	2.5	5			
P. aeruginosa	ND	ND	ND	ND	ND	ND	ND	ND	10	ND
	ND	ND	ND	ND	ND	ND	ND			
S. typhii	2.5	0.625	1.25	1.25	0.625	1.25	1.25	ND	10	ND
	5	2.5	5	2.5	2.5	2.5	5			
S. dysenteriae	2.5	1.25	1.25	0.625	0.625	0.625	1.25	ND	5	ND
	10	2.5	2.5	1.25	2.5	2.5	5			
K. pneumoniae	2.5	1.25	1.25	0.625	0.625	1.25	1.25	ND	5	ND
	5	5	5	2.5	2.5	2.5	2.5			
C. albicans	2.5	1.25	1.25	1.25	0.625	1.25	1.25	ND	ND	25
	10	5	5	2.5	2.5	5	5			
A. nigre	2.5	ND	ND	ND	ND	ND	ND	ND	ND	ND
	10	ND	ND	ND	ND	ND	ND			
T. rubrum	ND	ND	ND	ND	ND	ND	ND	ND	ND	25
	ND	ND	ND	ND	ND	ND	ND			

Upper values are MIC and lower values are MBC or MFC as the case may be. ND: not determined.

^{1}H-NMR (400 MHz, $CDCl_3$) δ 3.83 (3H, s, H-11), 6.21 (1H, d, J = 13.33 Hz, H-8), 6.42 (1H, d, J = 13.25 Hz, H-9), 7.11 (1.0H, t, J = 7.38 Hz, 4-H), 7.32 (2H, t, J = 7.60 Hz, H-3 and H 5), 7.64 (2H, d, J = 7.88 Hz, H-2 and H-6), 10.82 (1H, s, 1-NH). EI-MS: m/z 205 {[M]$^+$, 50%}. IR_{vmax} (neat) cm^{-1}: 3283.23 (N-H), 3079.91, 2953.75 (C-H), 1730.94, 1666.54 (C=O), 1206.53, 1165.92 (C-O).

3.3. IA/21/1/B (Ethyl N-Phenylamino Fumaramate). Light brown crystalline solid; melting point, 63°C.

^{13}C-NMR (400 MHz, $CDCl_3$). 166.7 (C7), 162.0 (C10), 137.9 (C1), 138.9 (C9), 126.0 (C8), 129.0 (C3 and C5), 124.6 (C4), 120.1 (C2 and C6), 61.9 (C11), 14.0 (C12).

^{1}H-NMR (400 MHz, $CDCl_3$) δ 1.30 (3H, t, J = 7.20 Hz, H-12), 4.26 (2H, *q*, J = 7.14 Hz, H-11), 6.16 (1H, *d*, J = 13.09 Hz, H-8), 6.39 (1H, *d*, J = 13.09 Hz, H-9), 7.10 (1H, *t*, J = 7.36 Hz, H-4), 7.31 (2H, *t*, J = 7.84 Hz, H-3 and H-5), 7.65 (2H, *d*, J = 7.96 Hz, H-2 and H-6), 10.7 (1H, s, 1-NH). EI-MS: m/z 219 {[M]$^+$, 50%}. IR_{vmax} (neat) cm^{-1}: 3285.55 (N-H), 3063.98, 2926.30 (C-H), 1716.89, 1667.67 (C=O), 1176.46, 1166.06 (C-O).

3.4. IA/22/1/B (n-propyl N-Phenylamino Fumaramate). Dark brown crystalline solid; melting point, 55°C.

^{13}C-NMR (400 MHz, $CDCl_3$). 166.7 (C7), 162.5 (C10), 137.9 (C1), 137.2 (C9), 126.6 (C8), 128.9 (C3 and C5), 124.5 (C4), 120.2 (C2 and C6), 67.3 (C11), 21.7 (C12) 10.3 (C13).

^{1}H-NMR (400 MHz, $CDCl_3$) δ 0.83 (3H, t, J = 7.42 Hz, H-13), 1.57 (2H, m, H-12), 4.04 (2H, t, J = 6.74 Hz), 6.04 (1H, d, J = 12.69 Hz, H-8), 6.31 (1H, d, J = 12.69 Hz, H-9), 7.00 (1H, t, J = 7.42 Hz, H-4), 7.21 (2H, d, J = 7.64 Hz, H-3 and H-5), 7.56 (2H, d, J = 7.76 Hz, H-2 and H-6), 10.3 (1H, s, 1-NH). EI-MS: m/z 233 {[M]$^+$, 50%}. IR_{vmax} (neat) cm^{-1}: 3259.72 (N-H), 3092.97, 2878.49 (C-H), 1719.46, 1662.54 (C=O), 1205.45, 1185.79 (C-O).

3.5. IA/23/1/B (Isopropyl N-Phenylamino Fumaramate). Dark brown crystalline solid; melting point, 70°C.

^{13}C-NMR (400 MHz, $CDCl_3$). 166.4 (C7), 161.6 (C10), 137.9 (C1), 140.3 (C9), 125.8 (C8), 129.0 (C3 and C5), 124.5 (C4), 120.1 (C2 and C6), 70.0 (C11), 21.7 (C12 and C13).

^{1}H-NMR (400 MHz, $CDCl_3$) δ 1.32 (6H, d, J = 6.28 Hz, H-12 and H-13), 5.15 (1H, m, H-11), 6.18 (1H, d, J =

13.36 Hz, H-8), 6.41 (1H, d, J = 13.45 Hz, H-9), 7.12 (1H, t, J = 7.38 Hz, H-4), 7.34 (2.1H, d, J = 15.81 Hz, H-3 and H-5), 7.67 (1.9H, d, J = 7.84 Hz, H-2 and H-6), 11.0 (1H, s, 1-NH). EI-MS: m/z 233 $\{[M]^+, 50\%\}$. IR_{vmax} (neat) cm^{-1}: 3238.13 (N-H), 3074.64, 2936.22 (C-H), 1729.86, 1660.48 (C=O), 1166.93, 1103.15 (C-O).

3.6. IA/19/1/B (n-Butyl N-Phenylamino Fumaramate). Yellow waxy solid; melting point, 42°C.

^{13}C-NMR (400 MHz, $CDCl_3$). 166.6 (C7), 162.8 (C10), 136.2 (C1), 138.0 (C9), 126.8 (C8), 128.8 (C3 and C5), 124.5 (C4), 120.2 (C2 and C6), 65.4 (C11), 30.3 (C12), 19.0 (C13), 13.6 (C14).

^{1}H-NMR (400 MHz, $CDCl_3$) δ 0.84 (3H, t, J = 7.60 Hz, H-14), 1.31 (2H, m, H-13), 1.56 (2H, m, H-12), 4.12 (2H, t, J = 6.78 Hz, H-11), 6.07 (1H, d, J = 12.69 Hz, H-8), 6.38 (1H, d, J = 12.53 Hz, H-9), 7.04 (1H, t, J = 7.40 Hz, H-4), 7.24 (2H, t, J = 7.64 Hz, H-3 and H-5), 7.63 (2H, d, J = 8.20 Hz, H-2 and H-6), 10.2 (1H, s, 1-NH). EI-MS: m/z 247 $\{[M]^+, 75\%\}$. IR_{vmax} (neat) cm^{-1}: 3298.98 (N-H), 2957.43, 2872.97 (C-H), 1720.16, 1659.63 (C=O), 1223.16, 1178.85 (C-O).

3.7. IA/20/1/B (2-Butyl N-Phenylamino Fumaramate). Golden brown oil.

^{13}C-NMR (400 MHz, $CDCl_3$). 166.4 (C7), 162.0 (C10), 138.0 (C1), 139.0 (C9), 126.3 (C8), 129.0 (C3 and C5), 124.5 (C4), 120.1 (C2 and C6), 74.4 (C11), 28.6 (C12), 19.2 (C14), 9.6 (C13).

^{1}H-NMR (400 MHz, $CDCl_3$) δ 0.87 (3H, t, J = 7.46 Hz, H-13), 1.23 (3H, d, J = 6.20 Hz, H-14), 1.58 (2H, m, H-12), 4.92 (1H, m, H-11), 6.11 (1H, d, J = 13.13 Hz, H-8), 6.35 (1H, d, J = 13.13 Hz, H-9), 7.06 (1H, t, J = 7.36 Hz), 7.27 (2H, d, J = 15.81 Hz), 7.62 (2H, d, J = 7.68 Hz), 10.8 (1H, s, 1-NH). EI-MS: m/z 247 $\{[M]^+, 50\%\}$. IR_{vmax} (neat) cm^{-1}: 3329.95 (N-H), 2972.12, 2878.77 (C-H), 1709.62, 1663.31 (C=O), 1245.78, 1175.08 (C-O).

The results show that the synthesized antibiotics had remarkable activity in the range 0.625–10 μg/mL against the microorganisms for which they were active against.

The derivatives showed higher activity than the analogue as was anticipated. This could be due to the reduction in polarity when the highly polar carboxylic functional group was converted to the less polar ester functional group. This has been shown to increase permeability across cell membranes [11–13]. All the synthesized compounds showed better activity than the standard drugs used for comparison. The antibacterial standard drug, sparfloxacin, was up to four-fold less active than most of the synthesized derivatives while the antifungal standard drug, fluconazole, was up to ten-fold less active than the synthesized analogue and derivatives for the microorganism against which they were active. All the synthesized compounds except (IA/01/1) and (IA/19/1/B) were able to inhibit the growth of *C. ulcerans* whereas the standard antibacterial could not. All the derivatives and the standard antifungal could not inhibit the growth of *A. nigre*, but the analogue (IA/01/1) was able to inhibit its growth. Two of the synthesized compounds (IA/19/1/B) and (IA/24/1/B) could also not inhibit the growth of *S. pyogenes*. All the synthesized compounds could not inhibit the growths of *P. aeruginosa* and *T. rubrum*.

4. Conclusion

The results show that the synthesized compounds have a clear advantage over the tested standard drugs, and this has opened up the possibility of their application in the treatment of various ailments for which the tested microorganism are responsible for. The possible applications of these compounds are endless and with further studies could compliment or even replace some of the antibiotic drugs currently in the market if they are found to be less toxic and better tolerated. Other applications for the new antibiotics are in the veterinary and agricultural fields where they could be useful in combatting some of the diseases that plague both animals and plants.

Some *in vivo* work is being carried out to establish the activity of the synthesized compounds as anticancer, anti-HIV, antimalarial, antitrypanosomiasis, antituberculosis, and so forth.

Acknowledgment

I. A. Bello would like to appreciate the ETF, Nigeria for sponsoring part of this research.

References

[1] S. M. Aronson, "The naming of antibiotics," *Medicine and Health*, vol. 80, no. 6, article 180, 1997.

[2] M. Radetsky, "The discovery of penicillin," *The Pediatric Infectious Disease Journal*, vol. 15, no. 9, pp. 811–818, 1996.

[3] S. A. Waksman, "What is an antibiotic or an antibiotic substance?" *Mycologia*, vol. 39, no. 5, pp. 565–569, 1947.

[4] F. von Nussbaum, M. Brands, B. Hinzen, S. Weigand, and D. Häbich, "Antibacterial natural products in medicinal chemistry—exodus or revival?" *Angewandte Chemie—International Edition*, vol. 45, no. 31, pp. 5072–5129, 2006.

[5] H. B. Maruyama, Y. Suhara, J. Suzuki Watanabe et al., "A new antibiotic, fumaramidmycin. I. Production, biological properties and characterization of producer strain," *Journal of Antibiotics*, vol. 28, no. 9, pp. 636–647, 1975.

[6] T. Hasegawa, M. Asai, K. Haibara, T. Yamano, H. Iwasaki, and M. Yoneda, "A new antibiotic, C-9154," *Journal of Antibiotics*, vol. 28, no. 9, pp. 713–716, 1975.

[7] Y. Suhara, H. B. Maruyama, Y. Kotoh et al., "A new antibiotic, fumaramidmycin. II. Isolation, structure and syntheses," *Journal of Antibiotics*, vol. 28, no. 9, pp. 648–655, 1975.

[8] J. Jumina, D. Siswanta, and A. K. Zulkarnain, "Sintesis dan Uji Aktivitas Biologis Turunan Antibiotik C-9154 Dari Vanilin," *Majalah Farmasi Indonesia*, vol. 12, no. 3, pp. 24–35, 2001.

[9] J. Jumina, I. Tahir, and A. K. Zulkarnain, "Synthesis and antimicrobe activity evaluation of ethyl salicyl fumarate and ethyl furfuryl fumarate," *Majalah Farmasi Indonesia*, vol. 13, no. 4, pp. 207–214, 2002.

[10] J. Jumina, A. K. Zulkarnain, and P. Mulyono, "Preparation and antibacterial activity of p-Anisyl ethyl fumarate and ethyl N-phenyl fumaramate," *Majalah Farmasi Indonesia*, vol. 16, no. 2, pp. 116–123, 2005.

[11] F. L. Strand, "The plasma membrane as a regulatory organelle," in *Physiology: A Regulatory Systems Approach*, pp. 49–67, MacMillan, New York, NY, USA, 2nd edition, 1983.

[12] R. Eckert, D. Randall, and G. Augustine, "Permeability and transport," in *Animal Physiology*, pp. 65–99, W. H. Freeman, New York, NY, USA, 3rd edition, 1988.

[13] B. Alberts, D. Bray, J. Lewis, M. Raff, K. Roberts, and J. D. Watson, "The plasma membrane," in *Molecular Biology of the Cell*, pp. 276–337, Garland, New York, NY, USA, 2nd edition, 1989.

2

Comparative Analysis of the Antioxidant Activity of *Cassia fistula* Extracts

Md. Irshad,[1,2,3] Md. Zafaryab,[1] Man Singh,[2] and M. Moshahid A. Rizvi[1]

[1] *Department of Biosciences, Jamia Millia Islamia, New Delhi 110025, India*
[2] *School of Chemical Sciences, Central University of Gujarat, Gujarat, Gandhinagar 382030, India*
[3] *Department of Medical Education, College of Medicine, King Saud University, Riyadh 11321, Saudi Arabia*

Correspondence should be addressed to M. Moshahid A. Rizvi, rizvijmi@gmail.com

Academic Editor: Hussein El-Subbagh

Antioxidant potential of various extracts of *Cassia fistula* was determined by the DPPH, FRAP, Fe^{3+} reducing power, and hydrogen peroxide scavenging assay. Methanolic extracts of *Cassia fistula* showed the highest amount of phenolic and flavonoid content and reducing capacity, whereas hexane extracts exhibited the lowest level of reducing capacity. The order of antioxidant activity in *Cassia fistula* extracts displayed from higher to lower level as methanolic extracts of pulp, methanolic extracts of seed, hexane extracts of pulp, and hexane extracts of seed. The antioxidant potential of *Cassia fistula* extracts significantly correlated ($P < 0.02$) with the phenolic content of the methanolic extracts. Ascorbic acid taken as control showed highest antioxidant power in the present study.

1. Introduction

Cassia fistula Linn. (*Caesalpiniaceae*) has great therapeutic implication in Indian system of medicine and exerts an antipyretic, analgesic, antiinflammatory, and hypoglycemic effects [1, 2]. Over the past few years, there has been an exponential growth in study of pharmacological properties of this plant [3–5]. Antioxidant components are micro-constituents that inhibit lipid oxidation by inhibiting the initiation or propagation of oxidizing chain reactions, and are also involved in scavenging of free radicals. Clinical approaches of antioxidants increased multifold during the recent time for the management and therapeutic implication of neurodegenerative disorders, aging, and chronic degenerative diseases. In view of the above, we designed the study to evaluate the antioxidant potential and phenolic content in *Cassia fistula*.

2. Materials and Methods

2.1. Reagent and Chemicals. All the chemicals and solvents were of analytical grade and obtained from Merck and HiMedia, Mumbai, India.

2.2. Preparation of Cassia fistula Extracts. The fresh ripe fruits of *Cassia fistula* were collected in June from the campus of Delhi College of Pharmaceutical Science and Research (DIPSAR), New Delhi, India, and properly authenticated. A voucher of specimen (PM 21) was stored in the laboratory for further reference. The fruit pulp and seed were separated and grounded to coarse powder. It was extracted with the hexane for 72 hrs and the same materials were reextracted with methanol for 72 hrs in soxhlet apparatus. The extract was filtered and dried in rotavapour. Hence, we obtained hexane extract of seed (HES), hexane extract of pulp (HEP), methanolic extract of seed (MES), and methanolic extract of pulp (MEP).

2.3. Estimation of Total Phenolic Content. The total phenol content was estimated in the methanolic extracts of seed and pulp using Folin-Ciocalteu reagent (FCR) according to the procedure reported by Singleton et al. [6], using standard Gallic acid ($Y = 1.8478x + 0.0702$, $R^2 = 0.9971$) and Tannic acid ($Y = 1.388x + 0.0472$, $R^2 = 0.9958$), curve standardized in the lab for the calculation of Gallic acid equivalent (GAE) and Tannic acid equivalent (TAE) per gram of extracts, respectively. The blue complex was formed by the reduction of reagent by phenolic compounds in extract. Briefly, aliquot of 400 μL of extract was added to 1.6 mL of sodium carbonate (7.5% in deionised water) and 2 mL of Folin-Ciocalteu reagent (diluted 10-fold in deionised water). After incubation of 1 hr at room temperature, absorbance was measured at 525 nm using LaboMed Inc. Spectrophotometer (USA). All determination was carried out in triplicate. Total phenolic content was expressed in mg GAE and TAE per gram of extracts, using calibration curve.

2.4. Estimation of Flavonoid Content. Total flavonoids contents were estimated both in methanolic extracts of pulp and seed by method of Zhishen et al. [7], using Quercetin standard. Briefly, 0.5 mL of aliquot of extract was added to 75 μL of 5% $NaNO_2$ solution. After 6 minutes, 150 μL of a 10% $AlCl_3$ $6H_2O$ solution was added and the mixture was allowed to stand another 5 minutes. Then, 0.5 mL of 1 mol/l NaOH and 2.5 mL of distilled water was added. The solutions were mixed and absorbance was measured at 510 nm using LaboMed Inc. Spectrophotometer (USA). All experiments were carried out in triplicate. Total flavonoid content was calculated as mgQE/g, using the following equation based on the calibration curve:

$$Y = 2.234x + 0.992, \qquad R^2 = 0.9927, \qquad (1)$$

where x was the absorbance and y was the mgQE/g.

2.5. DPPH Radicals Scavenging Activity. DPPH radicals scavenging of *Cassia fistula* extracts was estimated according to the method of Miliauskas [8, 9]. DPPH radicals absorbed maximum at 515 nm, which disappears with reduction by an antioxidant compound (s). Three milliliter (3 mL) DPPH solution in methanol (0.1 mM) was mixed with 100 μL of extracts (mg/mL). In control 100 μL methanol (without extracts) mixed with DPPH solution. The samples were incubated in a water bath for 20 min at 37°C, and the decrease in absorbance at 515 nm was measured. Ascorbic acid was used as positive reference. The experiment was carried out in triplicate. Radical scavenging activity was calculated using the following formula:

$$\%\,\text{inhibition} = \left[\frac{(\text{AC}-\text{AE})}{\text{AB}}\right] \times 100, \qquad (2)$$

where AC = absorbance of the control and AE = absorbance of tested samples.

2.6. Ferric Reducing Antioxidant Potential (FRAP) Assay. Ferric reducing power of *Cassia fistula* extracts were determined using FRAP assay [10, 11]. This method is based on the reduction of colourless ferric complex (Fe^{3+} tripyridyltriazine) to blue-colored ferrous complex (Fe^{2+} tripyridyltriazine) by the action of electron donating antioxidants at low pH. The reduction was monitored by measuring the change of absorbance at 593 nm. The working FRAP reagent was prepared by mixing 10 volumes of 300 mM acetate buffer, pH 3.6, with 1 volume of 10 mM TPTZ (2,4,6-tri(2-pyridyl)-*s*-triazine) in 40 mM HCl and with 1 volume of 20 mM ferric chloride. All the required solutions were freshly prepared before their uses. 100 μL of samples (mg/mL) were added to 3 mL of prepared FRAP reagent. The reaction mixture was incubated in a water bath for 30 min at 37°C. Then, the absorbance of the samples was measured at 593 nm. The difference between absorbance of sample and the absorbance of blank was calculated and used to calculate the FRAP value. FRAP value was expressed in terms of mmol Fe^{2+}/g of sample using ferric chloride standard curve $Y = 1.7057x - 0.2211$, $R^2 = 0.9904$. All measurements were calculated from the value obtained from triplicate assays.

2.7. Hydroxyl Radical Scavenging Activity Assay. The scavenging activity for hydroxyl radicals was measured with Fenton reaction [12, 13]. Reaction mixture contained 60 μL of 1.0 mM $FeCl_3$, 90 μL of 1 mM 1,10-phenanthroline, 2.4 mL of 0.2 M phosphate buffer (pH 7.8), 150 μL of 0.17 M H_2O_2, and 1.5 mL of extract at various concentrations. After incubation at room temperature for 5 min, the absorbance of reaction mixture was noted at 560 nm. The hydroxyl radicals scavenging activity was calculated according to the following equation and compared with ascorbic acid as standard:

$$\%\,\text{Inhibition} = \left(\frac{(\text{AB} - \text{AE})}{\text{AB}} \times 100\right), \qquad (3)$$

where AB was the absorbance of blank (without extract) and AE was the absorbance of tested samples.

2.8. Reducing Power Assay. The Fe^{3+} reducing power of extracts was determined by the method of Oyaizu [9, 11, 14]. The extract (0.75 mL) of various concentrations was mixed with 0.75 mL of phosphate buffer (0.2 M, pH 6.6) and 0.75 mL of potassium hexacyanoferrate ($K_3Fe(CN)_6$) (1%, w/v), followed by incubation at 50°C in a water bath for 20 min. The reaction was stopped by adding 0.75 mL of trichloroacetic acid (TCA) solution (10%) and mixture centrifuged at 800 g for 10 min. 1.5 mL of the obtained supernatant was mixed with 1.5 mL of distilled water and 0.1 mL of ferric chloride ($FeCl_3$) solution (0.1%, w/v) for 10 min. The absorbance of reaction mixture was taken at 700 nm. Higher value absorbance of the reaction mixture indicated greater reducing power.

3. Statistical Analysis

Results were expressed as mean value ± SD ($n = 3$). Regression analysis was performed to calculate the dose-response relation between the extracts. Linear regression analysis was performed to find out the correlation coefficient. Statistical significance was evaluated employing t-test and $P < 0.05$ which were considered to be significant.

4. Result and Discussion

4.1. DPPH Radical Scavenging. DPPH (1,1-diphenyl-2-picrylhydrazyl) analysis is one of the best-known, accurate, and frequently employed methods for evaluating antioxidant activity [15]. It is a stable free radical because of its

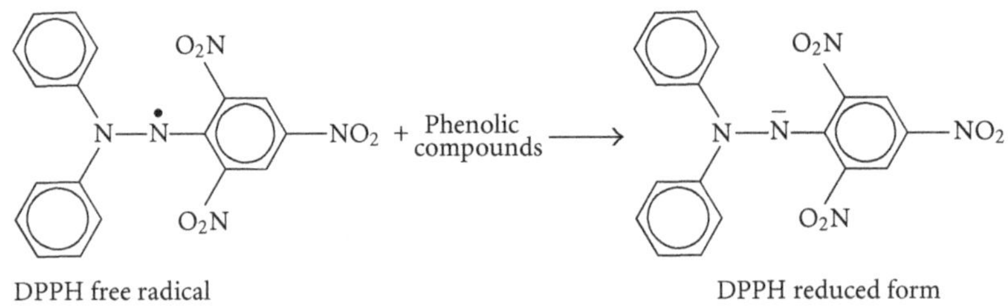

Scheme 1

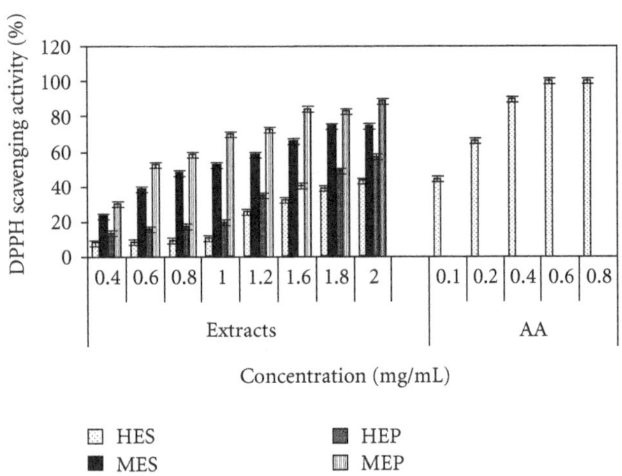

Figure 1: The graphical representation of DPPH scavenging activity of hexane extracts of seed (HES), pulp (HEP), methanolic extracts of seed (MES), and pulp (MEP). Values are expressed as mean ± standard deviation ($n = 3$). Ascorbic acid (AA) was used as a standard.

Table 1: Quantification of total phenolic and flavonoid contents in methanolic extract of pulp (MEP) and seed (MES).

Cassia fistula extracts	Total phenolic		Flavonoids**
	Quercetine*	Tannic acid**	
MEP	114.23 ± 1.05	107 ± 0.87	74.65 ± 1.12
MES	109 ± 1.05	96 ± 1.08	67.78 ± 0.98

*Total phenolic contents (mg gallic acid equivalent/g).
**Total phenolic contents (mg tannic acid equivalent/g).
***Flavonoid (mg quercetin equivalent/g).

spare electron delocalization over the whole molecule. The donation of H^+ to the DPPH radicals made a corresponding change from violet colour to pale yellow in the solution. The DPPH scavenging also made a proportionate decrease in its absorbance at 517 nm (see Scheme 1).

The order of DPPH scavenging against *Cassia fistula* extract was found to be in the order of MEP > MES > HEP > HES. Antioxidant activity of the methanol extracts of pulp and seed was also compared to total phenolic content and it was further found that radical scavenging effects of extracts were directly proportional to the phenolic content present in extracts (Table 1). Hence, methanol extract of pulp and seed showed the significant radical scavenging as the concentration increased, whereas hexane extract of pulp and seed showed low scavenging effect up to 1 mg/mL; thereafter, DPPH scavenging increased with the concentration of extract. The positive control, ascorbic acid showed maximum scavenging effect at very low concentration (Figure 1). A significant correlation coefficient (r, 0.978) was found between the antioxidant activity of the methanolic extract of pulp and seed. The proton radical scavenging action is known to be one of the important mechanisms for measuring antioxidant activity. This assay determines the scavenging of stable radical species *DPPH* by antioxidants compounds present in the extracts. The results showed the greater rate of DPPH scavenging activity of both the methanolic extracts as compared to thehexane extracts probably due to the presence of high content of phenolic compounds. EC_{50} value was determined from the plotted graph of scavenging activity against various concentrations of extracts, which is defined as the efficient concentration of antioxidant necessary to decrease the initial DPPH radicals concentration by 50% (Table 2). The lowest EC_{50} indicates the strongest ability of the extracts to act as DPPH radicals scavengers. Out of the all extracts, methanolic extract of pulp and seed showed the lowest EC_{50}, 0.915 and 1.088 mg/mL, whereas hexane extracts of pulp and seed showed 1.865 and 2.239 mg/mL, respectively (Table 2). Ascorbic acid showed highest DPPH radicals scavenging with EC_{50} of 0.102 mg/mL. Sun and Ho reported a significant correlation between total phenolics and scavenging ability of buckwheat extracts on DPPH radicals [16]. Our study clearly indicated that the methanol extracts of pulp and seed of *Cassia fistula* exhibited high content of phenolic compounds which was significantly correlated with the DPPH radical scavenging activity ($P < 0.005$).

4.2. Hydroxyl Radical Scavenging Activity. The hydroxyl radicals are extremely reactive oxygen species that can react with every possible molecule in living organisms, especially with proteins, DNA, and lipids [17]. Hydroxyl radicals are capable of rapid initiation of the lipid peroxidation process by extracting hydrogen atoms from unsaturated fatty acids [18]. The electron or proton donation capacities of *Cassia fistula* extracts were further confirmed by the Fenton reaction system. Under *in vitro*, condition very reactive OH^- radicals were formed through the Fenton reaction. The reversible reaction between Fe^{2+} and 1,10-Phenathroline

TABLE 2: EC_{50} value is defined as the effective concentration of antioxidant necessary to decrease the radical concentration by 50%. FRAP values represent as equivalent mmol of Fe^{2+}/gram sample.

Cassia fistula extracts	EC_{50} (DPPH) mg/mL	EC_{50} (hydroxyl peroxide) mg/mL	FRAP value
MEP	0.915	0.889	136.05
MES	1.088	1.058	112.02
HEP	1.865	1.723	75.09
HES	2.239	2.075	69.02
Ascorbic acid	0.102	0.105	—

formed $Fe(phen)_2^{2+}$, which react with H_2O_2 and formed OH^- free radicals [19, 20]:

$$\begin{aligned} Fe(phen)_3^{2+} &\leftrightarrows Fe(phen)_2^{2+} + phen, \\ Fe(phen)_2^{2+} + H_2O_2 &\longrightarrow Fe(phen)_2^{3+} + OH^- + {}^{\bullet}OH. \end{aligned} \quad (4)$$

Among the reactive oxygen species (ROs), H_2O_2 is the most reactive and predominant radical generating molecule endogenously during aerobic metabolism [21]. The results showed that methanolic extracts of pulp and seed of *Cassia fistula* have scavenging ability of OH^- free radicals in a dose-dependent manner (Figure 2). The highest activity was noted for the pulp extract of methanol followed by methanolic extract of seed at all concentrations. OH^- free radical scavenging potential of all the extracts was found to be lower than that of the reference compound ascorbic acid. The EC_{50} values of MEP, MES, HES, HEP, and ascorbic acid were 0.889, 1.058, 2.075, 1.723, and 0.105 mg/mL, respectively (Table 2). The trend obtained in OH^- free radical scavenging behaviour was similar to the DPPH free radical scavenging nature (Figure 2). The result of OH^- free radical scavenging activity was significantly correlated with the DPPH free radical scavenging ($P, 0.05$).

4.3. Reducing Power. The reducing power assay is often used to evaluate the ability of an antioxidant to donate an electron [22]. In this assay, the ability of extracts to reduce Fe^{3+} to Fe^{2+} was determined. The presence of antioxidants in the extracts resulted into reduction of the ferric cyanide complex (Fe^{3+}) to the ferrous cyanide form (Fe^{2+}). In reducing power assay, antioxidants cause the reduction of the Fe^{3+} into Fe^{2+}, thereby changing the solution into various shades from green to blue, depending on the reducing power of the compounds [23]. Strong reducing agents, however, formed Perl's Prussian blue colour and absorbed at 700 nm. Figure 3 showed the reducing activities of various extracts of *Cassia fistula* in comparison with ascorbic acid as standard. The higher the absorbance of the reaction mixture, the higher would be the reducing power. Methanolic extract of pulp and seed showed some degree of electron donation. Reducing power of different extracts increased with the concentration of the extract. Hexane extract of seed and pulp showed less degree of Fe^{3+} reduction than the methanolic extracts. The reducing power was found to be in order of MEP > MES > HEP > HES. Interestingly, the rate of reducing power of methanolic extracts of both pulp and seed (MEP and MES) first increased rapidly with concentration but later it decreased. The reducing power of reference compound (Ascorbic acid) was found to be higher than all the tested extracts. It has been reported that the reducing power of substances is probably because of their hydrogen-donating ability [24]. Methanolic extracts of pulp and seed might, therefore, contain high amount of reductones than hexane extracts of pulp and seed. Hence, methanolic extracts of fruit pulp and seed may act as electron donors and could react with free radicals to convert them into more stable products and then terminate the free radical chain reactions.

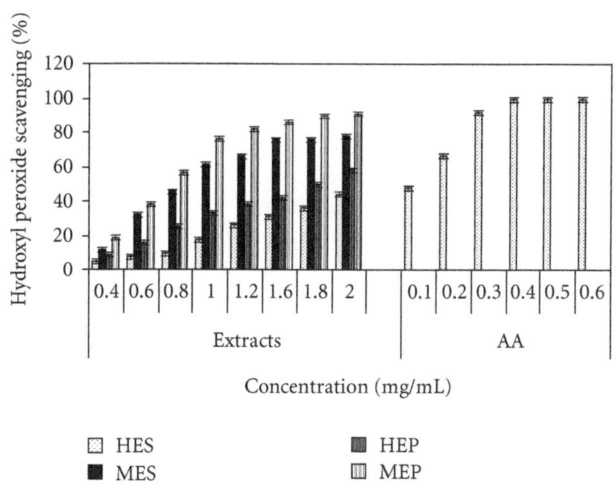

FIGURE 2: The graphical representation of hydroxyl radical scavenging activity of hexane extracts of seed (HES), pulp (HEP), methanolic extracts of seed (MES), and pulp (MEP). Values are expressed as mean ± standard deviation ($n = 3$). Ascorbic acid (AA) was used as a standard.

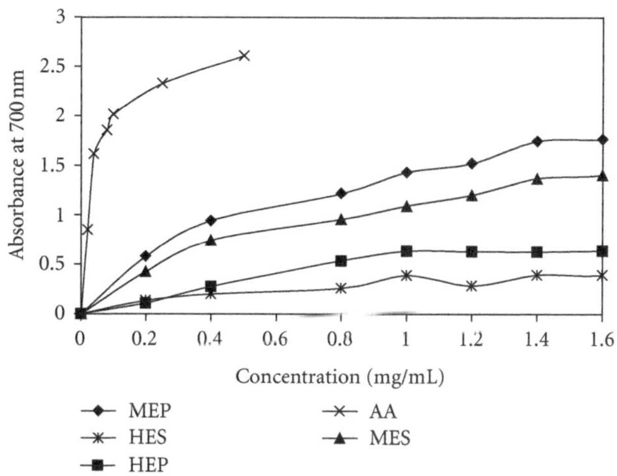

FIGURE 3: The graphical representation of reducing power of hexane extracts of seed (HES), pulp (HEP), methanolic extracts of seed (MES), and pulp (MEP). Values are expressed as mean ± standard deviation ($n = 3$). Ascorbic acid (AA) was used as a standard.

Bhalodia et al. [25] recently reported antioxidant activity of the *Cassia fistula* flower. They reported that the hydroalcoholic extract of flower demonstrated significant radical scavenging activity in ferric reducing power and DPPH assays. It was also correlated with the total phenols which could be responsible for the antioxidant activity [25].

The reducing power assay is often used to evaluate the ability of an antioxidant to donate an electron which is an important mechanism of phenolic antioxidant action [17]. Many reports have revealed that there is a direct correlation between antioxidant activities and reducing power of certain plant extracts [26, 27].

4.4. FRAP Assay. FRAP assay measures the reducing potential of an antioxidant reacting with a ferric tripyridyltriazine (Fe^{3+}-TPTZ) complex and producing a coloured ferrous tripyridyltriazine (Fe^{2+}-TPTZ) [28]. The reducing properties associated with the presence of compounds exert their action by breaking the free radical chain through donating a hydrogen atom [29, 30]. FRAP assay showed positive correlation between reducing power and phenolic content in *Cassia fistula* extracts (Table 2). Here also, methanolic extracts of pulp showed greater FRAP value as 136.05 equivalent mmol of Fe^{2+}/g sample. The other extracts MES, HEP, and HES showed FRAP value 112.02, 75.09, and 69.02 equivalent mmol of Fe^{2+}/g samples, respectively. The result was similar with Benzie and Szeto [27], who found a strong correlation between total phenolic content and FRAP assay. Rice-Evans et al. [30] reported that phenolic compounds have redox properties, which allow them to act as reducing agents, hydrogen donators, and singlet oxygen quenchers. The redox potential of phenolic compounds played an important role in determining the antioxidant potential.

5. Conclusion

Antioxidant potential of *Cassia fistula* extract demonstrated the highest reducing power in the methanolic extract of pulp and seed. It was concluded that the antioxidant activity of these extracts was directly proportional to the phenolic contents. The hexane extracts of seed and pulp, however, did not significantly reduce the free radicals under *in vitro* study.

Acknowledgment

The authors acknowledge Professor M. P. Sharma (Department of Botany, Jamia Hamdard, New Delhi, India) for the identification of plant materials.

References

[1] D. G. Patel, S. S. Karbhari, O. D. Gulati, and S. D. Gokhale, "Antipyretic and analgesic activities of *Aconitum spicatum* and *Cassia fistula*," *Pharmaceutical Biology*, vol. 157, no. 1, pp. 22–27, 1965.

[2] M. M. A. Rizvi, M. Irshad, G. El Hassadi, and S. B. Younis, "Bioefficacies of *Cassia fistula*: an Indian labrum," *African Journal of Pharmacy and Pharmacology*, vol. 3, no. 6, pp. 287–292, 2009.

[3] T. Bhakta, P. K. Mukherjee, K. Mukherjee, M. Pal, and B. P. Saha, "Studies on *in vivo* wound healing activity of *Cassia fistula* linn. Leaves (Leguminosae) in rats," *Natural Product Sciences*, vol. 4, no. 2, pp. 84–87, 1998.

[4] M. Irshad, M. Singh, and M. M. A. Rizvi, "Assessment of anthelmintic activity of *Cassia fistula* L," *Middle East Journal of Science and Research*, vol. 5, pp. 346–349, 2010.

[5] M. Irshad, S. Shreaz, N. Manzoor, L. A. Khan, and M. M. A. Rizvi, "Anticandidal activity of *Cassia fistula* and its effect on ergosterol biosynthesis," *Pharmaceutical Biology*, vol. 49, no. 7, pp. 727–733, 2011.

[6] V. L. Singleton, R. Orthofer, and R. M. Lamuela-Raventós, "Analysis of total phenols and other oxidation substrates and antioxidants by means of folin-ciocalteu reagent," *Methods in Enzymology*, vol. 299, pp. 152–178, 1998.

[7] J. Zhishen, T. Mengcheng, and W. Jianming, "The determination of flavonoid contents in mulberry and their scavenging effects on superoxide radicals," *Food Chemistry*, vol. 64, no. 4, pp. 555–559, 1999.

[8] G. Miliauskas, P. R. Venskutonis, and T. A. Van Beek, "Screening of radical scavenging activity of some medicinal and aromatic plant extracts," *Food Chemistry*, vol. 85, no. 2, pp. 231–237, 2004.

[9] W. Boonchum, Y. Peerapornpisal, D. Kanjanapothi et al., "Antioxidant activity of some seaweed from the Gulf of Thailand," *International Journal of Agriculture and Biology*, vol. 13, no. 1, pp. 95–99, 2011.

[10] S. Dudonné, X. Vitrac, P. Coutiére, M. Woillez, and J. M. Mérillon, "Comparative study of antioxidant properties and total phenolic content of 30 plant extracts of industrial interest using DPPH, ABTS, FRAP, SOD, and ORAC assays," *Journal of Agricultural and Food Chemistry*, vol. 57, no. 5, pp. 1768–1774, 2009.

[11] S. Luqman, S. Srivastava, R. Kumar, A. K. Maurya, and D. Chanda, "Experimental assessment of Moringa oleifera leaf and fruit for its antistress, antioxidant, and scavenging potential using in vitro and in vivo assays," *Evidence-Based Complementary and Alternative Medicine*, vol. 2012, Article ID 519084, 2012.

[12] L. Yu, S. Haley, J. Perret, M. Harris, J. Wilson, and M. Qian, "Free radical scavenging properties of wheat extracts," *Journal of Agricultural and Food Chemistry*, vol. 50, no. 6, pp. 1619–1624, 2002.

[13] G. P. Ganu, S. S. Jadhav, and A. D. Deshpande, "Antioxidant and antihyperglycemic potential of methanolic extract of bark of mimusops elengi l. In mice," *International Journal of Phytomedicine*, vol. 2, no. 2, pp. 116–123, 2010.

[14] M. Oyaizu, "Studies on products of browning reactions: antioxidant activities of products of browning reaction prepared from glucose amine," *Japanese Journal of Nutrition*, vol. 44, pp. 307–315, 1986.

[15] K. Zhou and L. Yu, "Effects of extraction solvent on wheat bran antioxidant activity estimation," *LWT—Food Science and Technology*, vol. 37, no. 7, pp. 717–721, 2004.

[16] T. Sun and C. T. Ho, "Antioxidant activities of buckwheat extracts," *Food Chemistry*, vol. 90, no. 4, pp. 743–749, 2005.

[17] H. Mohamed, M. Ons, E. T. Yosra, S. Rayda, G. Neji, and N. Moncef, "Chemical composition and antioxidant and radical-scavenging activities of Periploca laevigata root bark extracts," *Journal of the Science of Food and Agriculture*, vol. 89, no. 5, pp. 897–905, 2009.

[18] O. I. Aruoma, "Free radicals, oxidative stress, and antioxidants in human health and disease," *Journal of the American Oil Chemists' Society*, vol. 75, no. 2, pp. 199–212, 1998.

[19] J. Burgess and R. H. Prince, "1132. Kinetics of reactions of ligand-substituted tris-(2,2′- bipyridyl)iron(II) complexes," *Journal of the Chemical Society*, pp. 6061–6066, 1965.
[20] M. Cyfert, "Kinetics of reaction of $Fe(phen)_3^{2+}$ with hydrogen peroxide in neutral medium," *Inorganica Chimica Acta*, vol. 98, no. 1, pp. 25–28, 1985.
[21] C. Walling, "Fenton's reagent revisited," *Accounts of Chemical Research*, vol. 8, no. 4, pp. 125–131, 1975.
[22] A. Yildirim, A. Mavi, M. Oktay, A. A. Kara, O. F. Algur, and V. Bilaloglu, "Comparison of antioxidant and antimicrobial activities of Tilia (Tilia argentea Desf ex DC), sage (Salvia triloba L.), and Black tea (Camellia sinensis) extracts," *Journal of Agricultural and Food Chemistry*, vol. 48, no. 10, pp. 5030–5034, 2000.
[23] I. C. F. R. Ferreira, P. Baptista, M. Vilas-Boas, and L. Barros, "Free-radical scavenging capacity and reducing power of wild edible mushrooms from northeast Portugal: individual cap and stipe activity," *Food Chemistry*, vol. 100, no. 4, pp. 1511–1516, 2007.
[24] K. Shimada, K. Fujikawa, K. Yahara, and T. Nakamura, "Antioxidative properties of xanthan on the autoxidation of soybean oil in cyclodextrin emulsion," *Journal of Agricultural and Food Chemistry*, vol. 40, no. 6, pp. 945–948, 1992.
[25] N. R. Bhalodia, P. B. Nariya, R. N. Acharya, and V. J. Shukla, "Evaluation of in vitro antioxidant activity of flowers of *Cassia fistula* Linn," *International Journal of PharmTech Research*, vol. 3, pp. 589–599, 2011.
[26] I. I. Koleva, T. A. Van Beek, J. P. H. Linssen, A. De Groot, and L. N. Evstatieva, "Screening of plant extracts for antioxidant activity: a comparative study on three testing methods," *Phytochemical Analysis*, vol. 13, no. 1, pp. 8–17, 2002.
[27] I. F. F. Benzie and Y. T. Szeto, "Total antioxidant capacity of teas by the ferric reducing/antioxidant power assay," *Journal of Agricultural and Food Chemistry*, vol. 47, no. 2, pp. 633–636, 1999.
[28] M. H. Gordon, "The mechanism of antioxidant action in-vitro," in *Food Antioxidants*, B. J. F. Hudson, Ed., pp. 1–18, Elsevier Applied Science, London, UK, 1990.
[29] P. D. Duh, P. C. Du, and G. C. Yen, "Action of methanolic extract of mung bean hulls as inhibitors of lipid peroxidation and non-lipid oxidative damage," *Food and Chemical Toxicology*, vol. 37, no. 11, pp. 1055–1061, 1999.
[30] C. A. Rice-Evans, N. J. Miller, and G. Paganga, "Antioxidant properties of phenolic compounds," *Trends in Plant Science*, vol. 4, pp. 304–309, 1997.

3

Current Updates on Oxazolidinone and Its Significance

Neha Pandit,[1] Rajeev K. Singla,[2] and Birendra Shrivastava[1]

[1] *Department of Pharmaceutical Chemistry, School of Pharmaceutical Sciences, Jaipur National University, Jagatpura-Jaipur, Rajasthan 302025, India*
[2] *Sadbhavna College of Management & Technology, Jalaldiwal, Ludhiana-Barnala State Highway-13, Raikot (Ludhiana), Punjab, India*

Correspondence should be addressed to Rajeev K. Singla, rajeevsingla26@gmail.com

Academic Editor: Paola Fossa

Oxazolidinone is a five-member heterocyclic ring exhibiting potential medicinal properties with preferential antibacterial activity. Scientists reported various synthetic procedures for this heterocyclic structure. Current review articles tried to cover each and every potential aspect of oxazolidinone like synthetic routes, pharmacological mechanism of action, medicinal properties, and current research activities.

1. Introduction

The oxazolidinones are a new class of antimicrobial agents which have a unique structure and good activity against gram-positive pathogenic bacteria. Oxazolidinones are a class of compounds containing 2-oxazolidine in the structure. Oxazolidinones represent a new class of synthetic antibacterial agents active against multiple-resistant gram-positive pathogens, including methicillin-resistant *Staphylococcus aureus* (MRSA), penicillin-resistant streptococci, and vancomycin-resistant enterococci [1].

1.1. Chemical Structure [2]. Oxazolidinones are a class of azoles, oxazolidines with the carbon between the nitrogen and oxygen oxidized to a ketone, hence oxazolidinone (Figure 1).

1.2. Nomenclature. The antibacterial oxazolidinone template has a common nomenclature specially for aryl-5-(substituted) methyl-2-oxazolidinone. Throughout the paper, a consistent numbering system has been employed for description of the various oxazolidinone examined [3].

2. Synthetic Schemes

Earlier reviews have comprehensively covered the many synthetic approaches available for construction of the oxazolidinone ring. The most recent literature is replete with accounts of oxazolidinone templates that have seen extensive use as chiral auxiliaries.

(1) *Chiral Resolution Method.* Early work relied upon chiral resolution as a means to optically active oxazolidinone. The amino-diol resulting from reaction of an aniline with glycidol was resolved using (R)-mandelic acid. Diethylcarbonate effected cyclization to the 5-(R)-hydroxymethyl-3-phenyl-2-oxazolidinone. Numerous approaches had been employed for converting the 5-(R)-methylalcohol moiety to the 5-(S)-acetamidomethyl group. Here tosylation of the alcohol was followed by azide displacement, reduction, and acylation of the amine [4] (Figure 2).

(2) *Iodocyclocarbamation.* Many of the Upjohn SAR studies were carried out with racemic oxazolidinone, synthesized by iodocyclocarbamation. A key modification was the addition of pyridine, crucial for circumventing untoward alkylative side reaction [5] (Figure 3).

(3) *DuPONT Asymmetric Synthesis. The Herweh-Kauffmann/Speranza-Peppel Method.* This method involved high-temperature catalyst-mediated cyclization of an aryl isocyanate with an epoxide where LiBr catalyst was solubilized in refluxing xylene by tributylphosphine oxide. As demonstrated for DuP721 (R)-glycidol butyrate was cyclized with 4-acetylphenyl isocyanate, giving the oxazolidinone butyrate ester. Saponification provided the 5-(R)-hydroxymethyl oxazolidinone which was further

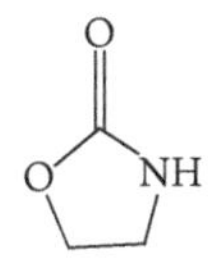

Figure 1: General Structure of oxazolidinone.

methylated and then converted to the acetamide, as described next for tosylate [5] (Figure 4).

(4) Synthesis of oxazolidin-2-ones derivatives was carried out starting from urea and ethanolamine reagents using microwave irradiation in a chemical paste medium in which a catalytic amount of nitromethane absorbed the microwaves and generated hot spots [6] (Figure 5).

(5) An efficient, versatile, and practical gram-scale preparation of oxazolidinone, imidazolidinone, and dioxolanone was achieved [6] (Figure 6).

(6) A mild and efficient gold (I)-catalyzed rearrangement of propargylic *tert*-butylcarbamates allowed the synthesis of various 5-methylene-1, 3-oxazolidin-2-ones, which would be less conveniently obtained using other methods [7] (Figure 7).

(7) Various *N*-Boc-protected alkynylamines were converted into the corresponding alkylidene 2-oxazolidinones or 2-oxazinones under very mild reaction conditions in the presence of a cationic Au (I) complex in high yield regardless of the substitution at nitrogen and alkyne terminus [8] (Figure 8).

(8) A nickel-catalyzed cycloaddition of aziridines with isocyanates proceeded smoothly to give iminooxazolidine derivatives in good yields. A longer reaction time allowed the isomerization of the iminooxazolidine to the corresponding imidazolidinone derivatives [9] (Figure 9).

(9) *Synthesis of Oxazolidinones by the Use of Halomethyloxirane, Primary Amine, and Carbonate Salt.* Primary amines reacted with carbonate salts (Na_2CO_3, K_2CO_3, Cs_2CO_3, and Ag_2CO_3) and halomethyloxiranes in the presence of a base such as DBU or TEA to give oxazolidinones in high yields. The use of K_2CO_3 among these carbonate gave the best yield in this synthesis. A reaction mechanism was proposed that the oxazolidinone was obtained from an oxazinanone intermediate via a bicyclo [2.2.1] intermediate. The present reaction can be widely applied to convenient synthesis of useful N-substituted oxazolidinones and chiral oxazolidinones [10] (Figure 10).

(10) *Direct Catalytic Asymmetric Amination of Aldehydes: Synthesis of Evans, Oxazolidinones and α-Amino Acids* [11] (Figure 11).

(11) *Conversion of Chiral α-Hydrazino Alcohols and N-Amino Oxazolidinone to Evans Auxiliaries [11](Figure 12).*

(12) *Solid-Phase Synthesis of Oxazolidinones by Cycloaddition of Resin-Bound Epoxides with Isocyanates.* The first solid-phase synthesis of oxazolidinones by cycloaddition of resin-bound epoxides with isocyanates is described. Synthesis of the title compounds was achieved by alkylation of resin-bound carbamates with glycidyl tosylate, followed by cycloaddition of the resulting epoxides with isocyanates at elevated temperature in high yields and purity. Because *N*-aryloxazolidinones have been known to possess various biological activities, this method is useful from the viewpoint of drug discovery [12] (Figure 13).

(13) *A Convenient Diastereoselective Synthesis Of Oxazolidinone: Approach to Unusual Amino Acid Statine* [13] (Figures 14 and 15).

(14) *Synthesis of* [5, 14, 14]*-Tricyclic Fused Oxazolidinone [15] (Figure 16).*

3. Pharmacological Background

Mechanism of Action. Oxazolidinones inhibit protein synthesis by binding at the P site at the ribosomal 50S subunit. Resistance to other protein synthesis inhibitors does not affect oxazolidinone activity; however rare development of oxazolidinone resistance cases, associated with 23S r-RNA alterations during treatment, has been reported. Linezolid (Figure 17), the first oxazolidinone available, has already taken its place in the clinic for treatment of gram-positive infections.

It selectively inhibits bacterial protein synthesis through binding to sites on the bacterial ribosome and prevents the formation of a functional 70S-initiation complex. Specifically, linezolid binds to a site on the bacterial 23S ribosomal RNA of the 50S subunit and prevents the formation of a functional 70S initiation complex (Figure 18), which is an essential component of the bacterial translation process [16].

4. Pharmacokinetic Parameter [17, 18]

(i) *Absorption.* It is rapidly and extensively absorbed after oral dosing. Maximum plasma concentrations are reached approximately 1 to 2 hours after dosing, and the absolute bioavailability is approximately 100%.

(ii) *Toxicity.* Clinical signs of acute toxicity lead to decreased activity, ataxia, vomiting and tremors.

(iii) *Protein Binding.* 31%.

(iv) *Biotransformation.* It is primarily metabolized by oxidation of the morpholine ring, which results in two inactive ring-opened carboxylic acid metabolites: the aminoethoxyacetic acid metabolite, and the hydroxyethyl glycine metabolite.

(v) *Half-Life.* 4.5–5.5 hours.

(vi) *Excretion.* Renal and fecal (Table 1).

4.1. Available Marketed Formulation [19]

4.1.1. Bacterial Resistace and the Search for Novel Antibacterial Structural Templates [5]

(i) The alarming escalation seen worldwide in the incidence of bacterial resistance to previously effective antibiotics continues to provide the impetus for the medicinal chemist to search for entirely new classes of antibacterial agents, that can cure bacterial infection by novel mechanism. As numerous bacteria have

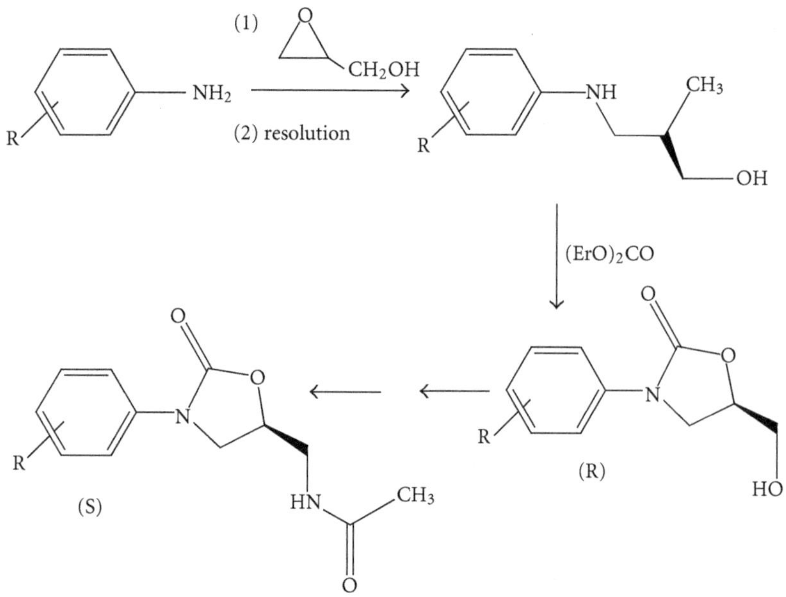

FIGURE 2: Chiral resolution method.

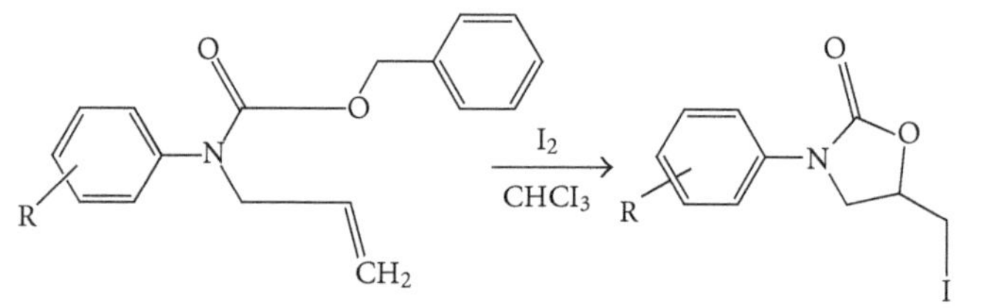

FIGURE 3: Iodocyclocarbamation.

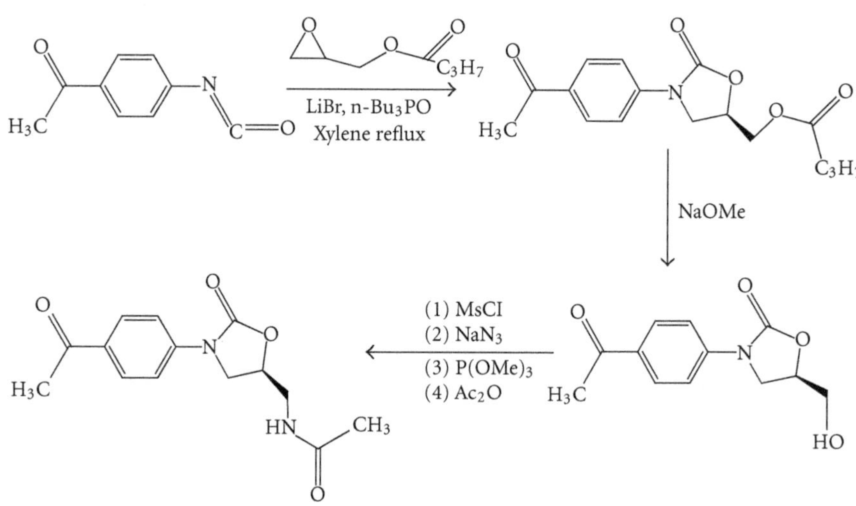

FIGURE 4: Du PONT asymmetric synthesis.

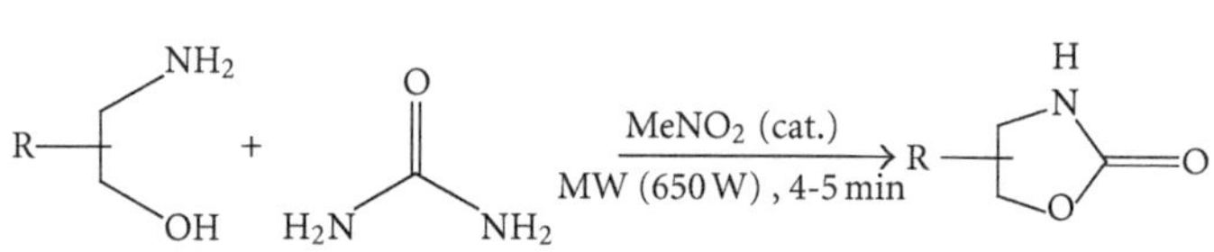

FIGURE 5: Oxazolidinones using urea and ethanolamine.

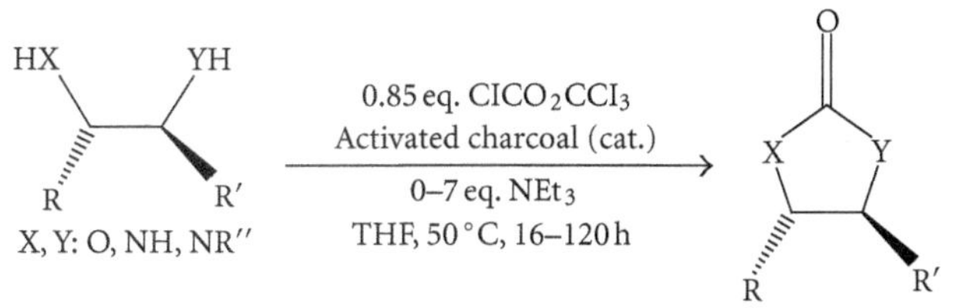

FIGURE 6: Efficient synthesis of oxazolidinone, imidazolidinone, and dioxolane.

TABLE 1: Available marketed formulation of oxazolidinone.

Brand name	Dosage form	Dosage	Company
Infulid	TAB	600 mg	Neiss
Linosept	TAB	600 mg	Micro Eros
Linospan	TAB	600 mg	Cipla
	IV	200 mg/100 mL	
Linox	TAB	600 mg	Unichem
Lizbid	TAB	600 mg	Piramal HC
Linezolid	TAB	600 mg	Glenmark (Integrace)
	IV BAG	600 mg	
Lizomed	TAB	600 mg	Aglowmed
	D-SYR	100 mg	

increasingly evidenced the evolution of multiple-antibiotic resistance, health care provides have been seriously challenged to provide effective therapy for the often life-threatening infection caused by these pathogens. Numerous reviews have recently appeared emphasizing the extent and severity of the resistance problem as it exists today, and the dim prospects envisioned for the future.

(ii) Some of the most problematic organism has been the multi-drug-resistant gram-positive bacteria. These include the highly virulent organism, methicillin-resistant *Staphylococcus aureus* MRSA, and penicillin- and cephalosporin-resistant *Streptococcus pneumoni.*

(iii) The enterococci have become increasingly of concern; there have been isolated problems of infection with vancomycin-resistant enterococci VREae. A major concern is that VRE will transfer the vancomycin-resistance genes encoded on a plasmid, to much more virulent organism *S. aureus,* as Nobel et al. have demonstrated to be feasible in an experimental setting.

Still another problem microorganism for which resistance has generated considerable medical concern is multidrug-resistant *M. tuberculosis* MDRTB.

This loss of antibacterial activity among drugs once efficacious against these pathogens has led to a summoning from numerous experts for the discovery and development of new antibiotic classes, so that health care practitioners will not be left benefit of effective therapeutic modalities. The oxazolidinone antibacterial agents are capable to provide an answer for these problems in new age.

5. Bacteriostatic versus Bactericidal Nature of Oxazolidinones [20]

Time-kill studies indicate that both Dup-721 (Figure 19) and Dup-105 (Figure 19) were bacteriostatic against all organism tested, expect the diphtheroids but U-100592 (Figure 19) and U-100766 (Figure 19) are bactericidal for *S. pneumoniae* and bacteriostatic for most staphylococci and enterococci.

5.1. Combination Therapy with Other Antibiotics [21]

(i) Antagonism between Dup-721 and ciprofloxacin or norfloxacin was reported for MRSA, *S. epidermidis* and *E. faecalis.*

(ii) In *In vivo* combination therapy studies, both drugs were found additive with vancomycin, gentamycin, or rifampin against *S. aureus*. Similarly, in a mixed *S. aureus* and *E. coli* infection model combination therapy with aztreonam or gentamycin as well as vancomycin, was found to be effective.

(iii) Oxazolidinones are a new class of totally synthetic antimicrobial agents against multidrug-resistant gram-positive bacteria, including methicillin-resistant *Staphylococcus aureus* (MRSA), *Staphylococcus epidermidis* (MRSE), penicillin-resistant *Streptococcus pneunomiae* (PRSP) and vancomycin-resistant enterococci (VRE) [22–25]. Oxazolidinone has:

(a) a novel mechanism of action,
(b) selectively binding,
(c) uniquely binding to the 50S ribosomal subunit,
(d) inhibiting bacterial translation at the initiation phase of protein synthesis,
(e) consequently, the drug would not show cross resistance with existing antibacterial agents.

The unique mechanism of action of oxazolidinones has attracted interest to develop derivatives with potent activity and broad spectrum.

6. Structure Activity Relationship (SAR) [26, 27]

To find out the most potent and comparatively less toxic compounds, we need to to do exhaustive study of SAR.

This paper cites the modifications directed for the different parts of the oxazolidinone template as shown in Figure 20. The "A" part manifests oxazolidinone ring, which bears aryl system on the 3rd position of the oxazolidinone ring termed as B region and the 4th position of the aryl group is extended by amine functionality that has been termed as C region. Oxazolidinone ring exclusively possess C5 side chain in the (*S*)-configuration, which has been optimized for the better efficacy.

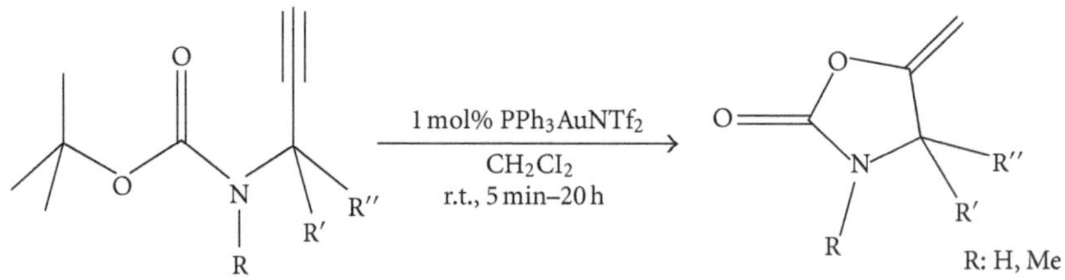

FIGURE 7: Gold catalyzed rearrangement in oxazolidinone synthesis.

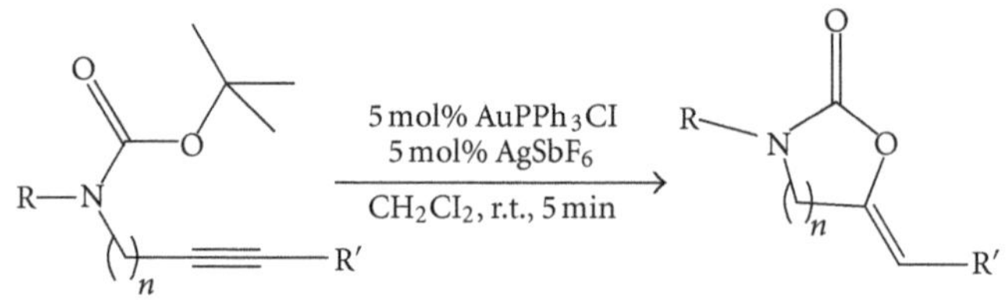

FIGURE 8: N-Boc protected oxazolidinone synthesis.

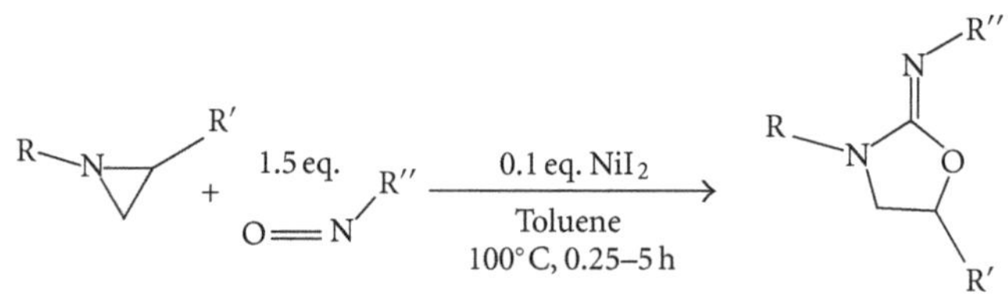

FIGURE 9: Nickel catalyzed cycloaddition in oxazolidinone synthesis.

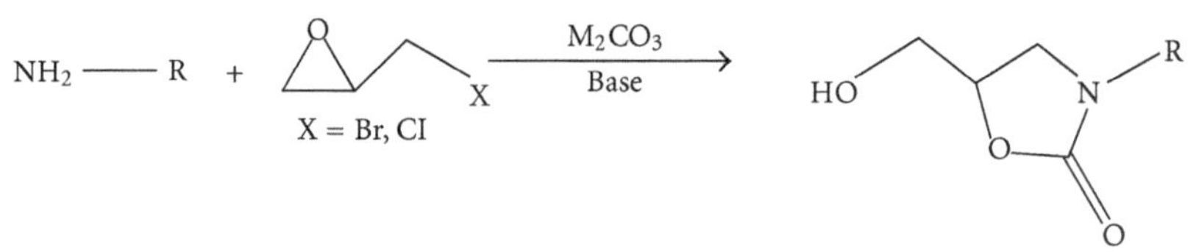

FIGURE 10: Use of halomethyloxirane, primary amine, and carbonate salt in oxazolidinone synthesis.

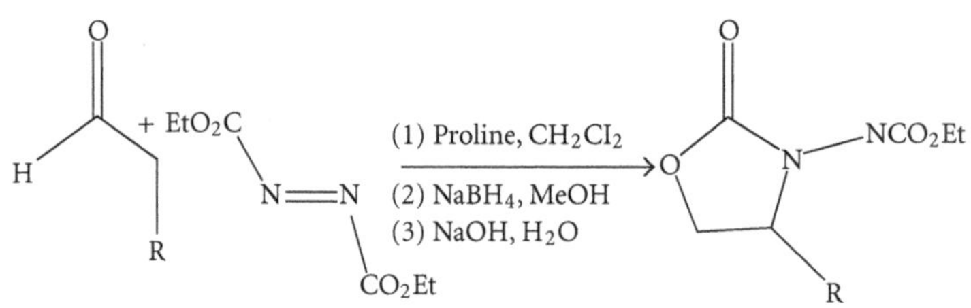

FIGURE 11: Direct catalytic asymmetric amination of aldehydes: synthesis of Evans, oxazolidinones and α-amino acids.

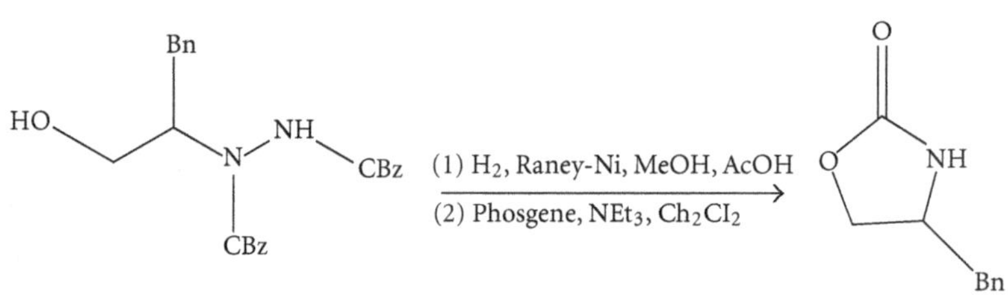

FIGURE 12: Conversion of chiral α-hydrazino alcohols and N-amino oxazolidinone to evans auxiliaries.

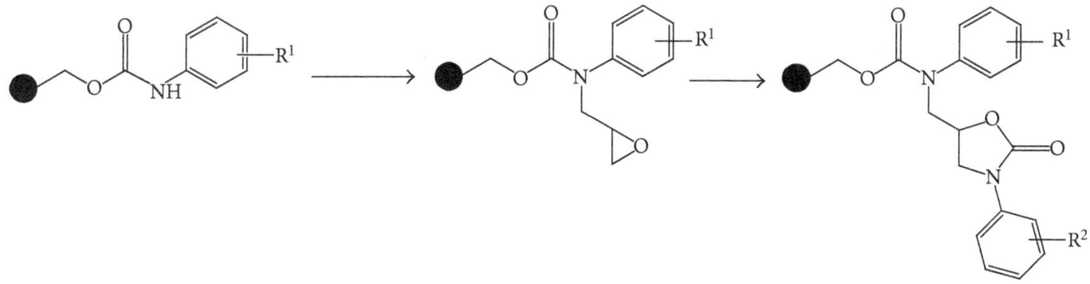

FIGURE 13: Solid-phase synthesis of oxazolidinones.

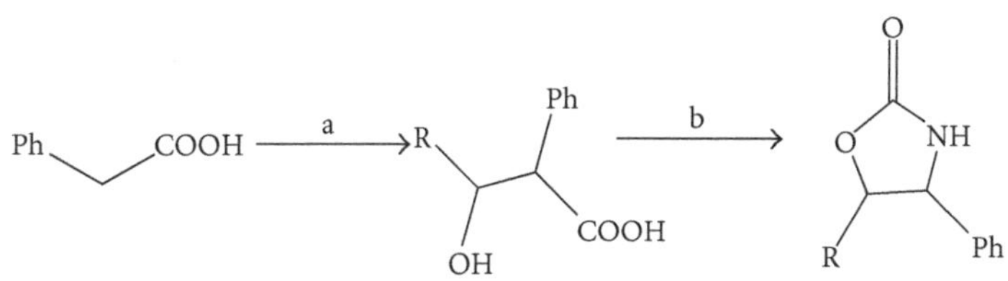

R = 2-thiophene
a = −78°C, THF
b = $(PhO)_2PON_3$, Et_3N, t-BuOH, 80°C

FIGURE 14: Diastereoselective synthesis of oxazolidinone part-I.

6.1. Synthesized Derivative of Oxazolidinone by Modification

(1) *Modifications in the Region A with C5*

(i) The thioanalogs of linezolid and eperezolid, **6** (Figure 21) and **7** (Figure 21), do not seem to bind to the 50S ribosomal subunit of the bacteria and thereby are unable to inhibit bacterial protein synthesis. The molecular modeling study supported this unexpected finding of oxazolidine-2-thione [16]. The similar trend of antibacterial activity has been reported in the early investigation for compound **8** (Figure 21), a thioanalogue of DuP 721 [28, 29].

(ii) Quesnelle et al. have reported isoxazolidinone agents having various substitutions at the C-region and most of the compounds have been found to be potent against gram-positive strains [30]. The compounds **9** (Figure 21) and **10** (Figure 21) have shown broad-spectrum antibacterial activity, additionally the sulfoxide **9** exhibited less toxicity compared to the linezolid in a seven-day study in rats.

(iii) Wang et al. have reported series of compounds with scaffolds **11, 12,** and **13** (Figure 21) containing chiral 1, 3 oxazinan-2-ones and oxazolidinones as basic core structures having tertiary amines core containing two aryl substituents [31].

(2) *Modifications in the Region B and at the C5* [32]. Bosch et al. have reported convenient synthesis of conformationally constrained analog of linezolid **14** (Figure 21) having a tricyclic moiety; however, no antibacterial activity has been disclosed for this compound.

Merck has described novel oxazolidinone derivatives **15** and **16** substituted with cyclopropyl moiety [33]. The compounds **15** (Figure 22) and **16** (Figure 22) exhibited impressive *in vitro* antibacterial activities against different strains **15**.

Selvakumar et al. have reported constrained analogues of linezolid such as hexahydroazolo-quinoxaline and tetrahydroazolo-benzothiazine compounds [34]. Amongst the compound of this class the tetrahydroazolo-benzothiazine and thiocarbamate at C5 side chain **17** (Figure 22) have shown antibacterial activity in the range of 0.25–1 μg/mL against resistant and sensitive gram-positive strains.

Choy et al. from Pfizer reported the synthesis of conformationally restricted oxazolidinone compounds [35]. The compound **18** (Figure 22) was found to be more potent and exhibited broad-spectrum antibacterial activity with MIC values in the range of <0.06–0.25 μg/mL for gram-positive organisms and 1-2 μg/mL for fastidious gram-negative organisms.

(3) *Modifications in the Region C with C5* [36]. A series of *N*-phenyl piperazinyl derivatives of oxazolidinone **19** (Figure 22) in which the nitrogen atom at 4-position of piperazinyl ring is substituted by different cinnamoyl groups. This optimization resulted in few potent compounds, which were found to be active against several gram-positive pathogens. In this class of compound some substituents were well tolerated on the phenyl ring of cinnamoyl group.

There is a reported synthesis of a few Mannich ketones of piperazinyl phenyl oxazolidinone derivatives and their antibacterial activity in various Gram-positive organisms such as *Bacillus subtilis, Staphylococcus aureus, Staphylococcus epidermidis,* and *Enterococcus faecalis* [37]. A moderately

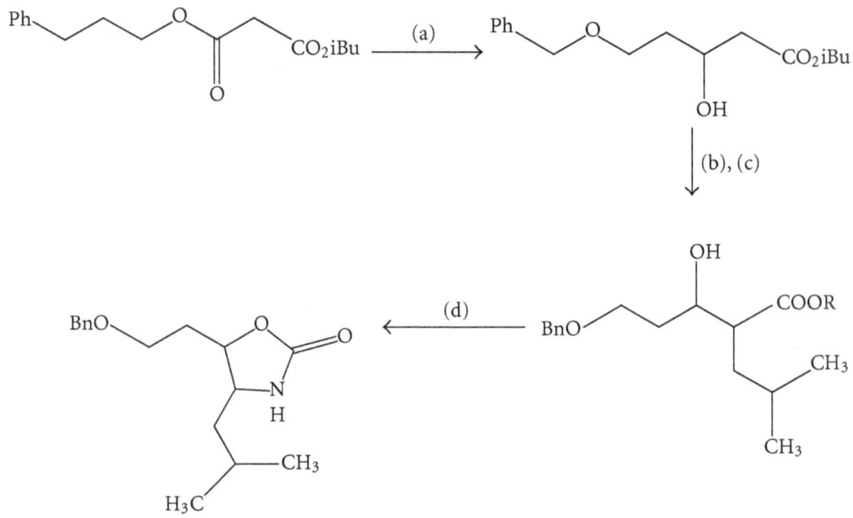

(a) = Baker's yeast, sucrose, 25°C, 3 days

(b) = −78°C, THF

(c) = NaOH, EtOH

(d) = $(PhO)_2PON_3$, Et_3N, tBuOH, 80°C

FIGURE 15: Diastereoselective synthesis of oxazolidinone part-II.

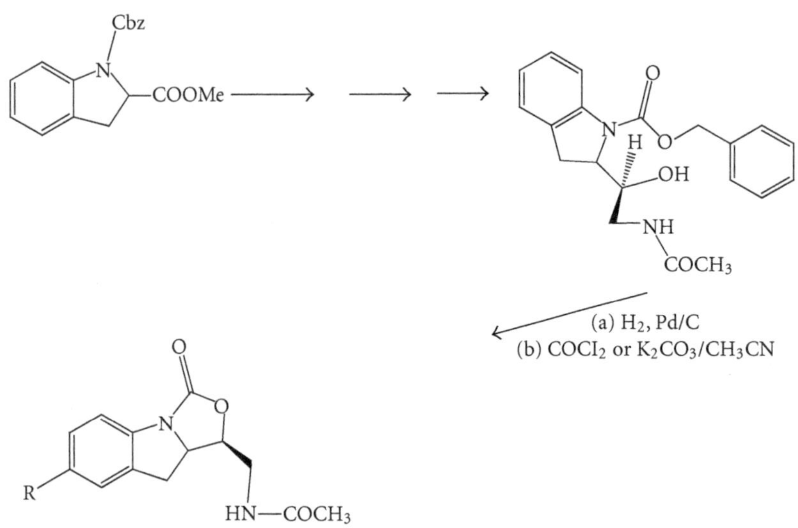

FIGURE 16: Diastereoselective synthesis of oxazolidinone part-II.

active compound **20** (Figure 22) has been transformed to active compound **21** (Figure 22). The compound **22** (Figure 22) having cyclohexanone has shown antibacterial activities against various gram-positive strains with MIC values in the range of 1–4 μg/mL, which is comparable to that of linezolid (0.5–4 μg/mL).

Ranbaxy has reported synthesis of oxazolidinones modified at the C region. The optimization of the series afforded a potent compound **23** (Figure 22) (Ranbezolid, RBx7644), which is under clinical development [38]. They have further modified the compound **23** to get the compound **24** (Figure 23), wherein nitrofuran ring is attached to the 2° nitrogen of the aminopiperidine.

In another embodiment a group of scientists from Merck has published a patent describing novel oxazolidinone derivatives possessing cyclopropyl moiety [33]. The disclosed compounds **25** and **26** (Figure 23) have shown excellent *in vitro* antibacterial activity against broader panel of both susceptible and resistant strains.

Scientists from Orchid have described novel oxazolidinones **27** (Figure 23) having variations in the regions B, C, and at the C5 side chain of oxazolidinone [39].

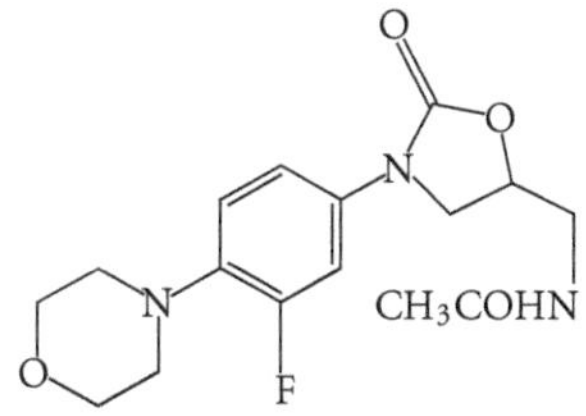

Figure 17

Pfizer has reported a series of oxazolidinone compounds **28** (Figure 23) containing dihydrothiopyran substituent at the C region and instead of acetamide other variations such as halogenated amide and thioamide derivatives were studied at the C5 side chain [40]. The *in vitro* activities of the C5 modified derivatives were found to be superior compared to the corresponding C5 acetamide derivative. The compounds of this class have shown better antibacterial activities towards Gram-positive *Staphylococcus pneumoniae, Enterococcus faecium,* and Gram-negative *Haemophilus influenzae* as compared to the compound **29** (Figure 23) (PNU-288034).

Zhai et al. have reported novel oxazolidinone derivatives having modified C-5 side chain. The compounds **30**, **31**, and **32** (Figure 24) have shown inferior *in vitro* antibacterial activities in MIC assay against various strains [41].

Trius Therapeutics has announced the initiation of Phase 1 clinical trials of TR-701 **33** (Figure 24) for the treatment of patients with serious Gram-positive bacterial infections, including those caused by methicillin-resistant *Staphylococcus aureus* and other drug-resistant strains [42].

A group of scientists from Ranbaxy has patented novel substituted phenyl oxazolidinone derivatives [43]. The compound **34** (Figure 24) of this series has shown significant *in vitro* antibacterial activity in the MIC assay against different strains such as methicillin-resistant *Staphylococcus aureus* (4.0 μg/mL; linezolid 2.0 μg/mL), vancomycin-resistant *Enterococci* (2.0 μg/mL; linezolid 2.0 μg/mL).

A series of arylcarbonyl and arylsulfonyl-piperizinyl-5-triazolylmethyl oxazolidinones **35** (Figure 24) with improved activity against Gram-positive bacterial clinical isolates than Gram-negative bacterial clinical isolates [44].

Fan et al. have reported triazolyl oxazolidinones as antibacterial agents. Most of the analogues displayed activity superior to the linezolid and the vancomycin in various Gram-positive bacteria [45]. In the antibacterial MIC assay, the compounds **36**, **37**, and **38** (Figure 24) were found to be potent and further studied for their *in vivo* efficacies in mice model; however none of the compounds showed *in vivo* activity.

Selvakumar et al. have reported novel chalcone oxazolidinone hybrids with antibacterial activity. Of these, the compound **39** (Figure 24) containing chalcone substituent in the C region showed significant antibacterial activities with MIC values of 4 μg/mL against methicillin-resistant *Staphylococcus aureus* strain [46]. The acetamide group at C-5 was converted to thiocarbamate to get compound **40** (Figure 25), which exhibited *in vitro* activity in the range of 0.25–2 ug/mL against resistant strains.

Reck et al. at AstraZeneca have reported substituted (pyridin-3-yl) phenyloxazolidinones as antibacterial agents with reduced activity against monoamine oxidase A [47]. The compound **41** (Figure 25) showed excellent activity against Gram-positive bacteria; however the compound **41** lacks the monoamine oxidase A inhibition and inhibits cytochrome P450 (CYP) due to its poor solubility.

Liu et al. have reported novel substituted oxotriazolylphenyl derivatives. Of the many compounds described, the compound **42** (Figure 25) has shown potent antibacterial activity against pathogens *Staphylococcus epidermidis, Streptococcus pneumoniae, Streptococcus enteritidis, and Streptococcus nonhemolyticus* at 0.10–6.25 μg/mL [48] O Vara Prasad et al. have reported new series of oxazolidinones with C5 carboxamide functionality, that is, reverse amides, which blocks the bacterial protein synthesis. These compounds have also exhibited less potency against monoamine oxidase enzymes, indicative for the lower side effects. The compound **43** (Figure 25) has shown reduction of myelotoxicity in a 14-day safety study in rodents and compared with linezolid [49].

Rudra et al. have reported series of heterocyclic oxazolidinones as antibacterial agents with identification of RBx 8700 as potent compound [50]. Systematic modification of the linker between the five-membered heterocycle and the piperazinyl ring of RBx 7644 and its thienyl analogue **44** (Figure 25) resulted in the identification of an expanded spectrum compound.

Reck et al. reported novel acyclic substituted (pyridin-3-yl) phenyl oxazolidinones **45**–**47** [51] (Figure 25). The acyclic 6-substituted pyridin-3-yl phenyl oxazolidinones **46** and **47** having bulky substituent demonstrated excellent antibacterial activity against Gram-positive organisms, including linezolid resistant *Streptococcus pneumonia* (1.0 μg/mL). The compounds with bulkier substituents showed reduced MAO-A activity (Ki = 40 μM; Ki = 19 μM) compared to the corresponding unsubstituted parent compound **48** (Figure 25) (Ki < 0.3 μM) and the improved solubility.

The high-molecular-weight novel oxazolidinone derivatives having variations in the regions B, C, and at the C5 side chain have been patented by a group of scientists from Ferrer International Company [52].

(4) *Hybrid Molecules* [53]. Morphochem AG has published a patent mentioning oxazolidinone-quinolone hybrid **49** (Figure 26). However, no discloser has been made for their antibacterial activities. Further, in another report they have described different variant of oxazolidinone-quinolone hybrids of type **50**. The disclosed derivatives **50** (Figure 26) possess the pharmacophore of quinolone and oxazolidinone linked together through a linker and the hybrid has been used as an antibacterial agent. However, there is no mention of the antibacterial activities for the disclosed oxazolidinone-quinolone hybrids.

(5) *Computational Studies.* Extensive studies on the syntheses of oxazolidinone antibacterial agents and their SAR have been reported; however only few QSAR studies have been published. The first QSAR study was reported in the year 1999, comprised a 3D-QSAR study on a small data

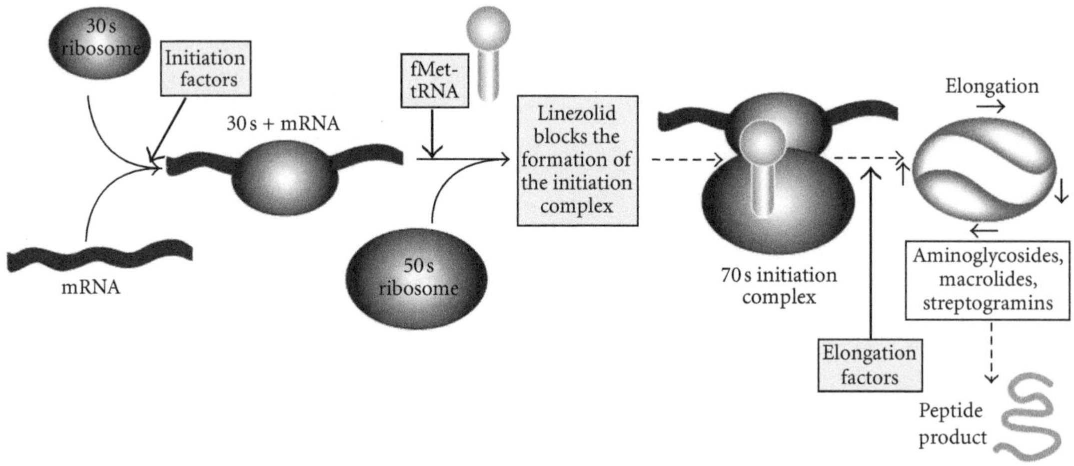

FIGURE 18

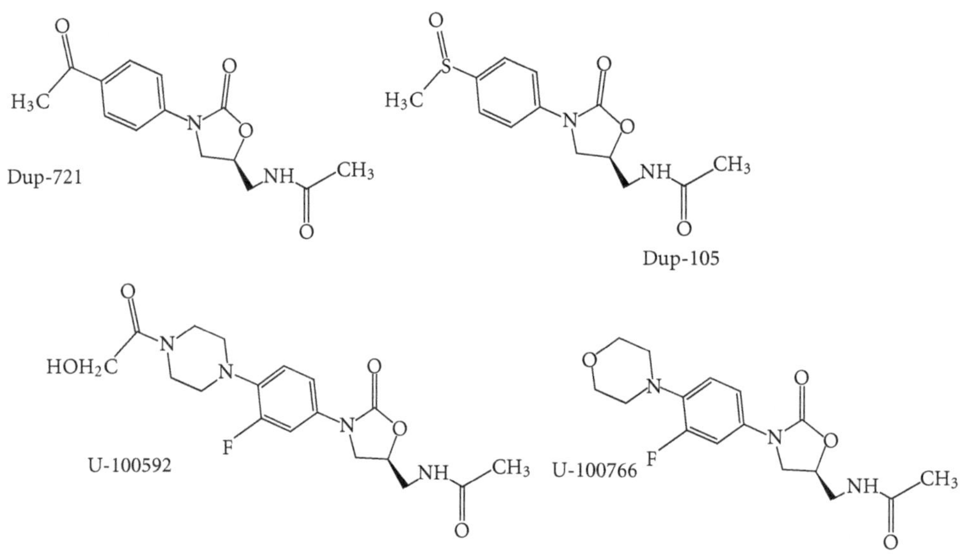

FIGURE 19

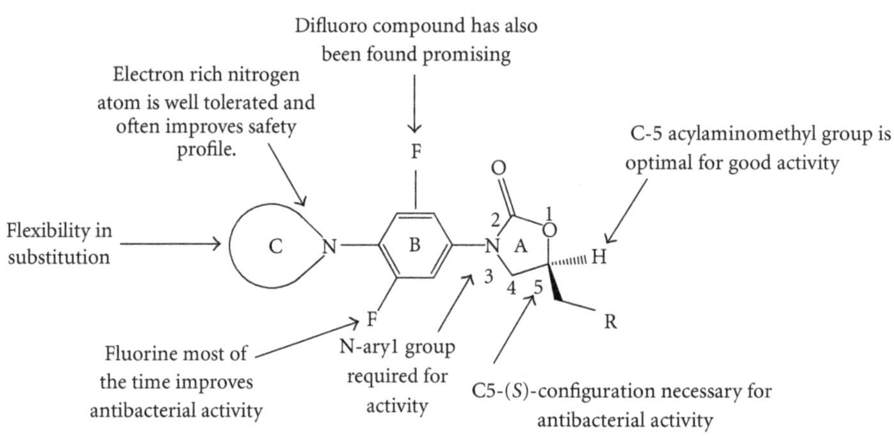

FIGURE 20

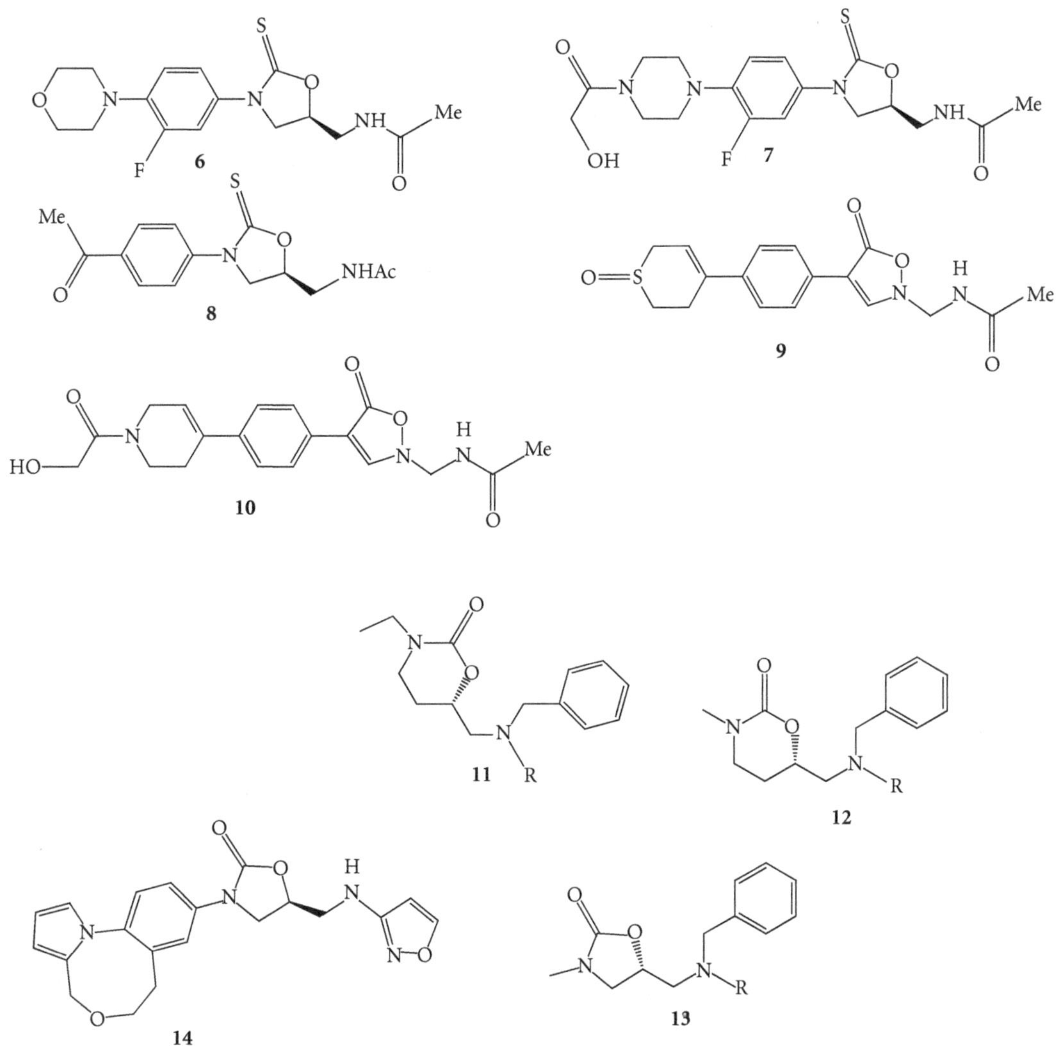

Figure 21

sets of two novel series of oxazolidinone antibacterial agents using "Comparative Molecular Field Analysis" (CoMFA), wherein a training set of 17 compounds with two reference compounds have been used. The CoMFA steric, electrostatic fields, and ClogP were used as descriptors, and the activities against methicillinresistant *Staphylococcus aureus* 88 (MRSA 88) as dependents were developed. The cross-validated r2cv (0.653) and conventional r2 (0.984) from the Partial Least Square (PLS) and CoMFA analyses indicated considerable reliability for predicting the antibacterial activities of oxazo lidinone antibacterial agents.

The second 3D-QSAR studies, reported by Karki and Kulkarni in the year 2001, used a genetic function algorithm with a data set of 60 compounds. The QSAR models were developed using a training set of 50 compounds and the *in vitro* MIC against *Staphylococcus aureus* SFCO-1. The r2 values reported for the models range from 0.629 to 0.732. The predictive ability of the QSAR model was evaluated with a test set of 10 compounds. The results obtained concluded that the antibacterial activity of the 3-aryloxazolidin-2-ones is strongly dependent on the electronic factor as expressed by lowest unoccupied molecular orbital energy (LUMO) and the spatial factor as expressed by density and thermodynamic factors accounted for molar refractivity and the heat of formation. The SAR study by Tokuyama et al. revealed that the antibacterial activity against Gram-positive bacteria including methicillin-resistant *Staphylococcus aureus* and vancomycin-resistant enterococci on 5 thiocarbonyl oxazolidinones was significantly affected by the lipophilicity, especially the calculated log *P* value, the balance between 5-hydrophilic (or hydrophobic) substituents and hydrophobic (or hydrophilic) substituents on the benzene ring (Table 2).

6.2. Oxazolidinone in Clinical Trials

6.2.1. Oxazolidinone: Biological Activity. The development of resistance by the antibiotics in the Gram-positive pathogenic bacteria over the last twenty years and continuing today

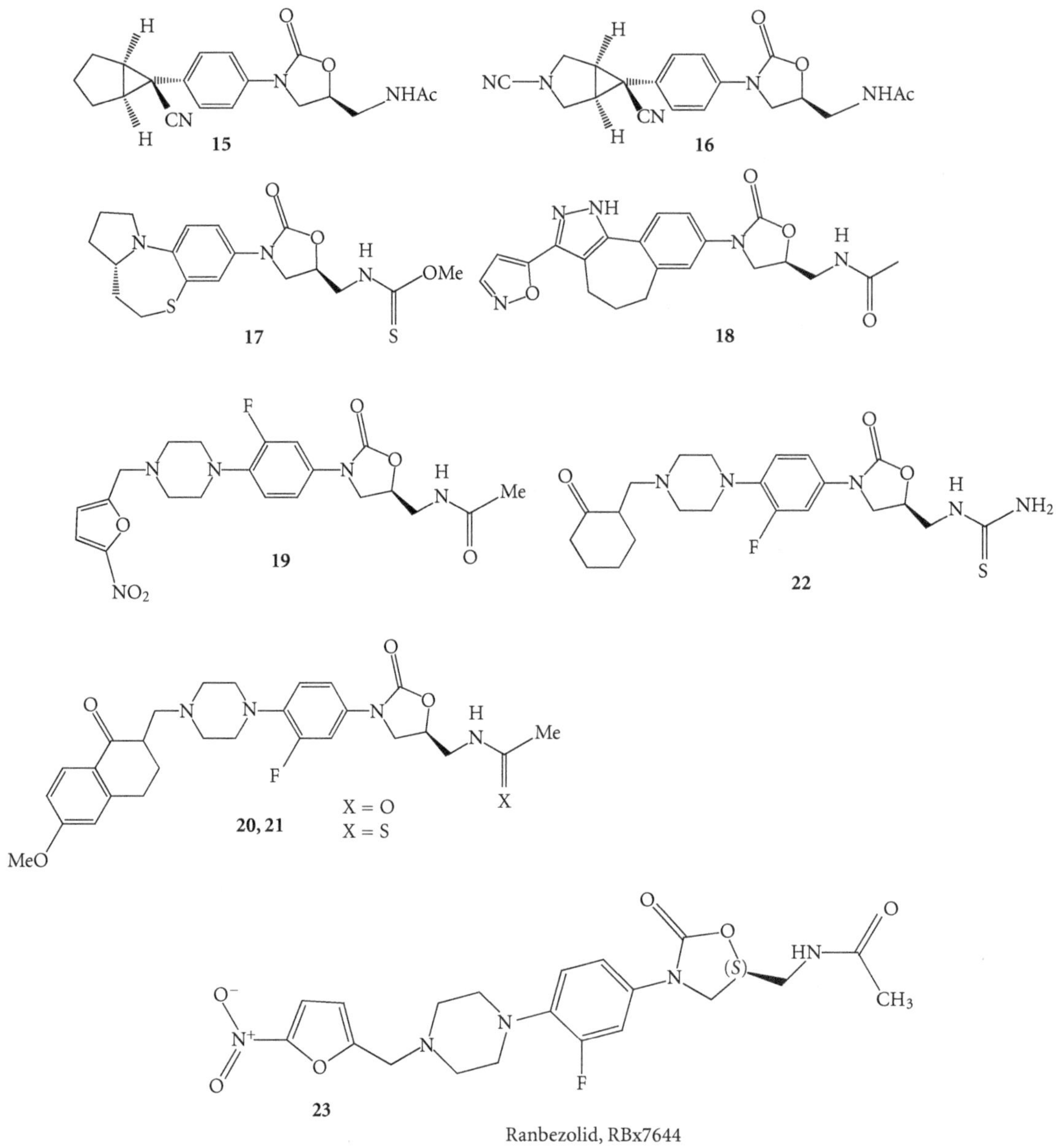

(a)

Figure 22

has created a need for new antibiotic classes, which may be unaffected by existing bacterial resistance. The oxazolidin-2-ones not only represent a new class with a novel mechanism of action, but also satisfy the requirement for overcoming the resistance mechanisms. Both linezolid and eperezolid, the first chemical candidates, arose from the piperazine subclass, with the first one being chosen further development because of its enhanced pharmacokinetic properties. The main attractive traits of the oxazolidinone series have encouraged further work in the area, and the patent literature reveals that extensive chemical investigation is currently being made. The unexpected early resistance development emphasizes the need for further exploration of features of the oxazolidinone to eliminate these deficiencies. Recently, several changes, involving the C5 side chain as well the N-phenyl heterocyclic ring, give promise for such improvement. Various biological activity show by oxazolidinone derivatives like the following.

(1) *Antibacterial Activity* [54]. Linezolid is an oxazolidinone developed by Pharmacia (formerly Pharmacia & Upjohn) for the treatment of multiresistant Gram-positive infections 187765, 317456. Linezolid resistance due to a 23S rRNA mutation may emerge in Enterococci during therapy with this antimicrobial and may be associated with clinical failure 368652. Following FDA approval, linezolid was launched in May 2000 368526, 368652. In April 2000, the FDA approved linezolid injection, tablets,

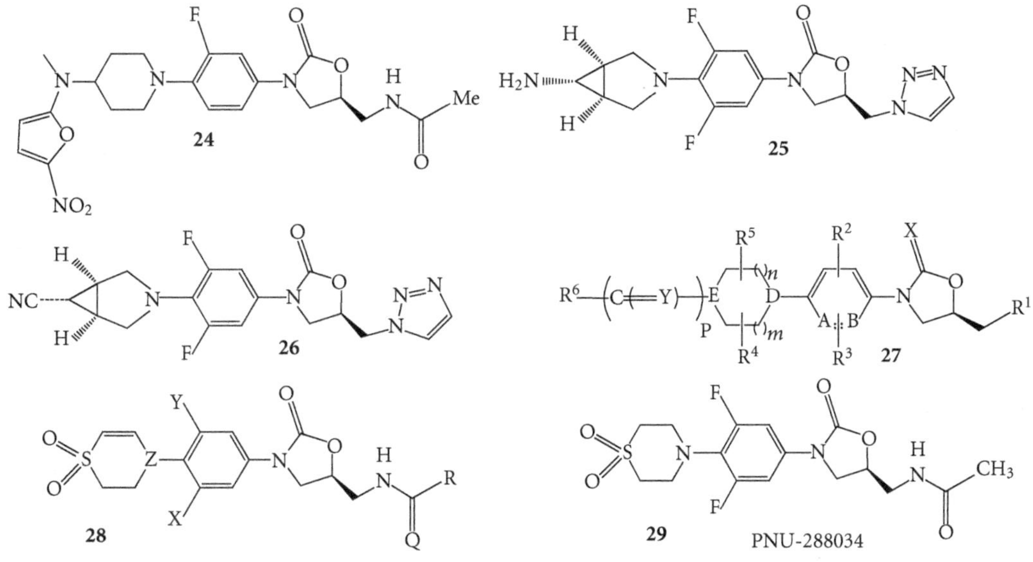

X = H, F;   Y = H, F
Z = CH, N; Q, S
R = CH_3, CHF_2, $CHCl_2$

Figure 23

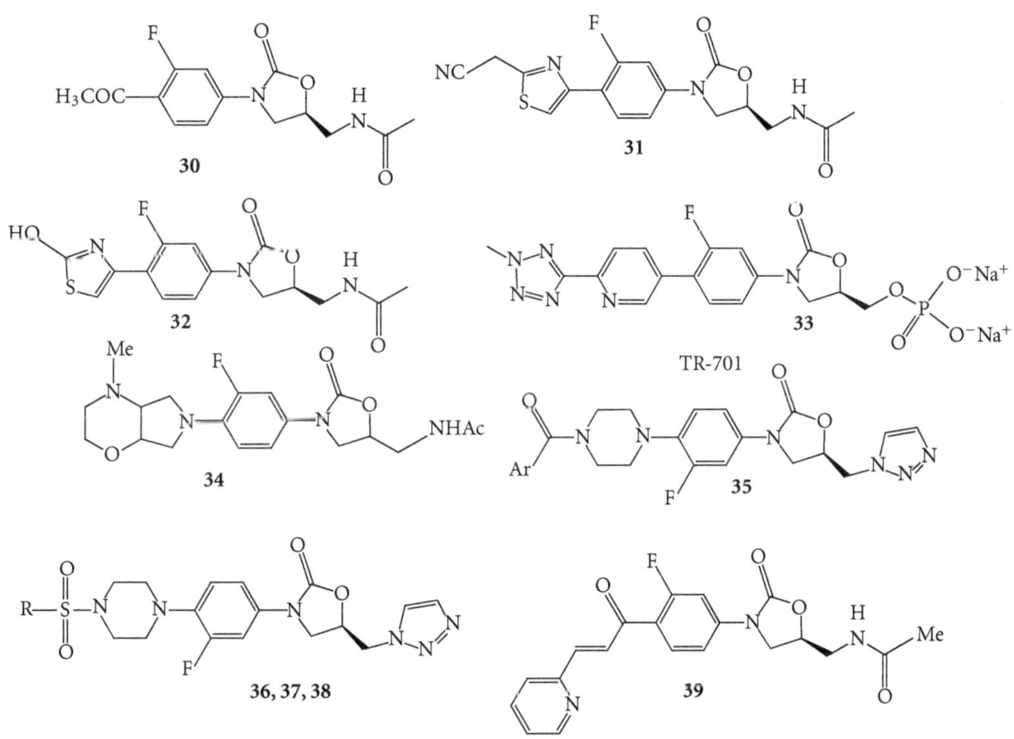

R = *m*-F-Ph
R = *m*-NO_2-Ph
R = *m*-NH_2-Ph

Figure 24

Table 2

S. no.	Compound name	Clinical phase	Sponser	Conditions
(1)	Rivaroxaban (BAY59-7939)	phase-III	Bayer's Johnson & Johnson Pharmaceutical Research Development	Deep vein thrombosis (DVT) or Pulmonary embolism (PE)
(2)	Radezolid	phase-II Complete	Rib-X pharmaceutics	Uncomplicated skin infection
(3)	Torezolid	phase-II	Trius Therapeutics	Complicated skin and skin-structure infections (cSSSI)
(4)	DA-7218	phase-III	Dong-A Pharmaceutical	Acute bacterial skin and skin-structure infections (ABSSSI)
(5)	MRX-1	phase-I	MicuRx Pharmaceutical	Bacterial infection
(6)	Radezolid	phase-II	Rib-X pharmaceutics	Community-acquired pneumonia
(7)	PNU-100480	phase- IIa	Pfizer pharmaceutical	Pulmonary Tuberculosis
(8)	RX-1741	phase- II	Rib-X pharmaceutics	Infectious skin disease

and oral suspension for the treatment of patients with infections caused by Gram-positive bacteria. It is indicated for adults in the treatment of nosocomial pneumonia, community-acquired pneumonia (CAP), complicated and uncomplicated skin and skin structure infections and vancomycin-resistant enterococcus (VRE) infections caused by methicillin-resistant *Staphylococcus aureus* (MRSA), VRE, faecium and penicillin-susceptible *Streptococcus pneumoniae* 363503.

Vancomycin resistance was reported in only 0.6 and 3.0% of *Enterococcus faecalis* and *E. faecium*, respectively. Penicillin resistance occurred in 25.1% of *Streptococcus pneumonia*, 4.9% at the high level (≥2 mg/L). The MIC_{90} for linezolid was 1 mg/L for streptococci and 2 mg/L for enterococci and staphylococci. Using the US FDA- and EUCAST-recommended susceptible breakpoints for linezolid, there were no confirmed reports of linezolid resistance minimum inhibitory concentration (MIC), ≥8 mg/L. The distribution of linezolid MIC values was unimodal and varied between 0.25 and 1 mg/L for streptococci (>90% of isolates) and between 1 and 2 mg/L for staphylococci (>90%) and enterococci (>95%).

Favorable clinical results shown by linezolid prompted many pharmaceutical industries and academic institutions to explore the possibilities of expansion of antibacterial spectrum of this class. A large number of publications, reviews, and patents testify to the interest of the various research groups in the oxazolidinone class of antibacterials. Furazolidone as shown in Figure 27 is the first member of the oxazolidinone class discovered in 1950 and appears to be the initial candidate responsible for the genesis of further work on oxazolidinone antibacterials.

YC-20 [55] (Figure 28) and linezolid were active against all Gram-positive organisms isolated, including strains resistant to other classes of antibiotics. YC-20 exhibited MIC50 and MIC90 values of <0.5 and 2 mg/L against all isolates. Overall, the activity of YC-20 is slightly superior to that of linezolid. The present study confirms and expands previous findings of the good *in vitro* activity of YC-20 against Gram-positive organisms. YC-20 has a potential role in the treatment of infections caused by Gram-positive pathogens, especially for multidrug-resistant Gram-positive bacteria (Table 3).

Radezolid (RX-1741) [56, 57] (Figure 28) is a novel oxazolidinone with broader spectrum of coverage and increased activity against Gram-positive organisms as compared to other oxazolidinones. Radezolid has recently completed successfully two Phase 2 clinical trials: one for community-acquired pneumonia (CAP) and the second for uncomplicated skin and skin structure infections (Usssi) (Table 4).

The spectrum of activity of torezolid (TR-700) (Figure 28), the active moiety of torezolid phosphate (TR-701), and proposes tentative MIC and disk diffusion breakpoints as well as quality control ranges [58]. The *in vitro* susceptibilities of 1,096 bacterial isolates, representing 23 different species or phenotypic groups, were determined for torezolid, linezolid, cefotaxime, and levofloxacin using Clinical and Laboratory Standards Institute (CLSI) broth microdilution MICs, minimum bactericidal concentrations (MBCs), agar dilution, and disk diffusion testing methods. Torezolid was very active against the majority of Gram-positive strains, including methicillin-susceptible and -resistant *Staphylococcus aureus* (MIC_{50} = 0.25 μg/mL, MIC_{90} ≤ 0.5 μg/mL), coagulase-negative staphylococci (CNS; MIC_{50} = 0.25 μg/mL, MIC_{90} ≤ 0.5 μg/mL), enterococci (MIC_{50} and MIC_{90} ≤ 0.5 μg/mL), and streptococci (MIC_{50} and MIC_{90} ≤ 0.25 μg/mL). Based upon MIC_{90}s, torezolid was 4-fold more active than linezolid against *S. aureus*, coagulase-negative staphylococci, and the enterococci and 8-fold more active than linezolid.

In 2002 AstraZeneca introduced posizolid (AZD2563) (Figure 28). Results indicate that posizolid has excellent, targeted bactericidal activity against all common gram-positive bacteria, regardless of resistance to other classes of antibiotics [59].

The *in vitro* activity of AZD2563, a novel oxazolidinone, was assessed against 595 Gram-positive cocci, comprising recent surveillance isolates and a collection of resistant

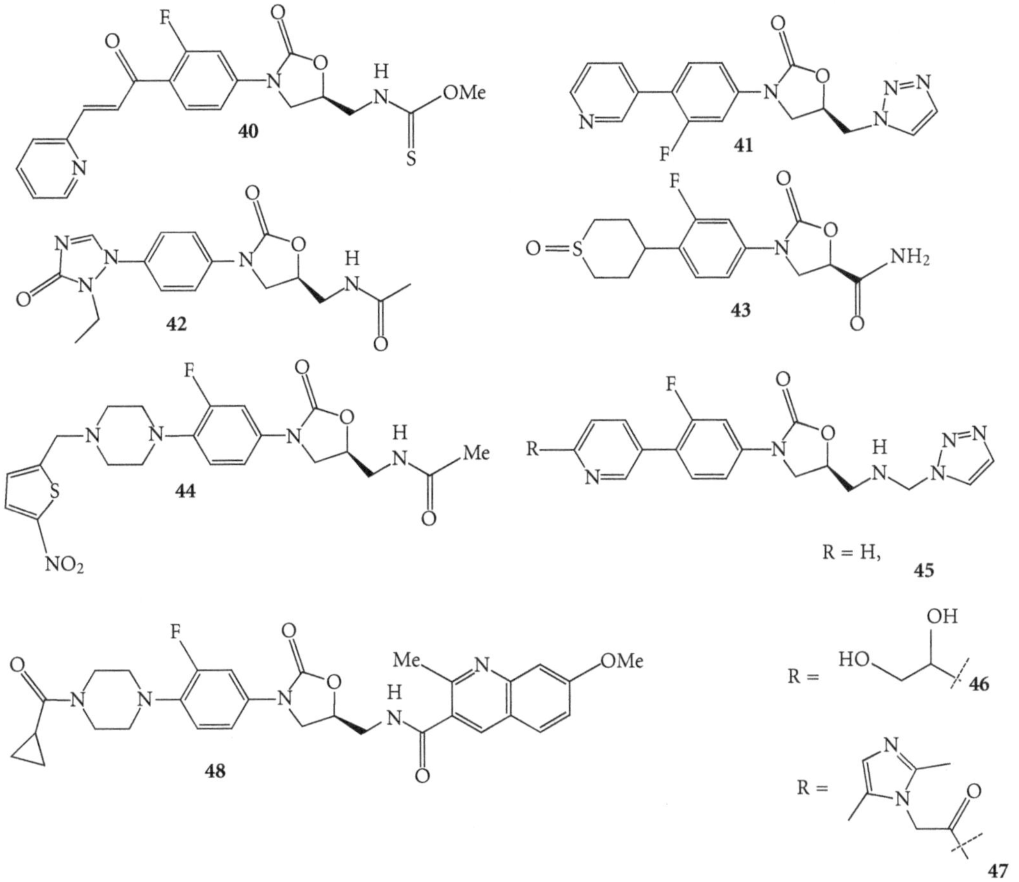

Figure 25

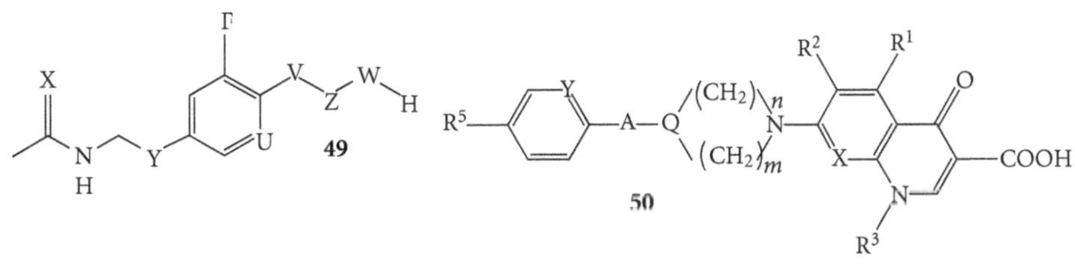

Figure 26

Table 3: Antimicrobial activity of YC-20 against 522 Gram-positive organisms from clinical samples.

Organism/antimicrobial agent	50% MIC (mg/L)	90% MIC (mg/L)	Range MIC (mg/L)
Staphylococcus aureus, MSSA (80)			
YC-20	0.25	0.5	0.06–0.5
Linezolid	0.25	1	0.25–1
Vancomycin	0.5	1	0.25–1
Ampicillin	1.0	2	0.25–4
Cefazolin	0.5	4	0.125–4
evofloxacin	0.125	1	0.06–2

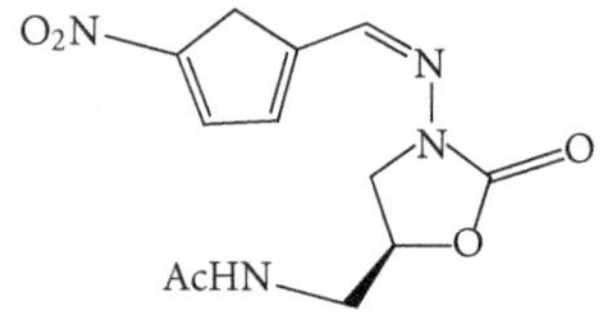

Furazolidone

Figure 27

Table 4: Antimicrobial activity of Radezolid against 522 Gram-positive organisms from clinical samples.

Bacteria	Linezolid (MICs in mg/L)	Radezolid (MICs in mg/L)
S. pneumonia	0.5–2	<0.25
S. pyogenes	2–4	0.03–0.125
E. faecalis	1–16	<0.25– 4
H. influenza	2–64	0.25–2
S. aureus	2–4	.5–4
L. pneumophila	4–16	1–4
C. trachomatis	8–16	.5–1

(including multiresistant), epidemiologically diverse isolates [60]. The MICs of AZD2563 for staphylococci, pneumococci, and enterococci had narrow ranges, 0.25–2 mg/L, with modal MICs of 1 mg/L for staphylococci and pneumococci, and 1–2 mg/L for enterococci. AZD2563 was equally active against the surveillance isolates and those that had been selected for their multiresistance to other agents. The MICs of AZD2563 were either the same as those of linezolid or twofold lower.

A few Mannich ketones of piperazinyl oxazolidinone derivatives have been synthesized and their antibacterial activities in various Gram-positive organisms such as *Bacillus subtilis, Staphylococcus aureus, Staphylococcus epidermidis,* and *Enterococcus faecalis* were evaluated by MIC determination [61]. The analog showed inferior activity than linezolid as well as eperezolid. Thus in an attempt to improve potency, we prepared several analogues by modifying the ketone part A. However, all such resulting compounds lost their *in-vitro* antibacterial activity. It has been reported that thioacetamide at the 5th position of the oxazolidinone improves the activity. MIC was determined by microbroth dilution technique and values reported in table represent the highest MIC value obtained in triplicate. S.a1, *Staphylococcus aureus* ZYABL 006; S.a2, *Staphylococcus aureus* ATCC 33591; S.e, *Staphylococcus epidermidis* ATCC 12228, B.s, *Bacillus subtilis* ATCC 6633; E.f, *Enterococcus faecalis* ATCC 29212.

The compound library (Figure 28) was screened using a disk diffusion assay on the Gram-positive bacterial reference strains, *M. smegmatis* ATCC 14468, *Bacillus subtilis* ATCC 6633, and *Enterococcus faecalis* ATCC [62]. *Mycobacterium smegmatis* was inoculated into Middlebrook 7H9 containing 0.2% glycerol and ADC enrichment while the other strains were inoculated into Mueller–Hinton broth. The inocula of all strains were allowed to incubate for approximately 24 h at 37°C and 180 rpm. Minimum inhibitory concentrations (MICs) of compound was also determined against the same strains using a broth macrodilution assay using either Middlebrook 7H9 containing 0.2% glycerol and ADC enrichment or Mueller–Hinton broth 20. The highest concentration for each compound tested was 32 lg/mL, and each subsequent tube was a twofold dilution of the previous. The M. smegmatis tubes were incubated for 48 h at 37°C while the other strains were incubated for 24 h and then analyzed. Thiazolyl blue tetrazolium bromide (MTT) was dissolved in MeOH (10 mg/mL) and added to the solution (20 lL) to aid in bacteria visualization.

The *in vitro* antibacterial activities of the compound and vancomycin, ciprofloxacin, and linezolid as the reference drugs were determined by the conventional agar dilution method using Mueller–Hinton agar medium [63]. The tested Gram-positive organisms included two clinical isolates of *S. aureus* resistant to methicillin (MRSA), *S. aureus* ATCC 29737, and *S. epidermidis* ATCC 12229. Gram-negative bacteria used in the study were *E. coli* ATCC 8739, *S. typhimurium* ATCC 1639 and *P. aeruginosa* ATCC 9027. The tested compounds, vancomycin and ciprofloxacin, were dissolved in DMSO while linezolid was dissolved in water. Suspensions of each of bacteria were prepared to contain approximately 106 colony-forming units (CFU/mL) and applied to plates with twofold serially diluted compounds to be tested in distilled water in concentration ranging from 0.01 to 100 mg/mL and incubated at 37°C for 18 h. To ensure that the solvent had no effect on bacterial growth, a control test was performed with test medium supplemented with DMSO at the same dilution as used in the experiments.

Minimum inhibitory concentrations (MICs, mg/mL) were determined on Mueller–Hinton (MH) agar with medium containing dilutions of antibacterial agents ranging from 0.12 to 64 mg/mL [64]. The test compounds were dissolved in 20% water in DMSO, while linezolid and vancomycin were dissolved in 40% water in ethanol and water, respectively. The tests were performed using MH agar plates for all staphylococci and enterococci and on MH agar plates supplemented with 5% sheep blood to facilitate the growth of *S. pneumoniae, H. influenzae,*and *M. catarrhalis.* The Gram-positive clinical isolates utilized in this study consisted of methicillin-resistant *S. aureus* (MRSA, n = 1/4 10), methicillin-susceptible S. aureus (MSSA, n = 1/4 10), methicillin-resistant coagulase-negative staphylococci (MR-CNS, n = 1/4 3), methicillin-sensitive coagulase-negative staphylococci (MS-CNS, n = 1/4 6), S. pneumoniae (n = 1/4 3), vancomycin-sensitive (VSE, n = 1/4 6) and vancomycin-resistant (VRE, n = 1/4 4) enterococci. The Gram-negative clinical isolates tested included *H. influenzae* (n = 1/4 4) and *M. catarrhalis* (n = 1/4 1). The reference strains, *S. aureus* ATCC 25923, *S. epidermidis* ATCC 12228 and *E. faecalis* ATCC 29212, *E. coli* ATCC 25922, and *H. influenzae* ATCC 49247 were used as controls. The final bacterial concentration for inocula was 107 CFU/mL and was incubated at 35°C for 18 h. The test compounds were also evaluated against *S. aureus* ATTC 25923 in MH broth supplemented with 50% human plasma to assess the extent of plasma binding and/or plasma instability. The MIC was defined as the lowest drug concentration that completely inhibited growth of the bacteria.

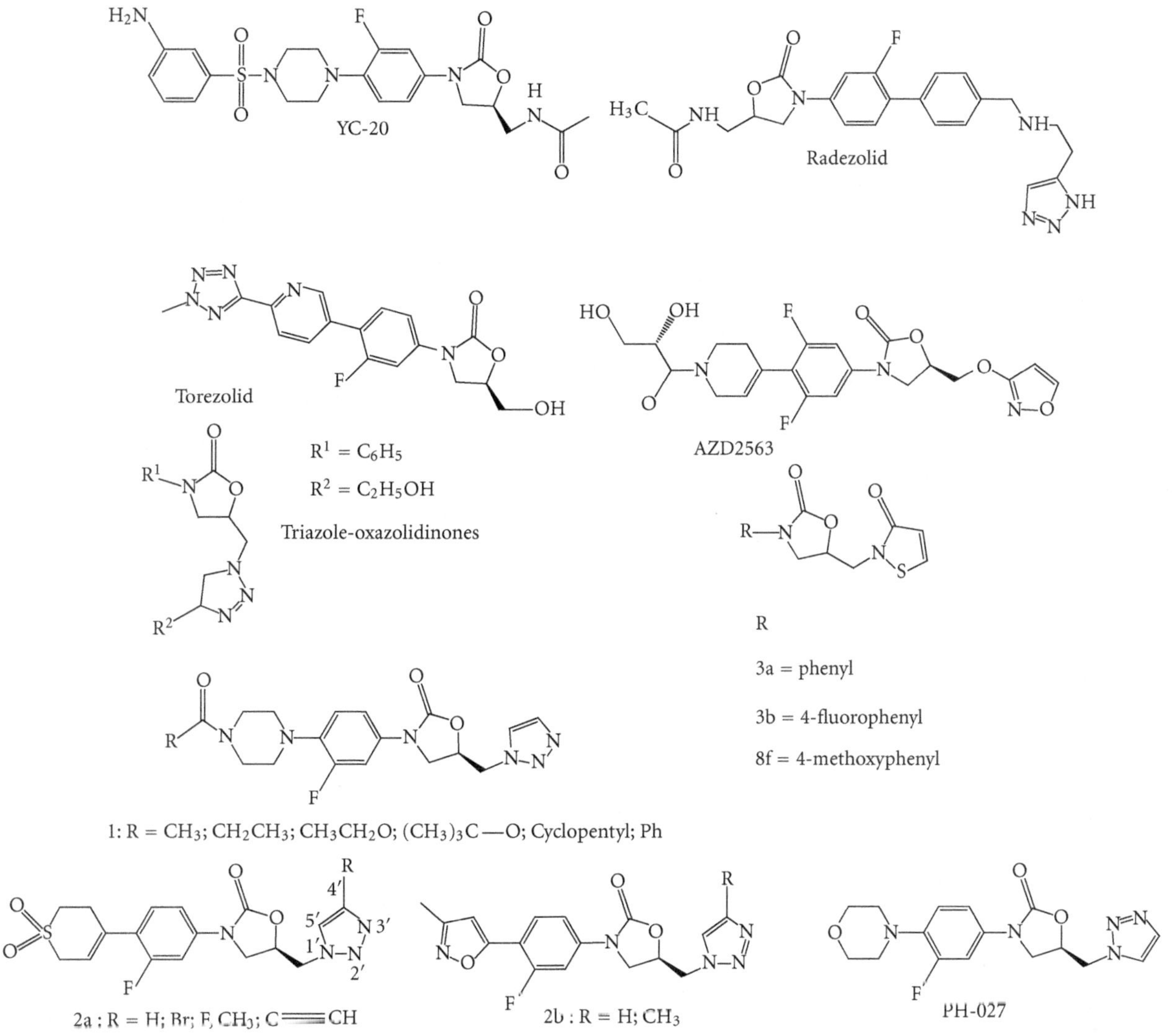

Figure 28

Against some of the gram-positive bacteria strains tested, the triazole oxazolidinone (Figure 28) PH-027 demonstrated MIC values comparable to or 1- to 2-fold lower that hose of linezolid and vancomycin [65]. In particular, the MIC values of PH-027 against vancomycin susceptible *E. faecium* (VSE), vancomycin-intermediate-resistant *E. faecalis* (VIRE), and vancomycin-resistant E. faecalis (VRE) were 0.5, 0.5, and 2 mg/mL, respectively, compared to an MIC of 2 mg/mL demonstrated by linezolid against these strains. However, the MIC's of vancomycin against the same strains were 4, 8 and >32 mg/mL, respectively. PH-027 demonstrated the most potent antibacterial activity against both sensitive and resistant gram-positive bacteria strains, including MRSA, MSSA, MS-CNS, MR-CNS, PRSP, and enterococci (*Enterococcus faecium* and *Enterococcus faecalis*).

The result of *in vitro* antibacterial activity against a spectrum of resistant and susceptible Gram-positive organisms [66] clearly shows that all compounds (Figure 29) bearing sulfonyl group have good antibacterial activity. Compound showed more potent antibacterial activity than linezolid and vancomycin. Obviously, the introduction of strong electron-withdrawing groups (e.g., CF3, F, and NO2) into the oxazolidinones can confer excellent antibacterial activity especially at the 3rd position of the phenylsulfonyl group, in addition, electron-donating groups (e.g., Cl).

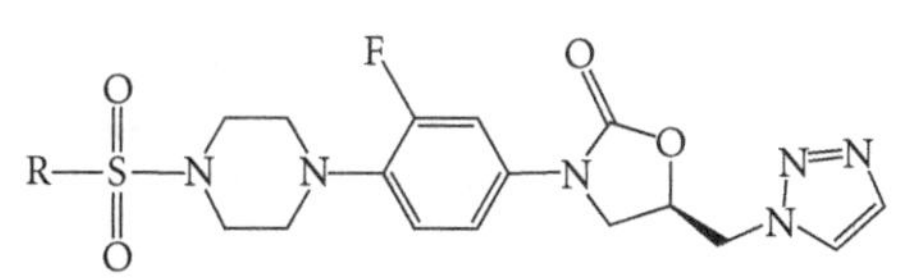

Figure 29

(2) *Anticoagulant Activity* [67, 68]. Rivaroxaban (BAY 59–7939) (Figure 30) is an oral anticoagulant invented and manufactured by Bayer; in a number of countries it is

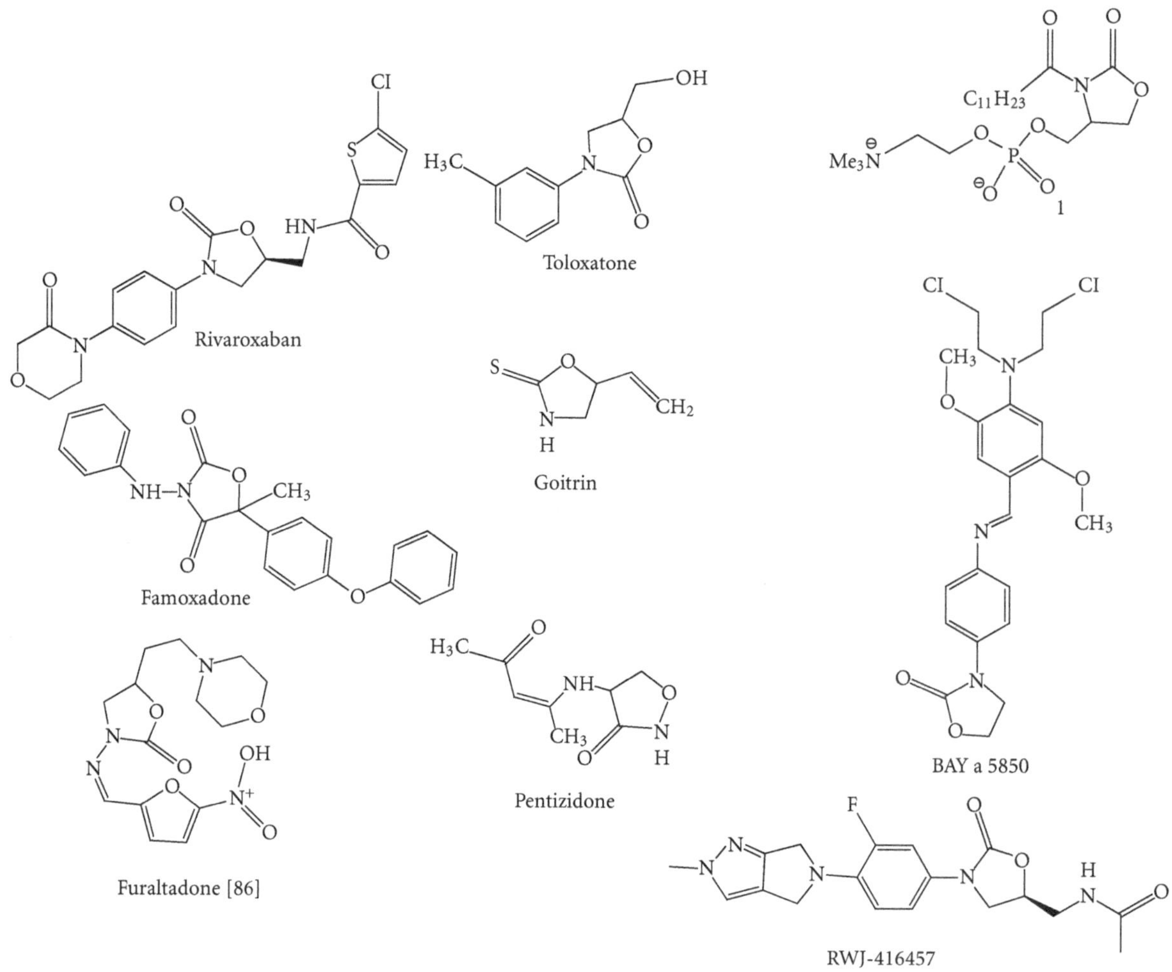

FIGURE 30

marketed as Xarelto. If approved by the United States FDA, it will be marketed by Ortho-McNeil Pharmaceutical. It is the first available orally active direct factor Xa inhibitor. Rivaroxaban is well absorbed from the gut and maximum inhibition of factor Xa occurs four hours after a dose. The effect lasts 8–12 hours, but factor Xa activity does not return to normal within 24 hours so once daily dosing is possible.

(3) *Antitubercular Activity.* During the course of investigation in the oxazolidinones antibacterial agent area was identified a subclass with especially potent in vitro activity against mycobacteria [69]. The salient structural feature of these oxazolidinone analogues, U-100480, U-101603, and U-101244, is their appended thiomorpholine. Potent activity against a screening strain of *M. tuberculosis* was determined by U-100480 and U-101603 (MIC = 0.125 mg/mL).

The activities of linezolid, eperezolid, and PNU-100480 were evaluated in a murine model of tuberculosis. Approximately 10 (7) viable *Mycobacterium tuberculosis* ATCC 35801 organisms were given intravenously to 4-week-old outbred CD-1 mice [70]. In the first study, treatment was started 1 day postinfection and was given by gavage for 4 weeks. Viable cell counts were determined from homogenates of spleens and lungs. PNU-100480 was as active as isoniazid. Linezolid was somewhat less active than PNU-100480 and isoniazid. Eperezolid had little activity in this model. In the next two studies, treatment was started 1 week postinfection. A dose-response study was performed with PNU-100480 and linezolid (both at 25, 50, and 100 mg/kg of body weight). PNU-100480 was more active than linezolid, and its efficacy increased with an escalation of the dose. Subsequently, the activity of PNU-100480 alone and in combination with rifampin or isoniazid was evaluated and was compared to that of isoniazid-rifampin. The activity of PNU-100480 was similar to that of isoniazid and/or rifampin in the various combinations tested. Further evaluation of these oxazolidinones in the murine test system would be useful prior to the development of clinical studies with humans.

(4) *Antidepressant or Psychotropic Activity.* Monoamine oxidase (MAO) inhibitors were developed as antidepressants but many drugs, including the novel oxazolidinone antibacterial agents, share similar molecular properties and have MAO inhibitory activity [71]. Factors important for binding antidepressants and modifications to decrease binding of oxazolidinones to avoid undesirable vascular effects are discussed [72].

In man, the antidepressant agent 3-(3-methylphenyl)-5-hydroxymethyl-2-oxazolidinone (toloxatone) (Figure 30) on oral dosing was mainly eliminated in urine (80% dose in 12 h).

Plasma concnentration of total radioactivity was max (5.8 μg equiv./mL) at 30 min to 1 h after administration and declined rapidly ($t_{1/2}$, 1.25 h). Unchanged drug accounted for 48, 32, and 13% of plasma radioactivity at 15 min, 1 h, and 6 h, respectively.

The drug was extensively metabolized. The major urinary metabolites were 3-(3-carboxyphenyl)-5-hydroxymethyl-2-oxazolidinone and a glucuronide of toloxatone. A minor urinary metabolite, characterized as a phenolic derivative, was also excreted conjugated.

On the basis of previous laboratory studies AS-8 (5-morpholinemethyl-3-(4-chlorobenzylideneamino)-2-oxazolidinone) was suggested to possess antidepressant-like activity. Forced swim test, learned helplessness, and conflict Vogel's test were performed after three prior administrations of AS-8 (24, 5, and 1 h before the test). The data have shown that AS-8 produces moderate antidepressant effect but did not induce anxiolytic-like action. Biochemical data revealed increased brain 5-HT and 5-HIAA levels following AS-8 administration [73]. The combined treatment of rats with AS-8 (100 mg/kg) and amitriptyline (5 mg/kg) or desipramine (1.25 mg/kg) significantly stimulated active behavior in the forced swim test above the level obtained with each of the drug given separately. The present data suggest the potential antidepressant efficacy of AS-8 in conjunction with small doses of tricyclic antidepressants.

(5) *Phospholipase Inhibitor* [74]. (S)- and (R)-3-dodecanoyl-4-phosphatidylcholinohydroxymethyl-2-oxazolidinone **(1)** (Figure 30), which are cyclic analogues of the amide phospholipid 7, were synthesized. The inhibitory activities of these analogues toward phospholipase A_2 were compared with that of the amide analogue 7. (S)- and (R)-3-dodecanoyl-4-phosphatidylcholinohydroxymethyl-2-oxazolidinone (1) were synthesized. The inhibitory activities of these analogues toward phospholipase A_2 were compared with that of the amide analogue.

(6) *Agriculture Fungicide* [75]. 5-Methyl-5-(4-phenoxyphenyl)-3-phenylamino-2,4-oxazolidinedione, DPX-JE874, is a new agricultural fungicide under development by DuPont. DPX-JE874 is a member of a new class of oxazolidinone fungicides which demonstrate excellent control of plant pathogens in the Ascomycete, Basidiomycete, and Oomycete classes which infect grapes, cereals, tomatoes, potatoes, and other crops. The synthesis, mode of action, and structure-activity relationships of two types of oxazolidinones, 2-thioxo-4-oxazolidinones and 2,4-oxazolidinediones, were discussed.

Famoxadone (3-anilino-5-methyl-5-(4-phenoxyphenyl)-1, 3-oxazolidine-2, 4-dione) (Figure 30), is a new agricultural fungicide recently commercialized by DuPont under the trade name Famoxate [76]. Famoxadone is a member of a new class of oxazolidinone fungicides that demonstrate excellent control of plant pathogens in the Ascomycete, Basidiomycete, and Oomycete classes that infect grapes, cereals, tomatoes, potatoes, and other crops.

Famoxadone is a preventative and curative fungicide recently commercialized for plant-disease control [77]. The molecule and its oxazolidinone analogs are potent inhibitors of mitochondrial ubiquinol: cytochrome c oxidoreductase (cytochrome bc_1) and they bind in the Q_0 site of the enzyme near the low potential heme of cytochrome b. Inhibitor binding constants for five mutant cytochrome bc_1 enzymes from *Saccharomyces cerevisiae* having single amino acid changes in their apocytochrome b located near the low potential heme were compared with their two parental wild-type enzymes. The five individual amino acid changes altered the inhibition constants for the inhibitors famoxadone, myxothiazol, azoxystrobin, and kresoxim-methyl in dissimilar fashion.

(7) *CNS Depressant* [78, 79]. A method of treating central nervous system diseases and disorders, the diseases and disorders being responsive to drugs possessing psychotropic activity. The method involves administering to a subject suffering from the disease or disorder a therapeutically effective amount of 5-morpholinomethyl-3-(4-chlorobenzylideneamine)-2-oxazolidinone or a pharmaceutically acceptable salt of 5-morpholinomethyl-3-(4-chlorobenzylideneamine)-2-oxazolidinone. The invention also relates to a pharmaceutical composition, the composition having pharmaceutically acceptable carriers and 5-morpholinomethyl-3-(4-chlorobenzylideneamine)-2-oxazolidinone and/or a pharmaceutically acceptable salt of 5-morpholinomethyl-3-(4-chlorobenzylideneamine)-2-oxazolidinone.

(8) *Centrally Acting Muscle Relaxants* [80]. The severity of anaemic decerebrate rigidity was quantitatively determined by measuring the frequency of electromyographic potentials in the rat. Some oxazolidinones markedly reduced the severity of this decerebrate rigidity in a dose-dependent manner, (4S,5R)-4-(2-methylpropyl)-3-[3-(perhydroazepin-1-yl)propyl]-5-phenyl-1,3-oxazolidin-2-one (MLV-6976) being the most potent. In addition to the oxazolidinones, an amino alcohol derivative, (1RS,2SR)-5-methyl-1-phenyl-2-(3-piperidinopropylamino)hexan-1-ol(MLV-5860) also reduced the rat decerebrate rigidity. In the oxazolidinone series, the optical isomers with absolute configuration (S) at the 4-position were more potent than the corresponding (4R)-isomers, while there was no significant difference in their LD50 values. Normal rats and mice receiving MLV-6976 at doses which reduced decerebrate rigidity showed no behavioural changes, impairment of motor coordination only appearing at extremely high doses. MLV-6976 and its derivatives did not affect spinal reflex potentials in cats. MLV-6976 reduced the severity of harmaline-induced tremor in mice in a dose-dependent manner, but slightly augmented tremorine-induced tremor. The frequency of the spike discharges induced by iontophoretically applied glutamate was reduced by MLV-6976 in a dose-dependent manner in rat cortical neurones. The amplitude of miniature endplate potentials of the rat diaphragm was decreased by MLV-6976 only at concentrations greater than 0.1 mM. It is concluded that MLV-6976 acts on the brainstem or/and higher levels of the brain rather than on the spinal cord

or the peripheral nervous system to reduce the excessive activities of the nervous system.

(9) *Antithyroid Agent* [81]. 5-Vinyloxazolidine-2-thione (VOT) administered orally to lactating rats was found to be efficiently transferred to the sucklings via the milk. In mothers the exposure to VOT resulted in an increased percentage of neutrophils, a decreased percentage of lymphocytes, and increases in the relative weights of liver and thyroid. Suckling rats showed a decreased number of leucocytes, increases in the relative weights of liver and thyroid and structural changes in the thyroid. Male sucklings were more affected than female pups. The antithyroid effects were clearly related to the maternally administered VOT doses.

RS-Goitrin (Figure 30) can be conveniently prepared by a simplification of the Ettlinger procedure. Goitrin is a moderate inhibitor of purified bovine adrenal dopamine beta-hydroxylase [82]. The administration of goitrin leads to a depression of brain norepinephrine and to an elevation of heart and adrenal dopamine.

Brassica vegetables are the major source of glucosinolates in the human diet. Certain glucosinolates are readily converted into goitrogenic species, notably 5-vinyloxazolidine-2-thione and thiocyanate ion [83]. The effect of dietary Brussels sprouts, a particularly rich source of such glucosinolates, on thyroid function has been examined.

(10) *Antiblastic Activity in Chemotherapy* [84]. Antiblastic activity means retardation of growth. 3-p-(2′, 5′-Dimethoxy-4′-(N,N-bis-(-chloroethyl)-amino)benzylideneamino)phenyl-2-oxazolidinone (GEA 29; BAY a 5850) (Figure 30) and its analogues use as antiblastic agent.

Some other oxazolidinone derivatives are used as anticancer in early clinical trial for example,

(i) 3-nitroso-5-methyl-2-oxazolidone,

(ii) 3-(2-hydroxy-3-(2-nitro-1H-imidazol-1-yl)propyl)-2-oxazolidinone,

(iii) 4-(4-(bis(2-chloroethyl)amino)-2,5-dimethoxy-phenyl) methylene aminophenyl)-2-oxazolidinone.

(11) *Use in Urinary Tract Infection* [85]. The antimicrobial activity of linezolid, a recently developed antibiotic agent active against Gram-positive bacteria, was tested against pathogens from three different collections. (1) Uropathogens from hospitalized urological patients (1990/1991) with complicated and/or hospital-acquired UTIs; Urologic Clinic, Hospital St. Elisabeth, Straubing. (2) Uropathogens from a multicentre study (1995/1996) comprising 37 urological centres throughout Germany. (3) MRSA isolates of patients and staff (1999/2000) within the Hospital St. Elisabeth, Straubing. Genotyping of the latter isolates was performed by pulsed-field electrophoresis. The minimal inhibitory concentrations (MICs) of linezolid determined by an agar (Isosensitest) dilution method using a multipoint inoculator and an inoculum of 104 cfu per point ranged for methicillin susceptible *Staphylococcus aureus* (MSSA) ($n = 27$) between 2 and 4 mg/L, for methicillin resistant *S. aureus* (MRSA) ($n = 35$) between 1 and 2 mg/L, for methicillin susceptible coagulase-negative staphylococci (CNS) (MSSE) ($n = 67$) between 0.5 and 4 mg/L, for methicillin resistant CNS (MRSE) ($n = 19$) between 0.25 and 2 mg/L, for *Enterococcus. faecalis* ($n = 184$) between 0.5 and 4 mg/L, for *E. faecium* ($n = 3$) 2 mg/L, and for *Streptococcus* spp. ($n = 4$) between 0.25 and 1 mg/L, indicating that all strains were susceptible. According to the *in vitro* activity, linezolid may be considered a promising antibacterial agent for the treatment of complicated UTI caused by Gram-positive uropathogens. Thus, linezolid should be evaluated in a clinical study (Table 5).

Table 5

Product name	Furaltadone (Figure 30)
Synonyms	5-(Morpholinomethyl)-3-((5-nitrofurfurylidene)amino)-2-oxazolidinon; 5-morpholinomethyl-3-(5-nitro-2-furfurylidine-amino)-2-oxazolidinone; altabactina; altafur; f-150; furazolin; furazoline; furmethanol
MF	C13H16N4O6
Use	Use in UTI

(12) *Cycloserine Analogues* [86]. (R)-4-[(1-Methyl-3-oxo-1-butenyl)amino]isoxazolidin-3-one is used as cycloserine analogues. Synonym of this compound is Pentizidone (Figure 30).

D-Cycloserine is a broad-spectrum antibiotic used with other antibiotics to treat various forms of tuberculosis [87]. Its prodrug sodium (R)-4-[(1-methyl-3-oxo-1-butenyl)amino]-3-isoxazolidinone hemihydrate, developed for better aqueous stability and solubility, is combined with another broad-spectrum antibiotic, fludalanine.

6.2.2. Current Research Work on Oxazolidinone. Mathur et al. demonstrated RBx 11760 MICs were in the range of 0.5–1 mg/L for *C. difficile* isolates, and it demonstrated concentration-dependent killing of *C. difficile* ATCC 43255 and *C. difficile* 6387 up to 2–4x MIC (1–2 mg/L). RBx 11760, at concentrations as low as 0.25–0.5 mg/L, resulted in a significant reduction in *de novo* toxin production as well as sporulation in different *C. difficile* isolates [88]. In contrast, vancomycin, metronidazole, and linezolid had little or no effect on toxin production and appeared to promote the formation of spores. In the hamster infection model, treatment with RBx 11760 resulted in prolonged survival of animals as compared with vancomycin or metronidazole, which correlated well with the histopathology results. Macromolecular labelling results suggest that RBx 11760 is a potent inhibitor of bacterial protein synthesis. RBx 11760 showed excellent *in vitro* and *in vivo* activity against *C. difficile*, and it could be a promising novel candidate for future drug development against *C. difficile* infection.

Skripkin et al. have reported new and improved antibiotics are urgently needed to combat the ever-increasing number of multidrug-resistant bacteria [89]. In this study, we characterized several members of a new oxazolidinone family, Rχ-01. This antibiotic family is distinguished by having *in vitro* and *in vivo* activity against hospital-acquired, as well as community-acquired, pathogens. We compared the 50S ribosome binding affinity of this family to that of the only marketed oxazolidinone antibiotic, linezolid, using chloramphenicol and puromycin competition binding assays. The competition assays demonstrated that several members of the Rχ-01 family displace, more effectively than linezolid, compounds known to bind to the ribosomal A site. We also monitored binding by assessing whether Rχ-01 compounds protect U2585 (*Escherichia coli* numbering), a nucleotide that influences peptide bond formation and peptide release, from chemical modification by carbodiimide. The Rχ-01 oxazolidinones were able to inhibit translation of ribosomes isolated from linezolid-resistant *Staphylococcus aureus* at submicromolar concentrations. This improved binding corresponds to greater antibacterial activity against linezolid-resistant enterococci. Consistent with their ribosomal A-site targeting and greater potency, the Rχ-01 compounds promote nonsense suppression and frameshifting to a greater extent than linezolid. Importantly, the gain in potency does not impact prokaryotic specificity as, like linezolid, the members of the Rχ-01 family show translation 50% inhibitory concentrations that are at least 100-fold higher for eukaryotic than for prokaryotic ribosomes.

Hilliard et al. have reported RWJ-416457 is an investigational pyrrolopyrazolyl-substituted oxazolidinone with activity against antibiotic-susceptible and -resistant gram-positive pathogens [90]. Efficacies of RWJ-416457, linezolid, and vancomycin against methicillin-susceptible *Staphylococcus aureus* (MSSA) and community-associated methicillin-resistant *S. aureus* (CA-MRSA) in murine skin and systemic infections were compared, as were efficacies against *Streptococcus pneumoniae* in a lower respiratory infection. In staphylococcal systemic infections, RWJ-416457 was equipotent to twofold more potent than linezolid, with 50% effective dose values ranging from 1.5 to 5 mg/kg of body weight/day. RWJ-416457 was two- to fourfold less potent than vancomycin against MSSA but up to fourfold more potent than vancomycin against CA-MRSA. In MSSA and CA-MRSA skin infections, RWJ-416457 demonstrated an efficacy similar to that of linezolid.

Locke et al. have reported that TR-700 (torezolid), the active moiety of the novel oxazolidinone phosphate prodrug TR-701, is highly potent against gram-positive pathogens, including strains resistant to linezolid (LZD) [91]. Here we investigated the potential of *Staphylococcus aureus* strains ATCC 29213 (methicillin-susceptible *S. aureus* MSSA) and ATCC 33591 (methicillin-resistant *S. aureus* MRSA) to develop resistance to TR-700. The spontaneous frequencies of mutation of MSSA 29213 and MRSA 33591 resulting in reduced susceptibility to TR-700 at 2x the MIC were 1.1×10^{-10} and 1.9×10^{-10}, respectively. These values are ~16-fold lower than the corresponding LZD spontaneous mutation frequencies of both strains. Following 30 serial passages in the presence of TR-700, the MIC for MSSA 29213 remained constant (0.5 μg/mL) while increasing eightfold (0.25 to 2.0 μg/mL) for MRSA 33591. Serial passage of MSSA 29213 and MRSA 33591 in LZD resulted in 64- and 32-fold increases in LZD resistance (2 to 128 μg/mL and 1 to 32 μg/mL, resp.). Domain V 23S rRNA gene mutations (*Escherichia coli* numbering) found in TR-700-selected mutants included T2500A and a novel coupled T2571C/G2576T mutation, while LZD-selected mutants included G2447T, T2500A, and G2576T. We also identified mutations correlating with decreased susceptibility to TR-700 and LZD in the *rplC* and *rplD* genes, encoding the 50S ribosomal proteins L3 and L4, respectively.

Resistance to linezolid (LZD) occurs through mutations in 23S rRNA and ribosomal proteins L3 and L4 or through methylation of 23S rRNA by Cfr [92]. Here we report novel L3 mutations, ΔSer145/His146Tyr and ΔMet169-Gly174, cooccurring with *cfr* in LZD-resistant *Staphylococcus aureus* isolates recovered from a hospital outbreak in Madrid, Spain. LZD MIC values (16, 32, or 64 μg/mL) correlated with the presence and severity of the L3 mutation. All isolates had TR-700 (torezolid) MIC values of ≤ 2 μg/mL.

The crystal structure of the antibiotic drug candidate RWJ-416457 (Figure 30) (systematic name: *N*-{(5*S*)-3-[4-(5,6-dihydro-2*H*,4*H*-2-methylpyrrolo [3,4-*c*]pyrazol-5-yl)-3-fluorophenyl]-2-oxo-1,3-oxazolidin-5-yl}acetamide), $C_{18}H_{20}FN_5O_3$, which belongs to the first new class of antibiotics discovered in the past 30 years, has been determined at 150 K [93]. Each molecule of this drug donates one hydrogen bond to a neighboring molecule and accepts one hydrogen bond to give $(O{=}C{-}R{-}N{-}H\cdots O{=}C{-}R{-}N{-}H\cdots)_n$ linkages along the *b*-axis direction. The compound contains a pyrrolopyrazole ring, which, owing to its uncommon structure, has been shown to have particular effectiveness against multidrug-resistant bacteria.

Locke et al. have reported that staphylococcal resistance to linezolid (LZD) is mediated through ribosomal mutations (23S rRNA or ribosomal proteins L3 and L4) or through methylation of 23S rRNA by the horizontally transferred Cfr methyltransferase [94]. To investigate the structural basis for oxazolidinone activity against LZD-resistant (LZD^r) strains, we compared structurally diverse, clinically relevant oxazolidinones, including LZD, radezolid (RX-1741), TR-700 (torezolid), and a set of TR-700 analogs (including novel CD-rings and various A-ring C-5 substituents), against a panel of laboratory-derived and clinical LZD^r *Staphylococcus aureus* strains possessing a variety of resistance mechanisms. Potency against all strains was correlated with optimization of C- and D-rings, which interact with more highly conserved regions of the peptidyl transferase center binding site. Activity against *cfr* strains was retained with either hydroxymethyl or 1,2,3-triazole C-5 groups but was reduced by 2- to 8-fold in compounds with acetamide substituents.

This phase 4, randomized, double-blind, multicenter trial compared the efficacy and safety of Zyvox with vancomycin in the treatment of nosocomial pneumonia proven to be caused by MRSA, a serious and difficult-to-treat bacterial infection that is resistant to many antibiotics [95]. The study randomized 1,225 patients between 2004 and 2010. The study was designed as a noninferiority study with nested superiority, meaning the primary endpoint would be tested for superiority if it met non-inferiority criteria. In the study, Zyvox was noninferior and statistically superior to vancomycin in achieving both clinical and microbiologic success. The primary endpoint was clinical outcome at end of study in the per-protocol population. Secondary analyses included clinical outcome at end of treatment in the per-protocol population, clinical outcomes in the modified intent-to-treat population at end of study and end of treatment, microbiologic outcomes at end of study and end of treatment in the per-protocol and modified intent-to-treat populations, and safety and tolerability in the intent-to-treat population. Patients were randomized to receive Zyvox IV 600 mg every 12 hours or vancomycin 15 mg/kg every 12 hours over the course of seven to 14 days; vancomycin doses could be titrated at the investigator's discretion based on creatinine clearance and vancomycin trough levels.

References

[1] J. Vara Prasad, "New oxazolidinones," *Current Opinion in Microbiology*, vol. 10, no. 5, pp. 454–460, 2007.

[2] B. Bozdogan and P. C. Appelbaum, "Oxazolidinones: activity, mode of action, and mechanism of resistance," *International Journal of Antimicrobial Agents*, vol. 23, no. 2, pp. 113–119, 2004.

[3] K. E. Brighty, W. Kohlbrenner, and P. R. McGuirk, "Chapter 15. Recent developments in antibacterial resistance mechanisms," *Annual Reports in Medicinal Chemistry*, vol. 28, pp. 141–150, 1993.

[4] Y. Luo, L. Liu, L. Li, and Q. Deng, "Chiral resolution of racemic 4-phenyl(benzyl)-2-oxazolidone by use of molecularly imprinted polymers," *Chromatographia*, vol. 65, no. 11-12, pp. 675–679, 2007.

[5] S. J. Brickner, "Oxazolidinone antibacterial agents," *Current Pharmaceutical Design*, vol. 2, no. 2, pp. 175–194, 1996.

[6] N. Alouane, A. Boutier, C. Baron, E. Vrancken, and P. Mangeney, "Remarkably efficient charcoal-promoted ring-closing carbonylations," *Synthesis*, no. 5, pp. 860–864, 2006.

[7] A. Buzas and F. Gagosz, "Gold(I)-catalyzed formation of 5-methylene-1,3-oxazolidin-2-ones," *Synlett*, no. 17, pp. 2727–2730, 2006.

[8] R. Robles-Machín, J. Adrio, and J. C. Carretero, "Gold-catalyzed synthesis of alkylidene 2-oxazolidinones and 1,3-oxazin-2-ones," *Journal of Organic Chemistry*, vol. 71, no. 13, pp. 5023–5026, 2006.

[9] T. Munegumi, I. Azumaya, T. Kato, H. Masu, and S. Saito, "[3+2] Cross-coupling reactions of aziridines with isocyanates catalyzed by nickel(II) iodide," *Organic Letters*, vol. 8, no. 3, pp. 379–382, 2006.

[10] Y. Osa, Y. Hikima, Y. Sato et al., "Convenient synthesis of oxazolidinones by the use of halomethyloxirane, primary amine, and carbonate salt," *Journal of Organic Chemistry*, vol. 70, no. 14, pp. 5737–5740, 2005.

[11] A. Bøgevig, K. Juhl, N. Kumaragurubaran, W. Zhuang, and K. A. Jørgensen, "Direct catalytic asymmetric amination of aldehydes: synthesis of evans oxazolidinones and a-AminoAcid," *Angewandte Chemie International Edition*, vol. 41, p. 1790, 2002.

[12] H. P. Buchstaller, "Solid-phase synthesis of oxazolidinones by cycloaddition of resin-bound epoxides with isocyanates," *Journal of Combinatorial Chemistry*, vol. 5, no. 6, pp. 789–793, 2003.

[13] K.-J. Hwang, C.-M. Yu, N.-K. Choi, and K.-H. Park, "A convenient diastereoselective synthesis of oxazolidinone: approach to unusal amino acid statine," *Bulletin of the Korean Chemical Society*, vol. 15, pp. 527–532, 1994.

[14] G. Bratulescu, "An excellent procedure for the synthesis of oxazolidin-2-ones," *Synthesis*, no. 20, pp. 3111–3112, 2007.

[15] D. M. Gleave, S. J. Brickner, P. R. Manninen et al., "Synthesis and antibacterial activity of [6,5,5] and [6,6,5] tricyclic fused oxazolidinones," *Bioorganic and Medicinal Chemistry Letters*, vol. 8, no. 10, pp. 1231–1236, 1998.

[16] D. L. Stevens, B. Dotter, and K. Madaras-Kelly, "A review of linezolid: the first oxazolidinone antibiotic," *Expert Review of Anti-Infective Therapy*, vol. 2, no. 1, pp. 51–59, 2004.

[17] F. Ntziora and M. E. Falagas, "Linezolid for the treatment of patients with atypical mycobacterial infection: a systematic review," *International Journal of Tuberculosis and Lung Disease*, vol. 11, no. 6, pp. 606–611, 2007.

[18] D. Andes, M. L. Van Ogtrop, J. Peng, and W. A. Craig, "In vivo pharmacodynamics of a new oxazolidinone (linezolid)," *Antimicrobial Agents and Chemotherapy*, vol. 46, no. 11, pp. 3484–3489, 2002.

[19] CIMS, *CMP Medica India Private Limited*, 2010.

[20] Zurenko et al., "In vitro antibacterial activity of U-100592 and U-100766, novel oxazolidinones antibiotics," in *Proceedings of the 35th Interscience Conference on Antimicrobial Agents and Chemotherapy (ICAAC '95)*, pp. 150–11, 1995.

[21] C. W. Ford et al., in *Proceedings of the 35th Internscience Conference on Antimicrobial Agents and Chemotherapy*, 1995.

[22] D. J. Abraham, Ed., *BURGER'S Medicinal chemistry and drug discovery*, vol. 5, Willey Interscience, 6th edition, 1997.

[23] D. J. Abraham, Ed., *BURGER'S Medicinal Chemistry and Drug Discovery*, vol. 2, Willey Interscience, 1996.

[24] D. J. Diekema and R. N. Jones, "Oxazolidinones: a review," *Drugs*, vol. 59, no. 1, pp. 7–16, 2000.

[25] V. A. Pankratov, T. M. Frenkel', and A. M. Fainleib, "2-oxazolidinone," *Russian Chemical Review*, vol. 52, no. 6, p. 576, 1983.

[26] B. K. Srivastava, R. Soni, J. Z. Patel, M. R. Jain, and P. R. Patel, "Oxazolidinone antibacterials and our experience," *Anti-Infective Agents in Medicinal Chemistry*, vol. 7, no. 4, pp. 258–280, 2008.

[27] M. R. Barbachyn and C. W. Ford, "Oxazolidinone structure-activity relationships leading to linezolid," *Angewandte Chemie*, vol. 42, no. 18, pp. 2010–2023, 2003.

[28] N. Gandhi, B. K. Srivastava, V. B. Lohray, and B. B Lohray, "Novel anti-infective compounds," *Tetrahedron Letters*, vol. 45, pp. 6269–6272, 2004.

[29] P. Senci, M. Caspani, F. Ripamonti, and R. Ciabatti, "Oxazolidinone antibacterials and our experience," *Journal of the Chemical Society, Perkin Transactions*, vol. 1, p. 2345, 1994.

[30] C. A. Quesnelle, P. Gill, S. Roy et al., "Biaryl isoxazolinone antibacterial agents," *Bioorganic and Medicinal Chemistry Letters*, vol. 15, no. 11, pp. 2728–2733, 2005.

[31] G. Wang, J. R. Ella-Menye, and V. Sharma, "Synthesis and antibacterial activities of chiral 1,3-oxazinan-2-one derivatives," *Bioorganic and Medicinal Chemistry Letters*, vol. 16, no. 8, pp. 2177–2181, 2006.

[32] R. Griera, C. Cantos-Llopart, M. Amat, J. Bosch, J.–C. del Castillo, and J.B Huguet, "New potential antibacterials: a synthetic route to *N*-aryloxazolidinone/3-aryl tetrahydroisoquinoline hybrids," *Bioorganic & Medicinal Chemistry Letters*, vol. 15, p. 2515, 2005.

[33] M. L. Hammond and Y. Fukuda, "Oxazolidinones antibiotics and derivatives," Pct International Application WO2005/005399, Merck & Co.-Kyorin Pharmaceutical, 2005.

[34] N. Selvakumar, B. Yadi Reddy, G. Sunil Kumar et al., "Synthesis of novel tricyclic oxazolidinones by a tandem SN_2 and SNAr reaction: SAR studies on conformationally constrained

analogues of Linezolid," *Bioorganic and Medicinal Chemistry Letters*, vol. 16, no. 16, pp. 4416–4419, 2006.

[35] A. L. Choy, J. V. N. Vara Prasad, F. E. Boyer, M. D. Huband, and M. R. Dermyer, "Synthesis and SAR of novel conformationally restricted oxazolidinones possessing Gram-positive and fastidious Gram-negative antibacterial activity. Part 2: Amino substitutions on heterocyclic D-ring system," *Bioorganic and Medicinal Chemistry Letters*, vol. 17, no. 16, pp. 4699–4702, 2007.

[36] B. B. Lohray, V. B. Lohray, B. K. Srivastava et al., "Oxazolidinone: search for highly potent antibacterial," *Bioorganic and Medicinal Chemistry Letters*, vol. 14, no. 12, pp. 3139–3142, 2004.

[37] B. K. Srivastava, P. B. Kapadnis, P. Pandya, and V. B. Lohray, "Novel Mannich ketones of oxazolidinones as antibacterial agents," *European Journal of Medicinal Chemistry*, vol. 39, no. 11, pp. 989–992, 2004.

[38] B. Das, S. Rudra, A. Yadav et al., "Synthesis and antibacterial activity of potent heterocyclic oxazolidinones and the identification of RBx 8700," *Bioorganic and Medicinal Chemistry Letters*, vol. 15, p. 4261, 2005.

[39] S. K. Agrawal, S. S. Pandey, M. M. Samuel, G. Singh, and S. Chithra, "New antibacterial agents," Pct International Application number WO2005/090339, Orchid Chemicals and Pharmaceuticals, 2005.

[40] A. R. Renslo, G. W. Luehr, S. Lam et al., "Synthesis and structure-activity studies of antibacterial oxazolidinones containing dihydrothiopyran or dihydrothiazine C-rings," *Bioorganic and Medicinal Chemistry Letters*, vol. 16, no. 13, pp. 3475–3478, 2006.

[41] X. Zhai, G. G. Zhang, D. H. Gou, J. P. Liu, and P. Gong, "Modification of novel oxazolidinone derivatives at C-5 side chain and their antibacterial activities," *Acta Pharmaceutica Sinica*, vol. 41, no. 10, pp. 985–989, 2006.

[42] R. Schaadt, D. Sweeney, D. Shinabarger, and G. Zurenko, "The *in vitro* activity of TR-701, a novel oxazolidinone antibacterial agent," in *Proceedings of the Interscience Conference on Antimicrobial Agents and Chemotherapy (ICAAC '07)*, Chicago, Ill, USA, September 2007.

[43] B. Das, S. Rudra, S. Sangita, M. Salman, and A. Rattan, "Oxazolidinone derivatives as antimicrobials," Pct International Application number WO2006/109156, Ranbaxy Laboratories, 2006.

[44] O. A. Phillips, E. E. Udo, A. A. M. Ali, and S. M. Samuel, "Structure-antibacterial activity of arylcarbonyl- and arylsulfonyl-piperazine 5-triazolylmethyl oxazolidinones," *European Journal of Medicinal Chemistry*, vol. 42, no. 2, pp. 214–225, 2007.

[45] H. Fan, Y. Chen, Z. Jiang et al., "Synthesis and biological evaluation of new N-linked 5-triazolylmethyl oxazolidinones," *European Journal of Medicinal Chemistry*, vol. 43, no. 8, pp. 1706–1714, 2008.

[46] N. Selvakumar, G. S. Kumar, A. Malar Azhagan et al., "Synthesis, SAR and antibacterial studies on novel chalcone oxazolidinone hybrids," *European Journal of Medicinal Chemistry*, vol. 42, no. 4, pp. 538–543, 2007.

[47] F. Reck, F. Zhou, C. J. Eyermann et al., "Novel substituted (pyridin-3-yl)phenyloxazolidinones: antibacterial agents with reduced activity against monoamine oxidase A and increased solubility," *Journal of Medicinal Chemistry*, vol. 50, no. 20, pp. 4868–4881, 2007.

[48] J. Liu, B. He, A. Yu, and W. Zhou, "Synthesis and antibacterial activity of substituted oxotriazolylphenyl oxazolidinones," *Chemical Biology and Drug Design*, vol. 70, no. 1, pp. 65–69, 2007.

[49] T. J. Poel, R. C. Thomas, W. J. Adams et al., "Antibacterial oxazolidinones possessing a novel C-5 side chain. (5R)-trans-3-[3-fluoro-4-(1-oxotetrahydrothiopyran-4-yl)phenyl] -2-oxooxazolidine-5-carboxylic acid amide (PF-00422602), a new lead compound," *Journal of Medicinal Chemistry*, vol. 50, no. 24, pp. 5886–5889, 2007.

[50] S. Rudra, A. Yadav, A. V. S. Raja Rao et al., "Synthesis and antibacterial activity of potent heterocyclic oxazolidinones and the identification of RBx 8700," *Bioorganic and Medicinal Chemistry Letters*, vol. 17, no. 24, pp. 6714–6719, 2007.

[51] F. Reck, F. Zhou, G. Zainoun, D. Carcanague, C. Eyermann, and M. Gravestock, "Novel acyclic substituted (pyridin-3-yl) phenyl oxazolidinones: antibacterial agents with reduced activity against monoamine oxidase A and improved solubility," in *Proceedings of the 17th European Congress of Clinical Microbiology and Infectious Diseases*, Munich, Germany, March-April 2007.

[52] M. Cano, A. Palomer, and A. Guglietta, "Oxazolidinone derivatives and use thereof as antibiotics," Pct International Application number WO2007/000432, Ferrer International S.A., 2007.

[53] B. K. Srivastava, R. Soni, J. Z. Patel, M. R. Jain, and P. R. Patel, "Oxazolidinone antibacterials and our experience," *Anti-Infective Agents in Medicinal Chemistry*, vol. 7, no. 4, pp. 258–280, 2008.

[54] J. F. Barrett, "Linezolid pharmacia corp," *Current Opinion in Investigational Drugs*, vol. 1, no. 2, pp. 181–187, 2000.

[55] Y. Cui, Y. Dang, Y. Yang, S. Zhang, and R. Ji, "Antibacterial activity of YC-20, a new oxazolidinone," *Journal of Antimicrobial Chemotherapy*, vol. 58, no. 2, pp. 452–454, 2006.

[56] S. Lemaire, K. Kosowska-Shick, P. C. Appelbaum, P. M. Tulkens, and F. Van Bambeke, "Radezolid (RX RX-1741), a novel oxazolidinone oxazolidinone, accumulates, extensively within human macrophages and PMNs and shows activity towards intracellular linezolid linezolid-sensitive and linezolid linezolid-resistant *Staphylococcus aureus*," in *Proceedings of the 26th International Congress of Chemotherapy and Infection*, Torranto, Canada, June 2009.

[57] http://www.rib-x.com/pipeline/rx_1741.

[58] S. D. Brown and M. M. Traczewski, "Comparative *in Vitro* antimicrobial activities of torezolid (TR-700), the active moiety of a new oxazolidinone, torezolid phosphate (TR-701), determination of tentative disk diffusion interpretive criteria, and quality control ranges," *Antimicrobial Agents and Chemotherapy*, vol. 54, no. 5, pp. 2063–2069, 2010.

[59] A. Wookey, P. J. Turner, J. M. Greenhalgh, M. Eastwood, J. Clarke, and C. Sefton, "AZD2563, a novel oxazolidinone: definition of antibacterial spectrum, assessment of bactericidal potential and the impact of miscellaneous factors on activity in vitro," *Clinical Microbiology and Infection*, vol. 10, no. 3, pp. 247–254, 2004.

[60] A. P. Johnson, M. Warner, and D. M. Livermore, "*In vitro* activity of a novel oxazolidinone, AZD2563, against randomly selected and multiresistant Gram-positive cocci," *Journal of Antimicrobial Chemotherapy*, vol. 50, no. 1, pp. 89–93, 2002.

[61] B. K. Srivastava, P. B. Kapadnis, P. Pandya, and V. B. Lohray, "Novel Mannich ketones of oxazolidinones as antibacterial agents," *European Journal of Medicinal Chemistry*, vol. 39, no. 11, pp. 989–992, 2004.

[62] J. A. Demaray, J. E. Thuener, M. N. Dawson, and S. J. Sucheck, "Synthesis of triazole-oxazolidinones via a one-pot reaction

and evaluation of their antimicrobial activity," *Bioorganic and Medicinal Chemistry Letters*, vol. 18, no. 17, pp. 4868–4871, 2008.

[63] N. Adibpour, A. Khalaj, and S. Rajabalian, "Synthesis and antibacterial activity of isothiazolyl oxazolidinones and analogous 3(2H)-isothiazolones," *European Journal of Medicinal Chemistry*, vol. 45, no. 1, pp. 19–24, 2010.

[64] O. A. Phillips, E. E. Udo, M. E. Abdel-Hamid, and R. Varghese, "Synthesis and antibacterial activity of novel 5-(4-methyl-1H-1,2,3-triazole) methyl oxazolidinones," *European Journal of Medicinal Chemistry*, vol. 44, no. 8, pp. 3217–3227, 2009.

[65] O. A. Phillips, E. E. Udo, A. A. M. Ali, and N. Al-Hassawi, "Synthesis and antibacterial activity of 5-substituted oxazolidinones," *Bioorganic and Medicinal Chemistry*, vol. 11, no. 1, pp. 35–41, 2003.

[66] H. Fan, G. Xu, Y. Chen et al., "Synthesis and antibacterial activity of oxazolidinones containing triazolyl group," *European Journal of Medicinal Chemistry*, vol. 42, no. 8, pp. 1137–1143, 2007.

[67] Oral anticoagulant rivaroxaban (Xarelto) cuts deep vein thrombosis risk by 82%: Bayer, http://www.dancewithshadows.com/pillscribe/oral-anticoagulant-rivaroxaban-xarelto-cuts-deep-vein-thrombosis-risk-by-82-bayer/.

[68] S. Roehrig, A. Straub, J. Pohlmann et al., "Discovery of the novel antithrombotic agent 5-chloro-*N*-((5S)-2-oxo-3-[4- (3-oxomorpholin-4-yl)phenyl]-1,3-oxazolidin-5-ylmethyl)thiophene-2-carboxamide (BAY 59-7939): an oral, direct factor Xa inhibitor," *Journal of Medicinal Chemistry*, vol. 48, no. 19, pp. 5900–5908, 2005.

[69] M. R. Barbachyn, D. K. Hutchinson, S. J. Brickner et al., "Identification of a novel oxazolidinone (U-100480) with potent antimycobacterial activity," *Journal of Medicinal Chemistry*, vol. 39, no. 3, pp. 680–685, 1996.

[70] M. H. Cynamon, S. P. Klemens, C. A. Sharpe, and S. Chase, "Activities of several novel oxazolidinones against Mycobacterium tuberculosis in a murine model," *Antimicrobial Agents and Chemotherapy*, vol. 43, no. 5, pp. 1189–1191, 1999.

[71] R. R. Ramsay and M. B. Gravestock, "Monoamine oxidases: to inhibit or not to inhibit," *Mini Reviews in Medicinal Chemistry*, vol. 3, no. 2, pp. 129–136, 2003.

[72] A. Malnoe and M. Strolin Benedetti, "Metabolic fate of 3-(3-methylphenyl)-5-hydroxymethyl-2-oxazolidone (toloxatone), a new antidepressant agent, in man," *Xenobiotica*, vol. 9, no. 5, pp. 281–288, 1979.

[73] W. Kostowski, A. Plaznik, A. Bidzinski, E. Rosnowska, M. Jessa, and M. Nazar, "Studies on antidepressant action of a new oxazolidinone derivative AS-8," *Polish Journal of Pharmacology*, vol. 46, no. 1-2, pp. 15–20, 1994.

[74] S. Katsumura, S. Iwama, T. Matsuda, T. Tani, S. Fujii, and K. Ikeda, "Synthesis of oxazolidinone phospholipid analogue as a new inhibitor of phospholipase A_2," *Bioorganic and Medicinal Chemistry Letters*, vol. 3, no. 12, pp. 2703–2706, 1993.

[75] J. A. Sternberg, D. Geffken, and J. B. Adams Jr., "Oxazolidinones: a new class of agricultural fungicides," in *Synthesis and Chemistry of Agrochemicals*, vol. 686 of *ACS Symposium Series*, chapter 22, pp. 216–227, 1998.

[76] J. A. Sternberg, D. Geffken, J. B. Adams et al., "Famoxadone: the discovery and optimisation of a new agricultural fungicide," *Pest Management Science*, vol. 57, no. 2, pp. 143–152, 2001.

[77] D. B. Jordan, R. S. Livingston, and J. J. Bisaha, "Oxazolidinones: a new chemical class of fungicides and inhibitors of mitochondrial cytochrome bc_1 function," *Pesticide Science*, vol. 55, no. 2, pp. 213–215, 1999.

[78] W. J. Welstead, G. C. Helsley, C. R. Taylor et al., "5-(2-Aminoethyl)-2-oxazolidinones with central nervous system depressant and antiinflammatory activity," *Journal of Medicinal Chemistry*, vol. 16, no. 10, pp. 1129–1132, 1973.

[79] "Pharmaceutical composition containing 5-morpholinomethyl-3-(4-chlorobenzylideneamine)-2-oxazolidinone and its use for the treatment of CNS disorders," United States Patent, 5643907.

[80] M. Masaki and H. Shinozaki, "A new class of potent centrally acting muscle relaxants: pharmacology of oxazolidinones in rat decerebrate rigidity," *British Journal of Pharmacology*, vol. 89, no. 1, pp. 219–228, 1986.

[81] H. J. Lewerenz, D. W. Bleyl, and W. Schnaak, "Effects of 5-vinyloxazolidine-2-thione on lactating rats and their sucklings," *Die Nahrung*, vol. 36, no. 4, pp. 388–393, 1992.

[82] N. Zenker, L. S. Hubbard, and J. Wright, "Inhibition of dopamine β-hydroxylase by goitrin, a natural antithyroid compound," *Journal of Natural Products*, vol. 51, no. 5, pp. 862–865, 1988.

[83] M. McMillan, E. A. Spinks, and G. R. Fenwick, "Preliminary observations on the effect of dietary Brussels sprouts on thyroid function," *Human Toxicology*, vol. 5, no. 1, pp. 15–19, 1986.

[84] M. Artico, G. De Martino, and R. Giuliano, "Research on compounds with antiblastic activity. XL. Synthesis of 3-p-(2',5'-dimethoxy-4'-(N,N-bis-(-chloroethyl)-amino)benzylideneamino)phenyl-2-oxazolidinone (GEA 29 ; BAY a 5850) and its analogues," *Il Farmaco; Edizione Scientifica*, vol. 26, no. 9, pp. 771–783, 1971.

[85] H. Onda, F. M. E. Wagenlehner, N. Lehn, and K. G. Naber, "In vitro activity of linezolid against Gram-positive uropathogens of hospitalized patients with complicated urinary tract infections," *International Journal of Antimicrobial Agents*, vol. 18, no. 3, pp. 263–266, 2001.

[86] http://www.chemicalbook.com/ChemicalProductProperty_EN_CB71179461.htm.

[87] D. G. Musson, S. M. Maglietto, and S. S. Hwang, "Simultaneous quantification of cycloserine and its prodrug acetylacetonylcycloserine in plasma and urine by high-performance liquid chromatography using ultraviolet absorbance and fluorescence after post-column derivatization," *Journal of Chromatography*, vol. 414, no. 1, pp. 121–129, 1987.

[88] T. Mathur, M. Kumar, T. K. Barman et al., "Activity of RBx 11760, a novel biaryl oxazolidinone, against *Clostridium difficile*," *Journal of Antimicrobial Chemotherapy*, vol. 66, no. 5, pp. 1087–1095, 2011.

[89] E. Skripkin, T. S. McConnell, J. DeVito et al., "Rχ-01, a new family of oxazolidinones that overcome ribosome-based linezolid resistance," *Antimicrobial Agents and Chemotherapy*, vol. 52, no. 10, pp. 3550–3557, 2008.

[90] J. J. Hilliard, J. Fernandez, J. Melton et al., "In vivo activity of the pyrrolopyrazolyl-substituted oxazolidinone RWJ-416457," *Antimicrobial Agents and Chemotherapy*, vol. 53, no. 5, pp. 2028–2033, 2009.

[91] J. B. Locke, M. Hilgers, and K. J. Shaw, "Novel ribosomal mutations in *Staphylococcus aureus* strains identified through selection with the oxazolidinones linezolid and torezolid (TR-700)," *Antimicrobial Agents and Chemotherapy*, vol. 53, no. 12, pp. 5265–5274, 2009.

[92] J. B. Locke, G. Morales, M. Hilgers et al., "Elevated linezolid resistance in clinical cfr-positive *Staphylococcus aureus* isolates

is associated with co-occurring mutations in ribosomal protein L3," *Antimicrobial Agents and Chemotherapy*, vol. 54, no. 12, pp. 5352–5355, 2010.

[93] E. Y. Cheung, M. L. Peterson, R. Chiarella, and B. M. Foxman, "A novel oxazolidinone anti-biotic: RWJ-416457," *Acta Crystallographica Section C*, vol. 66, no. 5, pp. o249–o251, 2010.

[94] J. B. Locke, J. Finn, M. Hilgers et al., "Structure-activity relationships of diverse oxazolidinones for linezolid-resistant *Staphylococcus aureus* strains possessing the cfr methyltransferase gene or ribosomal mutations," *Antimicrobial Agents and Chemotherapy*, vol. 54, no. 12, pp. 5337–5343, 2010.

[95] New Phase 4 Study Shows Higher Rates of Clinical and Microbiological Success for Zyvox Versus Vancomycin in MRSA Nosocomial Pneumonia, http://www.prnewswire.com/news-releases/new-phase-4-study-shows-higher-rates-of-clinical-and-microbiological-success-for-zyvox-versus-vancomycin-in-mrsa-nosocomial-pneumonia-105422943.html.

4

Aurones: A Promising Heterocyclic Scaffold for the Development of Potent Antileishmanial Agents

Marina Roussaki,[1] Sofia Costa Lima,[2] Anna-Maria Kypreou,[1] Panagiotis Kefalas,[3] Anabela Cordeiro da Silva,[2,4] and Anastasia Detsi[1]

[1] *Laboratory of Organic Chemistry, School of Chemical Engineering, National Technical University of Athens, Heroon Polytechniou 9, Zografou Campus, 15780 Athens, Greece*
[2] *Parasite Disease Group, IBMC-Institute for Molecular and Cell Biology, University of Porto, Rua do Campo Alegre, 823, 4150-180 Porto, Portugal*
[3] *Department of Food Quality and Chemistry of Natural Products, Mediterranean Agronomic Institute of Chania, International Centre for Advanced Mediterranean Agronomic Studies, 73100 Chania, Crete, Greece*
[4] *Department of Biological Sciences, Faculty of Pharmacy, University of Porto, Rua de Jorge Viterbo Ferreiera, 228, 4050-313 Porto, Portugal*

Correspondence should be addressed to Anastasia Detsi, adetsi@chemeng.ntua.gr

Academic Editor: O. Bruno

A series of (*Z*)-2-benzylidenebenzofuran-3-(2H)-ones (aurones) bearing a variety of substituents on rings A and B were synthesized and evaluated for their antiparasitic activity against the intracellular amastigote form of *Leishmania infantum* and their cytotoxicity against human THP1-differentiated macrophages. In general, aurones bearing no substituents on ring A (compounds **4a–4f**) exhibit higher toxicity than aurones with 4,6-dimethoxy substitution (compounds **4g–4l**). Among the latter, two aurones possessing a 2′-methoxy or a 2′-methyl group (compounds **4i** and **4j**) exhibit potent antileishmanial activity ($IC_{50} = 1.3 \pm 0.1\,\mu M$ and $IC_{50} = 1.6 \pm 0.2\,\mu M$, resp.), comparable to the activity of the reference drug Amphotericin B, whereas they present significantly lower cytotoxicity than Amphotericin B as deduced by the higher selectivity index.

1. Introduction

Leishmaniasis, an infectious disease caused by protozoan parasites belonging to the genus *Leishmania* (*L.*), is transmitted to humans through the bite of female phlebotomine sand flies infected with the parasite. This disease is manifested in three forms: cutaneous leishmaniasis (CL), which is the most common form, mucocutaneous leishmaniasis (MCL), and visceral leishmaniasis (VL) [1]. CL causes ulcers on the exposed parts of the human body which leave permanent scars when healed, whereas MCL causes the destruction of the mucous membranes of the nose, mouth, and throat cavities and surrounding tissues leading to serious disabling of the patient. Although they do not cause death of the patient, these two forms pose a serious social problem as a result of prejudice or even rejection of the disfigured patient from the community. VL is the most serious form of the disease which inevitably leads to death if left untreated [2]. Leishmaniasis is distributed in four continents and affects millions of people, especially in the poor developing countries where the lack of efficient medical care and resources, malnutrition, and weak immune system enhance the possibility of infection [3]. Moreover, co-infection with HIV is an emerging threat as the immunosuppressed HIV patients are especially vulnerable to VL infection. In fact, the majority of the VL cases reported in southern Europe involves HIV-infected patients [4]. In addition, in developed countries, where leishmaniasis is not endemic, the cases of infected population have increased over the last decade, in association with increasing international tourism, military operations, and the influx of immigrants from endemic countries [5].

Treatment of leishmaniasis lies exclusively on chemotherapy because no human vaccine is yet available [6]. The first-line drugs are the pentavalent antimonials meglumine antimoniate (Glucantime) and sodium stibogluconate (Pentostam), which are now obsolete in India due to resistance development although they are still used in other countries [2, 7]. As second-line treatment, liposomal formulations of the polyene antibiotic Amphotericin B have been developed in order to increase efficacy and reduce toxicity of the drug [8]. However, although it has high cure rates, this drug suffers from severe side-effects and high cost which renders its use unaffordable for patients in developing countries. New, cheaper formulations of Amphotericin B are currently being explored as alternatives [9]. The first oral treatment for leishmaniasis has been developed in 2002 using miltefosine, an alkylphospholipid initially used as an anticancer agent [10]. Unfortunately, miltefosine has exhibited teratogenic potential; therefore its use is prohibited in pregnant women [11]. Current advances in leishmaniasis chemotherapy include paromomycin, an aminoglycoside antibiotic, which has recently started to be used as a treatment for CL and VL and sitamaquine, a 8-aminoquinolone that has reached phase II clinical trials for the treatment of VL [10, 12, 13].

The development of new, more effective, less toxic, and affordable antileishmanial drugs is still a major goal in medicinal chemistry based not only on rational drug design [14, 15], but also on the exploitation of nature either as a source of phytochemical antiparasitics or as an inspiration for new molecular scaffolds [16–19]. The evaluation of the antiparasitic activity of plant extracts or isolated natural products constitutes an attractive approach for the development of new efficient drugs.

Aurones, (*Z*)-2-benzylidenebenzofuran-3-(2*H*)-ones (Figure 1), constitute a subclass of flavonoids which occur rarely in nature. They are responsible for the bright yellow color of some popular ornamental flowers and are biosynthesized from chalcones by the key enzyme aureusidin synthase. Although the studies of the biological activities of aurones are still limited, these natural products and their synthetic analogues have proved to be promising bioactive compounds with a broad spectrum of activity including anticancer [20–22] antimicrobial, [23] and antioxidant [24, 25] properties whereas they possess enzyme inhibitory [25, 26], or enzyme-inducing activity [27].

The antileishmanial activity of a series of aurones was first reported by Kayser and Kiderlen in 1999 [28]. In this study, five aurones and one aurone glucoside were tested against extracellular promastigotes of *L. donovani*, *L. infantum*, *L. enriettii*, and *L. major*, and against intracellular amastigote *L. donovani*, and the results revealed that the compounds were active against the parasites in concentrations ranging between 0.4 and 5.0 μg/mL while they showed moderate cytotoxicity. The same research group studied structurally diverse aurones and auronols for their inhibitory activity against the mitochondrial NADH-fumarate reductase of *L. major* promastigotes. The majority of the compounds were potent inhibitors of the enzyme in a dose- and structure-dependent manner [29].

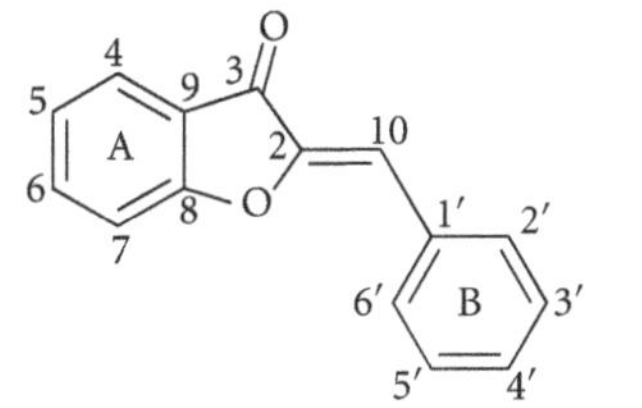

Figure 1: Chemical structure and numbering of the aurone scaffold.

As a continuation of our studies towards the synthesis and activity evaluation of novel bioactive natural product analogues, we have focused on the aurone scaffold in order to further explore its potential in the development of novel and efficient antileishmanial agents. In this work we describe the synthesis and structural characterization of aurone derivatives bearing various substituents on different positions of rings A and B as well as the evaluation of their activity against the intracellular form of *L. infantum* parasites.

2. Results and Discussion

2.1. Chemistry. Aurones **4a–4l** have been synthesized via the oxidative cyclization of the corresponding chalcones using mercury(II) acetate in pyridine. The synthesis and characterization of compounds **4a–4i** has been described in detail in one of our previous publications [25]. In order to accomplish the synthesis of aurones **4j–4l**, chalcones **3j–3l** were synthesized via the Claisen-Schmidt condensation reaction between 4′,6′-dimethoxy-2′-hydroxy-acetophenone (**1b**) and the appropriately substituted benzaldehyde in basic conditions (20% aqueous KOH in ethanol) (Scheme 1).

The oxidative cyclization of chalcones **3j–3l** proceeds smoothly by refluxing the corresponding chalcone with $Hg(OAc)_2$ in pyridine for 1 h (Scheme 1). The desired aurones **4j–4l** were isolated as yellow solids after acidification with aqueous HCl 10%, extraction with an appropriate solvent, and purification.

The structural characterization of the synthesized compounds was achieved using ^1H and ^{13}C NMR spectroscopy as well as MS/ESI(+) spectroscopy. The ^1H NMR spectra of chalcones **3j–3l** are characterized by a signal at a low field (~14 ppm) attributed to the proton of the 2′-OH group which is deshielded as a result of taking part in a strong intramolecular hydrogen bond with the neighbouring carbonyl group, and the pair of AB doublets owed to the vinylic protons, with J~15 Hz confirming the *E*-geometry of the double bond. In the case of chalcones **3j** and **3k**, the presence of a substituent on position 2 results in a significant downfield shift of the AB doublets (8.08 and 8.13 ppm for H_β and 7.81 and 7.87 ppm for H_α, resp.) as compared to the corresponding chemical shifts in the ^1H NMR spectrum of the naturally occurring chalcone

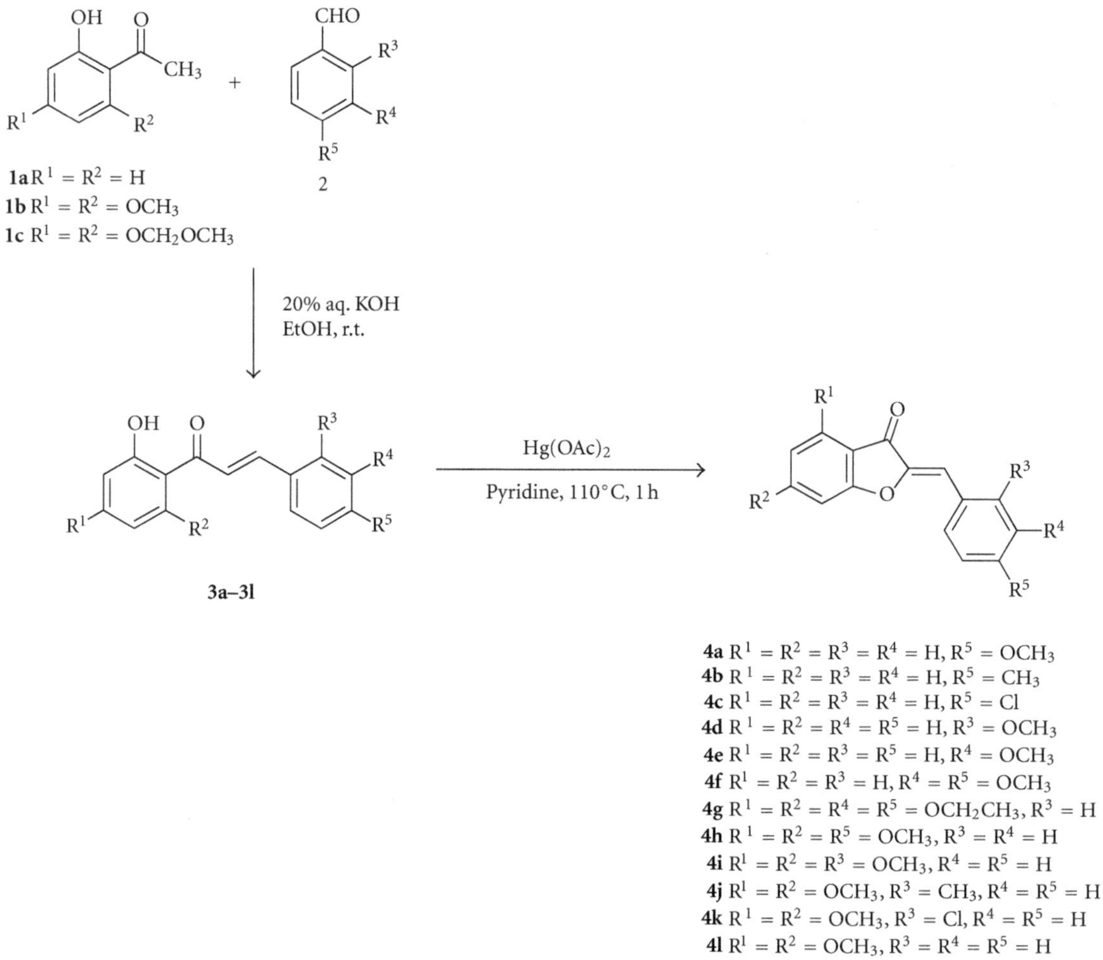

Scheme 1: Synthesis of chalcones **3a–3l** and of the corresponding aurones **4a–4l**.

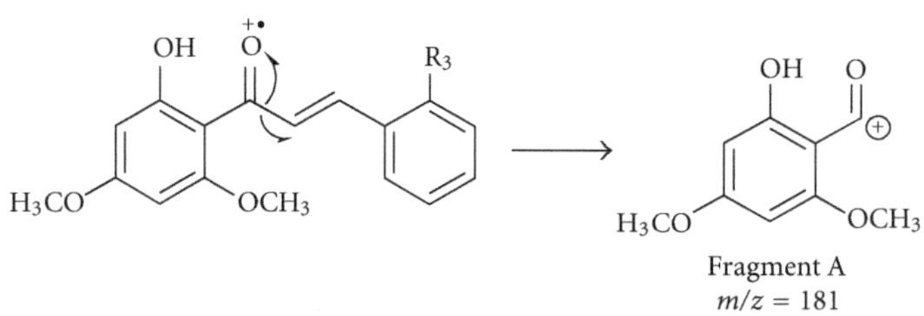

Scheme 2: Major fragmentation pathway of chalcones **3j–3l** in CID ESI-MS.

flavokawain B (**3l**) (7.91 ppm for H_β and 7.79 ppm for H_α). This difference should be attributed to the "*ortho*" effect, a phenomenon which renders the spectra of *ortho*-substituted aromatic compounds more complex than those of their *meta*- and *para*-substituted counterparts [30].

The mass spectra of chalcones **3j–3l** were obtained using the electrospray ionisation (ESI) technique and an ion trap detector. In the conditions of the ESI experiment, chalcones are protonated and the $[M+H]^+$ ion is detected instead of the molecular ion. In the collision-induced dissociation (CID) mass spectra of the $[M+H]^+$ cations of **3j–3l**, the fragmentation pattern observed involved a scission of the bond between the carbonyl carbon and the vinylic C_α, which produced Fragment A as the basic ion (Scheme 2).

In general, the ^1H NMR spectra of aurones are characterized by the presence of a singlet owed to the exocyclic vinylic proton, the chemical shift of which is indicative of the geometrical configuration of the molecule and is affected significantly by the position of the substituents on ring B and, to a lesser extent, by the presence of alkoxy substituents on ring A. In the case of the synthesized aurones **4a–4c** and **4e–4h, 4l,** this singlet appears in the range of 6.57–6.89 ppm [25], in accordance with the previous findings by us and other research groups [21, 24, 25]. The chemical shift of

TABLE 1: Energy calculation of selected energetically minimized aurones (kJ mol^{-1}).

Compound	Potential energy	Torsional energy
4a (*Z*-isomer)	88.08	−1.82
4a (*E*-isomer)	110.16	12.14
4d (*Z*-isomer)	107.57	13.54
4d (*E*-isomer)	117.81	13.18
4h (*Z*-isomer)	139.95	22.83
4h (*E*-isomer)	162.43	37.43
4i (*Z*-isomer)	132.81	0.76
4i (*E*-isomer)	169.79	38.41

the vinylic proton in the case of aurones **4d**, **4i–4k** which bear a substituent on position 2′ is observed downfield, at 7.00–7.49 ppm. Although some controversy was at first aroused as to if the downfield shift of vinylic H indicates the presence of the *E*-geometrical isomer of 2′-substituted aurones, the X-ray crystallographic analyses of two aurones reported recently proved that it is undoubtedly that the thermodynamically more stable *Z*-isomer always prevails [21, 31]. As a result, the difference in chemical shifts is attributed to the *ortho*-effect.

The observation of only one signal owed to the vinylic proton indicates that the oxidative cyclization procedure applied leads exclusively to the formation of a single geometrical isomer, in this case the thermodynamically more stable *Z*-aurone [24]. In order to further support this hypothesis, *in silico* approaches were employed to perform energy calculations for selected aurones (**4a**, **4d**, **4h**, and **4i**). Results indicate that the *Z* isomers exhibit lower potential energy thus higher stability than their *E* counterparts. Differences in energy values between *E*/*Z* isomers are mainly attributed to torsional energy contribution (Table 1).

2.2. Biological Assays. All compounds were tested *in vitro* with human THP1-differentiated macrophages and cell viability evaluated by MTT assay (Table 1). The ability to inhibit the growth of the parasite without affecting macrophage viability is a good indication of selectivity of the compound, a desirable property for a drug candidate. The less cytotoxic aurones on the mammalian cells ($IC_{50} > 20\,\mu M$) were then assessed against the intracellular amastigote stage of *L. infantum*. The selectivity index (SI) was determined based on ratio of IC_{50} values measured on the mammalian cells and for the intracellular parasite (IC_{50} for mammalian cell line/IC_{50} for parasite). This ratio provides an indication of the selectivity of these compounds against *L. infantum* amastigotes when compared to normal cells. The results are shown in Table 2.

Aurones **4a–4f** bearing no substituents on ring A generally exhibited higher toxicity against THP1 cells than aurones **4g–4l** which possess oxygenated substituents on the same ring. Among the nonsubstituted on ring A compounds, aurone **4e** did not show significant toxicity but it is not a very active antileishmanial agent either ($IC_{50} = 4.7 \pm 0.5\,\mu M$). The compounds possessing oxygenated substituents on positions 4 and 6 of ring A generally were less toxic against THP1 cells. However, aurone **4g**, one of the least toxic compound of the series, which possesses four methoxymethyl (MOM) groups, is a weak antiparasitic agent ($IC_{50} = 5.1 \pm 0.3\,\mu M$).

The position and electronic nature of the substituent on ring B seem to significantly affect the activity of the compounds against *L. infantum* parasite as well as their cytotoxicity: the most active and one of the less toxic compounds of the series, 2′,4,6-trimethoxyaurone **4i** bears a methoxy group on position 2′ of ring B and exhibits potent antileishmanial activity with $IC_{50} = 1.3 \pm 0.1\,\mu M$. Changing the electron-donating methoxy group to the nonoxygenated electron-donating methyl group on the same position results in compound **4j**, which exhibits comparable antileishmanial activity ($IC_{50} = 1.6 \pm 0.2\,\mu M$), with **4i** but slightly higher toxicity. On the other hand, when position 2′ of ring B is occupied by an electron-withdrawing chlorine atom as in aurone **4k**, antiparasitic activity is substantially lower ($IC_{50} = 12.2 \pm 1.4\,\mu M$) whereas the compound is not toxic to mammalian cells (in fact aurone **4k** is the least toxic compound of the series studied in this work). Aurone **4l** in which ring B is unsubstituted possesses significant antileishmanial activity ($IC_{50} = 2.1 \pm 0.9\,\mu M$) although lower than the 2′-substituted analogues **4i** and **4j**.

Compounds **4i** and **4j** exhibited higher selectivity towards the intracellular form of the *L. infantum* parasite, when compared to the toxicity exerted on the mammalian cells (SI = 57.5 and 43.4, resp.). Amphotericin B was used as a reference antileishmanial drug currently applied on the chemotherapy of leishmaniasis. Its activity on the intracellular amastigotes was very similar with the one obtained with compounds **4i** and **4j**, although Amphotericin B was more toxic on the human-differentiated macrophages (SI = 20.6).

3. Experimental

^1H NMR spectra and ^{13}C NMR spectra were recorded on a Varian Gemini 2000 300 MHz spectrometer or a Varian 600 MHz spectrometer. Coupling constants (J) are expressed in hertz (Hz). Chemical shifts (δ) of NMR are reported in parts per million (ppm) units relative to the solvent. Melting points were determined on a Gallenkamp MFB-595 melting point apparatus and are uncorrected.

The LC/MS analysis was performed using Varian LC/MS using the ElectroSpray Ionisation technique and an Ion Trap detector.

*3.1. General Procedure for the Synthesis of Chalcones **3j–3l**.* To a stirred solution of 2-hydroxy-4,6-dimethoxy-acetophenone (**1b**) (1eq) and a substituted benzaldehyde (1eq) in ethanol was added KOH (3eq, 20% w/v aqueous solution) and the mixture was stirred at room temperature for 24–72 h. The reaction mixture was cooled to 0°C (ice-water bath) and acidified with HCl (10% v/v aqueous solution). The precipitate formed was filtered and washed with 10% aqueous HCl solution.

TABLE 2: Antileishmanial activity and cytotoxicity of the synthesized aurones.

Compound	IC_{50} (μM)		Selectivity index (SI)[a]
	L. infantum intracellular amastigotes	THP1-differentiated macrophages	
4a	nd[b]	16.5 ± 3.1	nd
4b	nd	8.5 ± 0.8	nd
4c	nd	14.5 ± 2.1	nd
4d	nd	17.5 ± 3.1	nd
4e	4.7 ± 0.5	54.2 ± 7.3	11.5
4f	nd	20.5 ± 1.2	nd
4g	5.1 ± 0.3	62.5 ± 1.3	12.4
4h	nd	19.6 ± 2.5	nd
4i	1.3 ± 0.1	75.4 ± 4.7	57.5
4j	1.6 ± 0.2	68.1 ± 2.1	43.4
4k	12.2 ± 1.4	>100	>8.2
4l	2.1 ± 0.9	57.5 ± 3.4	26.9
Amphotericin B	1.2 ± 0.1	23.8 ± 2.3	20.6

[a]The selectivity index represents the ratio of IC_{50} on THP1-differentiated macrophages to the IC_{50} on intracellular parasite.
[b]nd: not determined, for compounds with IC_{50} values below 20 μM on THP1-differentiated macrophages.

2′-Hydroxy-4′,6′-dimethoxy-2-methyl-chalcone ***(3j)*** *[32].* Prepared following the general procedure starting from 2-hydroxy-4,6-dimethoxy-acetophenone (**1b**, 500 mg, 2.3 mmol) and o-methylbenzaldehyde (278 mg, 2.3 mmol), dissolved in 9 mL ethanol and KOH (20% aqueous solution, 2 mL). The solid was recrystallized from methanol to afford yellow crystals. Yield: 420 mg (61%). mp 117–120°C; ^{1}H NMR ($CDCl_3$, 300 MHz) δ: 14.26 (s, 1H), 8.08 (d, J = 15.3 Hz, 1H), 7.81 (d, J = 15.9 Hz, 1H), 7.64 (d, J = 6.6 Hz, 1H), 7.32–7.19 (m, 3H), 6.13 (s, 1H), 5.98 (s, 1H), 3.93 (s, 3H), 3.87 (s, 3H), 2.52 (s, 3H). ^{13}C NMR ($CDCl_3$, 75 MHz) δ ppm 192.87, 168.54, 166.37, 162.65, 140.12, 138.33, 134.65, 131.01, 129.94, 128.73, 126.74, 126.40, 106.48, 93.90, 91.39, 55.99, 55.74, 20.12. MS (ESI) *m/z* $[M+1]^+$ 299.

2′-Hydroxy-4′,6′-dimethoxy-2-chloro-chalcone ***(3k)***. Prepared following the general procedure starting from 2-hydroxy-4,6-dimethoxy-acetophenone (**1b**, 500 mg, 2.3 mmol) and o-chlorobenzaldehyde (328 mg, 2.3 mmol), dissolved in 9 mL ethanol and KOH (20% aqueous solution, 2 mL). The solid was recrystallized from methanol to afford yellow crystals. Yield: 518 mg (70%). mp 135–136°C (lit. [33] mp 136–137°C); ^{1}H NMR ($CDCl_3$, 600 MHz) δ: 14.19 (s, 1H), 8.13 (d, J = 15.6 Hz, 1H), 7.87 (d, J = 15.6 Hz, 1H), 7.69–7.67 (m, 1H), 7.43–7.42 (m, 1H), 7.31–7.28 (m, 2H), 6.11 (d, J = 1.8 Hz, 1H), 5.95 (d, J = 2.4 Hz, 1H), 3.89 (s, 3H, OCH_3), 3.83 (s, 3H, OCH_3). ^{13}C NMR ($CDCl_3$, 75 MHz) δ: 192.48, 168.62, 166.55, 162.62, 138.03, 135.53, 133.98, 130.83, 130.41, 130.18, 127.95, 127.12, 106.45, 93.94, 91.45, 56.03, 55.78. MS (ESI) *m/z* $[M+1]^+$ 319.

2′-Hydroxy-4′,6′-dimethoxy-chalcone *(Flavokawain B,* ***3l****).* Prepared following the general procedure starting from 2-hydroxy-4,6-dimethoxy-acetophenone (**1b**, 500 mg, 2.3 mmol) and benzaldehyde (247 mg, 2.3 mmol), dissolved in 9 mL ethanol and KOH (20% aqueous solution, 2 mL). The solid was recrystallized from methanol to afford orange crystals. Yield: 460 mg (72%). mp 74–76°C (lit. [34] mp 89–90°C); ^{1}H NMR ($CDCl_3$, 600 MHz): δ 14.27 (s, 1H), 7.91 (d, J = 15.6 Hz, 1H), 7.79 (d, J = 15.6 Hz, 1H), 7.61 (dd, J = 7.8 Hz, J = 1.2 Hz, 2H), 7.42–7.39 (m, 3H), 6.10 (d, J = 2.4 Hz, 1H,), 5.96 (d, J = 2.4 Hz, 1H), 3.90 (s, 3H), 3.84 (s, 3H). ^{13}C NMR ($CDCl_3$, 75 MHz) δ: 192.77, 168.54, 166.37, 162.63, 142.46, 135.68, 130.20, 129.01, 128.50, 127.65, 106.45, 93.90, 91.40, 55.99, 55.73. MS (ESI) *m/z* $[M+1]^+$285.

3.2. General Procedure for the Preparation of Z-Aurones ***4j–4k****.* To a solution of mercuric acetate (1.2eq) in pyridine was added chalcone (1eq) at room temperature and the mixture was stirred at 110°C for 1 h. The cooled reaction mixture was poured into ice cold water and acidified with HCl (10% aqueous solution). The precipitated solid was extracted with dichloromethane, the extracts were dried (Na_2SO_4) and the solvent was evaporated to give a solid which was further purified by recrystallization from methanol/hexane.

4, 6-Dimethoxy-2′-methylaurone *(****4j****).* Prepared following the general procedure starting from chalcone **3j** (200 mg, 0.67 mmol) in 7 mL pyridine. After recrystallization, the product was obtained as yellow crystals. Yield: 112 mg (48%). mp 180–183°C (lit. [21] mp 194–195°C); ^{1}H NMR ($CDCl_3$, 300 MHz) δ: 8.15 (d, J = 6 Hz, 1H), 7.32–7.23 (m, 4H,), 7.00 (s, 1H), 6.39 (d, J = 1.8 Hz, 1H), 6.15 (d, J = 1.8 Hz, 1 H), 3.98 (s, 3H), 3.93 (s, 3H), 2.52 (s, 3H). ^{13}C NMR ($CDCl_3$, 75 MHz)δ: 180.86, 169.30, 169.09, 159.57, 148.14, 138.96, 131.21, 130.90, 130.74, 129.42, 126.35, 124.73, 107.95, 105.41, 94.17, 89.33, 56.37, 56.28, 20.48.

4, 6-Dimethoxy-2′-chloro-aurone **(4k)**. Prepared following the general procedure starting from chalcone **3k** (250 mg, 0.78 mmol) in 7.3 mL pyridine. After recrystallization, the product was obtained as yellow crystals. Yield: 194 mg (78%). mp 191–193°C (lit. [27] mp 201–203°C); ^{1}H NMR ($CDCl_3$, 300 MHz) δ: 8.26 (dd, J = 7.5 Hz, J = 1.8 Hz, 1 H), 7.45 (dd, J = 7.8 Hz, J = 1.5 Hz, 1H), 7.38–7.28 (m, 2H), 7.22 (s, 1H), 6.39 (d, J = 1.8 Hz, 1H), 6.16 (d, J = 1.8 Hz, 1H), 3.98 (s, 3H), 3.93 (s, 3H). ^{13}C NMR ($CDCl_3$, 75 MHz) δ: 180.47, 169.28, 169.22, 159.74, 148.90, 135.70, 132.02, 130.87, 130.19, 130.08, 127.02, 106.07, 105.26, 94.33, 89.50, 56.41, 56.29. MS (ESI) m/z $[M+1]^+$ 317.

4, 6-Dimethoxyaurone (**4l**). Prepared following the general procedure starting from chalcone **3l** (200 mg, 0.70 mmol) in 6.6mL pyridine. After recrystallization, the product was obtained as yellow crystals. Yield: 88 mg (44%). mp 139–142°C (lit. [34] mp 152–153°C); ^{1}H NMR ($CDCl_3$, 300 MHz) δ: 7.87 (d, J = 7.6 Hz, 2 H), 7.46–7.36 (m, 3 H), 6.78 (s, 1H), 6.40 (d, J = 1.8 Hz, 1 H), 6.15 (d, J = 1.8 Hz, 1 H), 3.98 (s, 3 H), 3.94 (s, 3H); ^{13}C NMR ($CDCl_3$, 75 MHz) δ: 180.85, 169.25, 169.16, 159.60, 148.03, 132.74, 131.25, 131.15, 129.46, 128.92, 110.93, 105.40, 94.22, 89.41, 56.38, 56.26. MS (ESI) *m/z* $[M+1]^+$ 283.

3.3. Parasite and Cell Cultures

3.3.1. Parasites. A cloned line of *L. infantum* (MOM/MA671 TMAP263) promastigotes, stably expressing the luciferase gene (LUC), was grown in RPMI 1640 medium (Bio Whitaker, Belgium) supplemented with 10% heat-inactivated fetal bovine serum (FBS) (Bio Whitaker, Belgium), 2 mML-glutamine (Bio Whitaker, Belgium), 20 mM Hepes (Bio Whitaker, Belgium), 100 U/mL penicillin, and 100 μg/mL streptomycin (Bio Whitaker, Belgium). Selection of LUC positive parasites was done by adding geneticin-sulphate (Sigma-Aldrich, USA) to the culture media at a final concentration of 5 μg/mL. Parasites were maintained in culture, at 26°C by subpassage (10°C parasites/mL) at every 5 days.

L. infantum axenic amastigotes, stably expressing the LUC gene were derived from promastigotes by culturing them in MAA (Medium for Axenic Amastigotes) culture medium. MAA consisted of a modified medium 199 with Hanks balanced salt solution (Gibco-Invitrogen, Spain) supplemented with 0.5% soya broth trypto-casein (Bio-Rad, UK), 15 mM D-glucose (Panreac, Spain), 4 mM $NaHCO_3$ (Sigma-Aldrich, USA). The pH was adjusted to 5.8, and the media was 0.2 μm sterilized by filtration and further supplemented with 0.023 mM bovine hemin (Fluka, USA), 5 mM L-glutamine, and 25% of heat inactivated FBS. Amastigotes were maintained in culture, at 37°C in an atmosphere containing 5% CO_2 by subpassage (10^5 parasites/mL), at every 5 days.

3.3.2. Cell Lines. The human leukaemia monocyte THP1 cell line was cultured as a monolayer at 37°C in a humidified atmosphere containing 5% CO_2. Cells were grown in RPMI 1640 medium, supplemented with 10% heat-inactivated FBS, 2 mM L-glutamine, 100 U/mL penicillin, and 100 μg/mL streptomycin, and were maintained in culture, by subpassage every 3 days.

3.4. Toxicity of Aurones on THP1-Differentiated Macrophages. For macrophage differentiation, human leukaemia monocyte cell line (THP-1 cells) cells were incubated in the presence of 20 ng/ml phorbol 12-myristate 13-acetate (PMA, Sigma-Aldrich) for 18 h at 37°C, 5% CO_2,and left another 24 h with fresh medium containing no PMA to induce maturation. Serial dilutions of each aurone, ranging from 100 to 1.56 μM in culture media, were added to the wells, in quadruplicate, and incubated for 72 h at 37°C, 5% CO_2. Cell viability was assessed by the MTT assay, as described before [35].

The IC_{50} value, that is, the concentration of the aurone necessary to decrease cell viability to 50% of the untreated control, was determined by linear regression analysis.

3.5. Growth Inhibition Assays. THP1-differentiated macrophages were infected for a period of 4 h with stationary phase LUC expressing, *L. infantum* axenic amastigotes, at a 5 : 1 parasite to cell ratio. After the infection period, cells were washed with culture media to remove noninternalized parasites. Serial dilutions of the aurones, ranging from 50 to 0.78 μM in culture media, were added to the wells, in quadruplicate, and incubated for 72 h at 37°C, 5% CO_2. After the incubation period, the luciferase activity of intracellular amastigotes was determined as described before [36]. The percentage of growth inhibition was calculated as described above.

3.6. Molecular Modeling Calculations. Molecular Modeling calculations were performed using MacroModel module of Schrodinger package, (Schrodinger, LLC, New York, NY, 2011). *E* and *Z* isomers for the four selected aurones were initially energetically minimized using Molecular Mechanics with OPLS_2005 force field and $CHCl_3$ as solvent, simulating the environment of the NMR solvent. Minimization was performed with PRCG "Polak-Ribiere Conjugate Gradient" algorithm using 1000 iterations and an energy tolerance of 0.05 kcal/mol^{-1} Å^{-1}, to reach a local minimum.

4. Conclusions

A series of substituted aurones have been synthesized and tested for their antileishmanial activity against the intracellular form of *L. infantum* parasite as well as their cytotoxicity against mammalian cells. The position of the substituents as well as their electronic nature seem to play an important role in the antiparasitic activity. Among the twelve tested compounds, the aurones **4i** and **4j**, possessing methoxy groups at positions 4 and 6 of ring A and electron-donating groups on position 2′of ring B exhibited antiparasitic activity comparable to that of Amphotericin B. The significant antileishmanial activity of aurones **4i** and **4j** is combined with low cytotoxicity, thus rendering these compounds eligible for further studies toward the design and synthesis of novel antileishmanial agents possessing this naturally occurring "privileged" scaffold.

Acknowledgments

The present collaboration was promoted by COST action CM0801 "New drugs for neglected diseases." Sofia Costa Lima thanks Fundação para a Ciência e Tecnologia (FCT) and FSE (III Quadro Comunitário) for the Grant SFRH/BPD/37880/2007. The contribution of Dr. Panagiotis Zoumpoulakis in the molecular modeling calculations is gratefully acknowledged.

References

[1] http://www.who.int/leishmaniasis/burden/en/.

[2] H. C. Maltezou, "Drug resistance in visceral leishmaniasis," *Journal of Biomedicine & Biotechnology*, vol. 2010, Article ID 617521, 2010.

[3] J. Alvar, S. Yactayo, and C. Bern, "Leishmaniasis and poverty," *Trends in Parasitology*, vol. 22, no. 12, pp. 552–557, 2006.

[4] J. Alvar, P. Aparicio, A. Aseffa et al., "The relationship between leishmaniasis and AIDS: the second 10 years," *Clinical Microbiology Reviews*, vol. 21, no. 2, pp. 334–359, 2008.

[5] A. Pavli and H. C. Maltezou, "Leishmaniasis, an emerging infection in travelers," *International Journal of Infectious Diseases*, vol. 14, no. 12, pp. e1032–e1039, 2010.

[6] K. J. Evans and L. Kedzierski, "Development of vaccines against visceral leishmaniasis," *Journal of Tropical Medicine*, vol. 2012, Article ID 892817, 2012.

[7] S. Sundar, D. K. More, M. K. Singh et al., "Failure of pentavalent antimony in visceral leishmaniasis in India: report from the center of the Indian epidemic," *Clinical Infectious Diseases*, vol. 31, no. 4, pp. 1104–1107, 2000.

[8] C. Bern, J. Adler-Moore, J. Berenguer et al., "Liposomal amphotericin B for the treatment of visceral leishmaniasis," *Clinical Infectious Diseases*, vol. 43, no. 7, pp. 917–924, 2006.

[9] L. Ordóñez-Gutiérrez, R. Espada-Fernández, M. A. Dea-Ayuela, J. J. Torrado, F. Bolás-Fernandez, and J. M. Alunda, "*In vitro* effect of new formulations of amphotericin B on amastigote and promastigote forms of Leishmania infantum," *International Journal of Antimicrobial Agents*, vol. 30, no. 4, pp. 325–329, 2007.

[10] S. L. Croft, M. P. Barrett, and J. A. Urbina, "Chemotherapy of trypanosomiases and leishmaniasis," *Trends in Parasitology*, vol. 21, no. 11, pp. 508–512, 2005.

[11] B. L. Herwaldt, "Miltefosine—the long-awaited therapy for visceral leishmaniasis?" *The New England Journal of Medicine*, vol. 341, no. 24, pp. 1840–1842, 1999.

[12] T. S. Tiuman, A. O. Santos, T. Ueda-Nakamura, B. P. D. Filho, and C. V. Nakamura, "Recent advances in leishmaniasis treatment," *International Journal of Infectious Diseases*, vol. 15, no. 8, pp. e525–e532, 2011.

[13] T. Polonio and T. Efferth, "Leishmaniasis: drug resistance and natural products," *International Journal of Molecular Medicine*, vol. 22, no. 3, pp. 277–286, 2008.

[14] I. Papanastasiou, K. C. Prousis, K. Georgikopoulou et al., "Design and synthesis of new adamantyl-substituted antileishmanial ether phospholipids," *Bioorganic and Medicinal Chemistry Letters*, vol. 20, no. 18, pp. 5484–5487, 2010.

[15] V. P. Pandey, S. S. Bisht, M. Mishra et al., "Synthesis and molecular docking studies of 1-phenyl-4-glycosyl-dihydropyridines as potent antileishmanial agents," *European Journal of Medicinal Chemistry*, vol. 45, no. 6, pp. 2381–2388, 2010.

[16] M. L. Bello, L. D. Chiaradia, L. R. S. Dias et al., "Trimethoxy-chalcone derivatives inhibit growth of Leishmania braziliensis: synthesis, biological evaluation, molecular modeling and structure-activity relationship (SAR)," *Bioorganic and Medicinal Chemistry*, vol. 19, no. 16, pp. 5046–5052, 2011.

[17] J. A. Peixoto, M. L. A. E. Silva, A. E. M. Crotti et al., "Antileishmanial activity of the hydroalcoholic extract of miconia langsdorffii, isolated compounds, and semi-synthetic derivatives," *Molecules*, vol. 16, no. 2, pp. 1825–1833, 2011.

[18] R. Sen and M. Chatterjee, "Plant derived therapeutics for the treatment of Leishmaniasis," *Phytomedicine*, vol. 18, pp. 1056–1069, 2011.

[19] M. J. Chan-Bacab and L. M. Peña-Rodríguez, "Plant natural products with leishmanicidal activity," *Natural Product Reports*, vol. 18, no. 6, pp. 674–688, 2001.

[20] H. Cheng, L. Zhang, Y. Liu et al., "Design, synthesis and discovery of 5-hydroxyaurone derivatives as growth inhibitors against HUVEC and some cancer cell lines," *European Journal of Medicinal Chemistry*, vol. 45, no. 12, pp. 5950–5957, 2010.

[21] H. M. Sim, C. Y. Lee, P. L. R. Ee, and M. L. Go, "Dimethoxyaurones: potent inhibitors of ABCG2 (breast cancer resistance protein)," *European Journal of Pharmaceutical Sciences*, vol. 35, no. 4, pp. 293–306, 2008.

[22] H. M. Sim, K. Y. Loh, W. K. Yeo, C. Y. Lee, and M. L. Go, "Aurones as modulators of ABCG2 and ABCB1: synthesis and structure-activity relationships," *ChemMedChem*, vol. 6, no. 4, pp. 713–724, 2011.

[23] B. P. Bandgar, S. A. Patil, B. L. Korbad, S. C. Biradar, S. N. Nile, and C. N. Khobragade, "Synthesis and biological evaluation of a novel series of 2,2-bisaminomethylated aurone analogues as anti-inflammatory and antimicrobial agents," *European Journal of Medicinal Chemistry*, vol. 45, no. 7, pp. 3223–3227, 2010.

[24] S. Venkateswarlu, G. K. Panchagnula, A. L. Gottumukkala, and G. V. Subbaraju, "Synthesis, structural revision, and biological activities of 4′-chloroaurone, a metabolite of marine brown alga Spatoglossum variabile," *Tetrahedron*, vol. 63, no. 29, pp. 6909–6914, 2007.

[25] A. Detsi, M. Majdalani, C. A. Kontogiorgis, D. Hadjipavlou-Litina, and P. Kefalas, "Natural and synthetic 2′-hydroxy-chalcones and aurones: synthesis, characterization and evaluation of the antioxidant and soybean lipoxygenase inhibitory activity," *Bioorganic and Medicinal Chemistry*, vol. 17, no. 23, pp. 8073–8085, 2009.

[26] R. Haudecoeur, A. Ahmed-Belkacem, W. Yi et al., "Discovery of naturally occurring aurones that are potent allosteric inhibitors of hepatitis C virus RNA-dependent RNA polymerase," *Journal of Medicinal Chemistry*, vol. 54, no. 15, pp. 5395–5402, 2011.

[27] C. Y. Lee, E. H. Chew, and M. L. Go, "Functionalized aurones as inducers of NAD(P)H:quinone oxidoreductase 1 that activate AhR/XRE and Nrf2/ARE signaling pathways: synthesis, evaluation and SAR," *European Journal of Medicinal Chemistry*, vol. 45, no. 7, pp. 2957–2971, 2010.

[28] O. Kayser and A. F. Kiderlen, "Leishmanicidal activity of aurones," *Tokai Journal of Experimental and Clinical Medicine*, vol. 23, no. 6, pp. 423–426, 1998.

[29] O. Kayser, M. Chen, A. Kharazmi, and A. F. Kiderlen, "Aurones interfere with Leishmania major mitochondrial fumarate

reductase," *Zeitschrift fur Naturforschung C*, vol. 57, no. 7-8, pp. 717–720, 2002.

[30] Arunima and N. D. Kurur, "Ortho effect in fluorobenzenes: cross-correlated relaxation and quantum chemical studies," *Magnetic Resonance in Chemistry*, vol. 43, no. 2, pp. 132–138, 2005.

[31] M. Zhang, X.-H. Xu, Y. Cui, L.-G. Xiea, and C.-H. Kong, "Synthesis and herbicidal potentialof substituted aurones," *Pest Management Science*. In press.

[32] B. Srinivasan, T. E. Johnson, R. Lad, and C. Xing, "Structure—activity relationship studies of chalcone leading to 3-hydroxy-4,3′,4′,5′-tetramethoxychalcone and its analogues as potent nuclear factor κB inhibitors and their anticancer activities," *Journal of Medicinal Chemistry*, vol. 52, no. 22, pp. 7228–7235, 2009.

[33] P. Boeck, P. C. Leal, R. A. Yunes et al., "Antifungal activity and studies on mode of action of novel xanthoxyline-derived chalcones," *Archiv der Pharmazie*, vol. 338, no. 2-3, pp. 87–95, 2005.

[34] H. Sekizaki, "Synthesis of 2-benzylidene-3(2H)-benzofuran-3-ones (aurones) by oxidation of 2′-hydroxychalcones with mercury(II) acetate," *Bulletin of the Chemical Society of Japan*, vol. 61, pp. 1407–1409, 1988.

[35] T. Mosmann, "Rapid colorimetric assay for cellular growth and survival: application to proliferation and cytotoxicity assays," *Journal of Immunological Methods*, vol. 65, no. 1-2, pp. 55–63, 1983.

[36] G. Roy, C. Dumas, D. Sereno et al., "Episomal and stable expression of the luciferase reporter gene for quantifying Leishmania spp. infections in macrophages and in animal models," *Molecular and Biochemical Parasitology*, vol. 110, no. 2, pp. 195–206, 2000.

5

Microwave-Assisted Synthesis and Biological Evaluation of Dihydropyrimidinone Derivatives as Anti-Inflammatory, Antibacterial, and Antifungal Agents

Anjna Bhatewara, Srinivasa Rao Jetti, Tanuja Kadre, Pradeep Paliwal, and Shubha Jain

Laboratory of Heterocycles, School of Studies in Chemistry and Biochemistry, Vikram University, Ujjain, Madhya Pradesh 456010, India

Correspondence should be addressed to Shubha Jain; drshubhajain479@gmail.com

Academic Editor: Qidong You

A simple protocol for the efficient preparation of aryl and heteroaryl substituted dihydropyrimidinone has been achieved via initial Knoevenagel, subsequent addition, and final cyclization of aldehyde, ethylcyanoacetate, and guanidine nitrate in the presence of piperidine as a catalyst in solvent-free under microwave irradiation. The synthesized compounds showed a good anti-inflammatory, antibacterial, and antifungal activity.

1. Introduction

Pyrimidinones have been paid increasing attention, due to their various therapeutic and pharmacological properties, such as antiviral, antibacterial, antihypertensive, and antitumor effects [1]. More recently, they emerged as integral backbones of several calcium blockers, antihypertensive agents, α-1a-antagonists, and neuropeptide Y (NPY) antagonists [2]. Pyrimidinone derivatives are found as core units in many marine alkaloids (batzelladine and carambine), which have been found to be potent to HIV-gp-120 CD4 inhibitors [3].

Due to the remarkable biological utilization, the pyrimidinones attract many researchers as well as academicians. Recently, several methods improved the procedure using phosphorus pentoxide-methanesulfonic acid [4], potassium *ter*-butoxide (*t*-BuOK) [5], ammonium dihydrogen phosphate [6], silica-gel [7], mesoporous molecular sieve MCM-41 [8], cyanuric chloride [9], nano-$BF_3{\cdot}SiO_2$ [10], silica gel-supported polyphosphoric Acid [11], zirconium(IV) chloride [12], and indium(III) bromide [13] as catalysts. However, some of these one-pot procedures generally require strong protic or Lewis acids, prolonged reaction times, and high temperature. Consequently, there is a scope for further modification towards mild reaction condition, increased variation of the substituents, and improved yields.

Microwave promoted solvent-free reactions [14] are well known as environmentally benign methods that also usually provide improved selectivity, enhanced reaction rates, cleaner products, and manipulative simplicity [15]. However, these procedures are practically limited as the solvents in microwave oven at elevated temperatures create high pressures, which may cause explosion. To circumvent these problems, there is a need for the development of newer methods which proceed under mild and solvent free condition.

Nowadays solvent-free reactions gained much importance in organic synthesis because of the high yields and shorter reaction times. Earlier reported procedures for the synthesis of pyrimidine derivatives typically involved longer reaction time and fewer yields [16]. In the present communication, we would like to describe the advantages of dry reaction techniques coupled with microwave activation and their applications to organic synthesis [17].

2. Experimental

2.1. General. All reactions were carried out in an LG domestic unmodified microwave oven model MS-1947C/01. Reactions were routinely monitored by thin layer chromatography (TLC) on silica gel (precoated F 254 Merck plates) and visualized the products under UV light (254 nm). ^1H NMR spectra were determined in $CDCl_3$ or DMSO-d_6 solutions with a Bruker Advance II 400 MHz spectrometer and signals recorded in parts per million (δ) downfield from tetramethylsilane as internal standard. IR spectra were recorded on Perkin Elmer FT-IR spectrometer (Spectrum RX I) using KBr pellet technique. The elemental analyses were performed using thermo EA 2110 series. Melting points were recorded in open capillaries on LABINDIA melting point apparatus and were uncorrected. Mass spectra (ESI) were recorded on Waters Micromass Q-TOF Micro. Anti-inflammatory activity has been carried out in Institute of Pharmacy, Vikram University, Ujjain.

2.2. General Procedure for the Synthesis of 2-amino Dihydropyrimidinone Derivatives **4(a-h)**. A mixture of aldehyde **1** (1 mmol), ethyl cyanoacetate **2** (1.2 mmol), guanidine nitrate **3** (1.5 mmol), and 2-3 drops of piperidine was subjected to microwave irradiation at 60% power in 600 W microwave oven for 5 min. (Successive irradiation of 30–40 sec with cooling intervals of time, the temperature being 90–100°C). On completion of reaction, indicated by TLC, the mixture was cooled and quenched with water (3 × 10 mL). The solid product was separated and recrystallized from ethanol to afford pure products **4(a-h)** in good yields.

2.3. Spectral Data

2-amino-4-oxo-6-phenyl-1,4,5,6-tetrahydropyrimidine-5-carbonitrile **(4a)**. White crystals; Mp: 218–220°C; IR (KBr ν_{max}/cm^{-1}): 3478 (NH), 3090 (C–H), 2260 (CN), 1690 (C=O), 1617 (C=N), 1567 (C=C). ^1H NMR (300 MHz, $CDCl_3$) δ 8.56, 2.0 (s, 2H NH), 7.27–7.40 (m, 5H, Ar), 4.1 (d, CH, J = 8.4 Hz), 3.97 (d, CH, J = 11.5 Hz). ^{13}C NMR ($CDCl_3$) δ: 168.4, 153.3, 143.5, 128.5, 126.9, 126.7, 116.8, 43.2, 42.4. Molecular weight: 214.22; Mass (m/z): 214 (M^+); $C_{11}H_{10}N_4O$ (214.09); Cacld. C, 61.67; H, 4.71; N, 26.15; O, 7.47; Found. C, 60.07; H, 4.13; N, 25.87; O, 7.08.

2-amino-6-(4-methoxyphenyl)-4-oxo-1,4,5,6-tetrahydropyrimidine-5-carbonitrile **(4b)**. Light yellow crystals; Mp: 110–112°C; IR (KBr ν_{max}/cm^{-1}): 3318 (NH), 3055 (C–H), 2190 (CN), 1710 (C=O), 1620 (C=N), 1565 (C=C). ^1H NMR (300 MHz, $CDCl_3$) δ: 8.32 (s, 2H, NH_2), 6.94–7.18 (m, 4H, Ar), 3.9 (d, CH, J = 14.4 Hz), 3.72 (d, CH, J = 11.5 Hz), 3.50 (s, 3H, OCH_3), 2.0 (s, 2H NH). ^{13}C NMR ($CDCl_3$) δ: 166.7, 159.1, 153.5, 136.2, 127.4, 116.3, 115.0, 56.1, 43.6, 44.8. Molecular weight: 244; Mass (m/z): 243 (M^+); $C_{12}H_{12}N_4O_2$ (244.10); Cacld. C, 59.01; H, 4.95; N, 22.94; O, 13.10; Found. C, 58.87; H, 4.63; N, 22.65, O, 12.72.

2-amino-6-(3,4-dimethoxyphenyl)-4-oxo-1,4,5,6-tetrahydropyrimidine-5-carbonitrile **(4c)**. Yellow crystals; Mp: 127-128°C; IR (KBr ν_{max}/cm^{-1}): 3460 (NH), 3080 (arom. C–H), 2943 (methyl C–H), 1590 (C=C), 1290 (aryl OCH_3), 1670 (C=O), 1567 (C=N), 2310 (CN).^1H NMR (300 MHz, $CDCl_3$) δ: 6.74–6.96 (m, 3H, Ar), 2.0, 8.56 (s, 2H, NH), 3.97 (d, CH, J = 8.4 Hz), 4.1 (d, CH, J = 11.5 Hz), 3.83 (s, 2H, methylene proton). ^{13}C NMR ($CDCl_3$) δ: 168.4, 153.3, 149.6, 147.8, 136.8, 121.9, 118.9, 116.8, 109.8, 43.2, 42.7, 56.1. Molecular weight: 274.28; Mass (m/z): 248 (M^+); $C_{13}H_{14}N_4O_3$ (274.11); Cacld. C, 56.93; H, 5.14; N, 20.43; O, 17.50; Found. C, 56.14; H, 5.04; N, 21.98; O, 17.19.

2-amino-6-(4-nitrophenyl)-4-oxo-1,4,5,6-tetrahydropyrimidine-5-carbonitrile **(4d)**. Yellow crystals; Mp: 162–164°C; IR (KBr ν_{max}/cm^{-1}): 3440 (NH), 3080 (C–H), 2327 (CN), 1640 (C=O), 1560 (C=N), 1567 (C=C), 1523 (N=O). ^1H NMR (300 MHz, $CDCl_3$) δ: 7.55–8.21 (dd, 4H, Ar), 2.0, 8.56 (s, 2H, NH), 3.97 (d, CH, J = 11.5 Hz), 4.1 (d, CH, J = 8.4 Hz). ^{13}C NMR ($CDCl_3$) δ: 168.4, 153.3, 149.6, 145.9, 123.7, 123.4, 116.8, 43.2, 42.4. Molecular weight: 259.22; Mass (m/z): 259 (M^+); $C_{11}H_9N_5O_3$ (259.07); Cacld. C, 50.97; H, 3.50; N, 27.02; O, 18.52; Found. C, 50.73; H, 3.24; N, 26.92; O, 18.32.

2-amino-4-oxo-6-(1H-pyrrol-2-yl)-1,4,5,6-tetrahydropyrimidine-5-carbonitrile **(4e)**. Yellow crystals; Mp: 93–95°C; IR (KBr ν_{max}/cm^{-1}): 3420 (NH), 3111 (Ar C–H), 2360 (CN), 1720 (C=O), 1593 (C=N). ^1H NMR (300 MHz, $CDCl_3$) δ: 5.72–6.69 (m, 3H, Ar), 2.0, 8.56 (s, 2H, NH), 3.02 (m, 1H, CH), 3.9 (m, 1H, CH). ^{13}C NMR ($CDCl_3$) δ: 168.4, 153.3, 130.5, 118.0, 116.8, 108.5, 107.7, 43.9, 42.8. Molecular weight: 203.20; Mass (m/z): 203 (M^+); $C_9H_9N_5O$ (203.08); Cacld. C, 53.20; H, 4.46; N, 34.47; O, 7.87; Found. C, 52.84; H, 4.13; N, 33.94; O, 7.59.

2-amino-6-(furan-2-yl)-4-oxo-1,4,5,6-tetrahydropyrimidine-5-carbonitrile **(4f)**. Light yellow crystals; Mp: 76–78°C IR (KBr ν_{max}/cm^{-1}): 3427 (NH), 3121 (C–H), 2230 (CN), 1690 (C=O), 1615 (C=N), 1572 (C=C). ^1H NMR (300 MHz, $CDCl_3$) δ: 7.92 (s, 2H, NH_2), 6.54–7.67 (m, 3H, Ar), 4.1 (d, CH, J = 11.5 Hz), 3.82 (d, CH, J = 8.4 Hz), 2.0 (s, 2H, NH). ^{13}C NMR ($CDCl_3$) δ: 164.5, 153.6, 149.9, 142.1, 116.5, 111.1, 110.4, 44.3, 40.7. Molecular weight: 204; Mass (m/z): 203 (M^+); $C_9H_8N_4O_2$ (204.06); Cacld. C, 52.94; H, 3.95; N, 27.44; O, 15.67; Found. C, 52.81; H, 3.78; N, 26.97; O, 15.41.

2-amino-6-(1H-indol-3-yl)-4-oxo-1,4,5,6-tetrahydropyrimidine-5-carbonitrile **(4g)**. Yellow crystals; Mp: 190–192°C; IR (KBr ν_{max}/cm^{-1}): 3450 (NH), 3120 (C–H), 2310 (CN), 1690 (C=O), 1590 (C=N), 1534 (C=C). ^1H NMR (300 MHz, $CDCl_3$) δ: 10.8 (s, 1H, NH), 8.22 (s, 2H, NH_2), 7.11–7.69 (m, 5H, Ar), 3.7 (d, CH, J = 14.0 Hz), 3.4 (d, CH, J = 8.0 Hz), 2.0 (s, 2H, NH). ^{13}C NMR ($CDCl_3$) δ: 164.3, 154.1, 133.2, 126.4, 124.3, 122.5, 120.6, 114.7, 117.1, 110.9, 43.1, 45.4. Molecular weight: 253; Mass (m/z): 252 (M^+); $C_{13}H_{11}N_5O$ (253.10); Cacld. C, 61.65; H, 4.38; N, 27.65; O, 6.32; Found. C, 61.16; H, 3.99; N, 27.34; O, 5.89.

2-amino-4-oxo-6-(1-methyl-1H-pyrrol-2-yl)-1,4,5,6-tetrahydropyrimidine-5-carbonitrile **(4h)**. Yellow crystals; Mp: 148–150°C; IR (KBr ν_{max}/cm^{-1}): 3470 (NH), 3111 (arom. C–H), 2360 (CN), 1720 (C=O), 1542 (C=C), 1593 (C=N), 2953 (methyl C–H). ^1H NMR (300 MHz, $CDCl_3$) δ: 5.72–6.69 (m, 3H, Ar), 2.0, 8.56 (s, 2H, NH), 3.90 (s, 1H, CH_3), 3.97 (d, CH, J = 8.4 Hz), 4.1 (d, CH, J = 11.5 Hz). ^{13}C NMR ($CDCl_3$) δ: 168.4, 153.3, 132.1, 122.5, 116.8, 108.6, 108.4, 44.2, 40.3, 35.2. Molecular weight: 217.23; Mass (m/z): 217 (M^+); $C_{10}H_{11}N_5O$

(217.10); Cacld. C, 55.29; H, 5.10; N, 32.24; O, 7.37; Found. C, 54.94; H, 4.92; N, 31.91; O, 7.03.

2.4. Pharmacology. Colony bred healthy rats of Wistar strain and albino mice procured from local market from Ujjain were used for the study. They were housed in standard polypropylene cages under room temperature (24 ± 2°C), relative humidity (60%–70%), and exposed to 12 : 12 hours light : dark cycle. The rats were fed Nutrilab Rodent Feed and drinking water filtered through an Aqua guard water filter system *adlibitum.* They were allowed to acclimatize for 5 days prior to commencement of dosing. The protocol was ethically approved by IAEC of the institute.

2.5. Anti-Inflammatory Activity. The anti-inflammatory activity was determined in vivo [18] using the carrageenan-induced rat paw edema test [5, 11]. A solution of 0.1 mL of 1% carrageenan (Sigma-Aldrich, Dorset, UK) in saline was injected subplantarly in the right hind paw of the rats 1 h after IP administration of compounds. The paw thickness was measured from the ventral to the dorsal surfaces using a dial caliper immediately prior to carrageenan injection and then at each hour, up to 4 h after the subplanar injection. The edema was calculated as the thickness variation between the carrageenan and saline treated paw. Anti-inflammatory activity was expressed as the percent of inhibition of the edema when compared with the control group. The results are expressed as the mean ± SEM of n animals per group. The data was statistically analyzed by one way analysis of variance (ANOVA) followed by Tukey multicomparison test. Differences with $P < 0.05$ between experimental groups were considered statistically significant.

TABLE 1: Piperidine catalyzed synthesis of 2-amino dihydropyrimidinones.

Entry	Ar	Product[a]	Yield (%)[b]	M.P (°C)
1	C_6H_5	**4a**	92	218–220
2	4-(CH_3O)-C_6H_4	**4b**	93	110–112
3	3,4-(CH_3O)-C_6H_3	**4c**	87	127-128
4	4-NO_2-C_6H_4	**4d**	83	162–164
5	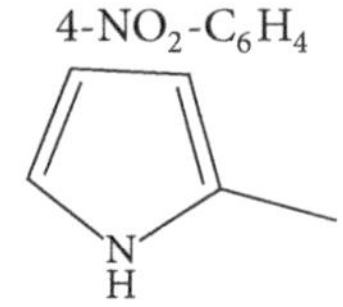	**4e**	79	93–95
6	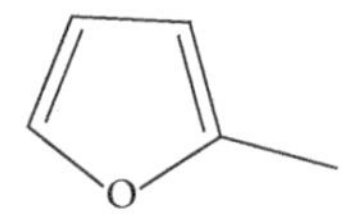	**4f**	91	76–78
7	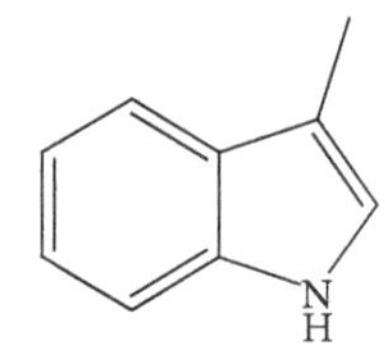	**4g**	90	190–192
8	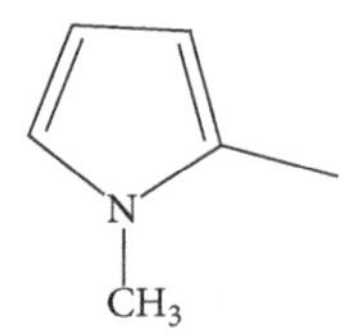	**4h**	87	148–150

[a]All compounds thus obtained were characterized by physical and spectral data. [b]Isolated yields.

2.6. Antibacterial Activity. Antibacterial activity of the prepared compounds **4d, 4e, 4f, 4g**, and **4h** were tested by the disk diffusion method [19]. Whattman no. 1 filter paper disks were sterilized by autoclaving for 1 h at 140°C. The sterile disks were impregnated with the test compounds (250 mg/mL). Agar plates were uniformly surface inoculated with fresh broth culture of *Staphylococcus aureus, Basillus subtilis, Escherichia coli*, and *Pseudomonas aeruginosa.* The impregnated disks were placed on the medium suitably spaced apart, and the plates were incubated at 30°C for 1 h to permit good diffusion and were then transferred to an incubator at 37 ± 2°C for 24 h. The zones of inhibition were measured on mm scale. The results of antimicrobial activity tests are listed in Table 3.

2.7. Antifungal Activity. Antifungal susceptibility test was done by disk diffusion method [20] using Sabouraud's dextrose agar medium. After sterilization, the medium was inoculated with *Candida albicans, Aspergillus flavus*, and *Aspergillus niger.* The standard antifungal agent clotrimazole (100 g/mL), solvent control (0.5% v/v Tween 80), and the newly synthesized compounds **4d**, **4e**, **4f**, **4g**, and **4h** in a concentration of 100 μg/mL were then added by sterile micropipette. The plates were then incubated at 37°C for 24 h and the diameter of zone of inhibition was measured and recorded in Table 3.

3. Results and Discussion

In the view of the above mentioned limitations of the reported method, pharmacological importance of dihydropyrimidinones and our ongoing endeavors to conduct organic synthesis under solvent free conditions [21], we describe an expeditious solvent less microwave accelerated approach for the rapid assembly of 2-amino dihydropyrimidinones. Aromatic aldehydes (**1a–h**, 1 mmol) on reaction with ethyl cyanoacetate (**2**, 1.2 mmol) and guanidine nitrate (**3**, 1.5 mmol) using dry conditions yielded corresponding 2-amino dihydropyrimidinones (Scheme 1).

As far as our interest in investigating the facile, rapid, and expeditious solvent-less methodology for 2-amino dihydropyrimidinone, we tried the reaction of benzaldehyde (**1a**, 1 mmol), with ethyl cyanoacetate (**2**, 1.2 mmol) and guanidine nitrate (**3**, 1.5 mmol) by varying microwave power from 150 watts to 750 watts. It was observed that by increase in power up to 600 watts, there was increase in yield and shortened reaction time. Beyond the 600 watts there was no significant change in reaction time and yield.

TABLE 2: Effects of compounds (**4d, 4e, 4f, 4g**, and **4h**) and indomethacin in the inhibition of carrageenan-induced rat paw edema.

Entry	Drug	Dose	Paw volume response at different time intervals in mean ± SEM			
			1 hour	2 hour	3 hour	4 hour
1	Control	10 mL/kg	0.5912 ± 0.005***	0.5930 ± 0.002***	0.6102 ± 0.002***	0.6303 ± 0.002***
2	Indomethacin (standard)	100 mg/kg	0.2700 ± 0.005** (54.33%)	0.3124 ± 0.005** (47.31%)	0.3133 ± 0.008** (48.65%)	0.3500 ± 0.005** (44.47%)
3	**4d**	20 mg/kg	0.4000 ± 0.005*** (32.34%)	0.4300 ± 0.005** (27.48%)	0.4500 ± 0.011*** (26.25%)	0.4711 ± 0.011*** (25.25%)
4	**4e**	40 mg/kg	0.3533 ± 0.003** (40.24%)	0.3833 ± 0.003** (35.36%)	0.3967 ± 0.003** (34.98%)	0.4200 ± 0.010** (33.36%)
5	**4f**	20 mg/kg	0.4133 ± 0.003** (30.09%)	0.4433 ± 0.003** (25.24%)	0.4667 ± 0.008*** (23.51%)	0.4933 ± 0.012*** (21.73%)
6	**4g**	40 mg/kg	0.3567 ± 0.003** (39.66%)	0.3867 ± 0.003** (34.78%)	0.4133 ± 0.003** (32.26%)	0.4520 ± 0.005*** (28.28%)
7	**4h**	20 mg/kg	0.3833 ± 0.003** (35.16%)	0.4133 ± 0.003*** (30.30%)	0.4433 ± 0.003** (27.35%)	0.4667 ± 0.008*** (25.95%)

Values are expressed as mean ± SEM; $n = 6$ in each group, $^{***}P < 0.001$, $^{**}P < 0.01$, compared to control. Data was analysed by one way ANOVA followed by Duneet's test. Formula % inhibition = $V_c - V_t/V_c \times 100$ (V-volume, C-control, t-test).

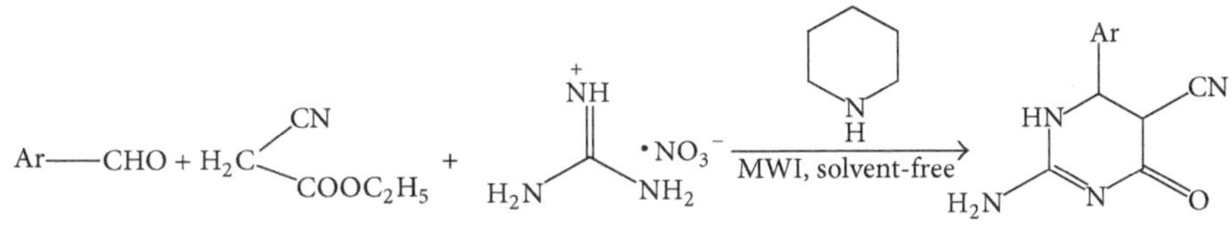

1(a–h) 2 3 4(a–h)

SCHEME 1

TABLE 3: Antimicrobial evaluation of synthesized compounds.

Compound	Norfloxacin	4d	4e	4f	4g	4h
Gram +ve bacteria						
S. aureus ATCC 25922	20	18	10	16	15	18
B. subtilis ATCC 6633	22	17	12	15	16	17
Gram −ve bacteria						
E. coli ATCC 25922	22	24	10	20	16	15
P. aeruginosa ATCC 27853	23	20	13	20	18	17
Compound	**Clotrimazole**	**4d**	**4e**	**4f**	**4g**	**4h**
Test fungi						
C. albicans ATCC 10231	20	21	16	18	17	17
A. flavus ATCC 204304	22	20	17	17	18	15
A. niger ATCC 16404	21	19	16	20	16	16

In order to evaluate the generality of this model reaction, we then prepared a range of 2-amino dihydropyrimidinone derivatives under the optimized reaction conditions. In all cases, aryl aldehydes and heteroaryl aldehydes with substituents carrying either electron-donating or electron-withdrawing groups reacted successfully and gave the expected products in good to excellent yields in relatively short reaction times. The kind of aldehyde has no significant effect on the reaction. The results are shown in Table 1.

In the presence of piperidine, reaction proceeds smoothly giving desired products in short time and in a good yield. The formation of the product takes place when aryl aldehydes were reacted with ethyl cyanoacetate to form arylmethylene ethylcyanoacetate, which subsequently added with guanidine followed by cyclization and tautomerization to form the desired product.

The results of the anti-inflammatory activity of compounds **4d, 4e, 4f, 4g,** and **4h** were summarized in Table 2. From the results, it is evident that compounds **4c, 4f**, and **4h** as well as indomethacin as the reference drug induced significant anti-inflammatory activity after 3 and 4 h in comparison to control and almost all of the tested compounds were shown moderate to good anti-inflammatory activity.

The MIC values of the test solutions are recorded in Table 3 which is recorded in zones of inhibition in mm for the bacteria and fungi.

The antimicrobial screening has shown that compounds **4d, 4f, 4g**, and **4f** have displayed moderate activity against gram +ve bacteria tested, that is, *S. aureus* and *B. subtilis*. Compounds **4d** and **4h** have shown broad spectrum activity against gram −ve bacteria tested, that is, *E. coli* and *P. aeruginosa* while compound **4e** has shown poor activity against both gram +ve and gram −ve bacteria tested.

Compound **4d** has shown good antifungal activity against all the tested fungi, that is, *C. albicans, A. flavus*, and *A. niger* while the remaining compounds **4e** and **4f** have shown good activity, and compounds **4g** and **4h** have shown moderate activity against all fungi tested.

4. Conclusion

In summary, we have described one-pot synthesis of 2-amino dihydropyrimidinone derivatives via a three component cycloaddition reaction under microwave irradiation. Another advantage of this method is excellent yields in shorter reaction time with high purity of the products. The synthesized compounds have shown good anti-inflammatory, antibacterial, and antifungal activities.

Conflict of Interests

The authors do not have any conflict of interests.

Acknowledgments

The authors are thankful to SAIF - IIT Bombay for providing elemental analysis and Department of Chemistry, Vikram University, Ujjain for extending laboratory facilities and IR data. The authors are also grateful to Anis Shaik, Institute of Pharmacy, Vikram University, Ujjain for providing microbiological facilities.

References

[1] C. O. Kappe, "100 years of the Biginelli dihydropyrimidine synthesis," *Tetrahedron*, vol. 49, no. 32, pp. 6937–6963, 1993.

[2] G. C. Rovnyak, S. D. Kimball, B. Beyer et al., "Calcium entry blockers and activators: conformational and structural determinants of dihydropyrimidine calcium channel modulators," *Journal of Medicinal Chemistry*, vol. 38, no. 1, pp. 119–129, 1995.

[3] A. D. Patil, N. V. Kumar, W. C. Kokke et al., "Novel alkaloids from the sponge *Batzella* sp.: inhibitors of HIV gp120-human CD4 binding," *Journal of Organic Chemistry*, vol. 60, no. 5, pp. 1182–1188, 1995.

[4] A. Borse, P. Mahesh, P. Nilesh, and S. Rohan, "A green, expeditious, one-pot synthesis of 3, 4-dihydropyrimidin-2(1*H*)-ones using a mixture of phosphorus pentoxide-methanesulfonic acid at ambient temperature," *ISRN Organic Chemistry*, vol. 2012, Article ID 415645, 6 pages, 2012.

[5] D. Abdelmadjid, C. Louisa, B. Raouf, and C. Bertrand, "A one-pot multi-component synthesis of dihydropyrimidinone/thione and dihydropyridine derivatives *via* Biginelli and Hantzsch condensations using *t*-BuOK as a catalyst under solvent-free conditions," *The Open Organic Chemistry Journal*, vol. 6, pp. 12–20, 2012.

[6] T. Reza, M. Behrooz, and G. Malihe, "Ammonium dihydrogen phosphate catalyst for one-pot synthesis of 3, 4-dihydropyrimidin-2(1*H*)-ones," *Chinese Journal of Catalysis*, vol. 33, no. 4–6, pp. 659–665, 2012.

[7] S. Agarwal, U. Aware, A. Patil et al., "Silica-gel catalyzed facile synthesis of 3, 4-dihydropyrimidinones," *Bulletin of Korean Chemical Society*, vol. 33, no. 2, pp. 377–378, 2012.

[8] R. Hekmatshoar, M. Heidari, M. M. Heravi, and B. Baghernejad, "Mesoporous molecular sieve MCM-41 catalyzed one-pot synthesis of 3,4-dihydro-2(1*H*)-pyrimidinones and -thiones under solvent-free conditions," *Bulletin of the Chemical Society of Ethiopia*, vol. 25, no. 2, pp. 309–313, 2011.

[9] J. A. Kumar, C. Shanmugam, and P. H. Babu, "One pot synthesis of dihydropyrimidinones catalyzed by Cyanuric chloride: an improved procedure for the Biginelli reaction," *Der Pharma Chemica*, vol. 3, no. 4, pp. 292–297, 2011.

[10] B. F. Mirjalili, A. Bamoniri, and A. Akbari, "One-pot synthesis of 3,4-dihydropyrimidin-2(1*H*)-ones (thiones) promoted by nano-$BF_3 \cdot SiO_2$," *Journal of the Iranian Chemical Society*, vol. 8, supplement 1, pp. S135–S140, 2011.

[11] M. Zeinali-Dastmalbaf, A. Davoodnia, M. M. Heravi, N. Tavakoli-Hoseini, A. Khojastehnezhad, and H. A. Zamani, "Silica gel-supported polyphosphoric acid (PPA-SiO_2) catalyzed one-pot multi-component synthesis of 3, 4-dihydropyrimidin-2(1*H*)-ones and -thiones: an efficient method for the Biginelli reaction," *Bulletin of the Korean Chemical Society*, vol. 32, no. 2, pp. 656–658, 2011.

[12] C. V. Reddy, M. Mahesh, P. V. K. Raju, T. R. Babu, and V. V. N. Reddy, "Zirconium(IV) chloride catalyzed one-pot synthesis of 3,4-dihydropyrimidin-2(1*H*)-ones," *Tetrahedron Letters*, vol. 43, no. 14, pp. 2657–2659, 2002.

[13] N. Y. Fu, Y. F. Yuan, Z. Cao, S. W. Wang, J. T. Wang, and C. Peppe, "Indium(III) bromide-catalyzed preparation of dihydropyrimidinones: improved protocol conditions for the Biginelli reaction," *Tetrahedron*, vol. 58, no. 24, pp. 4801–4807, 2002.

[14] A. Shaabani and A. Bazgir, "Microwave-assisted efficient synthesis of spiro-fused heterocycles under solvent-free conditions," *Tetrahedron Letters*, vol. 45, no. 12, pp. 2575–2577, 2004.

[15] M. M. Abelman, S. C. Smith, and D. R. James, "Cyclic ketones and substituted α-keto acids as alternative substrates for novel Biginelli-like scaffold syntheses," *Tetrahedron Letters*, vol. 44, no. 24, pp. 4559–4562, 2003.

[16] M. Kidwai, K. Singhal, and S. Kukreja, "One-pot green synthesis for pyrimido[4,5-d]pyrimidine derivatives," *Zeitschrift fur Naturforschung B*, vol. 62, no. 5, pp. 732–736, 2007.

[17] M. Kidwai, "Dry media reactions," *Pure and Applied Chemistry*, vol. 73, no. 1, pp. 147–151, 2001.

[18] M. B. Patil, S. S. Jalalpure, H. J. Pramod, and F. V. Manvi, "Anti-inflammatory activity of the leaves of Anacardium occidentale linn," *Indian Journal of Pharmaceutical Sciences*, vol. 65, no. 1, pp. 70–72, 2003.

[19] N. K. Jain, M. K. Chourasia, S. Jain, S. K. Jain, and N. K. Jain, "Development and characterization of engineered commensal bacteria for the delivery of enzymes," *Indian Journal of Pharmaceutical Sciences*, vol. 65, no. 2, pp. 113–121, 2003.

[20] A. R. Saundane, K. Rudresh, N. D. Satyanarayan, and S. P. Hiremath, "Pharmacological screening of 6H, 11H-indolo [3, 2-c] isoquinolin-5-ones and their derivatives," *Indian Journal of Pharmaceutical Sciences*, vol. 60, no. 6, pp. 379–383, 1998.

[21] T. Kadre, S. R. Jetti, A. Bhatewara, P. Paliwal, and S. Jain, "Green protocol for the synthesis of 3, 4-dihydropyrimidin-2(1*H*)-ones/thiones using TBAB as a catalyst and solvent free condition under microwave irradiation," *Archives of Applied Science Research*, vol. 4, no. 2, pp. 988–993, 2012.

6

Structural Stereochemistry of Androstene Hormones Determines Interactions with Human Androgen, Estrogen, and Glucocorticoid Receptors

Thomas L. Shaak,[1,2,3] Dayanjan S. Wijesinghe,[4,5] Charles E. Chalfant,[4,5,6] Robert F. Diegelmann,[2,5] Kevin R. Ward,[7] and Roger M. Loria[2,8]

[1] *Department of Integrative Life Sciences, Virginia Commonwealth University, Richmond, VA 23298, USA*
[2] *Virginia Commonwealth University Reanimation Engineering Science Center (VCURES), Virginia Commonwealth University, Richmond, VA 23298, USA*
[3] *Department of Biochemistry, Virginia Commonwealth University, 1101 E. Marshall Street, Sanger Hall, Room 2-004, Richmond, VA 23298, USA*
[4] *Hunter Holmes McGuire VA Medical Center, Richmond, VA 23249, USA*
[5] *Department of Biochemistry and Molecular Biology, Virginia Commonwealth University, Richmond, VA 23298, USA*
[6] *Massey Cancer Center, Richmond, VA 23298, USA*
[7] *Michigan Critical Injury and Illness Research Center, Department of Emergency Medicine, University of Michigan, Ann Arbor, MI 48109, USA*
[8] *Department of Microbiology, Immunology, Pathology and Emergency Medicine, Virginia Commonwealth University, Richmond, VA 23298, USA*

Correspondence should be addressed to Thomas L. Shaak; tshaak28@gmail.com

Academic Editor: Patrick Bednarski

DHEA, 17α-AED, 17β-AED, and 17β-AET exhibit strong biological activity that has been attributed to androgenic, estrogenic, or antiglucocorticoid activity in vivo and in vitro. This study compared DHEA, 17α-AED, 17β-AED, and 17β-AET for their ability to activate the human AR, ER, and GR and determine the relative androgenicity, estrogenicity, and glucocorticoid activity. The results show that, at the receptor level, these androstene hormones are weak AR and even weaker ER activators. Direct androstene hormone activation of the human AR, ERα, and ERβ may not be essential for their biological function. Similarly, these hormones indirectly activated the human GR, only in the presence of high dexamethasone concentrations. These results underscore the major difference between androstene hormone interactions with these nuclear receptors and their biological effects.

1. Introduction

DHEA, an androstene hormone, has been shown to possess a wide range of beneficial biological effects mainly attributed to immune system modulation [1]. DHEA is metabolized into more active metabolites, that is, 17β-AED and 17β-AET, as well as testosterone and estradiol [1, 2]. 17β-AED and 17β-AET have been reported to prevent the morbidity and mortality of otherwise lethal infections [3, 4], potentiate lymphocyte activation, and counteract the immune suppressive action of hydrocortisone [5–7], thus leading to beneficial effects in diverse human diseases including resistance to infection, neuroprotection, wound healing, diabetes, hepatic injury, cardiovascular disease, and cancer [8–10].

17α-AED mediates autophagy of glial and breast cancers and apoptosis of myeloid tumor cells [11–13]. 17β-AED and 17α-AED naturally exist in epimeric forms based on whether the hydroxyl group is above (β) or below (α) the Δ^5 cycloperhydrophenanthrene ring. Addition of a hydroxyl group at the C7 position to 17β-AED results in the formation of Δ^5-androstene-3β, 7β, 17β-triol (17β-AET). The biological activities of 17α-AED, 17β-AED, and 17β-AET have exhibited a structure-activity relationship that depends on the orientation and location of the hydroxyl groups [13].

Androstene hormones (AH) have been shown to promulgate their biological effects in many different animal models including mice, rats, monkeys, and some specific human tissues. Reports have associated the mechanism of action of androstene hormone metabolites with androgen, estrogen, and glucocorticoid receptor activity [14–16]. Adrenal hormones have been shown to activate both androgen and estrogen constructs. In this regard, it has been documented that 17β-AED can activate the AR in prostate tissue in the presence of commonly used antiandrogens [17]. Inhibitors of both the androgen receptor and the estrogen receptors demonstrated that AR and ERβ receptors combine to affect gene transcription [18]. Additionally, 17β-AED was recently shown to be a part of an anti-inflammatory mechanism that utilizes the ERβ [19]. 17β-AED and 17β-AET have been documented in vitro and in vivo to oppose the action of hydrocortisone indicating that there may be crosstalk with the GR [6, 20, 21].

DHEA has been shown to possess weak androgenicity and estrogenicity [22]. Because 17α-AED, 17β-AED, and 17β-AET are more potent metabolites of DHEA that exhibit strong biological activity that could be attributed to androgenic, estrogenic, or antiglucocorticoid activity in vivo and in vitro, it was advantageous to identify whether or not this is directly mediated by the human ER, AR, and GR at the cellular level. Additionally, androstenediol has been modeled as a chemical with a 3β-hydroxy and a saturated A ring which can act as an estrogen [23]. Consequently, we compared DHEA, 17α-AED, 17β-AED, and 17β-AET in the Indigo Biosciences nuclear receptor assay system for their ability to activate the human AR, ER, and GR and determine the relative androgenicity and estrogenicity of these androstene hormone derivatives.

2. Materials and Methods

2.1. Nuclear Receptor Transactivation Assays. Nuclear receptor transactivation assays were obtained from Indigo Biosciences (State College, PA, USA) and were utilized to assess the activation of human AR, ERβ, and ERα. receptors. Briefly, stocks of the compounds tested were prepared and diluted in a medium provided by the manufacturer. Cell medium was tested for hormone activity by mass spectrometry (Section 2.3). Frozen reporter cells provided in the assay kit were thawed and compound dilutions were added immediately. Cells were incubated for 24 hours and the activation response was measured on a luminometer (Perkin-Elmer, MA, USA). The cells consisted of non-human mammalian cells engineered by Indigo Biosciences to provide constitutive high-level expression of full length, unmodified human androgen receptor (NR3C4), human estrogen receptor 1 (NR3A1), human estrogen receptor 2 (NR3A2), and full length, human glucocorticoid receptor (NR3C1).

The nonhuman mammalian reporter cells included a luciferase reporter gene functionally linked to a human nuclear receptor-responsive promoter. The cells are engineered so that only interactions with the human receptor will induce luciferase expression in the treated reporter cells to quantitate nuclear receptor activation. Positive control ligand performance was measured by the manufacturer and provided in the technical manuals thus allowing accurate comparison for assay performance. Additionally, the control ligands of the receptors (testosterone, 17β-estradiol, and dexamethasone) were tested on the same test plates ($n = 3$ to allow statistical analysis) with the androstene hormones and controls.

2.2. Preparation of Stock Hormone Solutions. Stock solutions of 17β-AED, 17α-AED, and testosterone (Sigma-Aldrich, St. Louis, MO, USA) were prepared by dissolving the compounds into 100% ethanol to a final 50 mM stock solution concentration; 17β-AET was also a stock solution of 50 mM but was dissolved into 1 : 1 DMSO/ethanol because of solubility issues. Cyproterone acetate (Sigma-Aldrich) was diluted with 100% ethanol and used at a 10 uM concentration. All stock solutions were diluted to final concentrations using the dilution fluid provided in each kit. All tests were performed with negative controls on the same plate and contained media alone and media containing the same amount of ethanol utilized in the stock solutions. All assay control results were in accordance with the stated technical performance specifications.

2.3. Mass Spectrometry. LC-MS/MS analyses of the steroid hormones were carried out using a Shimadzu Nexera UPLC device coupled (Shimadzu Corp, Kyoto, Japan) to AB Sciex (Foster City, CA, USA) 5500 Hybrid Triple Quadrupole Linear Ion Trap Mass Spectrometer operating in multiple reaction monitoring mode. Nitrogen produced by a high-purity nitrogen generator (PEAK Scientific Instruments Ltd, Chicago, Il, USA) was used as curtain, nebulizer, and collision gases. Unit mass resolution was set in both mass-resolving quadrupole Q1 and Q3. Ionization of the analytes was carried out using an APCI source. Multiple MRM transitions were selected for each analyte to eliminate ambiguity in analyte identification. For all steroids other than AED, 25 microliters of the media was directly injected onto a 2.1 × 50 mm 2.6 μm C18 reverse phase column (Phenomenex) and was separated via a linear gradient of water : methanol 98 : 2 (Solvent A) to methanol : water 70 : 30 (Solvent B). Both solvents contained 5 mM ammonium formate with 1% formic acid. Separation of 17α-AED and 17β-AED was carried out using 2.1 × 150 mm 2.6 μm C18 reverse phase column (Phenomenex) and was done using same Solvent A as above and 98 : 2 methanol : water with 5 mM ammonium formate and 1% formic acid as Solvent B. Steroid hormones were detected using precursor product. MRM pairs are as follows: DHEA (271-213, 271-197), androstenetriol (307-158, 307-254), androstenediol (291-95, 291-159, 291-255), testosterone (289-97, 289-109), androstenedione (287-97, 287-109), and 17β-estradiol (273-107, 273-135, 273-77). Where there were multiple transitions, the dominant peak was used in the analysis.

All analytes demonstrated a minimal limit of detection of at least 0.6 nM.

2.3.1. Cellular Uptake of Androstene Hormones and Normalization of Transactivation Assay Results. It was relevant to determine the relative uptake of each androstene hormone in the Indigo Assay System. Cells and cell medium utilized in the assays were provided by the manufacturer (Indigo Biosciences). The following components were found to be below the limit of detection in the cells and cell medium: DHEA, 17β-estradiol, androstenedione, testosterone, 17α-AED, 17β-AED, and 17β-AET. LCMS data was utilized as a ratio between the background subtracted signal (area under the curve of the LCMS trace) at time zero and 24 hours for each analyte investigated. The ratios were then utilized to create normalization factors for the cellular uptake of 17α-AED, 17β-AED, and 17β-AET as compared to DHEA. The normalization factors were DHEA: 1, 17α-AED: 1.71, 17β-AED: 1.85, and 17β-AET: 2.40, respectively. Background activity due to cells, media, and vehicle was subtracted before the data were normalized.

2.3.2. Metabolism of Androstene Hormones. The cell medium was analyzed before and after incubation with assay cells. Mass spectroscopy (Section 2.3) was utilized to detect androstenedione, 17β-AED, 17α-AED, and 17β-AET, as well as testosterone and 17β-estradiol that were expressed in the medium. Approximately 9% of DHEA was metabolized to 17β-AED after a 24-hour incubation with assay cells. No other DHEA metabolites were detected. Mass spectroscopy did not detect any metabolites of the other androstene hormones in the media after a 24-hour incubation period. The mass spectrometry data show that DHEA was minimally metabolized to 17β-AED in this cell construct, but not to testosterone or 17β-estradiol (Section 2.3.2). 17β-AED, 17β-AET, and 17α-AED were not metabolized.

2.4. Statistical Analysis. All statistical analyses were performed using SigmaPlot version 12 (SSI, San Jose, CA, USA). Hormone EC_{50} level and the estrogen receptor alpha activation statistical analyses were performed with a Student's t-test while all other hormone activation statistical comparisons were performed with a one-way ANOVA. P value levels < 0.05 were considered significant. Statistics on test groups were done before normalization to cellular uptake and were performed between test groups and controls.

3. Results and Discussion

3.1. Androstene Hormone Structures. The hormones that were used in this study are listed in Figure 1. The structures demonstrate the similarities and unique characteristics of each androstene hormone. The main differences are the orientation of the hydroxyl group at position C17 for 17α-AED and 17β-AED, the orientation and position of the hydroxyl group at position C7 for 17β-AET, and the ketone group at position 17 for DHEA. 17α-AED and 17β-AED are chemically identical except for the placement of the hydroxyl group in relation to (above or below) the Δ^5 cycloperhydrophenanthrene ring. All adrenal hormones in this study, with the exception of DHEA, possess hydroxyl groups in the C3 and C17 position with 17α-AED having the C17 hydroxyl group in the (α) position. This position at C17 results in remarkable biological actions [11, 13] while the hydroxyl group at C3 was shown not to influence the biological activity [24].

3.2. Androstene Hormone Activation of the Human AR. The data demonstrate that both the orientation of the hydroxyl at position C17 and the addition of the hydroxyl at position C7 affected the ability of 17β-AED, 17β-AET, and 17α-AED to activate the human AR construct (Figure 2). The AR construct contains a luciferase reporter gene that is functionally linked to an AR responsive promoter. The luciferase reading is utilized as a surrogate measure for AR binding. The EC_{50} was calculated for testosterone as the 50% activation point. EC_{50} values for the androstene hormones and testosterone were calculated (Table 1) and compared utilizing the ratio of the androstene hormone EC_{50} to testosterone EC_{50} (relative androgenicity). All of the androstene hormones tested showed a significant ($P < 0.001$) reduced androgenicity when compared to testosterone (Table 1). 17β-AED had only 1/5th the ability of testosterone to activate the androgen receptor. Changing the orientation of the C17 hydroxyl group on 17α-AED resulted in a further reduction to 1/60th the activity as compared to testosterone. Addition of the hydroxyl group to the C7 position further reduced the ability to activate the human AR construct to 1/1326th as that of testosterone. The androstene hormone activation of the human AR was rank ordered based on strength of activation. The rank order was 17β-AED $\gg$ 17α-AED $\ggg$ 17β-AET (Figure 2). DHEA binding to the AR was excluded from these experiments since its androgenicity has been reported previously [22].

3.3. Androstene Hormone Activation of the Human ERβ and ERα Receptors. The data demonstrate that both the orientation of the hydroxyl at position C17 and the addition of the hydroxyl at position C7 affected the ability of DHEA, 17β-AED, 17β-AET, and 17α-AED to activate the human ERβ construct (Figure 3). The ER construct contains a luciferase reporter gene that is functionally linked to an ER responsive promoter. The luciferase reading is utilized as a surrogate measure for ER binding. The EC_{50} was calculated for 17β-estradiol as the 50% activation point. EC_{50} values for the androstene hormones and 17β-estradiol were calculated (Table 2) and compared utilizing the ratio of the androstene hormone EC_{50} to 17β-estradiol EC_{50} (relative estrogenicity). All of the tested androstene hormones demonstrated a significantly ($P < 0.001$) decreased estrogenicity compared to 17β-estradiol (Table 2). 17β-AED had only 1/282nd the ability of 17β-estradiol to activate the ERβ. The orientation change of the hydroxyl group at position C17 of 17β-AED to the (α) position resulted in 17α-AED possessing 1/7609th the ability of 17β-estradiol to activate the ERβ receptor as 17β-estradiol. This represents a drastic decrease in estrogenicity from the 17β-AED epimer. The presence of the hydroxyl group at C7 of 17β-AET resulted in 1/587th the ability to activate the ERβ.

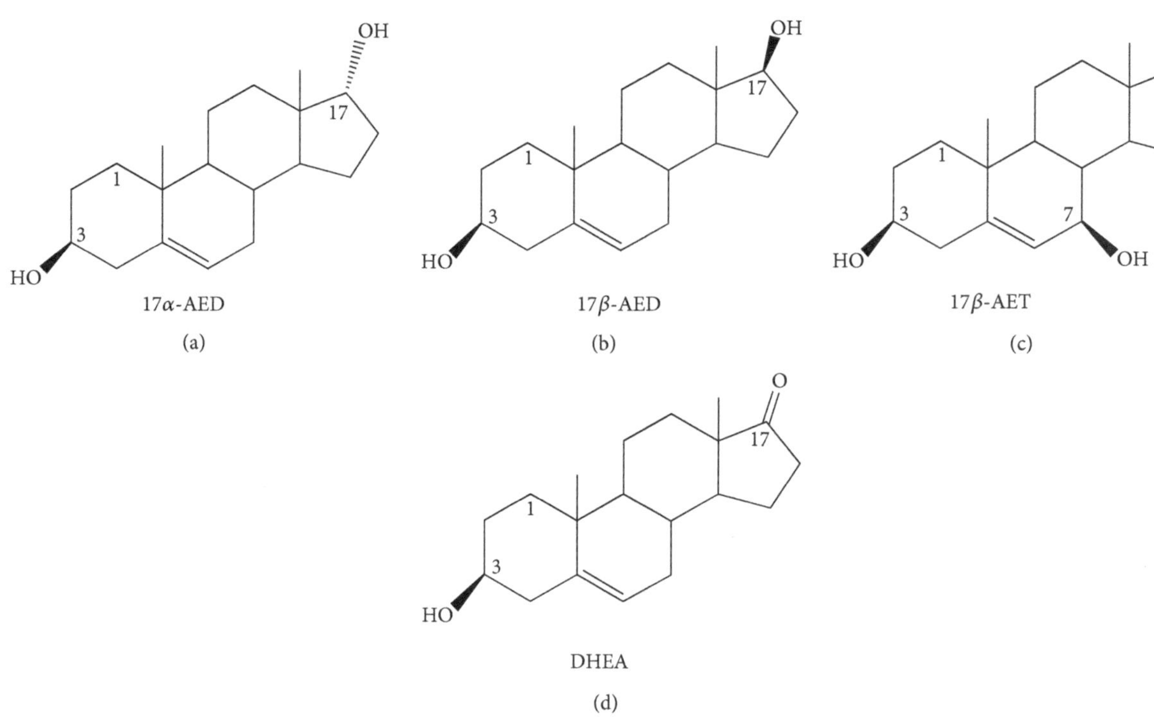

FIGURE 1: The structures of the androstene hormones. The androstene hormones are shown with the Δ^5 cycloperhydrophenanthrene ring. All steroids have a C3 hydroxyl group in the (β) beta position. The C7 hydroxyl group of androstenetriol is in the β-position. The C17 hydroxyl of androstenediol epimers are in either the (α) alpha or (β) beta position.

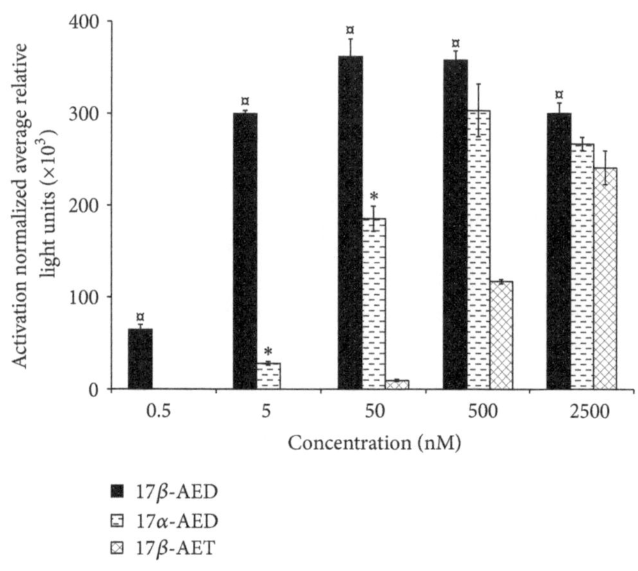

FIGURE 2: Androstene hormone activation of the human androgen receptor. Reporter cells were treated with androstene hormone metabolites ($n = 3$), incubated for 24 hours, and then assayed for luciferase activity. Androstene hormone activity was normalized to cellular uptake (Section 2.3.1). Error bars, ±1 SD. Statistical significance, $P < 0.001$ from all less activating androstene hormone metabolites (¤), $P < 0.001$ from less activating androstene hormone metabolites (∗).

TABLE 1: Relative androgenicity of androstene hormones.

Hormone	EC_{50} (nM)	Androgenicity	P value
Testosterone	0.35	1	1
Δ^5-Androstene-3β, 17β-diol	1.8	1/5	<0.001
Δ^5-Androstene-3β, 17α-diol	21	1/60	<0.001
Δ^5-Androstene-3β, 7β, 17β-triol	464	1/1326	<0.001

The EC_{50} was calculated as the 50% activation point. The androgenicity is the testosterone EC_{50} divided by the androstene hormone EC_{50}. This data does not normalize the cellular uptake of androstene hormones.

DHEA, with a ketone group in the C17 position, possessed 1/3543rd the ability to activate the ERβ.

The rank order of androstene hormone activation on the human ERβ receptor can be displayed as follows: 17β-AED > 17β-AET > DHEA > 17α-AED (Figure 3). These androstene hormones also specifically activated the human ERβ receptor and demonstrated the crucial effect of the (β) C17 hydroxyl group. 17β-AED and 17α-AED activated the ERβ receptor 2 and 3 orders of magnitude lower, respectively, than 17β-estradiol (Table 2). The 17β-AED demonstrated an estrogenicity of 1/1176 when assayed on the human ERα. Activation of the ERα receptor by 17β-AED did not become apparent until the concentration reached 25 nM (Figure 4) which was 3 orders of magnitude lower than 17β-estradiol, further demonstrating the weak estrogenicity displayed by these hormones at the level of the ERs. Finally, it should be

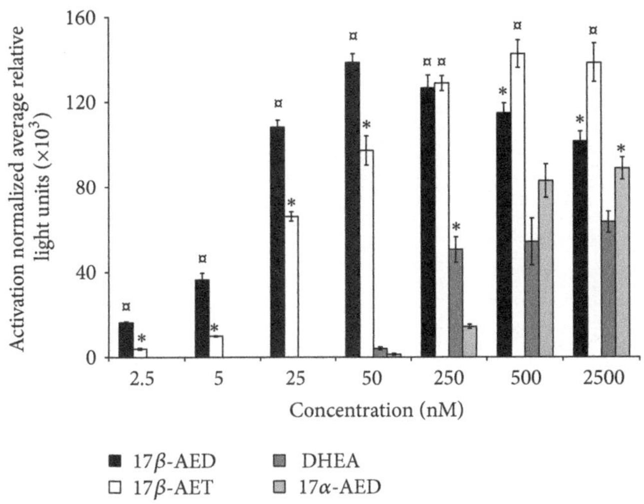

FIGURE 3: Androstene hormone activation of the human estrogen receptor beta. Reporter cells were treated with androstene hormones ($n = 3$), incubated for 24 hours, and then assayed for luciferase activity. Androstene hormone activity was normalized to cellular uptake (Section 2.3.1). Error bars, ±1 SD. Statistical significance, $P < 0.001$ (¤) versus androstene hormone metabolites in the same treatment group, $P < 0.001$ (∗) versus lower reacting androstene hormones in the same treatment group.

TABLE 2: Relative estrogenicity of androstene hormones.

Hormone	EC_{50} (nM)	Estrogenicity	P value
17β-Estradiol	0.046	1	1
Δ^5-Androstene-3β, 17β-diol	13	1/282	<0.001
Δ^5-Androstene-3β, 7β, 17β-triol	27	1/587	<0.001
DHEA	163	1/3543	<0.001
Δ^5-Androstene-3β, 17α-diol	350	1/7609	<0.001

The EC_{50} was calculated as the 50% activation point. The estrogenicity is the 17β-estradiol EC_{50} divided by the androstene hormone EC_{50}. This data does not normalize the cellular uptake of androstene hormones.

noted that the androstene hormones only weakly activated the AR and were even weaker activators of the human ERs.

3.4. Androstene Hormones Activation of the Human Glucocorticoid Receptor (GR). 17β-AED and especially 17β-AET are known to produce significantly affected glucocorticoid activity in vivo [6, 20, 21]. Therefore, the human GR construct response to 17β-AED, 17β-AET, DHEA, and 17α-AED alone and in combination with dexamethasone was evaluated. The results showed that dexamethasone alone activated the human GR while 17β-AED, 17β-AET, DHEA, and 17α-AED alone were negative at all concentrations tested (Figure 5). The human GR was then tested for activity with androstene hormones in the presence of dexamethasone. Unexpectedly, at suprapharmacological levels of dexamethasone there was a considerable activation of the human GR by 1 μM of each of the androstene hormones, which was greater than with dexamethasone alone with a higher activity when dexamethasone concentration was increased from 333 pM to 1000 pM (Figure 6).

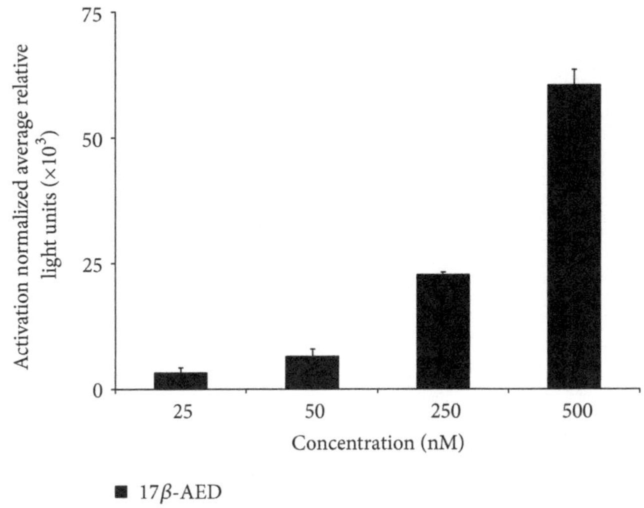

FIGURE 4: 17β-AED activation of the human estrogen receptor alpha. Reporter cells were treated with 17β-AED ($n = 3$), incubated for 24 hours, and then assayed for luciferase activity. 17β-AED activity was normalized to cellular uptake (Section 2.3.1), error bars, ±1 SD.

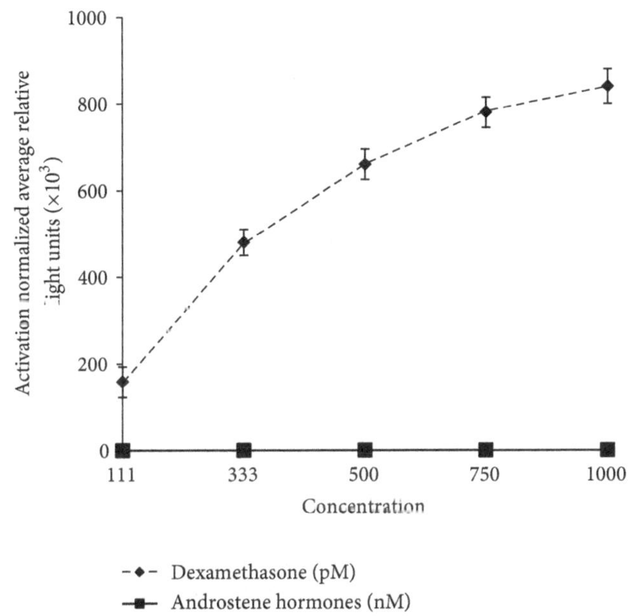

FIGURE 5: Dexamethasone or androstene hormone activation of the human glucocorticoid receptor. Reporter cells were treated with dexamethasone or androstene hormones alone ($n = 3$), incubated for 24 hours, and then assayed for luciferase activity. Androstene hormones are DHEA, 17β-AED, 17α-AED, and 17β- AET. Error bars, ±1 SD.

The rank order of activation of the dexamethasone-bound human GR in the presence of the androstene hormones is as follows: 17β-AET > 17β-AED > 17α-AED > DHEA. These data demonstrate that the C7 hydroxyl present in 17β-AET produced the strongest activation of the dexamethasone-bound human GR. The (β) C17 hydroxyl of 17β-AED produced a stronger activation than did the (α) C17 hydroxyl of 17α-AED. Thus, while 17β-AED and 17α-AED produced

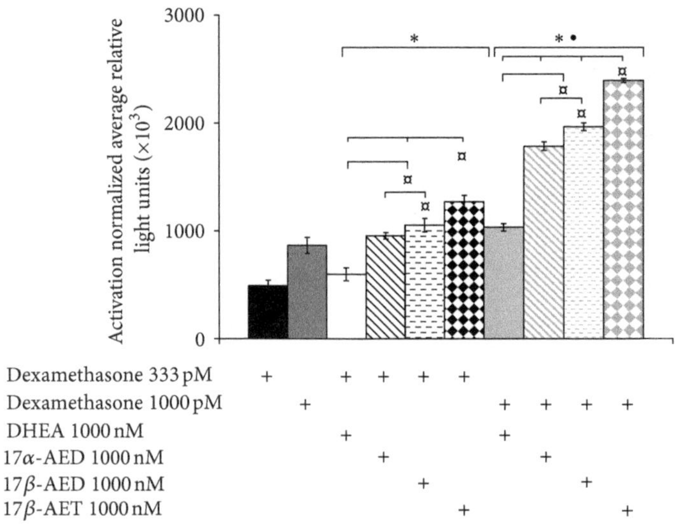

FIGURE 6: Androstene hormones and dexamethasone activation of the human GR. Reporter cells were treated with dexamethasone alone or a combination of dexamethasone and androstene hormone, incubated for 24 hours, and then assayed for luciferase activity ($n = 3$). Androstene hormone activity was normalized to cellular uptake (Section 2.3.1). Error bars, ±1 SD. Statistical significance, P is at least < 0.05 from other androstene hormone metabolites within the dexamethasone concentration treatment group (¤), P is at least < 0.05 versus control (∗), P is at least < 0.05 from 333 pM dexamethasone treatment group (•).

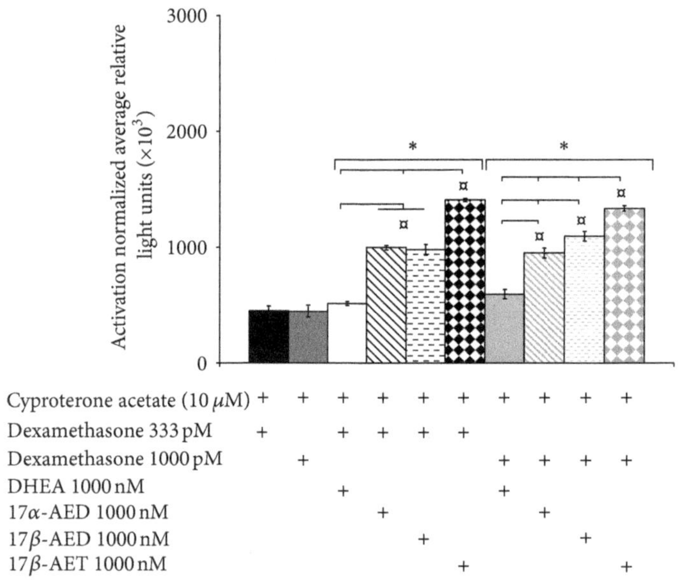

FIGURE 7: Androstene hormone and dexamethasone activation of the human GR in the presence of cyproterone acetate. Reporter cells were treated with dexamethasone and cyproterone acetate alone or androstene hormones with dexamethasone and cyproterone acetate ($n = 3$), incubated for 24 hours, and then assayed for luciferase activity. Androstene hormone activity was normalized to cellular uptake (Section 2.3.1). Error bars, ±1 SD. Statistical significance, P is at least < 0.05 from other androstene hormone metabolites within the treatment group (¤), $P < 0.05$ from controls (∗).

unique activation of the dexamethasone-bound human GR, the effect of the C17 hydroxyl group conformation was less apparent. DHEA, which has the keto group in the C17 position, possessed the least ability to activate the dexamethasone-bound human GR.

We next tested these effects with the dexamethasone inhibitor, cyproterone acetate, to see if we could reduce or eliminate the dexamethasone and androstene hormone activation of the human GR. Cyproterone acetate was selected as the inhibitor because of its unique glucocorticoid receptor inhibiting properties [25]. Cyproterone effectively inhibited the activation of the human GR by dexamethasone (Figure 7). Cyproterone acetate at a concentration of 10 uM was tested in the presence of the androstene hormones alone (1.0 uM) and there was no activation detected (data not shown). Androstene hormones, however, in the presence of cyproterone and dexamethasone exhibited different levels of activation that were significantly increased above the dexamethasone/cyproterone alone controls (Figure 7). These results are of particular clinical significance because it demonstrates that high-dose dexamethasone alters the human GR to interact with other biologically active hormones at the receptor level. Importantly, dexamethasone is known to cause adverse effects in humans [26].

The rank order of activation of the dexamethasone/cyproterone acetate-bound human GR in the presence of these androstene hormones is as follows: 17β-AET > 17β-AED > 17α-AED > DHEA. This rank order of activation on the inhibited human GR was the same as that of the uninhibited human GR indicating that the interaction of the androstene hormones and the dexamethasone/cyproterone-bound human GR was not disrupted. Since cyproterone acetate is a passive inhibitor of the human GR and opposes dexamethasone through an overlapping steroid scaffold mechanism, this suggests that the androstene hormone activation is mediated by an interaction that occurs outside the dexamethasone/cyproterone acetate-bound complex [25]. Additionally, the presence of dexamethasone-bound human GR is required to observe activation by the androstene hormones while cyproterone acetate alone does not mediate this effect. Together these data suggest an indirect activation of the ligand-bound human GR by 17β-AET, 17β-AED, 17α-AED, and DHEA.

4. Conclusion

This report indicates that the position of the hydroxyl group at C17 and/or the addition of the hydroxyl group at position C7 significantly affected the ability of 17β-AET, 17β-AED, 17α-AED, and DHEA to interact with the human estrogen, androgen, and ligand-bound glucocorticoid receptors. 17β-AET, 17β-AED, 17α-AED, and DHEA were shown to interact either directly or indirectly with the human AR, ER, and GR. Importantly, 17β-AET, 17β-AED, 17α-AED, and DHEA were shown to possess weak androgenicity and even weaker estrogenicity at the receptor level. Clinically, this is beneficial because the biological effects can be realized without unwanted androgenic or estrogenic effects.

In a stark contrast to the minimal receptor activation of AR, ER, and GR, these same androstene hormones produce striking biological effects in vitro and in vivo which have been attributed to activity with the AR, ER, or GR. Clearly, these effects may not be mediated by the direct androstene hormone interaction with the human AR, ERα, ERβ, and GR. Indeed, the biological mechanism, may not require AR or ER to achieve significant effects [27, 28]. Furthermore, the interaction with dexamethasone is indirect, occurs at high doses, and is not abolished by cyproterone acetate. Taken together, the data shows that interactions of 17β-AET, 17β-AED, 17α-AED, and DHEA with the human AR, ERα, ERβ, and GR are directed by the structure activity of these androstene hormones with minimal androgenic, estrogenic, or glucocorticoid effects and accentuates the need to further uncover the implied yet unidentified main mechanism(s) of action of these important adrenal hormones.

Abbreviations

17β-AED:	Δ^5-androstene-3β, 17β-diol
17α-AED:	Δ^5-androstene-3β, 17α-diol
17β-AET:	Δ^5-androstene-3β, 7β, 17β-triol
Androstene hormones:	17β-AED, 17α-AED, 17β-AET, DHEA
Androstene hormone metabolites:	17β-AED, 17α-AED, 17β-AET
DHEA:	Dehydroepiandrosterone
AR:	Androgen receptor
ERα:	Estrogen receptor alpha
ERβ:	Estrogen receptor beta
GR:	Glucocorticoid receptor
(α):	Alpha position
(β):	Beta position.

Acknowledgments

This work was performed in conjunction with the Virginia Commonwealth University Reanimation Engineering Science Center (VCURES), Virginia Commonwealth University, Richmond, VA, USA. This work was supported by the US Army Medical Research and Materiel Command under Grant no. W81XWH-07-1-08-1-0483. As well as a gift to Roger M. Loria from the SCDR Cancer Research Fund, this work was also supported by a Career Development Award (CDA1) from the Department of Veterans Affairs (DSW), research grants from the Veteran's Administration (VA Merit Award BX001792 to C.E.C. and a Research Career Scientist Award to C. E. Chalfant); from the National Institutes of Health via HL072925 (C. E. Chalfant), CA154314 (C.E.C.); from the US-Israel Binational Science Foundation via BSF#2011380 (C.E.C) and NHIC06-RR17393 (to Virginia Commonwealth University for renovation). The views expressed in this paper are those of the author and do not reflect the official policy or position of the United States Air Force, Department of Defense, or the US Government.

References

[1] J. S. Dillon, "Dehydroepiandrosterone, dehydroepiandrosterone sulfate and related steroids: their role in inflammatory, allergic and immunological disorders," *Current Drug Targets*, vol. 4, no. 3, pp. 377–385, 2005.

[2] R. Oberbeck and P. Kobbe, "DEhydroepiandrosterone (DHEA): a steroid with multiple effects. Is there any possible option in the treatment of critical illness?" *Current Medicinal Chemistry*, vol. 17, no. 11, pp. 1039–1047, 2010.

[3] R. M. Loria and D. A. Padgett, "Androstenediol regulates systemic resistance against lethal infections in mice," *Annals of the New York Academy of Sciences*, vol. 685, pp. 293–295, 1993.

[4] R. M. Loria and D. A. Padgett, "Mobilization of cutaneous immunity for systemic protection against infections," *Annals of the New York Academy of Sciences*, vol. 650, pp. 363–366, 1992.

[5] R. M. Loria, D. A. Padgett, and P. N. Huynh, "Regulation of the immune response by dehydroepiandrosterone and its metabolites," *Journal of Endocrinology*, vol. 150, pp. S209–S220, 1996.

[6] R. M. Loria, "Antiglucocorticoid function of androstenetriol," *Psychoneuroendocrinology*, vol. 22, supplement 1, pp. S103–S108, 1997.

[7] D. A. Padgett, R. M. Loria, and J. F. Sheridan, "Endocrine regulation of the immune response to influenza virus infection with a metabolite of DHEA-androstenediol," *Journal of Neuroimmunology*, vol. 78, no. 1-2, pp. 203–211, 1997.

[8] F. Svec and J. R. Porter, "The actions of exogenous dehydroepiandrosterone in experimental animals and humans," *Proceedings of the Society for Experimental Biology and Medicine*, vol. 218, no. 3, pp. 174–191, 1998.

[9] C. D. Santos, M. P. A. Toldo, F. H. Santello, M. D. V. Filipin, V. Brazão, and J. C. do Prado Júnior, "Dehydroepiandrosterone increases resistance to experimental infection by *Trypanosoma cruzi*," *Veterinary Parasitology*, vol. 153, pp. 238–243, 2008.

[10] V. R. Feeser, N. B. Menke, K. R. Ward, R. M. Loria, and R. F. Diegelmann, "Androstenediol reverses steroid-inhibited wound healing: biomedical hypothesis article," *Wound Repair and Regeneration*, vol. 17, no. 5, pp. 758–761, 2009.

[11] P. Huynh and R. Loria, "Contrasting effects of alpha- and beta-androstenediol on oncogenic myeloid cell lines *in vitro*," *Journal of Leukocyte Biology*, vol. 62, no. 2, pp. 258–2267, 1997.

[12] M. R. Graf, W. Jia, and R. M. Loria, "The neuro-steroid, 3β androstene 17α diol exhibits potent cytotoxic effects on human malignant glioma and lymphoma cells through different programmed cell death pathways," *British Journal of Cancer*, vol. 97, no. 5, pp. 619–627, 2007.

[13] M. R. Graf, W. Jia, M. L. Lewbart, and R. M. Loria, "The anti-tumor effects of androstene steroids exhibit a strict structure-activity relationship dependent upon the orientation of the hydroxyl group on carbon-17," *Chemical Biology and Drug Design*, vol. 74, no. 6, pp. 625–629, 2009.

[14] H. Miyamoto, S. Yeh, H. Lardy, E. Messing, and C. Chang, "δ5-Androstenediol is a natural hormone with androgenic activity in human prostate cancer cells," *Proceedings of the National Academy of Sciences of the United States of America*, vol. 95, no. 19, pp. 11083–11088, 1998.

[15] R. Hackenberg, I. Turgetto, A. Filmer, and K. D. Schulz, "Estrogen and androgen receptor mediated stimulation and inhibition of proliferation by androst-5-ene-3β,17β-diol in human mammary cancer cells," *Journal of Steroid Biochemistry and Molecular Biology*, vol. 46, no. 5, pp. 597–603, 1993.

[16] C. C. Head, M. J. Farrow, J. F. Sheridan, and D. A. Padgett, "Androstenediol reduces the anti-inflammatory effects of restraint stress during wound healing," *Brain, Behavior, and Immunity*, vol. 20, no. 6, pp. 590–596, 2006.

[17] P. Marwah, A. Marwah, H. A. Lardy, H. Miyamoto, and C. Chang, "C19-Steroids as androgen receptor modulators: design, discovery, and structure-activity relationship of new steroidal androgen receptor antagonists," *Bioorganic and Medicinal Chemistry*, vol. 14, no. 17, pp. 5933–5947, 2006.

[18] J. T. Arnold, X. Liu, J. D. Allen, H. Le, K. K. McFann, and M. R. Blackman, "Androgen receptor or estrogen receptor-β blockade alters DHEA-, DHT-, and E2-induced proliferation and PSA production in human prostate cancer cells," *Prostate*, vol. 67, no. 11, pp. 1152–1162, 2007.

[19] K. Saijo, J. G. Collier, A. C. Li, J. A. Katzenellenbogen, and C. K. Glass, "An ADIOL-ERβ-CtBP transrepression pathway negatively regulates microglia-mediated inflammation," *Cell*, vol. 145, no. 4, pp. 584–595, 2011.

[20] D. A. Padgett and R. M. Loria, "*In vitro* potentiation of lymphocyte activation by dehydroepiandrosterone, androstenediol, and androstenetriol," *The Journal of Immunology*, vol. 153, no. 4, pp. 1544–1552, 1994.

[21] R. M. Loria and D. A. Padgett, "Androstenediol regulates systemic resistance against lethal infections in mice," *Archives of Virology*, vol. 127, no. 1–4, pp. 103–115, 1992.

[22] F. Chen, K. Knecht, E. Birzin et al., "Direct agonist/antagonist functions of dehydroepiandrosterone," *Endocrinology*, vol. 146, no. 11, pp. 4568–4576, 2005.

[23] M. E. Baker, "Co-evolution of steroidogenic and steroid-inactivating enzymes and adrenal and sex steroid receptors," *Molecular and Cellular Endocrinology*, vol. 215, no. 1-2, pp. 55–62, 2004.

[24] R. M. Loria and M. R. Graf, "17α-androstenediol-mediated oncophagy of tumor cells by different mechanisms is determined by the target tumor," *Annals of the New York Academy of Sciences*, vol. 1262, no. 1, pp. 127–133, 2012.

[25] C. Honer, K. Nam, C. Fink et al., "Glucocorticoid receptor antagonism by cyproterone acetate and RU486," *Molecular Pharmacology*, vol. 63, no. 5, pp. 1012–1020, 2003.

[26] K. Matsumoto, H. Yamasaki, S. Akazawa et al., "High-dose but not low-dose dexamethasone impairs glucose tolerance by inducing compensatory failure of pancreatic β-cells in normal men," *Journal of Clinical Endocrinology and Metabolism*, vol. 81, no. 7, pp. 2621–2626, 1996.

[27] P. N. Huynh, W. H. Carter, and R. M. Loria, "17 Alpha androstenediol inhibition of breast tumor cell proliferation in estrogen receptor-positive and -negative cell lines," *Cancer Detection and Prevention*, vol. 24, no. 5, pp. 435–444, 2000.

[28] R. Loria, M. Graf, W. Jia, and T. Powell, "5-androstene 3β,17α diol induces autophagic cell death of human breast cancer cells and enhances radiation cytotoxicity," *Journal of Cancer Therapeutics and Research*, vol. 1, no. 1, 25 pages, 2012.

7

Development of 5-Substituted *N*-Methylmorphinan-6-ones as Potent Opioid Analgesics with Improved Side-Effect Profile

Helmut Schmidhammer and Mariana Spetea

Department of Pharmaceutical Chemistry, Institute of Pharmacy and Center for Molecular Biosciences Innsbruck (CMBI), University of Innsbruck, Innrain 52a, 6020 Innsbruck, Austria

Correspondence should be addressed to Helmut Schmidhammer, helmut.schmidhammer@uibk.ac.at

Academic Editor: Severo Salvadori

One of the most important functions of the opioid system is the control of pain. Among the three main opioid receptor classes (μ, δ, κ), the μ (MOR) is the main type targeted for pharmacotherapy of pain. Opioid analgesics such as morphine, oxycodone and fentanyl are agonists at the MOR and are the mainstay for the treatment of moderate-to-severe pain. However, adverse effects related to opioid use are severe and often lead to early discontinuation and inadequate analgesia. The development of more effective and safer medications for the management of pain still remains a major direction in pharmaceutical research. Chemical approaches towards the identification of novel MOR analgesics with reduced side effects include structural modifications of 14-alkoxy-*N*-methylmorphinan-6-ones in key positions that are important for binding, selectivity, potency, and efficacy at opioid receptors. This paper describes a representative strategy to improve the therapeutic usefulness of opioid analgesics from the morphinan class of drugs by targeting position 5. The focus is on chemical and biological studies and structure-activity relationships of this series of ligands. We report on 14-alkoxymorphinan-6-ones having a methyl and benzyl group at position 5 as strong opioid antinociceptive agents with reduced propensity to cause undesired effects compared to morphine although interacting selectively with MORs.

1. Introduction

The analgesic action of extracts of the opium poppy plant *Papaver somniferum* has been recognized for centuries. Morphine (Figure 1), the primary active component of opium, was isolated in 1805 by the German pharmacist Friedrich Sertürner, and more than 120 years elapsed when Gulland and Robinson proposed its correct structure [1]. Today, opioid analgesics play a central role in pain control and are generally considered to be highly effective in the management of moderate-to-severe pain [2, 3]. They can be classified into three classes: natural derivatives occurring in opium such as morphine and codeine; partially synthetic derivatives, including hydromorphone, oxycodone, oxymorphone, and buprenorphine; and synthetic compounds such as levorphanol, butorphanol, fentanyl, sufentanil, and the recently introduced tapentadol (Figure 1) [3–5].

Opioids act on three G-protein-coupled receptors (GPCRs) that is, μ (MOR), δ (DOR), and κ (KOR) [6], but it appears that the analgesic action of the most commonly used opioid analgesics is mediated primarily via the MOR. Activation of MORs, widely expressed in the nervous system and in peripheral tissues [7–10], is responsible not only for beneficial (analgesia) effects but also for a number of several undesired effects, which limits their clinical usefulness [3, 4, 11]. Adverse effects associated with opioid analgesics include respiratory depression, nausea, sedation, dizziness, vomiting, hypotension, and constipation. Long-term opioid use can cause tolerance, and thus complicating optimal pain treatment. Another concern with prolonged use of opioids is physical dependence and development of addictive disorders [4, 11]. Drug deaths from opioids are a serious and increasing issue [4]. On this basis, the development of more effective and safer medications for the management of pain, especially severe and chronic pain, still remains a major direction in pharmaceutical research.

Chemical approaches towards the identification of novel MOR analgesics with reduced side effects are represented by

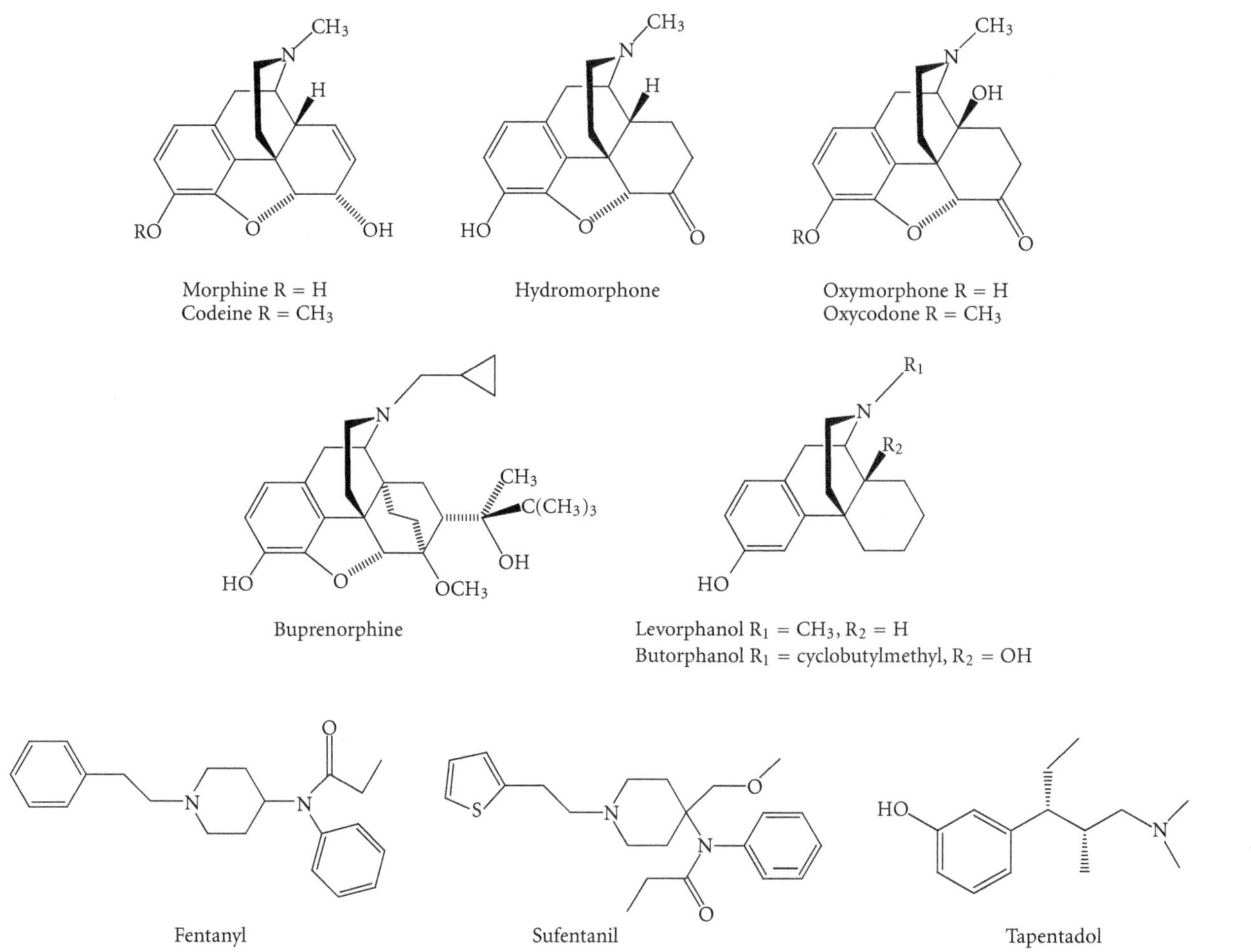

Figure 1: Examples of clinically used opioid analgesics.

structural modifications of morphinan-6-ones in key positions that are important for binding, selectivity, potency and efficacy at opioid receptors. A representative example is the development of the 14-*O*-methyl-substituted derivative of the clinically used MOR analgesic oxymorphone, namely, 14-*O*-methyloxymorphone (**1**) (Figure 2) [12]. It was reported that substitution of the hydroxyl group with a methoxy group in position 14 not only increases affinity to opioid receptors, while retaining the MOR selectivity of oxymorphone, but also markedly enhances the antinociceptive potency [12]. However, this MOR agonist induces the classical opioid unwanted effects of the conventional MOR analgesics [12–14]. An overview on synthesis and structure-activity relationships (SARs) on different 14-alkoxy-substituted morphinans has been recently published [15].

An alternative strategy to improve the therapeutic usefulness of opioid analgesics from the morphinan class of drugs is to target position 5. The 5-methyl-substituted *N*-methylmorphinan-6-one, metopon (Figure 3), which shows higher analgesic potency than morphine, had also an improved side-effect profile concerning respiratory depression, physical dependence, mental dullness, tolerance, and nausea in patients [16, 17]. It was shortly used in an oral formulation for chronic pain relief in hospitalized cancer patients in the late 1940s [16]. Due to the low synthetical availability of metopon it was never introduced to the market as an analgesic drug. The positive preclinical and clinical findings on metopon represented a stimulating aspect for the design of differently substituted derivatives of metopon. Herein, we review recent advances on the development of 14-alkoxy substituted analogues of metopon, and also of a 5-benzylated analogue with agonist action at the MOR, as

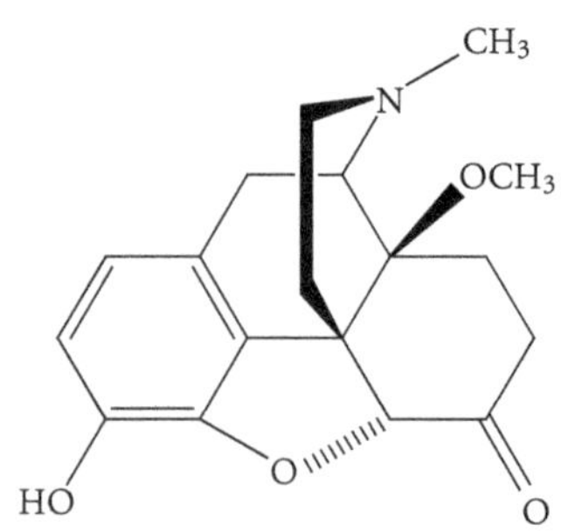

Figure 2: Structure of 14-*O*-methyloxymorphone (**1**).

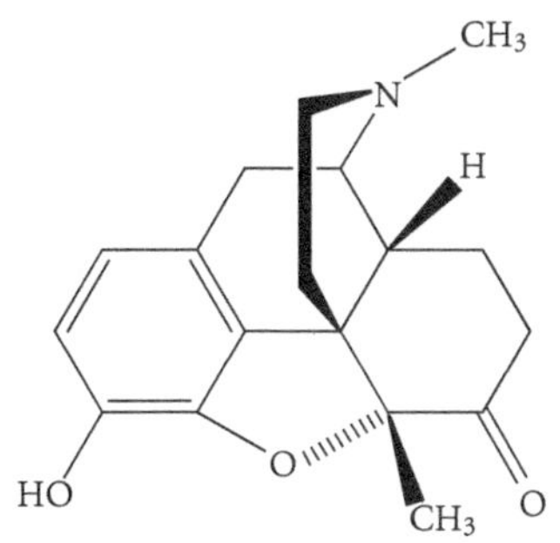

FIGURE 3: Structure of metopon.

potent analgesics with reduced adverse effects. The focus is on SAR and pharmacological studies as well as on the description of synthetic procedures developed and used in the preparation of such derivatives.

2. Synthesis of 5-Substituted *N*-Methylmorphinan-6-ones

The starting point of extensive research in the area of *N*-methylmorphinan-6-ones with new substitution patterns was represented by the synthesis of 14-*O*-methyloxymorphone (**1**). While 14-*O*-methyloxymorphone (**1**) can be prepared starting from oxymorphone in four synthetic steps [12], 14-alkoxy-*N*-methylmorphinan-6-ones substituted in position 5 have to be prepared from thebaine (**2**) as starting material. Introduction of an alkyl substituent in position 5 of thebaine (**2**) can be accomplished by formation of the thebaine anion using *n*-butyllithium in THF at low temperature [18], followed by alkylation with the respective alkylating agent (methyl fluorosulfonate, dimethyl sulfate, or benzyl chloride), yielding 5-methylthebaine (**3**) and 5-benzylthebaine (**4**), respectively [18–21]. Treatment with performic acid afforded 14-hydroxy-5-methylcodeinone (**5**) [19, 20] and its 5-benzyl analogue **6** (Scheme 1) [14]. The *β*-orientation of the 14-hydoxy group was proved by X-ray analysis [20]. 14-O-Alkylation of **5** with dimethyl or diethyl sulfate in DMF in the presence of NaH gave the respective 14-alkoxycodeinones **7** and **8** which were catalytically hydrogenated to afford **10** and **11**. Ether cleavage of both compounds with 48% HBr solution yielded 77% of 14-methoxymetopon (**13**) and 75% of 14-ethoxymetopon (**14**) [19, 20]. Analogously, 5-benzyl-14-*O*-methyloxymorphone (**15**) (75% from **12**) was obtained from 14-hydroxy-5-benzylcodeinone (**6**) via intermediates **9** and **12** (Scheme 1) [14].

14-Benzyloxymetopon (**18**) was prepared from 14-hydroxy-5-methylcodeinone (**5**). Compound **5** was first 3-O-demethylated using 48% HBr solution to give phenol **16** which was 3,14-bis-O-benzylated to afford compound **17**. Concomitant hydrogenation of the 7,8 double bond and hydrogenolysis of the 3-*O*-benzyl ether over Pd/C yielded 73% of **18** (Scheme 2) [13]. 14-Hydoxy-5-methylcodeinone (**5**) was also used for the synthesis of 14-phenylpropoxymetopon (PPOM, **21**) and its 3-*O*-methyl ether **20**. 14-O-Alkylation with cinnamyl bromide in DMF in the presence of NaH gave compound **19** which was catalytically hydrogenated to afford 73% of 3-O-methylated PPOM (**20**). Ether cleavage using 48% HBr solution yielded 88% of PPOM (**21**) (Scheme 3) [22].

3. Biological Activities of 5-Substituted *N*-Methylmorphinan-6-ones and Structure-Activity Relationship Studies

The applied strategy to obtain MOR analgesics in the *N*-methylmorphinan-6-one class exhibiting more favorable pharmacological features was initially based on the introduction of a small alkyl group such as methyl at the oxygen in position 14 of oxymorphone leading to 14-*O*-methyloxymorphone (**1**) [12]. This oxymorphone analogue displayed improved binding affinities at all three opioid receptors, while maintaining the MOR selectivity of the parent molecule, together with a considerable increase in in vitro and in vivo agonist potency (Table 1). In different in vitro functional assays including mouse vas deferens (MVD), guinea pig brain (GPI), and rat vas deferens (RVD) bioassays, and the [^{35}S]GTPγS binding assay using rat brain preparations, 14-*O*-methyloxymorphone proved to be a more potent MOR agonist than morphine and oxymorphone (Tables 1 and 2) [13, 14, 23, 24]. Compound **1** was also reported to possess up to 40 times higher antinociceptive potency than oxymorphone [12, 24], and it was up to 800 times more potent than morphine after subcutaneous (s.c.) administration in different pain tests in mice and rats (Tables 1 and 2) [12–14, 24]. Although it induced effective analgesia, this oxymorphone analogue **1** also produced the typical opioid adverse actions in mice after s.c. administration such as respiratory depression [12], physical dependence [12], inhibition of the gastrointestinal tract [13], and locomotor impairment [14].

Metopon (5-methyldihydromorphone; Figure 3) was reported to be about three times more potent than morphine as analgesic with lower tendency to produce nausea, sedation, respiratory depression, development of tolerance and dependence [16, 17]. Further chemical derivatization in the class of *N*-methylmorphinan-6-ones using 14-*O*-methyloxymorphone (**1**) as the lead, targeted position 5 by introducing a methyl group, giving rise to a new opioid compound, 14-methoxymetopon (**13**) [27]. 14-Methoxymetopon maintained the high affinity at the MOR in the subnanomolar range of its 5-unsubstituted analogue **1**, while DOR and KOR affinities were reduced by two to three times, resulting in higher MOR selectivity (Table 1). In in vitro bioassays, derivative **13** was a potent agonist in the GPI (IC_{50} = 6.1 nM) [13], MVD (IC_{50} = 24.4 nM [13] and 12.7 nM [23]) and RVD preparations (IC_{50} = 268 nM) [24], showing comparable potency to 14-*O*-methyloxymorphone (**1**), and it was several times more active as agonist than morphine (Table 2). 14-Ethoxymetopon (**14**) was also described as MOR selective (Table 2), and a potent agonist in the GPI bioassay displaying similar potency to its 14-methoxy-substituted analogue **13**, and about 50 times greater

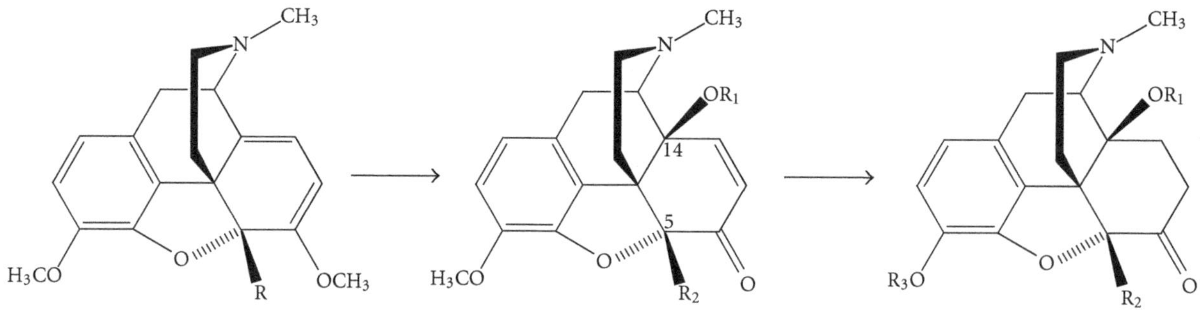

2 R = H
3 R = CH_3
4 R = CH_2Ph

5 R_1 = H, R_2 = CH_3
6 R_1 = H, R_2 = CH_2Ph
7 R_1 = R_2 = CH_3
8 R_1 = C_2H_5, R_2 = CH_3
9 R_1 = CH_3, R_2 = CH_2Ph

10 R_1 = R_2 = R_3 = CH_3
11 R_1 = CH_2H_5, R_2 = R_3 = CH_3
12 R_1 = R_3 = CH_3, R_2 = CH_2Ph
13 R_1 = R_2 = CH_3, R_3 = H
14 R_1 = CH_2H_5, R_2 = CH_3, R_3 = H
15 R_1 = CH_3, R_2 = CH_2Ph, R_3 = H

Scheme 1: Preparation of 14-methoxymetopon (**13**), 14-ethoxymetopon (**14**), and 5-benzyl-14-*O*-methyloxymorphone (**15**).

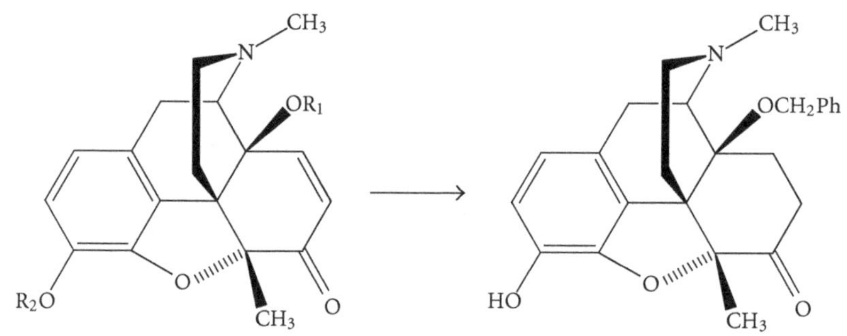

5 R_1 = H, R_2 = CH_3
16 R_1 = R_2 = H
17 R_1 = R_2 = CH_2Ph

18

Scheme 2: Preparation of 14-benzyloxymetopon (**18**).

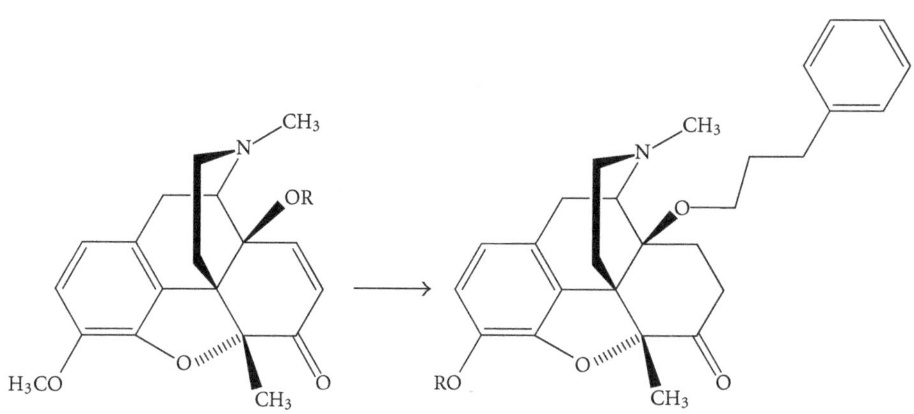

5 R = H
19 R = $CH_2CH = CHPh$

20 R = CH_3
21 R = H

Scheme 3: Preparation of 14-phenylpropoxymetopon (PPOM; **21**) and its 3-*O*-methyl ether (**20**).

Table 1: In vitro and in vivo opioid activity and SAR study on the variation of the substituent in position 5 in 14-alkoxy-substituted *N*-methylmorphinan-6-ones.

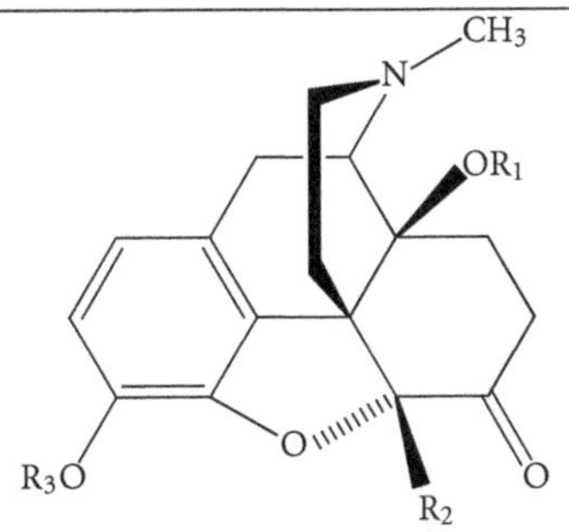

	R_1, R_2, R_3	Binding affinity (K_i, nM) MOR[a]	DOR[b]	KOR[d]	Selectivity DOR/MOR	KOR/MOR	Agonist potency (relative to morphine) MVD[e]	GPI[f]	Antinociceptive potency (relative to morphine)
Oxymorphone	H, H, H	0.97	80.5	61.6	83	63	0.4	ND	10[h] 18[i] 13[j]
1	Me, H, H	0.10	4.80	10.2	48	102	331	156	126[h] 300[i] 810[j]
13	Me, Me, H	0.15	13.3	25.2	89	168	202	52	94[h] 82[i] 99[j]
14	Et, Me, H	0.46	12.2[c]	43.2	26	94	ND	49[g]	46[h]
15	Me, Bz, H	0.31	13.1	22.8	42	73	ND	ND	50[i] 53[j]
18	Bz, Me, H	0.18	3.67	2.46	20	14	696	167	103j
20	PhPr, Me, Me	0.62	6.33	25.0	10	40	ND	ND	327[h] 436[i] 235[j]
21	PhPr, Me, H	0.20	0.14	0.40	0.7	2	ND	ND	8,500[h] 24,000[i] 2,500[j]

Bz: benzyl; Et: ethyl; Me: methyl; PhPr: phenylpropyl; K_i: inhibition constant; ND: not determined.
[a]Binding against [^{3}H]DAMGO in rat brain membranes [13, 14, 22, 23, 25, 26].
[b]Binding against [^{3}H][Ile^{5,6}]deltorphin II or [c][^{3}H]DSLET in rat brain membranes [13, 14, 22, 23, 25, 26].
[d]Binding against [^{3}H]U69,593 in rat or guinea pig brain membranes [13, 14, 22, 23, 25, 26].
[e]Determined in the MVD [13, 23, 24].
[f]Determined in the GPI [13, 23].
[g]Relative to normorphine [25].
[h]Determined in the hot-plate test in mice after s.c. administration [12–14, 22].
[i]Determined in the tail-flick test in mice after s.c. administration [14, 22, 24].
[j]Determined in the writhing test in mice after s.c. administration [20, 22, 25].

TABLE 2: Comparison of pharmacological and physicochemical properties of morphine, 14-*O*-methyloxymorphone (**1**), 14-methoxymetopon (**13**), and its 5-benzyl-substituted analogue **15**.

	Agonist activity[a] [^{35}S]GTPγS (EC_{50}, nM; % stim)	Antinociception ED_{50} (mg/kg)	Motor coordination[d] dose (mg/kg)	Log P[g]
Morphine	462; 85	2.63[b] 2.29[c]	10[e]	0.88
1	23.7; 108	0.017[b] 0.014[c]	0.06[e]	0.60
13	63.0; 110	0.028[b] 0.028[c]	0.1[e]	1.12
15	13.7; 85	0.053[b] 0.043[c]	0.2[f]	1.49

EC_{50}: effective concentration necessary to produce a 50% effect; ED_{50}: effective analgesic dose to produce a 50% effect; log P: partition coefficient.
[a]Determined in [^{35}S]GTPγS functional assays in rat brain membranes; Data as % stimulation relative to DAMGO [14].
[b]Determined in the hot-plate test in mice after s.c. administration [14].
[c]Determined in the tail-flick test in mice after s.c. administration [14].
[d]Determined in the rotarod test in mice after s.c. administration [14].
[e]Significant decrease [14].
[f]No significant effect [14].
[g]The experimental log P determined in octanol/water [14].

than that of normorphine [25]. The potent MOR agonist activity of 14-methoxymetopon was also established using [^{35}S]GTPγS functional assays in rat brain [14, 26], calf striatum [28], and Chinese hamster ovary (CHO) cells expressing mouse MOR splice variants [28]. When compared to 14-*O*-methyloxymorphone (**1**), the 5-methyl analogue **13** showed less than three times lower agonist potency, while equal efficacy based on stimulation of [^{35}S]GTPγS binding in rat brain preparation (Table 2), signifying that methylation in position 5 did not considerably alter opioid agonist activity in vitro. 14-Methoxymetopon is available in tritium-labeled form ([^{3}H]14-methoxymetopon) [26], which was described to have high affinity and selectivity for both native and recombinant MORs [26, 28]. In vivo agonist activities of 14-methoxymetopon (**13**) and its 14-*O*-ethyl analogue 14-ethoxymetopon (**14**) were also reported. Introduction of a methyl substituent in position 5 of 14-*O*-methyloxymorphone (**1**) resulted in a somewhat reduced analgesic potency in the mouse hot-plate and tail-flick tests after s.c. administration [13, 14, 24]. 14-Methoxymetopon and 14-ethoxymetopon were significantly more potent than morphine in producing antinociception (Table 1). A large number of studies reported on the analgesic properties of 14-methoxymetopon (**1**) in various pain models in mice, rats and dogs. These experiments were performed with different routes of administration, including s.c., intraperitoneal (i.p.), intravenous (i.v.), intracerebroventricular (i.c.v.), and intrathecal (i.t.), and employed a wide range of pain stimuli, that is, thermal, chemical, electrical and mechanical. The pain tests consisted of models of acute nociceptive pain such as hot-plate [13, 14, 25, 29, 30], tail-flick [14, 25, 30–32], tail-electrical stimulation [33], and skin-twitch [34], visceral pain (acetic acid- [25, 27] and paraphenylquinone (PPQ-) induced writhing [30]), and inflammatory pain (carrageenan-induced inflammatory hyperalgesia [35]). 14-Methoxymetopon induced effective analgesia in all models of pain in rodents, showing different degrees of higher potency than morphine, depending on the analgesic paradigm and route of administration used (Table 3). In dogs, it had similar antinociceptive efficacy to sufentanil in the skin-twitch test after i.v. application [33]. 14-Methoxymetopon (**13**) significantly reduced pain-behavior in response to heat and mechanical stimulation in the inflamed paw of rats with carrageenan-induced inflammatory hyperalgesia [35].

14-Methoxymetopon (**13**) was generally described to produce less severe adverse effects than traditional MOR analgesics. The gastrointestinal inhibitory activity of this derivative after s.c. administration was relatively weak, its maximal inhibition was only approximately 65%, whereas morphine completely blocked the transit in the charcoal test in mice [31]. Intravenous administration of 14-methoxymetopon (**13**) does not induce respiratory depression compared to sufentanil when given in equianalgesic doses to dogs [34]. A dose of 12 μg/kg sufentanil decreases oxygen and carbon dioxide tension (PaO_2 and $PaCO_2$) by 41% and 57%, respectively, while at the same dose of 14-methoxymetopon, gas levels remain almost unchanged [34]. In the same study using conscious dogs, it was shown that the MOR opioid agonist **13** evokes significantly reduced bradycardia and hypotension, and produces less sedation than sufentanil [34]. The maximal bradycardic effect was 19% after 14-methoxymetopon and 42% after sufentanil, and the maximal decrease in the mean arterial blood pressure was 20% after sufentanil, and only 6% after derivative **13** at the highest i.v. dose in dogs [34]. The physical dependence liability of 14-methoxymetopon in the naloxone-induced withdrawal jumping test in mice was much lower than that of morphine when administered i.p. in equianalgesic doses with 78% response showed by morphine-treated animals and 38% by mice receiving derivative **13** [25]. Similar observations on the induction of minimal physical dependence compared to morphine were made in rats not only for 14-methoxymetopon (**13**) but also for its 14-ethoxy

TABLE 3: Overview of the analgesic activity of 14-methoxymetopon (**13**) in various models of pain in rodents.

Pain model	Route	ED_{50}	Potency versus morphine	Reference
Hot-plate test (mouse)				
50°C	s.c.	54 μg/kg	24	[29]
52°C	s.c.	28 μg/kg	94	[14]
56°C	s.c.	30 μg/kg	28	[30]
Hot-plate test 55°C (rat)	s.c.	15 μg/kg	313	[25]
Tail-flick test (mouse)	s.c.	30 μg/kg	63	[30]
	s.c.	28 μg/kg	82	[14]
	s.c.	7.6 μg/kg	500	[31]
	i.c.v.	0.29 fg/animal	>1,000,000	[31]
	i.t.	0.31 fg/animal	>1,000,000	[31]
Tail-flick test (rat)	s.c.	7.2 μg/kg	250	[25]
	i.p.	40 μg/kg	125	[25]
Tail electric stimulation test (rat)	s.c.	30 μg/kg[a]	167	[33]
Writhing test (mouse)				
acetic acid	s.c.	7 μg/kg	99	[25]
PPQ	s.c.	9 μg/kg	44	[30]
Carrageenan-induced inflammatory pain (rat)	s.c.	20 μg/kg[a]	100	[35]

[a] Significant effect.

substituted analogue, 14-ethoxymetopon (**14**) in naloxone-induced abstinence, both MOR agonists show negligible withdrawal syndromes [25]. The rate of development of tolerance to analgesia was markedly slower than that of morphine. In the tail-flick test in rats, a much lower degree of tolerance was reported after chronic treatment with compound **13** for 7 and 11 days than morphine when given i.p. in equianalgesic doses [25, 32]. Also, 14-ethoxymetopon (**14**) failed to induce significant tolerance to the antinociceptive action [25]. Additional behavioral studies described that derivative **13** is more effective in rats in reducing the emotive/affective component of pain and in producing an anxiolytic effect than morphine in the elevated-plus maze [33]. Despite 14-methoxymetopon's classification as a highly potent and selective MOR agonist which is strongly supported by in vitro binding [25, 26, 28, 31] and functional studies [13, 14, 24–26, 28], antagonism of its in vitro [14, 23, 26] and in vivo effects by naloxone, naltrexone and selective MOR antagonists [14, 25, 31, 34], and antisense mapping studies [31], it is evident that 14-methoxymetopon has pharmacological and functional profiles distinct from those of traditional MOR agonists.

The effect of the replacement of the 5-methyl group in 14-methoxymetopon (**13**) by a benzyl group resulting in analogue **15** was examined aiming for understanding the role of the substitution pattern in position 5 in *N*-methylmorphinan-6-ones on the interaction with opioid receptors [14]. Exchanging the 5-methyl group in **13** with a benzyl group in **15**, left binding affinities at DOR and KOR largely unaffected, while retaining the high affinity at the MOR (K_i of 0.15 nM for **13** versus 0.31 nM for **15**), and the MOR selectivity (Table 1) [14]. On the basis of in vitro [^{35}S]GTPγS functional findings, the presence of a 5-benzyl group in compound **15** yielded a new MOR agonist, showing a five times increase in potency compared to its 5-methyl substituted derivative **13**; while it was slightly less efficacious in stimulating [^{35}S]GTPγS binding, but displaying similar efficacy to morphine (Table 2) [14]. The binding affinity of analogue **15** for opioid receptors is comparable to that of 14-methoxymetopon **13**; it is hence likely that the presence of an arylalkyl group such as benzyl at position 5 leads to an increase in agonist potency. A similar profile was noted when comparing in vitro agonist properties of the 5-benzyl analogue **15** to its 5-unsubstituted derivative 14-*O*-methyloxymorphone (**1**) (Tables 1 and 2). Compound **15** produced effective and naloxone-sensitive analgesia in the hot-plate and tail-flick tests in mice after s.c. administration [14]. Its antinociceptive potency was less than two times lower than that of 14-methoxymetopon **13**, and 50 times higher than that of morphine (Tables 1 and 2) [14]. Substitution of the 5-methyl group in **13** with a benzyl group appeared to be well tolerated leading to a highly potent and efficacious MOR antinociceptive agent. Besides, the introduction of a benzyl substituent at position 5 in 14-*O*-methyloxymorphone (**1**), giving rise to compound **15**, produced a three times decrease in antinociceptive potency (Table 2). Further experiments performed with this new opioid molecule showed that contrary to morphine, 14-*O*-methyloxymorphone (**1**) and 14-methoxymetopon (**13**), no significant alteration in motor coordination was induced by derivative **15** in the mouse rotarod test at any of the s.c. analgesic doses (Table 2) [14]. Results from pharmacological investigations have shown that exchanging the 5-methyl with a benzyl group in 14-methoxymetopon (**13**) afforded a potent MOR antinociceptive agent with decreased propensity to cause locomotor dysfunction.

TABLE 4: Comparison of antinociceptive potencies of 14-methoxymetopon (**13**) and its 14-phenylpropoxy substituted analogues **20** and **21**, with morphine and etorphine.

	ED_{50} (μg/kg)[a]		
	Hot-plate test	Tail-flick test	PPQ writhing test
20	2.6	4.4	1.7
21	0.10	0.08	0.16
13	30	30	9.0
Morphine	850	1,920	400
Etorphine	1.0	2.0	0.40

[a]Determined in mice after s.c. administration [22].

Subsequent synthetical and biological work targeted 14-arylalkyloxymetopon derivatives of 14-methoxymetopon (**13**), resulting in 14-benzyloxymetopon (**18**) [13] and 14-phenylpropoxy-substituted analogues **20** and PPOM (**21**) [22]. The benzyloxy or phenylpropoxy substitution in position 14 gave rise to new structures, **18** and **21**, respectively, which had markedly enhanced binding affinities at both DOR and KOR, while the high affinity at MORs remains unaffected compared to the parent compound **13** (Table 1). As a result, the MOR selectivity was markedly reduced for the benzyloxy derivative **18**, with a complete loss of MOR selectivity for PPOM (**21**), while some MOR selectivity was still depicted by its 3-methoxy analogue **20**. All three 5-methyl, 14-arylalkyloxy-substituted compounds, **18**, **20**, and **21**, showed high antinociceptive activity when administered s.c. to mice in pain models including hot-plate, tail-flick, and writhing tests (Table 1) [13, 22]. Notable was the observation, that in vivo, PPOM (**21**) was an extremely potent opioid agonist exhibiting substantially augmented analgesic potency compared not only to its 14-methoxy analogue **13** and morphine, but it was even more effective in inducing analgesia than etorphine (Table 4), a MOR morphinan used in veterinary medicine for anesthesia of large animals and wildlife species [36, 37]. Moreover, the 3-*O*-methyl ether of PPOM (**20**) also showed several hundredfolds greater analgesic potency than morphine, although compared to PPOM (**21**), its potency was 10 to 55 times lower depending upon the applied analgesic assay and displaying similar potency to etorphine (Table 4) [22].

Assessment of physicochemical properties has gained increased relevance in drug development, particularly in understanding the behavior of bioactive molecules and correlation with their biological profiles [38, 39]. The lipophilicity of some *N*-methylmorphinan-6-ones including 14-*O*-methyloxymorphone (**1**), 14-methoxymetopon (**13**) and its 5-benzyl analogue **15** was experimentally evaluated in comparison to morphine (Table 2) and oxymorphone [24]. While similar lipophilicity is shown by 14-*O*-methyloxymorphone (**1**) and its parent molecule oxymorphone ($\log P$ values of 0.60 and 0.67, resp.) and also morphine, the presence of a 5-methyl group in compound **13** leads to a more lipophilic compound (Table 2). Further increase in lipophilicity resulted after replacement of the 5-methyl with a 5-benzyl group in derivative **13** (Table 2), making such structures feasible candidates for oral and/or transdermal delivery.

4. Conclusion

The summarized research reports on the development of novel opioid analgesics from the class of *N*-methylmorphinan-6-ones have highlighted the spectrum of chemical strategies, biological and pharmacological properties and ligand-based SARs, directed towards the discovery of more effective and safer pain medications. Targeting position 5 in 14-alkoxy-*N*-methylmorphinas-6-ones represents a promising approach for tuning activities and influencing interaction with opioid receptors in this class of compounds. We reported that the presence of methyl and benzyl groups at position 5 gives rise to strong opioid antinociceptive agents with reduced propensity to induce undesired effects compared to morphine although interacting selectively with MORs. The promising experimental results represent a useful and valuable aspect for design and optimization of existing structural templates increasing the chance of identifying clinically useful analgesics for superior management of pain.

Abbreviations

$[^{35}S]GTP\gamma S$:	Guanosine-5′-O-(3-$[^{35}S]$thio)-triphosphate
DMF:	*N,N*-dimethylformamide
DOR:	δ opioid receptor
EC_{50}, IC_{50}:	Concentration necessary to produce a 50% effect
ED_{50}:	Effective analgesic dose necessary to elicit a 50% effect
GPI:	Guinea pig ileum
i.c.v.:	Intracerebroventricular
i.p.:	Intraperitoneal
i.t.:	Intrathecal
i.v.:	Intravenous
K_i:	Inhibition constant
KOR:	κ opioid receptor
MOR:	μ opioid receptor
MVD:	Mouse vas deferens
NaH:	Sodium hydride
PPOM:	14-Phenylpropoxymetopon

PPQ: Paraphenylquinone
RVD: Rat vas deferens
SAR: Structure-activity relationship
s.c.: Subcutaneous.

Acknowledgments

The paper was supported by the Austrian Science Fund (FWF): P12668 and TRP19-B16, European Community (EPILA, QLK6-1999-02334), and Drug Evaluation Committee of the College on Problems of Drug Dependence of the USA (N01DA-1-7725).

References

[1] S. Benyhe, "Morphine: new aspects in the study of an ancient compound," *Life Sciences*, vol. 55, no. 13, pp. 969–979, 1994.

[2] C. J. Woolf, "Overcoming obstacles to developing new analgesics," *Nature Medicine*, vol. 16, no. 11, pp. 1241–1247, 2010.

[3] C. E. Inturrisi, "Clinical pharmacology of opioids for pain," *Clinical Journal of Pain*, vol. 18, no. 4, pp. S3–S13, 2002.

[4] R. Benyamin, A. M. Trescot, S. Datta et al., "Opioid complications and side effects," *Pain Physician*, vol. 11, no. 2, pp. S105–S120, 2008.

[5] I. Power, "An update on analgesics," *British Journal of Anaesthesia*, vol. 107, no. 1, pp. 19–24, 2011.

[6] G. W. Pasternak, Ed., *The Opioid Receptors*, Humana Press, New York, NY, USA, 2010.

[7] A. Mansour, C. A. Fox, H. Akil, and S. J. Watson, "Opioid-receptor mRNA expression in the rat CNS: anatomical and functional implications," *Trends in Neurosciences*, vol. 18, no. 1, pp. 22–29, 1995.

[8] L. McCarthy, M. Wetzel, J. K. Sliker, T. K. Eisenstein, and T. J. Rogers, "Opioids, opioid receptors, and the immune response," *Drug and Alcohol Dependence*, vol. 62, no. 2, pp. 111–123, 2001.

[9] C. Stein, "Opioid receptors on peripheral sensory neurons," *Advances in Experimental Medicine and Biology*, vol. 521, pp. 69–76, 2003.

[10] P. Holzer, "Opioid receptors in the gastrointestinal tract," *Regulatory Peptides*, vol. 155, no. 1–3, pp. 11–17, 2009.

[11] L. J. Crofford, "Adverse effects of chronic opioid therapy for chronic musculoskeletal pain," *Nature Reviews Rheumatology*, vol. 6, no. 4, pp. 191–197, 2010.

[12] H. Schmidhammer, L. Aeppli, L. Atwell et al., "Synthesis and biological evaluation of 14-alkoxymorphinans. 1. Highly potent opioid agonists in the series of (−)-14-methoxy-*N*-methylmorphinan-6-ones," *Journal of Medicinal Chemistry*, vol. 27, no. 12, pp. 1575–1579, 1984.

[13] R. Lattanzi, M. Spetea, F. Schüllner et al., "Synthesis and biological evaluation of 14-alkoxymorphinans. 22. Influence of the 14-alkoxy group and the substitution in position 5 in 14-alkoxymorphinan-6-ones on *in vitro* and *in vivo* activities," *Journal of Medicinal Chemistry*, vol. 48, no. 9, pp. 3372–3378, 2005.

[14] M. Spetea, C. R. Bohotin, M. F. Asim, K. Stübegger, and H. Schmidhammer, "*In vitro* and *in vivo* pharmacological profile of the 5-benzyl analogue of 14-methoxymetopon, a novel μ opioid analgesic with reduced propensity to alter motor function," *European Journal of Pharmaceutical Sciences*, vol. 41, no. 1, pp. 125–135, 2010.

[15] H. Schmidhammer and M. Spetea, "Synthesis of 14-alkoxymorphinan derivatives and their pharmacological actions," in *Topics in Current Chemistry*, H. Nagase, Ed., pp. 63–91, Springer, Heidelberg, Germany, 2011.

[16] N. B. Eddy, "Metopon hydrochloride (methyldihydromorphinone hydrochloride)," *Canadian Medical Association Journal*, vol. 58, pp. 79–80, 1948.

[17] N. B. Eddy, "Pharmacology of metopon and other new analgesic opium derivatives," *Annals of the New York Academy of Sciences*, vol. 51, no. 1, pp. 51–58, 1948.

[18] R. M. Boden, M. Gates, S. P. Ho, and P. Sundararaman, "Derivatives of the thebaine anion. 1. Structure of metopon. A direct demonstration," *Journal of Organic Chemistry*, vol. 47, no. 7, pp. 1347–1349, 1982.

[19] H. Schmidhammer, F. Fritsch, W. P. Burkard, L. Eggstein-Aeppli, F. Hefti, and M. I. Holck, "(−)-*N*-[(cyclopropyl)methyl]-3,4-dimethoxy-5-methylmorphinan-6-one, an opioid agonist with preference for kappa opioid receptors," *Helvetica Chimica Acta*, vol. 71, no. 3, pp. 642–647, 1988.

[20] H. Schmidhammer, J. B. Deeter, N. D. Jones, J. D. Leander, D. D. Schoepp, and J. K. Swartzendruber, "Synthesis, structure elucidation, and pharmacological evaluation of 5-methyl-oxymorphone (= 4,5α-epoxy-3,14-dihydroxy-5,17-dimethylmorphinan-6-one)," *Helvetica Chimica Acta*, vol. 71, no. 7, pp. 1801–1804, 1988.

[21] M. Gates, R. M. Boden, and P. Sundararaman, "Derivatives of the thebaine anion. 2. 5-methylmorphine, 5-methylcodeine, 5-methylheroin, and some related compounds," *Journal of Organic Chemistry*, vol. 54, no. 4, pp. 972–974, 1989.

[22] J. Schütz, M. Spetea, M. Hoch et al., "Synthesis and biological evaluation of 14-alkoxymorphinans. 20.1 14-phenylpropoxymetopon: an extremely powerful analgesic," *Journal of Medicinal Chemistry*, vol. 46, no. 19, pp. 4182–4187, 2003.

[23] M. Spetea, T. Friedmann, P. Riba et al., "*In vitro* opioid activity profiles of 6-amino acid substituted derivatives of 14-O-methyloxymorphone," *European Journal of Pharmacology*, vol. 483, no. 2-3, pp. 301–308, 2004.

[24] P. Riba, T. Friedmann, K. P. Király et al., "Novel approach to demonstrate high efficacy of μ opioids in the rat vas deferens: a simple model of predictive value," *Brain Research Bulletin*, vol. 81, no. 1, pp. 178–184, 2010.

[25] Z. Fürst, B. Búzás, T. Friedmann, H. Schmidhammer, and A. Borsodi, "Highly potent novel opioid receptor agonist in the 14-alkoxymetopon series," *European Journal of Pharmacology*, vol. 236, no. 2, pp. 209–215, 1993.

[26] M. Spetea, F. Tóth, J. Schütz et al., "Binding characteristics of [3H]14-methoxymetopon, a high affinity mu-opioid receptor agonist," *European Journal of Neuroscience*, vol. 18, no. 2, pp. 290–295, 2003.

[27] H. Schmidhammer, A. Schratz, and J. Mitterdorfer, "Synthesis and biological evaluation of 14-alkoxymorphinans. Part 8. 14-methoxymetopon, an extremely potent opioid agonist," *Helvetica Chimica Acta*, vol. 73, no. 6, pp. 1784–1787, 1990.

[28] L. Mahurter, C. Garceau, J. Marino, H. Schmidhammer, G. Tóth, and G. W. Pasternak, "Separation of binding affinity and intrinsic activity of the potent μ-opioid 14-methoxymetopon," *Journal of Pharmacology and Experimental Therapeutics*, vol. 319, no. 1, pp. 247–253, 2006.

[29] G. Zernig, A. Saria, R. Krassnig, and H. Schmidhammer, "Signal transduction efficacy of the highly potent mu opioid agonist 14- methoxymetopon," *Life Sciences*, vol. 66, no. 19, pp. 1871–1877, 2000.

[30] E. Greiner, M. Spetea, R. Krassnig et al., "Synthesis and biological evaluation of 14-alkoxymorphinans. 18.1 *N*-substituted 14-phenylpropyloxymorphinan-6-ones with unanticipated agonist properties: extending the scope of common structure-activity relationships," *Journal of Medicinal Chemistry*, vol. 46, no. 9, pp. 1758–1763, 2003.

[31] M. A. King, W. Su, C. L. Nielan et al., "14-methoxymetopon, a very potent μ-opioid receptor-selective analgesic with an unusual pharmacological profile," *European Journal of Pharmacology*, vol. 459, no. 2-3, pp. 203–209, 2003.

[32] K. P. Király, P. Riba, C. D'Addario et al., "Alterations in prodynorphin gene expression and dynorphin levels in different brain regions after chronic administration of 14-methoxymetopon and oxycodone-6-oxime," *Brain Research Bulletin*, vol. 70, no. 3, pp. 233–239, 2006.

[33] L. Urigüen, B. Fernández, E. M. Romero et al., "Effects of 14-methoxymetopon, a potent opioid agonist, on the responses to the tail electric stimulation test and plus-maze activity in male rats: neuroendocrine correlates," *Brain Research Bulletin*, vol. 57, no. 5, pp. 661–666, 2002.

[34] E. Freye, H. Schmidhammer, and L. Latasch, "14-methoxymetopon, a potent opioid, induces no respiratory depression, less sedation, and less bradycardia than sufentanil in the dog," *Anesthesia and Analgesia*, vol. 90, no. 6, pp. 1359–1364, 2000.

[35] I. Bileviciute-Ljungar, M. Spetea, Y. Guo, J. Schütz, P. Windisch, and H. Schmidhammer, "Peripherally mediated antinociception of the μ-opioid receptor agonist 2-[(4,5α-epoxy-3-hydroxy-14β-methoxy-17-methylmorphinan-6β-yl) amino]acetic acid (HS-731) after subcutaneous and oral administration in rats with carrageenan-induced hindpaw inflammation," *Journal of Pharmacology and Experimental Therapeutics*, vol. 317, no. 1, pp. 220–227, 2006.

[36] G. F. Stegmann, "Etorphine-halothane anaesthesia in two five-year-old African elephants (Loxodonta africana)," *Journal of the South African Veterinary Association*, vol. 70, no. 4, pp. 164–166, 1999.

[37] J. Sterken, J. Troubleyn, F. Gasthuys, V. Maes, M. Diltoer, and C. Verborgh, "Intentional overdose of large animal immobilon," *European Journal of Emergency Medicine*, vol. 11, no. 5, pp. 298–301, 2004.

[38] A. Avdeef and B. Testa, "Physicochemical profiling in drug research: a brief survey of the state-of-the-art of experimental techniques," *Cellular and Molecular Life Sciences*, vol. 59, no. 10, pp. 1681–1689, 2002.

[39] B. Faller, "Physicochemical profiling in early drug discovery: new challenges at the age of high-throughput screen and combinatorial chemistry," in *Chemistry and Molecular Aspects of Drug Design and Action*, E. A. Rekka and P. N. Kourounakis, Eds., pp. 303–312, CRC Press, 2008.

8

Synthesis and Anticonvulsant Activity of Various Mannich and Schiff Bases of 1,5-Benzodiazepines

Surendra N. Pandeya and Neha Rajput

Department of Pharmaceutical Sciences, Saroj Institute of Technology and Management, Sultanpur Road, Lucknow 226002, India

Correspondence should be addressed to Neha Rajput, nr.17neha@gmail.com

Academic Editor: Jochen Lehmann

Benzodiazepines have a various behavioral effects in addition to their anxiolytic action. There is every reason to believe that the BZ/GABA receptor complex is involved in these effects, since GABAmimetic manipulations modify the effect of BZ in tests of convulsive activity, motor function, and appetitive behavior. 1,5-Benzodiazepines are biologically important molecules and are extensively used clinically as analgesic, hypnotic, sedative, and antidepressive agents. Hence, 1,5-Benzodiazepines were synthesized by condensation of o-phenylenediamine and ketones, for example, cyclohexanone and acetone in presence of sulfated zirconia (catalyst). Mannich bases were synthesized with acetophenone, p-nitroacetophenone, p-chloroacetophenone, and formaldehyde. Schiff bases were synthesized using Mannich base of 1,5-benzodiazepines with p-chloroaniline and p-chlorophenylsemicarbazide in the presence of glacial acetic acid. All the synthesized compounds were characterized by ^{1}H NMR and IR spectral analyses. All the synthesized derivatives were evaluated at the dose of 30 mg/kg b.w for anticonvulsant activity by isoniazid induced convulsion model, and the compounds NBZD-3 and NBZD-8 were found to be the most active among all compounds. Among all the synthesized derivatives, compounds NBZD-13 and NBZD-17 were found to be the most active among all compounds using thiosemicarbazide induced model. Although NBZD-8, NBZD-10, and NBZD-18 are the compounds which had shown good anticonvulsant activity and have an advantage over that, they were not sedative.

1. Introduction

A benzodiazepine is a psychoactive drug whose core chemical structure is the fusion of a benzene ring and a diazepine ring. The first benzodiazepine, chlordiazepoxide (Librium), discovered accidentally by Leo Sternbach in 1955 and made available in 1960 by Hoffmann La Roche, which has also marketed diazepam (Valium) since 1963 [1]. 1,5-Benzodiazepines constitute an important class of psychopharmaca [2], in particular as tranquilizers and also as potent Virucides and nonnucleoside inhibitors of HIV-1 reverse transcriptase [3].

Benzodiazepine has a traditional place in antiepileptic therapy. The clinical use of BZDs can be divided into two categories. First, in the acute treatment of seizures as drugs of choice in status epilepticus and also in some cases of febrile seizures. Second, the BZDs are utilized in long-term therapy of certain seizures types primarily in the pediatrics' population [4].

There are some differences between the effects of 1,5- and 1,4-benzodiazepines. A greater therapeutic potential and lower incidence of side effects were described for 1,5-BZDs when compared to 1,4-BZDs. 1,5-BZD is used as adjuvant therapy in resistant cases of epilepsies [5]. BZDs exhibit potent anticonvulsant actions in a wide variety of animal seizures models. They are particularly effective against seizures induced by electroshock [6], and various chemoconvulsants, in kindled seizures and in absence seizures [7].

Beside this, 1,5-benzodiazepines show antifungal, antibacterial [8], antifeedant [9], anti-inflammatory analgesic [10], and anticonvulsant activities [11]. The benzodiazepines nucleus is a well-studied traditional pharmacophoric scaffold that has emerged as a core structure unit of various biological activities [12].

Although, the first benzodiazepine was introduced as a drug nearly 35 years ago, the research in this area is still very active and is directed towards the synthesis of compounds with enhanced pharmacological activity [13]. The chemical structure of the benzodiazepines seems at first sight to be unique among the various types of central depressant drugs [14]. 1,5-Benzodiazepines derivatives show a large number of pharmacological properties such as they acted as sedatives [15], Cerebrovasodialators [16], neuroleptics [17], antispasmodic [18], anticonvulsant [19], tranquilizing agents [20], antibacterial [21], and psoriasis [22] and used for the treatment of small pox [23].

2. Experimental Procedure

Starting material and reagents were procured from commercial chemical suppliers. All the chemicals and solvents used were of laboratory grade. Melting points were determined in open capillary tubes and are uncorrected. IR spectra (KBr, cm^{-1}) were recorded on Perkin Elmer Spectrometer, 1H NMR-(δ, ppm) spectra were recorded on a Brucker 300 MHz NMR spectrometer using TMS as an internal standard. The purity of compounds and progress of the reaction were checked by TLC using silica gel-G as adsorbent.

2.1. Procedure for Preparation of Fused Ring Benzodiazepine Nucleus

2.1.1. Synthesis of Fused Ring Benzodiazepine Nucleus. Synthesis of fused ring benzodiazepines in presence of sulfated zirconia involves 2 steps which are as follows [9].

Preparation of Catalyst. 25 gm of zirconium oxychloride was dissolved in doubly distilled water (pH-2). Dilute aq. ammonia was then added dropwise from a burette with vigorous (pH = 8). Precipitate was washed with distilled water several times and dried for 24 h. Sample was ground to fine powder and immersed in an 0.5 M H2SO4 solution (30 mL) for 30 min. Excess water was evaporated on water bath, and the resulting sample was oven dried.

Synthesis of Benzodiazepines. 1 : 2.5 mole ratio mixture of o-phenylenediamine and ketone (cyclohexanone (Scheme 1)) with catalytic amount of sulfated zirconia was taken in round bottom flask (RBF) with stirring at ambient condition for 2-3 h. 10 mL of CH2Cl2 was added to reaction mixture, and catalyst was recovered by filtration.

2.2. Procedure for Preparation of Mannich Base Derivatives

2.2.1. Synthesis of Various Mannich Base Derivatives of Fused Ring Benzodiazepine (Scheme 2). Equimolar quantity of fused ring benzodiazepine (**NBZD-1**, 0.01 M), formaldehyde, and various acetophenones (i.e., acetophenone, p-nitroacetophenone, and p-chloroacetophenone) were taken in RBF, and mixture was refluxed for 2.30 h. Completion of reaction was monitored by TLC analysis for several times. Then, reaction mixture was evaporated on water bath and dried. Melting point, Rf value, and % yield were noted. Various Mannich base derivatives are shown in Scheme 2.

2.3. Procedure for Preparation of Schiff Base Derivatives

2.3.1. Synthesis of Various Schiff Base Derivatives of Fused Ring Benzodiazepines. Equimolar quantities of Mannich base derivatives (0.01 M, **NBZD-3**, **NBZD-4**, and **NBZD-5**), in individual reactions, were dissolved in glacial acetic acid and added with p-chloroaniline (Scheme 3) or p-chlorophenylsemicarbazide (Scheme 4) and taken in RBF and mixture was refluxed for 3 h, respectively. Completion of reaction was monitored by TLC analysis for several times in chloroform : ethanol (1 : 1). Then, reaction mixture was evaporated on water bath and dried. Melting point, Rf value, and % yield were noted.

2.4. Procedure for Preparation of 1,5-Benzodiazepine Nucleus

2.4.1. Synthesis of 1,5-Benzodiazepine Nucleus. Synthesis of 1,5-benzodiazepines in presence of sulphated zirconia involves 2 steps which are as follows [9].

Synthesis of Benzodiazepines. 1 : 2.5 mole ratio mixture of o-phenylenediamine and acetone (Scheme 5) with catalytic amount of sulfated zirconia was taken in round bottom flask (RBF) with stirring at ambient condition for 2-3 h. 10 mL of CH2Cl2 was added to reaction mixture, and catalyst was recovered by filtration.

2.5. Procedure for Preparation of Mannich Base Derivatives

2.5.1. Synthesis of Various Mannich Base Derivatives of 1,5-Benzodiazepine (Scheme 6). Equimolar quantities of fused ring benzodiazepine (**NBZD-12**, 0.01 M), formaldehyde, and various acetophenones (i.e., acetophenone, p-nitroacetophenone, and p-chloroacetophenone) were taken in RBF, and mixture was refluxed for 2.30 h. Completion of reaction was monitored by TLC analysis for several times. Then, reaction mixture was evaporated on water bath and dried. Melting point, Rf value, and % yield were noted. Various Mannich base derivatives are shown in Scheme 6.

2.6. Procedure for Preparation of Schiff Base Derivatives

2.6.1. Synthesis of Various Schiff Base Derivatives of 1,5-Benzodiazepines. Equimolar quantities of Mannich base derivatives (0.01 M, **NBZD-13**, **NBZD-14**, and **NBZD-15**), in individual reactions, were dissolved in glacial acetic acid and added with p-chloroaniline (Scheme 7) or p-chlorophenylsemicarbazide (Scheme 8), and were taken in RBF, and mixture was refluxed for 3 h, respectively. Completion of reaction was monitored by TLC analysis for several times in chloroform : ethanol (1 : 1). Then, reaction mixture was evaporated on water bath and dried. Melting point, Rf value, and % yield were noted.

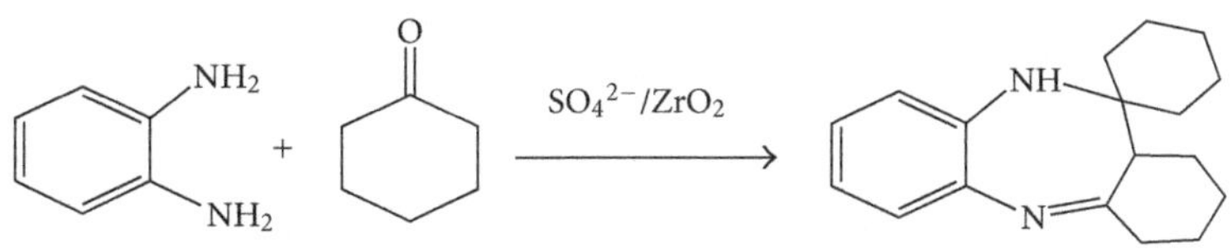

O-phenylenediamine | Cyclohexanone | **NBZD-1**

SCHEME 1: Synthesis of fused ring benzodiazepine.

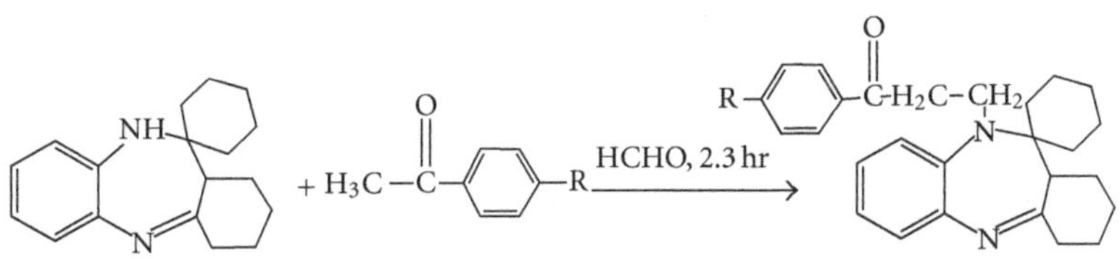

R = H
R = NO2
R = Cl

R = H (**NBZD-3**),
R = NO2 (**NBZD-4**),
R = Cl (**NBZD-5**)

NBZD-1 | Ketone | Mannich bases

SCHEME 2: Synthesis of various Mannich base derivatives of fused ring benzodiazepines.

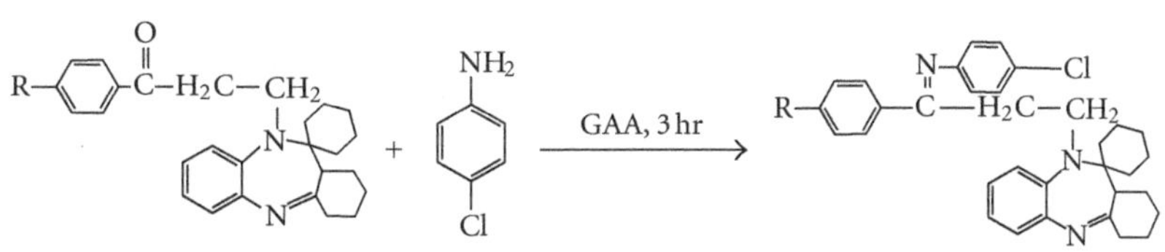

R = H (**NBZD-3**),
R = NO2 (**NBZD-4**),
R = Cl (**NBZD-5**)

R = H (**NBZD-6**),
R = NO2 (**NBZD-7**),
R = Cl (**NBZD-8**)

Mannich base of 1,5-BZD | p-chloroaniline | Schiff base of 1,5-BZD

SCHEME 3: Synthesis of Schiff base derivative of fused ring benzodiazepine from p chloroaniline.

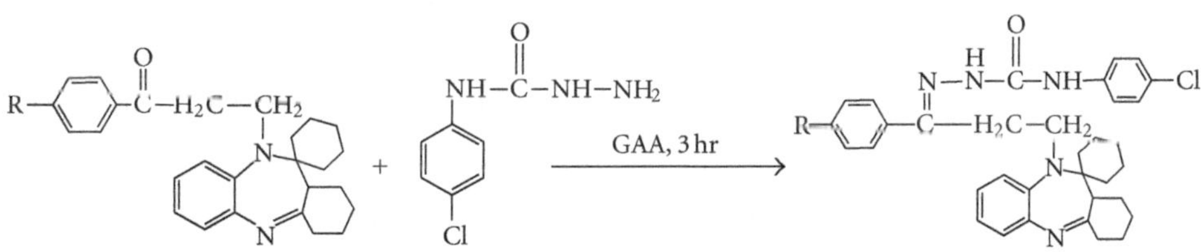

R = H (**NBZD-3**),
R = NO2 (**NBZD-4**),
R = Cl (**NBZD-5**)

R = H (**NBZD-9**),
R = NO2 (**NBZD-10**),
R = Cl (**NBZD-11**)

Mannich base of fused ring BZD | p-chlorophenylsemicarbazide | Schiff base of fused ring BZD

SCHEME 4: Synthesis of Schiff base derivative of fused ring benzodiazepine from p-chlorophenylsemicarbazide.

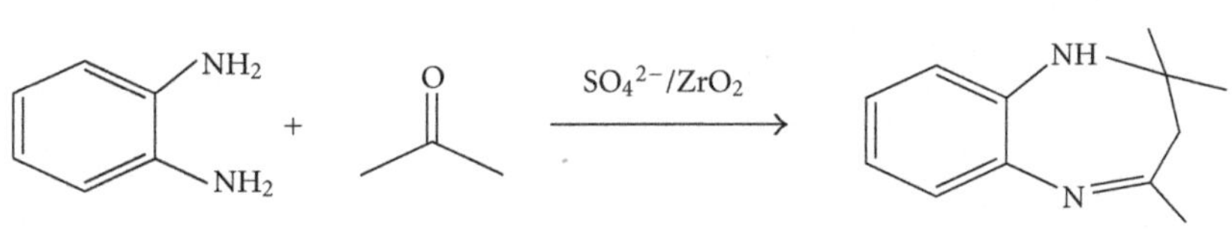

O-phenylenediamine | Acetone | **NBZD-12**

SCHEME 5: Synthesis of 1,5-benzodiazepine (**NBZD-12**).

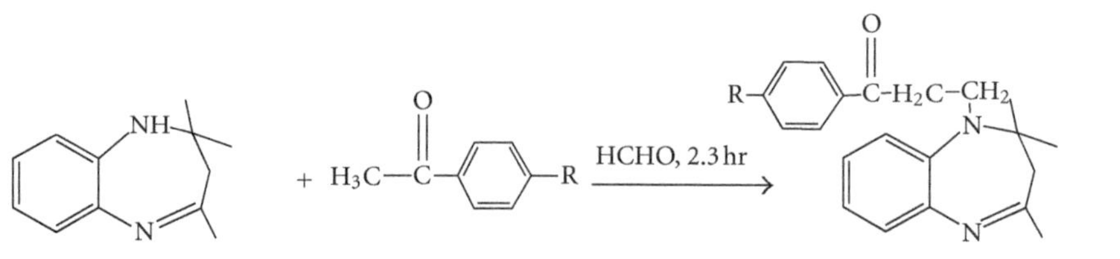

R = H
R = NO_2
R = Cl

R = H (**NBZD-13**),
R = NO_2 (**NBZD-14**),
R = Cl (**NBZD-15**)

NBZD-12 Ketone Mannich bases

Scheme 6: Synthesis of various Mannich base derivatives of 1,5-benzodiazepines.

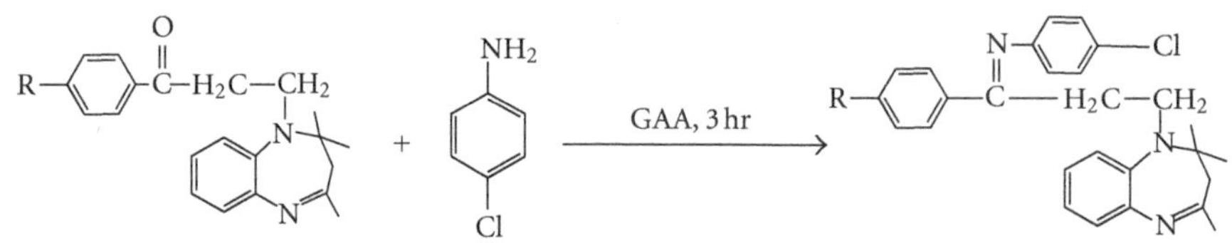

R = H (**NBZD-13**),
R = NO_2 (**NBZD-14**),
R = Cl (**NBZD-15**)

R = H (**NBZD-16**),
R = NO_2 (**NBZD-17**),
R = Cl (**NBZD-18**)

Mannich base of 1,5-BZD p-chloroaniline Schiff base of 1,5-BZD

Scheme 7: Synthesis of Schiff base derivative of 1,5-benzodiazepine from p-chloroaniline.

3. Anticonvulsant Activity

3.1. Chemical Induced Model. Ten mice of either sex with a weight of 22–25 g were treated with the test compounds (30 mg/kg b.w) or the standard (e.g., diazepam 10 mg/kg b.w) by i.p. administration. Controls received the vehicle only. 30 min after i.p. treatment, the animals were injected with a subcutaneous dose of (300 mg/kg, s.c) isoniazid, thiosemicarbazide (20 mg/kg, s.c). The occurrence of clonic seizures, tonic seizures, and death or recovery were recorded after 0.5 hr, 1 hr, 2 hr, and 4 hr, respectively, for isoniazid induced convulsion (Table 3) and also Thiosemicarbazide induced convulsion (Table 4).

3.2. Neurotoxicity Screen. Minimal motor impairment was measured in mice by the rotarod test. The mice were trained to stay on an accelerating rotarod that rotates at 20 revolutions per minute. The rod diameter was 3.2 cm. Neurotoxicity was indicated by the inability of the animal to maintain equilibrium on the rod for at least 1 min in each of the three trials. The dose at which the animals were unable to grasp the rotarod was determined (Tables 3 and 4).

DOSE: Test drug: 30 mg/Kg b.w i.p.

3.3. Sedative-Hypnotic Activity. This test was performed with the test substances in a dose of 30 mg/kg by phenobarbitone induced narcosis in rats. The compounds in PEG (polyethylene glycol) were administered i.p to a group of six rats. After 30 min, rats were then placed on their back and loss of righting reflex was taken as onset of sleep. The time taken by the rats to awake was noted. A control was also performed after pretreatment with test substances vehicle (PEG) and injected phenobarbitone (Table 5).

4. Results

4.1. Physicochemical Characterization. See Table 1.

4.2. Elemental Analysis. See Table 2.

4.3. Representative Spectral Analysis

(1) 10-Spirocyclohexane-1,2,3,9,10,10a-hexahydrobenzo[b]cyclohexane [e][1, 4] diazepine (**NBZD-1**):

^{1}H NMR (300 MHz, δ): CH(m, 6.4–7.0, 4H, phenyl), NH(s, 4.1, 1H), CH_2(m, 1.2–1.6, 18H, cyclohexane), CH(s, 2.7, 1H, diazepine ring).

IR (KBr): NH(Ar, 3030 cm^{-1}, str), CH(Ar, 3180 cm^{-1}, str), CH(Ar, 800 cm^{-1}, bend), C=N (1618 cm^{-1}, Str) CH_2(1490 cm^{-1}, str), C–C(Ar, 1600 cm^{-1}), C=C(Ar, 1410, 1500, 1580 cm^{-1}).

(2) 1-Phenyl-3-(10-spirocyclohexane-1,2,3,9,10,10a-hexahydrobenzo[b]cyclohexane[e][1, 4]diazepine-1-yl) propan-1-one (**NBZD-3**):

^{1}H NMR (300 MHz, δ): CH(m, 7.3–7.9, 5H, acetophenone), CH_2(s, 2.8, 2H, $-COCH_2$), CH_2(s, 3.5,

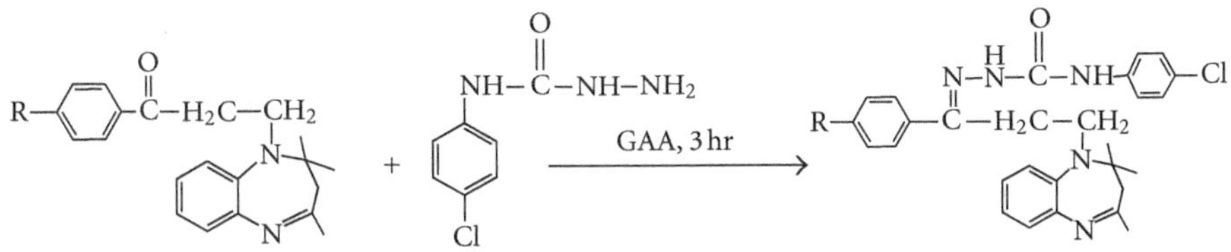

R = H (**NBZD-13**),
R = NO_2 (**NBZD-14**),
R = Cl (**NBZD-15**)

R = H (**NBZD-19**),
R = NO_2 (**NBZD-20**),
R = Cl (**NBZD-21**)

Mannich base of 1,5-BZD p-chlorophenylsemicarbazide Schiff base of 1,5-BZD

SCHEME 8: Synthesis of Schiff base derivative of 1,5-benzodiazepine from p-chlorophenylsemicarbazide.

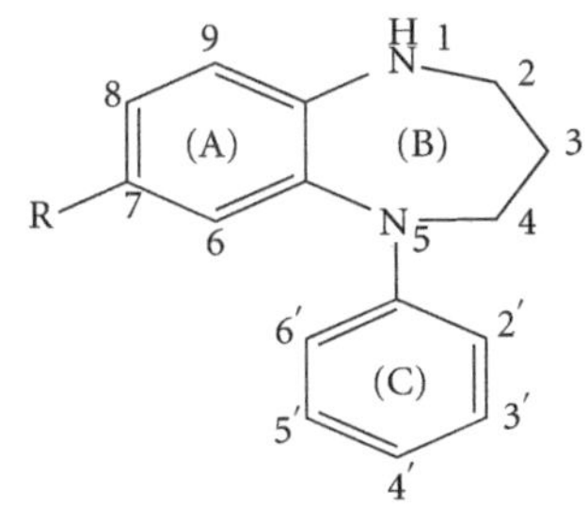

General structure of 1,5-benzodiazepine

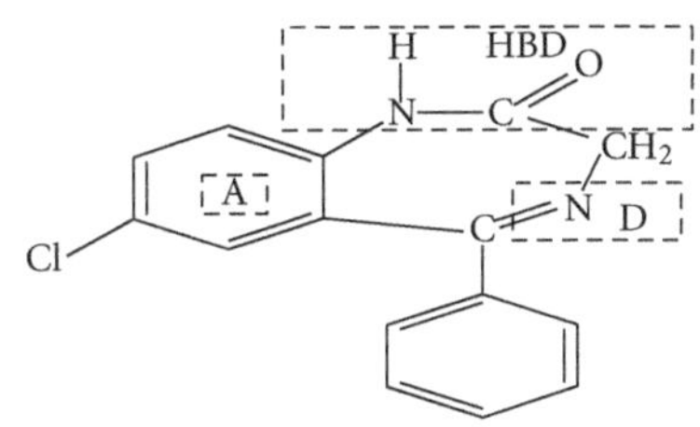

N-desmethyl diazepam
A = hydrophobic unit, HBD = hydrogen bonding domain, D = electron donor

SCHEME 9

TABLE 1: Physicochemical data of synthesized compounds.

Compound code	Molecular formula	Molecular weight	Melting point (°C)	% yield	Rf value	Log P value
NBZD-1	$C_{18}H_{24}N$	268.19	110	76.48	0.689	4.25
NBZD-3	$C_{27}H_{32}N_2O$	400.25	92	85.7	0.78	6.25
NBZD-4	$C_{33}H_{36}N_3Cl$	510.12	132	83.3	0.73	6.15
NBZD-5	$C_{34}H_{28}ClN_5O$	568.15	140	72	0.83	6.81
NBZD-6	$C_{27}H_{31}N_3O_3$	445.55	98	84.3	0.63	8.93
NBZD-7	$C_{27}H_{31}ClN_2O$	435.00	98	79.2	0.82	6.70
NBZD-8	$C_{33}H_{35}ClN_4O_2$	554.24	130	67.2	0.57	9.49
NBZD-9	$C_{34}H_{37}ClN_6O_3$	612.26	126	56.98	0.52	8.07
NBZD-10	$C_{33}H_{35}Cl_2N_3$	543.22	136	65.8	0.54	9.56
NBZD-11	$C_{34}H_{37}Cl_2N_5O$	601.24	130	66	0.66	8.62
NBZD-12	$C_{12}H_{16}N_2$	188.60	96	72	0.68	2.22
NBZD-13	$C_{21}H_{24}N_2O$	320.3	98	72.32	0.72	4.22
NBZD-14	$C_{21}H_{23}N_3O_3$	365.43	102	92.1	0.72	4.83
NBZD-15	$C_{21}H_{24}N_2O$	320.43	110	57.6	0.75	4.78
NBZD-16	$C_{21}H_{23}ClN_2O$	354.87	108	82.9	0.73	6.90
NBZD-17	$C_{27}H_{28}ClN_3$	429.98	140	90.4	0.78	7.20
NBZD-18	$C_{27}H_{27}ClN_4O_2$	474.98	136	64.06	0.75	7.46
NBZD-19	$C_{27}H_{27}Cl_2N_3$	464.43	142	85.6	0.63	6.04
NBZD-20	$C_{27}H_{27}ClN_4O_2$	474.18	138	67.9	0.55	—
NBZD-21	$C_{28}H_{29}ClN_6O_3$	532.20	142	65.2	0.76	6.59

Rf value: solvent system; chloroform : methanol 1 : 1.

TABLE 2: Elemental analysis of synthesized compounds.

Compound code	Calculated value			Observed value		
	C	H	N	C	H	N
NBZD-1	80.55	9.01	10.44	80.31	8.90	10.24
NBZD-3	80.96	8.05	6.99	80.56	8.00	6.79
NBZD-4	72.78	7.01	9.43	72.70	6.89	9.23
NBZD-5	74.55	7.18	6.44	74.23	7.04	6.25
NBZD-6	77.70	7.11	8.24	77.56	7.01	8.16
NBZD-7	71.40	6.26	10.09	71.19	6.16	10.00
NBZD-8	72.78	6.48	7.72	72.66	6.32	7.61
NBZD-9	71.88	6.74	12.33	71.34	6.66	12.21
NBZD-10	66.60	6.08	13.71	66.45	6.00	13.62
NBZD-11	67.77	6.19	11.62	67.44	6.02	11.52
NBZD-12	76.55	8.57	14.88	76.25	8.34	14.54
NBZD-13	78.71	7.55	8.74	78.59	7.56	8.70
NBZD-14	69.02	6.34	11.50	68.66	6.30	11.43
NBZD-15	71.07	6.53	7.89	71.00	6.45	7.68
NBZD-16	75.42	6.56	9.77	75.21	6.36	9.53
NBZD-17	68.27	5.73	11.80	68.20	5.60	11.70
NBZD-18	69.83	5.83	9.05	69.70	5.78	8.90
NBZD-19	68.91	6.20	3.28	68.78	6.09	3.10
NBZD-20	63.09	5.48	15.77	63.01	5.32	15.57
NBZD-21	64.37	5.59	13.40	64.30	5.40	13.28

TABLE 3: Anticonvulsant activity using Isoniazid induced convulsion model.

Compound code	Isoniazid induced model			Neurotoxicity study	
	0.5 hr	1 hr	2 hr	1 hr	4 hr
				Dose: 30 mg/kg	
NBZD-1	30 mg	30 mg	Not protected	NN	NN
NBZD-3	30 mg	30 mg	30 mg	NN	NN
NBZD-4	30 mg	Not protected	Not protected	NN	NN
NBZD-5	30 mg	Not protected	Not protected	NN	NN
NBZD-6	30 mg	30 mg	Not protected	NN	NN
NBZD-7	30 mg	Not protected	Not protected	NN	NN
NBZD-8	30 mg	30 mg	30 mg	NN	NN
NBZD-9	30 mg	Not protected	Not protected	NN	NN
NBZD-10	30 mg	30 mg	Not protected	NN	NN
NBZD-11	30 mg	30 mg	Not protected	NN	NN
Control	—	—	—		

Symbol (NN) indicates no neurotoxicity at 30 mg/kg b.w.

2H, $-NHCH_2$), CH(m, 6.4–7.0, 4H, phenyl), CH_2(m, 1.2–1.5, 18H, cyclohexane), CH(s, 2.5, 1H, diazepine ring).

IR (KBr): C=O(1700 cm^{-1}, str), CH(Ar, 3180 cm^{-1}, str), CH(Ar, 810 cm^{-1}, bend), C=N(1618 cm^{-1}, Str), CH_2(1490 cm^{-1}, str), C–C(Ar, 1600 cm^{-1}), C=C(Ar, 1410, 1500, 1580 cm^{-1}).

(3) 1-(4-Nitrophenyl)-3-(10-spirocyclohexane-1,2,3,9, 10,10a-hexahydrobenzo[b]cyclohexane[e][1,4] diazepine-1-yl) propan-1-one (**NBZD-4**).

^{1}H NMR (300 MHz, δ): CH(m, 8.1-8.2, 4H, p-nitro-acetophenone), CH_2(s, 2.7, 2H, $-COCH_2$), CH_2(s, 3.5, 2H, $-NHCH_2$), CH(m, 6.4–7.0, 4H, phenyl),

Table 4: Anticonvulsant activity using thiosemicarbazide induced convulsion model.

Compound code	Thiosemicarbazide induced model			Neurotoxicity study Dose: 30 mg/kg	
	0.5 hr	1 hr	2 hr	1 hr	4 hr
NBZD-12	30 mg	30 mg	Not protected	NN	NN
NBZD-13	30 mg	30 mg	30 mg	NN	NN
NBZD-14	30 mg	Not protected	Not protected	NN	NN
NBZD-15	30 mg	Not protected	Not protected	NN	NN
NBZD-16	30 mg	30 mg	Not protected	NN	NN
NBZD-17	30 mg	Not protected	Not protected	NN	NN
NBZD-18	30 mg	30 mg	30 mg	NN	NN
NBZD-19	30 mg	Not protected	Not protected	NN	NN
NBZD-20	30 mg	Not protected	Not protected	NN	NN
NBZD-21	30 mg	30 mg	Not protected	NN	NN
Control	—	—	—		

Symbol (NN) indicates no neurotoxicity 30 mg/kg of b.w.

Table 5: Sedative activity of synthesized compound.

Compounds code	Sleeping time (mean ± SEM) (min)
NBZD-1	120 ± 9.00**
NBZD-3	138 ± 10.53**
NBZD-4	140 ± 11.92**
NBZD-7	141 ± 11.21**
NBZD-8	68 ± 12.6 NS
NBZD-10	63 ± 9.05 NS
NBZD-11	148 ± 12.15**
NBZD-12	124 ± 10.12**
NBZD-14	110 ± 11.41 **
NBZD-15	100 ± 10.98**
NBZD-17	148 ± 11.54**
NBZD-18	76 ± 10.26 NS
NBZD-20	112 ± 9.62**
NBZD-21	157 ± 12.09**
Phenobarbitone (control)	56 ± 11.47

Values represent the mean ± SEM of six animals for each group.
*Significant at $P < 0.05$ and **significant at $P < 0.01$ (Dunnett's test).
Test drug (30 mg/kg) and Phenobarbitone (40 mg/kg).
NS denotes not significant at $P < 0.01$ (student's t-test).

NH(s, 4.0, 1H,), CH_2(m, 1.22–1.59, 18H, cyclohexane), CH(s, 2.7, 1H, diazepine ring).

IR (KBr): C=O(1710 cm^{-1}, str), CH(Ar, 3150 cm^{-1}, str), CH(Ar, 800 cm^{-1}, bend), C=N(1658 cm^{-1}, Str), CH_2 (1490 cm^{-1}, str), C–C(Ar, 1610 cm^{-1}), C=C(Ar, 1410, 1560, 1580 cm^{-1}), N–O(1350 cm^{-1}, str).

(4) (4-Chloro-phenyl)-[1-chlorophenyl-3-10-spirocyclohexane-1,2,3,9,10,10a-hexahydrobenzo[b]cyclohexane[e][1,4]diazepine-1yl)-propylidene]-amine (**NBZD-8**).

^1H NMR (300 MHz, δ): CH(m, 7.30–7.5, 4H, p-chloroacetophenone), CH(m, 7.2-7.3, 4H, p-chloroaniline), CH_2(s, 1.6, 2H, $-COCH_2$), CH_2(s, 3.4, 2H, $-NHCH_2$), CH(m, 6.6–7.1, 4H, phenyl), CH_2(m, 1.3–1.5, 18H, cyclohexane), CH(s, 2.7, 1H, diazepine ring).

IR (KBr): C=N(1569 cm^{-1}, str), C–Cl(727 cm^{-1}, str), C–Cl(760 cm^{-1}), C–H(2975 cm^{-1}, str assym), CH (1383.9 cm^{-1}, def sym.), C–H(Ar, 3072 cm^{-1}, str), CH(Ar, 3150 cm^{-1}, str), CH(Ar, 860 cm^{-1}, bend), C=N(1678 cm^{-1}, Str) CH_2(1490 cm^{-1}, str), C–C(Ar, 1600 cm^{-1}), C=C(Ar, 1410, 1500, 1580 cm^{-1}).

(5) (4-Chlorophenylhyrazinecarboxamide)[1-nitrophenyl-3-(10-spirocyclohexane-1,2,3,9,10,10a-hexahydrobenzo[b]cyclohexane [e][1,4] diazepine-1-yl)-propylidene]-amine (**NBZD-10**).

^1H NMR (300 MHz, δ): CH(m,7.9–8.2, 4H, p-nitroacetophenone), NH(s, 7.0, 1H, =NNH, p-chlorophenylsemicarbazide), NH(s, 6.0, 1H, $-NHC_6H_4Cl$, p-chlorophenylsemicarbazide), CH (m, 7.2–7.6, 4H, p-chlorophenylsemicarbazide), CH (m, 6.6–7.1, 4H, phenyl), CH_2(m, 1.3–1.5, 18H, cyclohexane), CH(s, 2.3, 1H, diazepine ring), CH_2(s, 1.6, 2H, $-COCH_2$), CH_2(s, 3.4, 2H, $-NHCH_2$).

IR (KBr): C=N(1569 cm^{-1}, str), C–Cl(728 cm^{-1}, str), C–H(2970 cm^{-1}, str assym), CH(1353.9 cm^{-1}, def sym.), C–H(Ar, 3062 cm^{-1}, str), CH(Ar, 3180 cm^{-1}, str), CH(Ar, 810 cm^{-1}, bend), C=N(1638 cm^{-1}, Str) CH_2(1490 cm^{-1}, str), C–C(Ar, 1680 cm^{-1}), C=C(Ar, 1410, 1500, 1580 cm^{-1}), NO(1380 cm^{-1}, str).

(6) (4-Chlorophenylhyrazinecarboxamide)[1-nitrophenyl-3-(10-spirocyclohexane-1,2,3,9,10,10a-hexahydrobenzo[b]cyclohexane [e][1,4] diazepine-1-yl)-propylidene]-amine (**NBZD-11**):

^1H NMR (300 MHz, δ): CH(m,7.6-7.7, 4H, p-chloroacetophenone), NH(s, 9.0, 1H, =NNH, p-chlorophenylsemicarbazide), NH(s, 6.0, 1H, $-NHC_6H_4Cl$, p-chlorophenylsemicarbazide), CH(m, 7.2–7.6,4H, p-chlorophenylsemicarbazide), CH (m, 6.6–7.1, 4H, phenyl), CH_2(m, 1.3–1.6, 18H, cyclohexane), CH(s, 2.7, 1H, diazepine ring), CH_2(s, 1.6, 2H, $-COCH_2$), CH_2(s, 3.4, 2H, $-NHCH_2$).

IR (KBr): C=N(1559 cm^{-1}, str), C–Cl(787 cm^{-1}, str), C–Cl(769 cm^{-1}), C–H(2975 cm^{-1}, str assym), CH(1353.9 cm^{-1}, def sym.), C–H(Ar, 3062 cm^{-1}, str), CH(Ar, 3180 cm^{-1}, str), CH(Ar, 880 cm^{-1}, bend), C=N(1638 cm^{-1}, Str) CH_2(1490 cm^{-1}, str), C–C(Ar, 1600 cm^{-1}), C=C(Ar, 1410, 1500, 1580 cm^{-1}).

(7) (1-Phenyl-3-(2,2,4-trimethyl-2,3-dihydrobenzo[b][1,4] diazepin-1-yl)-propan-1-one) (**NBZD-13**).

^1H NMR (300 MHz, δ): CH(m, 7.3–7.8, 5H, Acetophenone), CH_2(s, 2.7, 2H, —$COCH_2$), CH_2(s, 3.5, 2H, $-NHCH_2$), CH(m, 6.6–7.1, 4H, phenyl), 2xCH_3(s, 1.28, 6H), CH_3(s, 0.9, 3H), CH_2(s, 2.5, 2H, diazepine ring).

IR (KBr): C=O(1700 cm^{-1}, str), NH(Ar, 3230 cm^{-1}, str), CH(Ar, 3180 cm^{-1}, str), CH(Ar, 800 cm^{-1}, bend), C=N(1618 cm^{-1}, Str), CH_3(2980 cm^{-1}, str), C–C(Ar, 1610 cm^{-1}), C=C(Ar, 1410, 1500, 1580 cm^{-1}).

(8) 1-(4-Chloro-phenyl)-3-(2,2,4-trimethyl-2,3-dihydrobenzo[b][1,4]diazepin-1-yl)-propan-1-one (**NBZD-15**).

^1H NMR (300 MHz, δ): CH(m, 7.3–7.8, 4H, p-chloroacetophenone), CH_2(s, 2.78, 2H, —$COCH_2$), CH_2(s, 3.5, 2H, $-NHCH_2$), CH(m, 6.6–7.1, 4H, phenyl), 2xCH_3(s, 1.2, 6H), CH_3(s, 0.9, 3H), CH_2(s, 2.4, 2H, Diazepine ring).

IR (KBr): C=O(1710 cm^{-1}, str), NH(Ar, 3030 cm^{-1}, str), CH(Ar, 3280 cm^{-1}, str), CH(Ar, 800 cm^{-1}, bend), C=N(1618 cm^{-1}, Str), CH_3(2990 cm^{-1}, str), C–C(Ar, 1600 cm^{-1}), C=C(Ar, 1410, 1500, 1580 cm^{-1}), C–Cl(760 cm^{-1}).

(9) (4-Chloro-phenyl)-[1-(4-nitrophenyl)-3-(2,2,4-trimethyl-2,3-dihydro-benzo[b][1,4] dizepine-1-yl)-propylidene (**NBZD-17**).

^1H NMR (300 MHz, δ): CH(m, 7.8–8.2, 4H, p-nitroacetophenone), CH(m, 7.2-7.3, 4H, p-chloroaniline), CH_2(s, 1.6, 2H, $-N=C-CH_2$), CH_2(s, 3.4, 2H, $-NHCH_2$), CH(m, 6.6–7.1, 4H, phenyl), 2xCH_3(s, 1.2, 6H), CH_3(s, 0.9, 3H), CH_2(s, 2.5, 2H, diazepine ring).

IR (KBr): C=N(1599 cm^{-1}, str), C–Cl(728 cm^{-1}, str), C–H(2985 cm^{-1}, str assym), CH(1353.9 cm^{-1}, def sym.), C–H(Ar, 3062 cm^{-1}, str), CH(Ar, 3180 cm^{-1}, str), CH(Ar, 800 cm^{-1}, bend), C=N(1618 cm^{-1}, Str), CH_3(2990 cm^{-1}, str), C–C(Ar, 1600 cm^{-1}), C=C(Ar, 1410, 1500, 1580 cm^{-1}), NO(1350 cm^{-1}, str)

(10) 4-Chlorophenylhyrazinecarboxamide) [1-nitrophenyl-3-(2,2,4-trimethyl-2,3-dihydrobenzo[b][1,4] diazepine-1-yl)-propylidene]-amine (**NBZD-20**).

^1H NMR (300 MHz, δ): CH(m, 7.9–8.2, 4H, p-nitroacetophenone), NH(s, 9.0, 1H, =NNH, p-chlorophenylsemicarbazide), NH(s, 6.0, 1H, $-NHC_6H_4Cl$, p-chlorophenylsemicarbazide), CH(m, 7.2–7.5,4H, p-chlorophenylsemicarbazide), CH_2(s, 1.6, 2H, $-N=C-CH_2$), CH_2(s, 3.4, 2H, $-NHCH_2$), CH(m, 6.6–7.1, 4H, phenyl), 2xCH_3(s, 1.28, 6H), CH_3(s, 0.9, 3H), CH_2(s, 2.5, 2H, diazepine ring).

IR (KBr): C=N(1570 cm^{-1}, str), C–Cl(730 cm^{-1}, str), C–H(2975 cm^{-1}, str assym), CH(1353.9 cm^{-1}, def sym.), C–H(Ar, 3062 cm^{-1}, str), CH(Ar, 3180 cm^{-1}, str), CH(Ar, 800 cm^{-1}, bend), C=N(1618 cm^{-1}, Str), CH_3(2990 cm^{-1}, str), C–C(Ar, 1600 cm^{-1}), C=C(Ar, 1410, 1500, 1580 cm^{-1}), NO(1350 cm^{-1}, str).

4.4. Anticonvulsant Activity Using Chemical Induced Method. See Tables 3 and 4.

4.5. Sedative Activity. See Table 5.

5. Conclusion

All the synthesized derivatives were evaluated at the dose of 30 mg/kg b.w for anticonvulsant activity by isoniazid induced convulsion model and the compounds **NBZD-3** and **NBZD-8** were found to be most active among all compounds. Among all the synthesized derivatives, compounds **NBZD-13**, and **NBZD-17** were found to be most active among all compounds using thiosemicarbazide induced model. Activity of the drugs interfering with motor coordination was checked by the rotarod test. None of the synthesized compounds were found to be neurotoxic at a dose of 30 mg/kg b.w among all the tested compounds. The compounds **NBZD-1**, **NBZD-3**, **NBZD-4**, **NBZD-7**, **NBZD-11**, **NBZD-12**, **NBZD-14**, **NBZD-15**, **NBZD-17**, **NBZD-20**, **NBZD-21** were found to cause sedation. Although **NBZD-8**, **NBZD-10**, and **NBZD-18** are the compounds which had shown good anticonvulsant activity and have an advantage over that, they were not sedative.

Sulfated zirconia is one of the important agent that has attracted much attention recently because of its superacidity, nontoxicity, and low cost. Sulfated zirconia catalyzes many reactions under very mild condition in vapor as well as liquid phase.

6. Structure Activity Relationship

6.1. Anticonvulsant Scheme 9

(1) Highly active: **NBZD-3**, **NBZD-8**, **NBZD-13**, and **NBZD-18**.

(2) Moderately active: **NBZD-1**, **NBZD-6**, **NBZD-10**, **NBZD-11**, **NBZD-12**, and **NBZD-16**, **NBZD-21**.

(3) Less active: **NBZD-4**, **NBZD-7**, **NBZD-9**, **NBZD-14**, **NBZD-15**, **NBZD-17**, **NBZD-19**, and **NBZD- 20**.

Most functional subtypes of the $GABA_A$ receptor contain α, β, subunits, with the different benzodiazepine binding site ligands. BZ-binding site ligands act through mechanisms which modulate the inhibiting effects of GABA.

(1) In the basic structure of benzodiazepine, early SAR studies indicated that the seven-membered imino ring B was essential for its affinity towards the BZ-binding site.

(2) 4-5 Carbimino double bond has also been shown to substantially contribute to the binding affinity of compound. Saturation leads to complete less of activity. It acts as a two-electron donor site.

(3) The primary chemical moieties of the compounds which contribute to high receptor binding affinity are restricted to positions 7, 2, 1.

  (a) Position 2 is the most effective place. Presence of an electrophilic and bulky substituent at position 2 results in strong increase in receptor binding affinity of the corresponding compounds.

  (b) Compounds **NBZD-3** and **NBZD-8** had shown good anticonvulsant activity as they have cyclohexane ring at position 2.

(4) Molar refractivity is the most important parameter at position 1, suggesting that the molecular size of the substituent needs to be restricted at position 1 for effective ligand binding. Compounds **NBZD-1** and **NBZD-8** have less substituent as compared to other and hence more active.

(5) Compounds **NBZD-6**, **NBZD-10**, and **NBZD-11** were found to be moderately active anticonvulsant action, hence this shows that chloro-substituted derivatives are rather good anticonvulsant agent as compared to nitro-substituted derivatives.

(6) Among these synthesized compounds which have methyl groups at the 2nd and 4th position, **NBZD-13**, and **NBZD-18** were found to be most active. Hence it shows that good activity compounds are preferred with less substitution the at 1st position.

Acknowledgments

The authors are thankful to Central Drug Research Institute, Lucknow, India. They express their thanks to the Head and the Department of Pharmaceutical Sciences, Saroj Institute of Technology and Management, Lucknow, India, for their assistance during the course of study.

References

[1] C. Page, C. Michael, M. Sutter, M. Walker, and B. Hoffman, *Integrated Pharmacology*, Mosby, 2nd edition, 2005.

[2] G. A. Archer and L. H. Sternbach, "The chemistry of benzodiazepines," *Chemical Reviews*, vol. 68, no. 6, pp. 747–784, 1968.

[3] R. H. Smith Jr., W. L. Jorgensen, J. Tirado-Rives et al., "Prediction of binding affinities for TIBO inhibitors of HIV-1 reverse transcriptase using Monte Carlo simulations in a linear response method," *Journal of Medicinal Chemistry*, vol. 41, no. 26, pp. 5272–5286, 1998.

[4] M. A. Rogowaski and R. J. Porter, "Antiepileptic drugs: pharmacological mechanisms and clinical efficacy with consideration of promising developmental stage compounds," *Pharmacological Reviews*, vol. 42, no. 3, pp. 223–286, 1990.

[5] R. Fisher and D. Blum, "Clobazam, oxcarbazepine, tiagabine, topiramate, and other new antiepileptic drugs," *Epilepsia*, vol. 36, no. 2, pp. S105–S114, 1995.

[6] L. Steru, R. Chermat, and B. Millet, "Comparative study in mice of ten 1,4-benzodiazepines and of clobazam: anticonvulsant, anxiolytic, sedative, and myorelaxant effects," *Epilepsia*, vol. 27, no. 1, pp. S14–S17, 1986.

[7] A. K. Shenoy, J. T. Miyahara, E. A. Swinyard, and H. J. Kuperberg, "Comparative anticonvulsant activity and neurotoxicity of clobazam, diazepam, phenobarbital, and valproate in mice and rats," *Epilepsia*, vol. 23, no. 4, pp. 399–408, 1982.

[8] K. P. Jadhav and D. B. Ingled, "Synthis of 2,4-diaryl-2,3-dihyrdo-1,5-benzothiazeoines and thier 1,1-dioxides as antibacterial agents," *Indian Journal of Chemistry*, vol. B22, p. 180, 1983.

[9] R. J. Reddy, D. Ashok, and P. N. Sharma, "Synthesis of 4.6-bis(2′-substitutcri-2-3′-dihydro-1.5- benzothiazepin-4′-yl)resorcinols as potential antifeedants," *Indian Journal of Chemistry B*, vol. 32, pp. 404–406, 1993.

[10] K. Satyanarayana and M. N. A. Rao, "Synthesis of 3-[4-[2t 3-dihydro-2-(substituted aryl)-1,5- benzothiazcpin-4-yl]phcnyl]sydnones as potential antiinflammatory agents," *Indian Journal of Pharmaceutical Sciences*, vol. 55, pp. 230–233, 1993.

[11] G. Dessarro, A. Chimirri, A. Dessaro, and R. Gitto, *Chemical Abstracts*, vol. 106, p. 11222g, 1995.

[12] J. K. Landquist, "Application as pharmaceuticals," in *Comprehensive Heterocyclic Chemistry*, A. R. Kalritzky and C. W. Rees, Eds., vol. 1, pp. 143–183, Peramon, Oxford, UK, 1984.

[13] L. H. Sternbach, "The benzodiazepine story," *Journal of Medicinal Chemistry*, vol. 22, no. 1, pp. 1–7, 1979.

[14] P. Danneberg and K. H. Weber, "Chemical structure and biological activity of the diazepines," *British Journal of Clinical Pharmacology*, vol. 16, no. 2, pp. 231S–243S, 1983.

[15] A. Walser and R. I. Fryer, *Chemical Abstracts*, vol. 89, p. 59906, 1978.

[16] T. Hashiyama, H. Inoue, M. Takeda, S. Murtates, and T. Nagao, "Reactions of 3-phenylglycidic esters. V. Reaction of methyl 3-(4-methoxyphenyl) glycidate with 2-nitroaniline and synthesis of 1,5-benzodiazepine derivatives," *Chemical & Pharmaceutical Bulletin*, vol. 33, pp. 2348–2358, 1985.

[17] J. K. Chakrabarti, L. Horsman, T. M. Hotten, I. A. Pullar, D. E. Tupper, and F. C. Wright, "4-Piperazinyl-10H-thieno[2,3-b][1,5]benzodiazepines as potential neuroleptics," *Journal of Medicinal Chemistry*, vol. 23, no. 8, pp. 878–884, 1980.

[18] S. Caccia, G. Grossi, C. Fracasso et al., "Acid-catalysed hydrolysis and benzodiazepine-like properties of 5-(dialkylamino)- and 5-(alkylthio)-substituted 8-chloro-6-phenyl-6H-[1,2,4]triazolo[4,3-a][1,5]benzodiazepines in mice," *Journal of Pharmacy and Pharmacology*, vol. 50, no. 7, pp. 723–728, 1998.

[19] A. Bauer, K. H. Weber, and P. Danneberg, *Chemical Abstracts*, vol. 82, p. 4331, 1975.

[20] R. Unny, P. Joshi, M. P. Dobhal, and Y. C. Joshi, "Synthesis of 1,3-diketone and its reaction with different *N*-nucleophiles (part II)," *Heterocyclic Communications*, vol. 9, no. 2, pp. 171–174, 2003.

[21] N. Fokialakis, P. Magiatis, I. Chinou, S. Mitaku, and F. Tillequin, "Megistoquinones I and II, two quinoline alkaloids with antibacterial activity from the bark of Sarcomelicope megistophylla," *Chemical and Pharmaceutical Bulletin*, vol. 50, no. 3, pp. 413–414, 2002.
[22] R. Pili, *Proceedings of the American Association for Cancer Research*, vol. 43, p. 5363, 2002.
[23] N. K. Raju, T. S. Seniappan, S. Thangaraju, and S. M. Palathurai, *Heterocycles*, vol. 52, no. 2, p. 357, 2000.

Synthesis and Anticancer Properties of Silver(I) Complexes Containing 2,6-Bis(substituted)pyridine Derivatives

Korany A. Ali,[1] Mokhles M. Abd-Elzaher,[2] and Khaled Mahmoud[3]

[1] *Applied Organic Chemistry Department, National Research Centre, Dokki, Cairo 12622, Egypt*
[2] *Inorganic Chemistry Department, National Research Centre, Dokki, Cairo 12622, Egypt*
[3] *Department of Pharmacognosy, National Research Centre, Dokki, Cairo 12622, Egypt*

Correspondence should be addressed to Korany A. Ali; kornykhlil@yahoo.com

Academic Editor: Feng Liang

Several new 2,6-bis(substituted)pyridine ligands and 2,6-bis(substituted)pyridine Ag(I) nitrate complexes were synthesized and characterized spectroscopically. The newly synthesized ligands include pyridine-2,6-bis(3-oxopropanenitrile) (**1**), pyridine-2,6-bis(2-cyano-*N*-phenyl-3-oxopropanethioamide) (**2**), and pyridine-2,6-bis((*E*)-2-(2-phenylhydrazono)-3-oxopropanenitrile) (**3**). The newly synthesized ligands and silver(I) complexes were evaluated for their *in vitro* anticancer activity against four human cancer cell lines including hepatocellular carcinoma (HePG2), lung adenocarcinoma (A549), colon carcinoma (HT29), and breast adenocarcinoma (MCF7). Most of the newly synthesized silver(I) complexes exhibited better activity than the ligands, and the results have been compared with doxorubicin as a reference drug.

1. Introduction

Chemical and biological activity of the organic complexes is not possessed only by the metal or organic ligand but also this activity can be fine-tuned by subtle changes in the electronic and steric properties of the complexes or by variation of the oxidation state of the metal. These features provide a versatile platform for drug design that is now being exploited in several areas. For centuries, silver compounds have been known to possess interesting biological properties that show potent antibacterial properties [1–4]. Also, they were popular remedies for tetanus and rheumatism in the 19th century and for colds and gonorrhea before the advent of antibiotics in the early part of the 20th century [5]. Additionally, silver compounds used for treating mental illness, epilepsy, and nicotine addiction [6, 7]. Furthermore, silver compounds have reemerged as a viable treatment option for infections encountered in burns, open wounds, and chronic ulcers [8–11]. On the other hand, functionalized pyridine derivatives are gaining a great deal of interest in medicinal and organic synthesis, where some of pyridine derivatives are used as bactericides [12], fungicides [13], and anticancer agents [14–17]. In view of these observations and in continuation of our current interest in the synthesis of organic compounds for biological evaluations [18–25] and our interest in the chemistry of 2,6-disubstituted pyridine derivatives [26–29], we described herein a facile synthesis of novel silver complex with some of the newly synthesized 2,6-disubstituted pyridine ligands. The newly synthesized compounds were evaluated for their *in vitro* anticancer activity against four human cancer cell lines including hepatocellular carcinoma (HePG2), lung adenocarcinoma (A549), colon carcinoma (HT29), and breast adenocarcinoma (MCF7).

2. Results and Discussion

2.1. Synthesis and Characterization of 2,6-bis(substituted) pyridine Ligands. Three ligands have been prepared, pyridine-2,6-bis(3-oxopropanenitrile) (**1**), pyridine-2,6-bis(2-cyano-*N*-phenyl-3-oxopropanethioamide) (**2**), and pyridine-2,6-bis((*E*)-2-(2-phenylhydrazono)-3-oxopropanenitrile) (**3**).

Structures of the synthesized ligands were confirmed on the basis of their elemental analysis and spectral data. The elemental analyses of the prepared ligands are consistent with the calculated results from the empirical formula of each one. Ligand **1** was prepared using ethyl 2,6-pyridine dicarboxylate and acetonitrile in dry THF and in presence of sodium hydride [26]. The IR spectrum of ligand **1** revealed two bands at 2264 and 1720 cm^{-1} which were assigned to CN and C=O groups, respectively. The band that appeared at 1566 cm^{-1} was assigned to C=N of the pyridine moiety. ^{1}H NMR of **1** was carried out using DMSO-d_6 as a solvent. The signal of the two CH_2 groups was appeared at 3.74 ppm, whereas the signals of the pyridine moiety were appeared at 8.05–8.31 ppm.

To increase solubility and hydrogen bonds capabilities of the synthesized ligands we have used pyridine-2,6-bis(3-oxopropanenitrile) (**1**) as starting material for the preparation of Ligand **2** and Ligand **3**. Ligand **2** was prepared as brown powder in a good yield using pyridine-2,6-bis(3-oxopropanenitrile) (**1**) and phenyl isothiocyanate in DMF in the presence of KOH with stirring for 6 h (Scheme 1).

Four essential bands were found in the IR spectra of Ligand **2** at 3191, 2210, 1655, and 1555 cm^{-1}. These bands were assigned to (NH), (CN), (C=O), and (C=N in pyridine), respectively. ^{1}H NMR was carried out using DMSO-d_6 at room temperature. Signal of the two CH group was appeared at 3.60 ppm, whereas the aromatic and the pyridine protons were appeared in the expected region at 7.25–7.43 and 7.74–8.02 ppm, respectively, in addition to singlet signal at 12.02 ppm corresponding to 2NH protons. Elemental analysis, ^{13}C NMR, and also mass spectra of **2** were consistent with the proposed structure.

Compound **1** was coupled with benzene diazonium salt to yield coupling product pyridine-2,6-bis((*E*)-2-(2-phenylhydrazono)-3-oxopropanenitrile) (**3**) (Scheme 1).

Characterization of ligand **3** was confirmed on the basis of its elemental analysis and spectral data. For example, four bands in IR spectrum at 3187, 2213, 1674, and 1549 cm^{-1} correspond to NH, CN, C=O, and C=N group, respectively. Its ^{1}H NMR spectrum revealed signals at the range 7.10–7.28 ppm corresponding to the phenyl protons and multiplet at 7.97–8.20 ppm for pyridine protons in addition to singlet signal at 12.63 ppm corresponding to 2NH protons. Mass spectrum and elemental analysis confirmed also the structure of **3**. All the ligands were isolated in a good yield, and they are stable in air and light.

2.2. Synthesis and Characterization of 2,6-bis(substituted) pyridine Ag(I) Nitrate Complexes. In general, all of the ligands have good ability to form complexes due to the presence of several donor atoms in each ligand. So, preparations of the complexes were carried out easily and gave good yields. Formations of the complexes were observed from the change in colour of the ligand in the mixture as well as from the precipitation of the complexes (in the most cases). The silver complex **4** was prepared as yellow solid in good yield from the reaction of ligand **1** with silver nitrate (1 : 1 molar ratio), in ethanol, at room temperature (Scheme 2).

The silver complex **5** was prepared using ligand **2** with silver nitrate (1 : 1 molar ratio), in methanol, under reflux condition (Scheme 3). The yield was good, and the colour of the prepared complex was brown.

Refluxing of ligand **3** with $AgNO_3$ in methanol affords yellow precipitate of complex **6** in good yield (Scheme 4).

Characterization of the synthesized complexes was carried out using IR, NMR, and elemental analysis. All the complexes are stable in air and light.

The elemental analyses data of the complexes are consistent with the calculated results from the empirical formula of each compound. The IR spectra of the metal complexes were recorded in KBr.

It was found that C=O frequency of each of the complexes (**4–6**) was shifted to lower values compared with that of the ligands (in the range 1625–1609 cm^{-1}) indicating the participation of the C=O group in a coordinated bond with the metal ions. Confirmation of this coordinated bond was observed in the IR spectra; it was found that a new band in the complexes appeared at 460–561 cm^{-1}; this band was not found in the free ligand and was assigned to C=O–Ag bond. Also, the band at approx. 1565 cm^{-1} which was assigned to the frequency of the C=N in the pyridine moiety was slightly shifted to lower frequencies (11–52 cm^{-1}) in the spectra of the metal complexes. This shift suggests that the coordination takes place also through the pyridine nitrogen to the metal ion. These results were confirmed also from the presence of a new band at about 440 cm^{-1} in the complexes which was assigned to the Ag–N bond. This band was observed only in the spectra of the metal complexes and not in the spectra of the ligands, thus confirming the participation of the nitrogen atom in the formed complex [30–33].

The ^{1}H NMR spectra of the complexes **4–6** were recorded at room temperature using DMSO-d_6 as a solvent. The spectra of the complexes showed nearly the same signals appeared in the spectra of ligands with slightly downfield shift, which may be due to complex formation of the C=O and the pyridinyl nitrogen with metal ions.

2.3. In Vitro Anticancer Screening. In vitro anticancer activity evaluation of the newly synthesized compounds was carried out against four human cancer cell lines including hepatocellular carcinoma (HePG2), lung adenocarcinoma (A549), colon carcinoma (HT29), and breast adenocarcinoma (MCF7) using MTT method [34]. Doxorubicin HCl is one of the most effective anticancer agents was used as a reference drug in this study. The relationship between drug concentrations and cell viability was plotted to calculate IC_{50} (μM) the value which corresponds to the concentration required for 50% inhibition of cell viability. Table 1 shows the *in vitro* cytotoxic activity of the newly synthesized compounds against the tested cancer cell lines; where some compounds revealed significant activity compared to doxorubicin. From Table 1, it was observed that most of the synthesized ligands showed low to moderate activity, while most of the prepared silver complexes exhibit excellent cytotoxic activity against tested cancer cell lines.

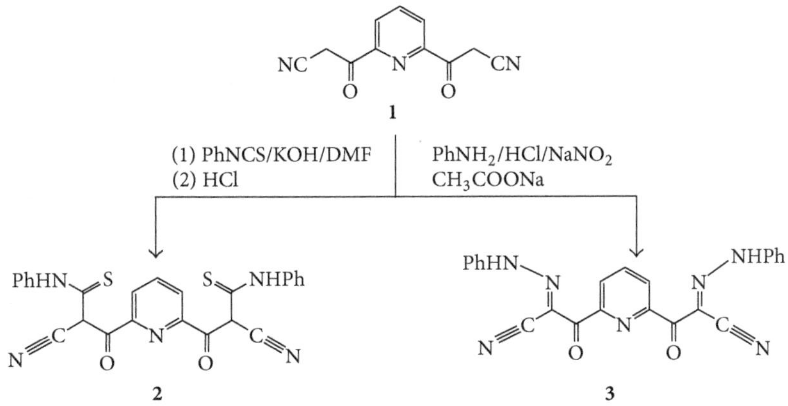

SCHEME 1: Synthetic pathway for the formation of ligands **2** and **3**.

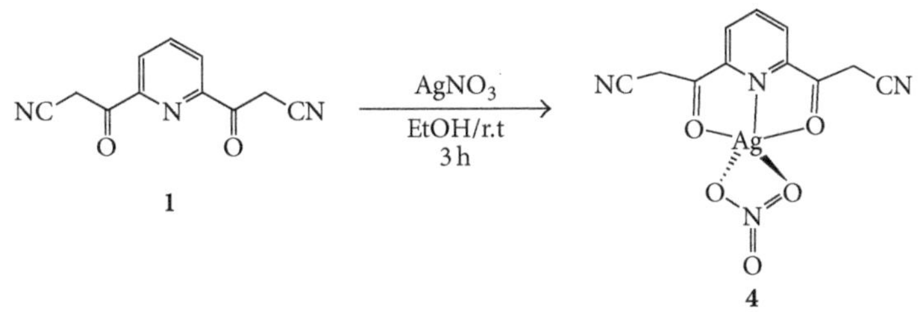

SCHEME 2: Synthetic pathway for the formation of complex **4**.

TABLE 1: Anticancer screening of the newly synthesized ligands and Ag(I) nitrate complexes against the tested human cancer cell lines.

Compound	Cytotoxicity[a,b] (IC_{50}) (μM)			
	HEPG2	A549	HT29	MCF7
1	45.11	NA	NA	NA
2	NA	NA	NA	NA
3	NA	34.55	49.82	29.5
4	1.52	1.41	1.75	3.67
5	6.12	1.08	8.85	10.72
6	1.62	3.71	52.65	3.66
Doxorubicin	4.61	3.51	5.22	2.78

[a] IC_{50}, compound concentration required to inhibit tumor cell proliferation by 50%.
[b] The same experiment is carried out in triplicate.
Sample concentration range (100—0.78 μM) using MTT assay.
NA: compounds having IC_{50} more than 60 μM.

In HePG2 silver complexes **4** (IC_{50} = 1.52 μM) and **6** (IC_{50} = 1.62 μM) revealed significant activity compared to doxorubicin (IC_{50} = 0.4.61 μM). In A549 silver complexes **4** (IC_{50} = 1.41 μM), **5** (IC_{50} = 1.08 μM), and **6** (IC_{50} = 3.71 μM) revealed significant activity compared to doxorubicin (IC_{50} = 3.51 μM). In HT29 silver complex **4** (IC_{50} = 1.75 μM) revealed significant activity compared to doxorubicin (IC_{50} = 5.22 μM). In MCF7 silver complexes **4** (IC_{50} = 3.67 μM) and **6** (IC_{50} = 3.66 μM) revealed moderate activity compared to doxorubicin (IC_{50} = 2.78 μM).

3. Conclusions

In the present study we have described the synthesis and characterizations of new silver complexes of 2,6-disubstituted pyridine ligands. The synthesized ligands and complexes were tested with doxorubicin on four human cancer cell lines including hepatocellular carcinoma (HePG2), lung adenocarcinoma (A549), colon carcinoma (HT29), and breast adenocarcinoma (MCF7) using MTT method. The entire synthesized complex displayed significant activity more than the corresponding ligands. While most of the prepared silver complexes exhibit excellent cytotoxic activity against tested cancer cell lines more than the reference drug. In the near future we will conduct some studies related to the mechanism of actions of the synthesized Ag(I) complexes, for example, assays against potential targets, particularly DNA: DNA binding properties of the proposed complexes by viscosity measurements, electronic absorption, fluorescence, and/or DNA electrophoresis.

4. Experimental

4.1. Chemistry. General: all melting points were measured on a Gallenkamp melting point apparatus. The infrared spectra were recorded in potassium bromide discs on a Pye Unicam SP 3-300 and Shimadzu FT IR 8101 PC infrared spectrophotometers. The NMR spectra were recorded on a Varian Mercury VXR-300 NMR spectrometer. ^{1}H NMR (300 MHz) and ^{13}C NMR (75.46 MHz) were run in deuterated dimethyl sulfoxide (DMSO-d_6). Chemical shifts were related to that

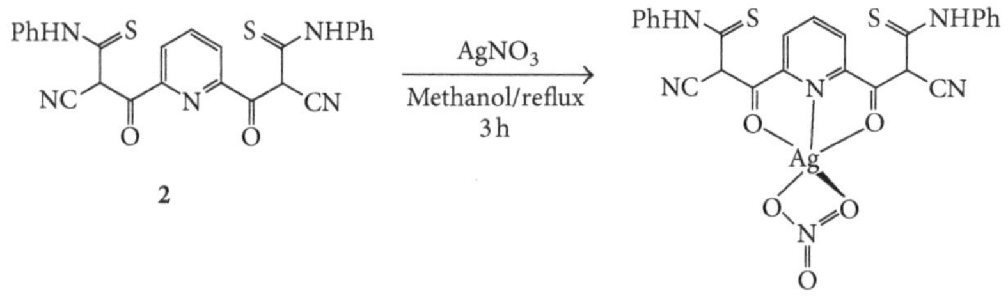

5

SCHEME 3: Synthetic pathway for the formation of complex **5**.

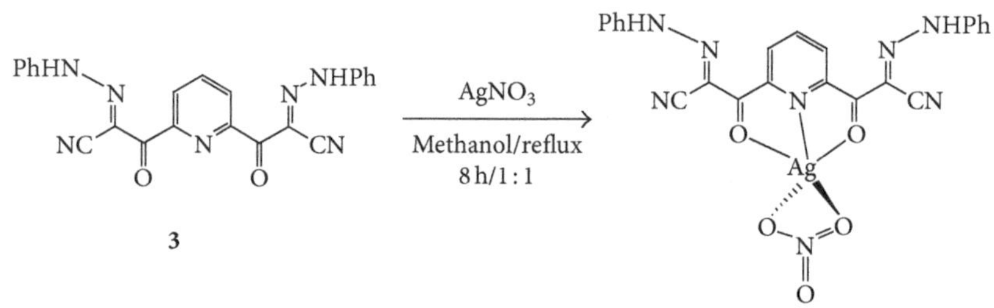

6

SCHEME 4: Synthetic pathway for the formation of complex **6**.

of the solvent. Mass spectra (EI, 70 ev) were recorded on a Shimadzu GCMS-QP1000 mass spectrometer. Elemental analyses were carried out at the Microanalytical Centre of Cairo University, Giza, Egypt, and recorded on Elementar-Vario EL (Germany) automatic analyzer. All reactions were followed by TLC (Silica gel, Aluminium Sheets 60 F_{254}, Merck).

4.2. Preparation of 2,6-bis(substituted)pyridine Ligands

*4.2.1. Pyridine-2,6-bis(3-oxopropanenitrile) (**1**).* To a mixture of ethyl pyridine-2,6-dicarboxylate (4.45 g, 20 mmol) and acetonitrile (2.70 mL, 50 mmol), in dry THF (5 mL), sodium hydride (2 g, 50%) was added. The reaction mixture was refluxed for 2 h and then allowed to cool. The precipitated product was filtered off, washed with diethyl ether, and dried. The crude product was dissolved in water (30 mL), and the resulting alkaline solution was treated with concentrated hydrochloric acid until it becomes slightly acidic (pH 5). The precipitated solid product was filtered off, washed with water, dried, and finally recrystallized from MeOH to afford brown crystals of pyridine-2,6-bis(3-oxopropanenitrile) (**1**) [26]. Yield (3.41 g, 80%); mp: 185-186°C; IR (KBr pellet, cm^{-1}): 2264, 1720, 1566, 1339, 1206, 1024, 927, 835. ^1H NMR (300 MHz, DMSO-d_6): δ 3.74 (s, 4H, $2CH_2$), 8.05–8.31 (m, 3H, pyridine). ^{13}C NMR (75 MHz, DMSO-d_6): δ 29.6, 116.4, 123.7, 126.4, 165.7, 190.7. MS (*m/z*, %): 213 [M^+] (75), 145 (100). Anal. Calcd. for $C_{11}H_7N_3O_2$ (213.19): C, 61.97; H, 3.31; N, 19.71. Found: C, 61.94; H, 3.34; N, 19.75.

*4.2.2. Pyridine-2,6-bis(2-cyano-N-phenyl-3-oxopropanethioamide) (**2**).* To a stirred solution of potassium hydroxide (1.12 g, 20 mmol) in DMF (20 mL) pyridine-2,6-bis(3-oxopropanenitrile) (**1**) (2.13 g, 10 mmol) was added. The mixture was stirred for 30 min, and then phenyl isothiocyanate (2.70 g, 20 mmol) was added. Stirring was continued for 6 h, and then the mixture was poured over crushed ice containing hydrochloric acid (2 N, 10 mL). The obtained solid product was filtered off, washed with water, dried, and finally recrystallized from ethanol to afford brown powder of pyridine-2,6-bis(2-cyano-*N*-phenyl-3-oxopropanethioamide) (**2**). Yield (0.41 g, 85%); mp: 114-115°C. IR (KBr pellet, cm^{-1}): 3191, 2210, 1655, 1560 1468, 1380, 760, 695, 557. ^1H NMR (300 MHz, DMSO-d_6): δ 3.60 (s, 2H, 2CH), 7.25–7.43 (m, 10H, Ar-H), 7.74–8.02 (m, 3H, pyridine-H), 12.02 (s, 2H, 2NH, D_2O-exchangeable). ^{13}C NMR (75 MHz, DMSO-d_6): δ 57.21, 118.70, 119.24, 123.50, 124.17, 124.95, 128.96, 129.23, 139.98, 150.21, 180.14. MS (*m/z*, %): 483 [M^+] (17), 317 (30), 197 (14), 144 (19), 105 (55), 78 (45), 60 (100). Anal. Calcd. for $C_{25}H_{17}N_5S_2O_2$ (483.56): C, 62.09; H, 3.54; N, 14.48. Found: C, 62.16; H, 3.48; N, 14.53.

*4.2.3. Pyridine-2,6-bis((E)-2-(2-phenylhydrazono)-3-oxopropanenitrile) (**3**).* To a stirred solution of pyridine-2,6-bis(3-oxopropanenitrile) (**1**) (2.13 g, 10 mmol) in EtOH (100 mL), sodium acetate trihydrate (3 g) was added. After stirring for 10 min, the mixture was cooled to −4°C and treated with aniline diazonium salt solution (prepared by diazotizing aniline (10 mmol) in hydrochloric acid (6 N, 6 mL) with sodium nitrite solution (0.7 g, 10 mmol), in 5 mL water). Addition of the diazonium salt was carried out with rapid stirring over a period of 30 min. The reaction mixture was stirred for further 2 h at −4°C and then was left for 6 h, in a refrigerator. The resulting solid was collected by filtration, washed thoroughly with water, dried, and finally recrystallized from EtOH/dioxane to give pyridine-2,6-bis((*E*)-2-(2-phenylhydrazono)-3-oxopropanenitrile) (**3**). Yield (3.5 g, 83%, yellow crystals); mp: 221–223°C. IR (KBr pellet, cm^{-1}):

3187, 2213, 1674, 1549, 1478, 1430 1357, 1266, 916, 755, 683 cm^{-1}. ^1H NMR (300 MHz, DMSO-d_6): δ 7.10–7.28 (m, 10H, ArH's), 7.97–8.20 (m, 3H, pyridine-H's) 12.63 (s, 2H, 2NH, D_2O-exchangeable). ^{13}C NMR (75 MHz, DMSO-d_6): δ 115.11, 120.11, 125.11, 126.14, 138.05, 142.31, 150.4, 154.12, 195.17. MS *m/z* (%): 421 [M^+] (13), 315 (46), 212 (30), 167 (27), 105 (30), 92 (25), 77 (86), 51 (100). Anal. Calcd. for $C_{23}H_{15}N_7O_2$ (421.41): C, 65.55; H, 3.59; N, 23.27. Found: C, 65.47; H, 3.64; N, 23.21.

4.3. Synthesis of 2,6-bis(substituted)pyridine Ag(I) Nitrate Complexes

4.3.1. [$AgL_1(NO_3)$] (4). A mixture of $AgNO_3$ (0.17 g, 1 mmol) and pyridine-2,6-bis(3-oxopropanenitrile) (**1**) (0.21 g, 1 mmol) in EtOH (20 mL) was stirred at room temperature for 3 h. The precipitated product was filtered off, washed with water, dried, and finally recrystallized from EtOH/dioxane to afford yellow powder of complex **4**.

Yield (0.27 g, 70%); IR (KBr pellet, cm^{-1}): 3430, 2165, 1611, 1571, 1517, 1375, 561, 460. ^1H NMR (300 MHz, DMSO-d_6): δ 4.21 (s, 4H, $2CH_2$), 7.95–8.15 (m, 3H, pyridine-H). ^{13}C NMR (75 MHz, DMSO-d_6): δ 28.2, 115.4, 125.7, 126.3, 155.7, 181.7 Anal. Calcd. for $C_{11}H_7AgN_4O_5$ (387.07): C, 34.49; H, 1.84; N, 14.63. Found: C, 35.01; H, 2.01; N, 14.43.

4.3.2. [$AgL_2(NO_3)$] (5). A mixture of $AgNO_3$ (0.17 g, 1 mmol) and pyridine-2,6-bis(2-cyano-*N*-phenyl-3-oxopropanethioamide) (**2**) in MeOH was refluxed for 4 h. The resultant solution was allowed to evaporate slowly at room temperature for several days to give yellow powder of complex **5**. The precipitated product was filtered off, washed with water/MeOH, dried, and finally recrystallized from EtOH/DMF to afford yellow powder of complex **5**. Yield (0.51 g, 78%); IR (KBr pellet, cm^{-1}): 3386, 2198, 1621, 1553, 1473, 1321, 754, 693, 561. ^1H NMR (300 MHz, DMSO-d_6): δ 7.25–8.16 (m, 13H, Ar-H, pyridine-H), 12.11 (s, 2H, 2NH, D_2O-exchangeable). ^{13}C NMR (75 MHz, DMSO d_6): δ 116.04, 121.31, 125.23, 125.81, 138.11, 141.33, 150.4, 160.17, 185.85. Anal. Calcd for $C_{25}H_{17}AgN_6O_5S_2$ (653.44): C, 45.95; H, 2.62; N, 12.86. Found: C, 46.12; H, 2.75; N, 12.45.

4.3.3. [$AgL_3(NO)$] (6). To a stirred solution of $AgNO_3$ (0.17 g, 1 mmol) in EtOH (20 mL), pyridine-2,6-bis((*E*)-2-(2-phenylhydrazono)-3-oxopropanenitrile) (**3**) (0.42 g, 1 mmol) was added. The reaction mixture was refluxed for 8 h and then allowed to cool. The formed solid product was filtered off, washed with ethanol/water, dried, and recrystallized from EtOH/dioxane to afford complex **6**. Yield (0.41 g, 70%); IR (KBr pellet, cm^{-1}): 3430 (NH), 2167 (CN), 1707 (C=O), 1512 (C=N pyridine moiety), 460 (C=O–Ag), 441 (Ag–N). ^1H NMR (300 MHz, DMSO-d_6): δ 7.22–7.47 (m, 10H, ArH's), 8.21–8.42 (m, 3H, pyridine-H), 12.10 (br s, 2H, 2NH, D_2O-exchangeable). ^{13}C NMR (75 MHz, DMSO-d_6): δ 57.21, 117.15, 122.25, 125.50, 126.17, 127.95, 128.86, 129.52, 138.45, 150.52, 194.14.Anal. Calcd for $C_{23}H_{15}AgN_8O_5$ (591.28): C, 46.72; H, 2.56; N, 18.95. Found: C, 46.52; H, 2.36; N, 18.65.

4.4. In Vitro Assay for Anticancer Activity. The synthesized compounds were supplied to the Bioassay-Cell Culture Laboratory, National Research Centre, Cairo, Egypt, for *in vitro* antitumor screening on Lung adenocarcinoma (A549), hepatocellular carcinoma (HePG2), colon carcinoma (HT29), and caucasian breast adenocarcinoma (MCF7) (American Type Culture Collection). Cell viability was assessed by the mitochondrial-dependent reduction of yellow MTT (3-(4,5-dimethylthiazol-2-yl)-2,5-diphenyl tetrazolium bromide) to purple formazan [34, 35].

Procedure. All the following procedures were done in a sterile area using a Laminar flow cabinet biosafety class II level (Baker, SG403INT, Sanford, ME, USA). HePG2 cell line was cultured in RPMI-1640, and MCF7 cell line was cultured in DMEM. Cells were plated in 96-well plates (having about 10000 cells/well). The plates are then incubated for 24 h in 37°C incubation and 5% CO_2 atmosphere before treatment with the compounds to allow attachment of cell to the wall of the plate. Tested compounds were dissolved in DMSO and different concentrations of the compounds under test were added to the cell monolayer. Triplicate wells were prepared for each individual concentration. Then the plate was incubated for 48 h in 37°C incubator. After the completion of compound exposure, 40 μL of MTT solution (2.5 mg/mL) was added into each well for an additional 4 h. Formazan was dissolved in 200 μL (10%) sodium dodecyl sulphate, and the absorbance at $\lambda = 495$ nm was measured. The concentration of DMSO as a solvent for the different compounds was 0.1% in the culture medium used and was without any effect on cell growth. Cell viability at given compound concentration was calculated as the percentage of absorbance in wells with the compound-treated cells to that of vehicle control cells (100%).

Acknowledgment

Financial support for this work was provided by the program of the Science and Technology Development Fund (STDF), Egypt (Project no. 170). The program is financed through the Ministry of State for Scientific Research and is administered by Higher Council for Science and Technology.

References

[1] J. B. Wright, D. L. Hansen, and R. E. Burrell, "The comparative efficacy of two antimicrobial barrier dressings: *in vitro* examination of two controlled release of silver dressings," *Wounds*, vol. 10, no. 6, pp. 179–188, 1998.

[2] J. W. Richard, B. A. Spencer, L. F. McCoy et al., "Acticoat versus Silverlon: the truth," *Journal of Burns and Surgical Wound Care*, vol. 1, no. 1, pp. 11–19, 2002.

[3] A. D. Russell and W. B. Hugo, "7 antimicrobial activity and action of silver," *Progress in Medicinal Chemistry*, vol. 31, pp. 351–370, 1994.

[4] F. Fu-Ren and J. F. Allen, "Chemical, electrochemical, gravimetric, and microscopic studies on antimicrobial silver films," *The Journal of Physical Chemistry B*, vol. 106, no. 2, pp. 279–287, 2002.

[5] S. M. Mirsattari, R. R. Hammond, M. D. Sharpe, F. Y. Leung, and G. B. Young, "Myoclonic status epilepticus following repeated oral ingestion of colloidal silver," *Neurology*, vol. 62, no. 8, pp. 1408–1410, 2004.

[6] M. R. Alidaee, A. Taheri, P. Mansoori, and S. Z. Ghodsi, "Silver nitrate cautery in aphthous stomatitis: a randomized controlled trial," *British Journal of Dermatology*, vol. 153, no. 3, pp. 521–525, 2005.

[7] F. Tanweer and J. Hanif, "Re: silver nitrate cauterisation, does concentration matter?" *Clinical Otolaryngology*, vol. 33, no. 5, pp. 503–504, 2008.

[8] B. G. A. Lansdown, "The role of silver.," *European Tissue Repair Society Bulletin*. In press.

[9] J. B. Wright, K. Lam, A. G. Buret, M. E. Olson, and R. E. Burrell, "Early healing events in a porcine model of contaminated wounds: effects of nanocrystalline silver on matrix metalloproteinases, cell apoptosis, and healing," *Wound Repair Regeneration*, vol. 10, no. 3, pp. 141–151, 2002.

[10] H. N. Paddock, G. S. Schultz, and K. J. Perrin, "Clinical assessment of silver-coated antimicrobial dressing on MMPs and cytokine levels in non-healing wounds," in *Proceedings of the Annual Meeting of the Wound Healing Society*, Baltimore, Md, USA, July 2002.

[11] H. Wu and H., "Synthesis, structure, DNA-binding properties and antioxidant activity of silver(I) complexes containing V-shaped bis-benzimidazole ligands," *Dalton Transactions*, vol. 41, no. 29, pp. 8829–8838, 2012.

[12] G. Malicorne, J. Bompart, L. Giral, and E. Despaux, "Synthesis and antibacterial activity of 4,7-dihydro-4-ethyl-7-oxothieno(3,2-*b*)pyridine-6-carboxylic acids," *European Journal of Medicinal Chemistry*, vol. 26, no. 1, pp. 3–11, 1991.

[13] S. L. Hargreaves, B. L. Pilkington, S. E. Russell, and P. A. Worthington, "The synthesis of substituted pyridylpyrimidine fungicides using palladium-catalysed cross-coupling reactions," *Tetrahedron Letters*, vol. 41, no. 10, pp. 1653–1656, 2000.

[14] E. Szoko and T. Tabi, "Analysis of biological samples by capillary electrophoresis with laser induced fluorescence detection," *Journal of Pharmaceutical and Biomedical Analysis*, vol. 53, no. 5, pp. 1180–1192, 2010.

[15] H. Zhao, L. Shi, J. Q. Cao, W. Li, X. Wen, and Y. Q. Zhao, "A new triterpene saponin from *Panax japonicus* C. A. Meyer var *major* (Burk.) C. Y. Wu et K. M. Feng," *Chinese Chemical Letters*, vol. 21, no. 10, pp. 1216–1218, 2010.

[16] B. K. Banik and F. F. Becker, "Novel 6,12-disubstituted chrysene as potent anticancer agent: synthesis, *in vitro* and *in vivo* study," *European Journal of Medicinal Chemistry*, vol. 45, no. 10, pp. 4687–4691, 2010.

[17] M. T. Rodrigues, J. C. Gomes, J. Smith, and F. Coelho, "Simple and highly diastereoselective access to 3,4-substituted tetrahydro-1,8-naphthyridines from Morita-Baylis-Hillman adducts," *Tetrahedron Letters*, vol. 51, no. 38, pp. 4988–4990, 2010.

[18] G. Mlostoń, A. Pieczonka, K. A. Ali, A. Linden, and H. Heimgartner, "A new approach to morpholin-2-one derivatives via the reaction of β-amino alcohols with dicyanofumarates," *Arkivoc*, pp. 181–192, 2012.

[19] A. M. Farag, A. S. Mayhoub, T. M. Abdalla et al., "Synthesis and structure-activity relationship studies of pyrazole-based heterocycles as antitumor agents," *Archiv der Pharmazie*, vol. 343, no. 7, pp. 384–396, 2010.

[20] A. E. Amr, K. A. Ali, and M. M. Abdalla, "Cytotoxic, antioxidant activities and structure activity relationship of some newly synthesized terpenoidal oxaliplatin analogs," *European Journal of Medicinal Chemistry*, vol. 44, no. 2, pp. 901–907, 2009.

[21] A. M. Farag, Y. M. Elkholy, and K. A. Ali, "Regioselective synthesis of diazaspiro[4.4]nona- and tetrazaspiro[4.5]deca-2,9-diene-6-one derivatives," *Journal of Heterocyclic Chemistry*, vol. 45, no. 1, pp. 279–283, 2008.

[22] S. A. Ghozlan, A. A. Mohammed, A. E. Amr, K. A. Ali, and A. A. Abd El-Wahab, "Synthesis and antimicrobial activity of some heterocyclic 2,6-bis(substituted)-1,3,4-thiadiazolo-, oxadiazolo-, and oxathiazolidino-pyridine derivatives from 2,6-pyridine dicarboxylic acid dihydrazide," *Journal of Heterocyclic Chemistry*, vol. 48, no. 5, pp. 1103–1110, 2011.

[23] K. A. Ali, E. A. Ragab, T. A. Farghaly, and M. M. Abdalla, "Synthesis and structure-activity relationship studies of pyrazole-based heterocycles as antitumor agents," *Acta Poloniae Pharmaceutica*, vol. 68, pp. 237–247, 2011.

[24] A. M. Farag, K. A. K. Ali, T. M. A. El-Debss et al., "Design, synthesis and structure-activity relationship study of novel pyrazole-based heterocycles as potential antitumor agents," *European Journal of Medicinal Chemistry*, vol. 45, no. 12, pp. 5887–5898, 2010.

[25] K. A. Ali, H. M. Hosni, E. A. Ragab, and S. I. Abd El-Moez, "Synthesis and antimicrobial evaluation of some new cyclooctanones and cyclooctane-based heterocycles," *Archiv der Pharmazie*, vol. 345, pp. 231–239, 2012.

[26] K. A. Ali, "A new convenient synthesis of some novel 2,6-disubstituted-pyridine derivatives," *Arkivoc*, vol. 2010, no. 11, pp. 55–63, 2010.

[27] K. A. Ali, M. A. Elsayed, and H. S. Abdalghfar, "Synthesis and reactions of 2,6-bis[3-oxo-3-propanenitrile-2-(*N*, *N*-dimethylamino) methylene]pyridine," *Arkivoc*, vol. 2011, no. 2, Article ID 37179, 2010.

[28] K. A. Ali, M. A. Elsayed, and A. M. Farag, "Synthesis of some new pyridine-2,6-bis-heterocycles," *Heterocycles*, vol. 85, no. 8, pp. 1913–1923, 2012.

[29] K. A. Ali, "Synthesis of pyridine-2,6-bis-((*E*)-2-benzylidene-3-oxopropanenitrile) and Its behavior towards nitrogen binucleophiles," *Heterocycles*, vol. 85, no. 8, pp. 1975–1986, 2012.

[30] M. T. Casey, P. Guinan, A. Canavan, M. McCann, C. Cardin, and N. B. Kelly, "Reaction of 1,1'-diacetylferrocene with hydrazine hydrate: synthesis and X-ray crystal structures of bis(hydrazine)bis(hydrazinecarboxylato-N',O)iron(II), $[Fe(N_2H_4)_2(O_2CNHNH_2)_2]$, and the cyclic biferrocene diazine, $[N(Me)CC_5H_4FeC_5H_4C(Me)N]_2$," *Polyhedron*, vol. 10, no. 4-5, pp. 483–489, 1991.

[31] M. M. Abd-Elzaher, "Synthesis and spectroscopic characterization of some tetradentate Schiff bases and their nickel, copper and zinc complexes," *Synthesis and Reactivity in Inorganic and Metal-Organic Chemistry* , vol. 30, no. 9, pp. 1805–1816, 2000.

[32] S. Sarawat, G. S. Srivastava, R. C. Mehrotra, and R. C., "Schiff base complexes of organotin(IV). Reactions of trialkyltin(IV) chlorides and alkoxides with *N*-substituted salicylideneimines," *Journal of Organometallic Chemistry*, vol. 129, no. 2, pp. 155–161, 1977.

[33] M. M. Abd-Elzaher, "Synthesis and spectroscopic characterization of some ferrocenyl Schiff bases containing pyridine moiety and their complexation with cobalt, nickel, copper and zinc," *Journal of the Chinese Chemical Society*, vol. 51, pp. 499–504, 2004.

[34] T. Mosmann, "Rapid colorimetric assay for cellular growth and survival: application to proliferation and cytotoxicity assays,"

Journal of Immunological Methods, vol. 65, no. 1-2, pp. 55–63, 1983.

[35] B. S. El-Menshawi, W. Fayad, K. Mahmoud et al., "Screening of natural products for therapeutic activity against solid tumors," *Indian Journal of Experimental Biology*, vol. 48, no. 3, pp. 256–264, 2010.

10

In Silico Inhibition Studies of Jun-Fos-DNA Complex Formation by Curcumin Derivatives

Anil Kumar[1] and Utpal Bora[1,2]

[1] *Computational Biology Laboratory, Department of Biotechnology, Indian Institute of Technology Guwahati, Assam, Guwahati 781039, India*
[2] *Biotech Hub, Centre for the Environment, Indian Institute of Technology Guwahati, Assam, Guwahati 781039, India*

Correspondence should be addressed to Utpal Bora, ubora@iitg.ernet.in

Academic Editor: Armando Rossello

Activator protein-1 (AP1) is a transcription factor that consists of the Jun and Fos family proteins. It regulates gene expression in response to a variety of stimuli and controls cellular processes including proliferation, transformation, inflammation, and innate immune responses. AP1 binds specifically to 12-O-tetradecanoylphorbol-13-acetate (TPA) responsive element 5′-TGAG/CTCA-3′ (AP1 site). It has been found constitutively active in breast, ovarian, cervical, and lung cancers. Numerous studies have shown that inhibition of AP1 could be a promising strategy for cancer therapeutic applications. The present in silico study provides insights into the inhibition of Jun-Fos-DNA complex formation by curcumin derivatives. These derivatives interact with the amino acid residues like Arg155 and Arg158 which play a key role in binding of Jun-Fos complex to DNA (AP1 site). Ala151, Ala275, Leu283, and Ile286 were the residues present at binding site which could contribute to hydrophobic contacts with inhibitor molecules. Curcumin sulphate was predicted to be the most potent inhibitor amongst all the natural curcumin derivatives docked.

1. Introduction

Activator protein-1 (AP1) is a transcription factor that consists of either homo- or heterodimers of the Jun and Fos family proteins [1]. It regulates gene expression in response to a variety of stimuli, including environmental stresses, UV radiation, cytokines, and growth factors. AP1 in turn controls a number of cellular processes including proliferation, transformation, inflammation, and innate immune response. The Jun and Fos proteins share similar amino acid sequences that comprise the basic DNA-binding sequence and the adjacent leucine zipper region by which these proteins dimerize [2–4]. The AP1 transcription factor binds specifically to 12-O-tetradecanoylphorbol-13-acetate (TPA) responsive element 5′-TGAG/CTCA-3′ which is commonly referred to as the AP1 site [5, 6]. *C-fos* and *c-jun* genes are autoregulated; the transcription of *c-jun* is stimulated by its own product, and in contrast *c-fos* is negatively autoregulated [7–9]. AP1 has been found constitutively active in many cancers including breast, ovarian, cervical, and lung. Numerous studies have shown that inhibition of AP1 has a profound effect on the behavior of cancer cells and tumors suggesting that AP1 could be a promising target for cancer therapy [10].

Curcumin, a dietary spice derived from the plant Turmeric (*Curcuma longa*), is used as a traditional medicine for inflammatory conditions [11]. Further, curcumin has been reported to have anti-inflammatory, anti-oxidant, and anticancer effects [12–15]. In vivo administration of curcumin was found to reduce the incidence and size of tumors in mice [16–19]. Moreover, curcumin was reported to inhibit proliferation and cell cycle progression in cancer cells [20]. Curcumin suppresses constitutive AP1 activity in HL-60, Raji, and prostate cancer cell lines (LNCaP, PC-3, and DU145) [21–25]. Curcumin was also reported to suppress LPS-induced cyclooxygenase-2 gene expression by inhibiting AP1 DNA binding in BV2 microglial cells [26]. It was confirmed that curcumin directly interacts with Jun-Fos dimer and inhibits its binding to DNA (AP1 site) [27]. Some synthetic curcumin derivatives have been discovered

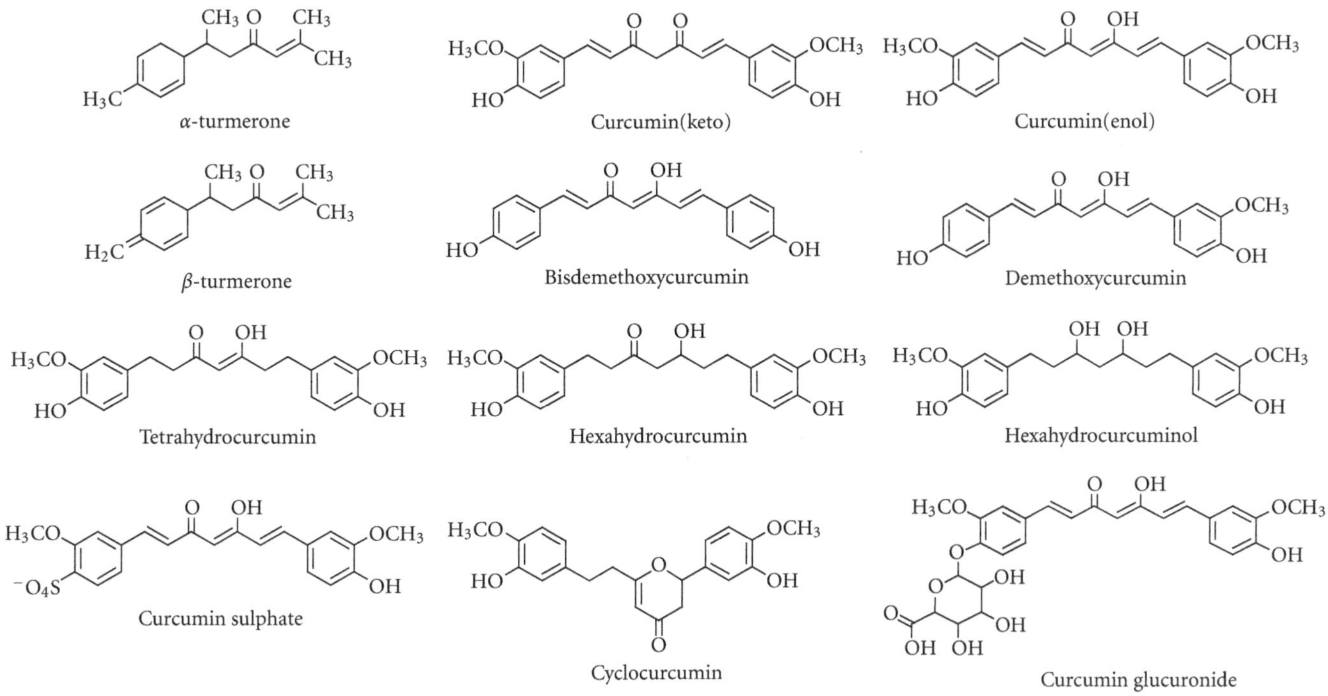

FIGURE 1: Chemical structures of natural curcumin derivatives.

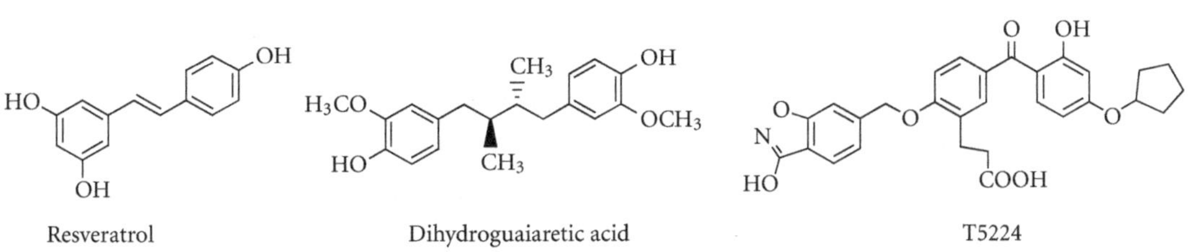

FIGURE 2: Known inhibitors of Jun-Fos-DNA complex formation used in the study.

as inhibitors of Jun-Fos-DNA complex formation [28–30]. However, no information on the site of interaction is reported yet. In the present study we investigate the interaction of curcumin derivatives with Jun-Fos complex by molecular docking studies.

2. Methodology

To investigate the interaction with Jun-Fos complex, curcumin natural derivatives (Figure 1), synthetic curcumin-based inhibitors (Table 1), and other known inhibitors of Jun-Fos-DNA complex formation (Figure 2) were drawn and 3D optimized using MarvinSketch (Free Academic License) and saved in Protein Data Bank (PDB) file format [31]. These molecules were prepared for molecular docking by merging nonpolar hydrogens, assigning Gastegier charges, and saving them in PDBQT file format using AutoDock Tools (ADT) 1.5.6 [32]. X-ray crystal structure of Jun-Fos-DNA complex (PDB ID: FOS1) was obtained from the Protein Data Bank (http://www.rcsb.org/pdb). For molecular docking DNA and other heteroatoms (water, ions, etc.) were removed using PyMOL 0.99. Gasteiger charges were assigned, and Jun-Fos complex was saved in PDBQT file format using ADT.

Grid and docking parameter files were prepared using ADT, and molecular docking was performed with AutoDock 4.2.1 (Scripps Research Institute, USA) considering all the rotatable bonds of curcumin derivatives as rotatable and Jun-Fos complex as rigid [33]. Grid box size of 90 × 90 × 90 Å with 0.375 Å spacing was selected that include the whole basic DNA-binding sequence and the adjacent leucine zipper region of Jun-Fos complex. Empirical-free energy function and Lamarckian genetic algorithm, with an initial population of 150 randomly placed individuals, a maximum number of 2,500,000 energy evaluations, a mutation rate of 0.02, and a crossover rate of 0.80, were used to perform molecular docking. Hundred independent docking runs were performed for each molecule. Curcumin derivative-Jun-Fos complex for lowest free energy of binding (ΔG) confirmation from the largest cluster was written in PDBQT format and converted to PDB file format using UCSF Chimera 1.6.1. Further, these complexes were analyzed using PyMOL 0.99 for possible polar and hydrophobic interactions. All the docking studies were performed at Intel(R) Xeon(R) CPU (3.2 GHz) with Linux-based operating system Fedora 15.

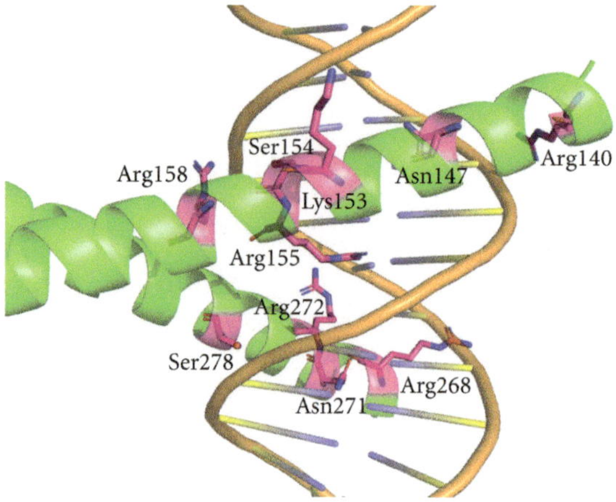

Figure 3: X-ray crystal structure of Jun-Fos-DNA complex (PDB ID: FOS1) showing amino acid residues (magenta) which form hydrogen bond with DNA (AP1 site).

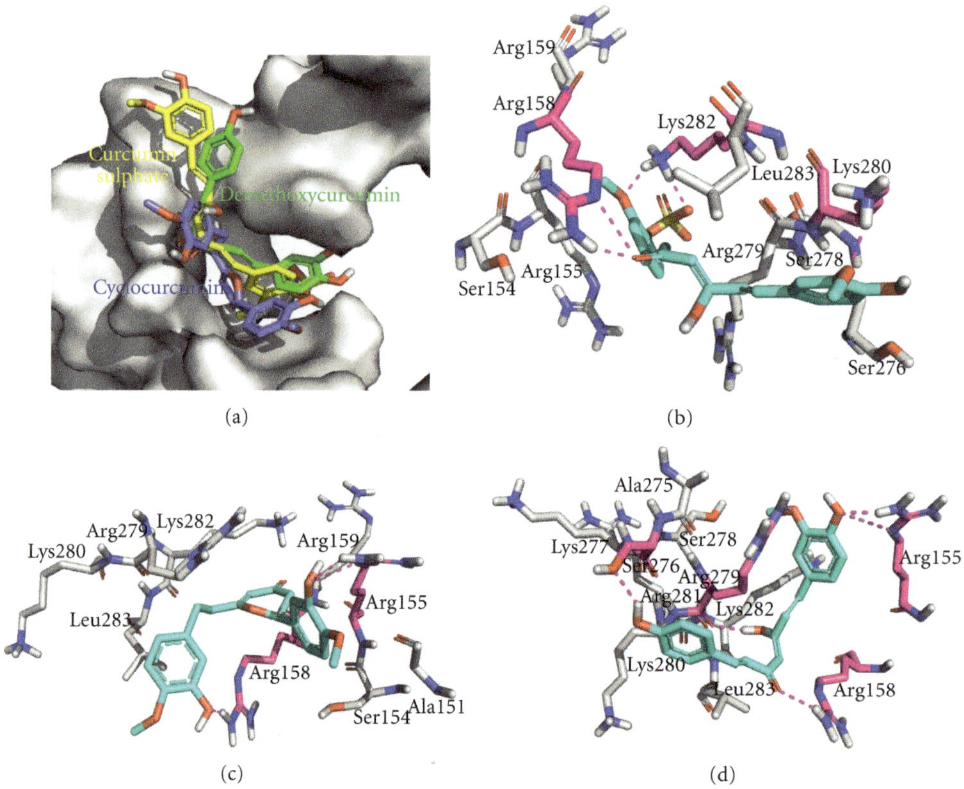

Figure 4: Binding modes of natural curcumin derivatives. (a) Curcumin sulphate (yellow), cyclocurcumin (blue), and demethoxycurcumin (green) docked to DBR of Jun-Fos complex; (b) curcumin sulphate (cyan) showing polar contacts with Arg158, Lys280, and Lys282 (magenta) (c) cyclocurcumin showing polar contacts with Arg155 and Arg158 (magenta); (d) demethoxycurcumin showing polar contacts with Arg155, Arg158, Ser276, and Arg279 (magenta).

3. Results and Discussions

X-ray crystal structure of Jun-Fos-DNA complex shows that Arg140, Asn147, Lys153, Ser154, Arg155, Arg158, Arg268, Asn271, Arg272, and Ser278 are the key residues by which Jun-Fos complex binds to DNA through hydrogen bonding (Figure 3). To predict the interaction of curcumin derivatives with Jun-Fos complex, natural curcumin derivatives and other known inhibitors of Jun-Fos-DNA complex formation were docked over DNA-binding region (DBR) of Jun-Fos complex, and results were summarized in Table 2.

Amongst all the natural curcumin derivatives docked to Jun-Fos complex curcumin sulphate bound with ΔG of -8.20 kcal/mol and predicted KI of 976.64 nM followed by cyclocurcumin and demethoxycurcumin which bound with ΔG of -5.75 and -5.72 kcal/mol and predicted KI of 61.42

Table 1: Synthetic curcumin-based inhibitors of Jun-Fos-DNA complex formation.

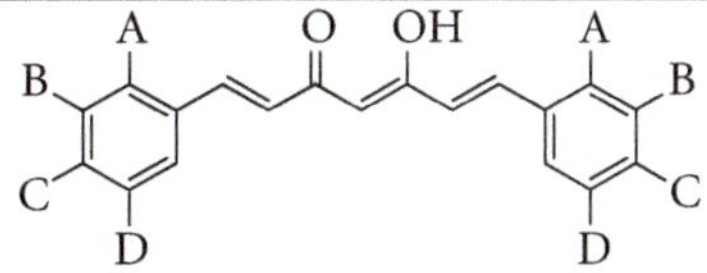

Compounds	A	B	C	D	References
CHC001	H	OCH_3	H	H	
CHC002	H	OCH_3	OCH_3	OCH_3	
CHC003	H	OCH_3	H	OCH_3	
CHC004	H	OCH_3	OCH_3	H	
CHC005	H	H	OCH_3	H	
CHC006	H	H	H	H	[28, 29]
CHC007	H	NO_2	OH	H	
CHC008	H	OH	H	H	
CHC009	H	NO_2	H	H	
CHC010	NO_2	H	H	H	
CHC011	H	H	NO_2	H	
BJC003	H	H	CH_3	H	
BJC004	H	NO_2	CH_3	H	[28, 30]
BJC005	H	NO_2	OH	OCH_3	

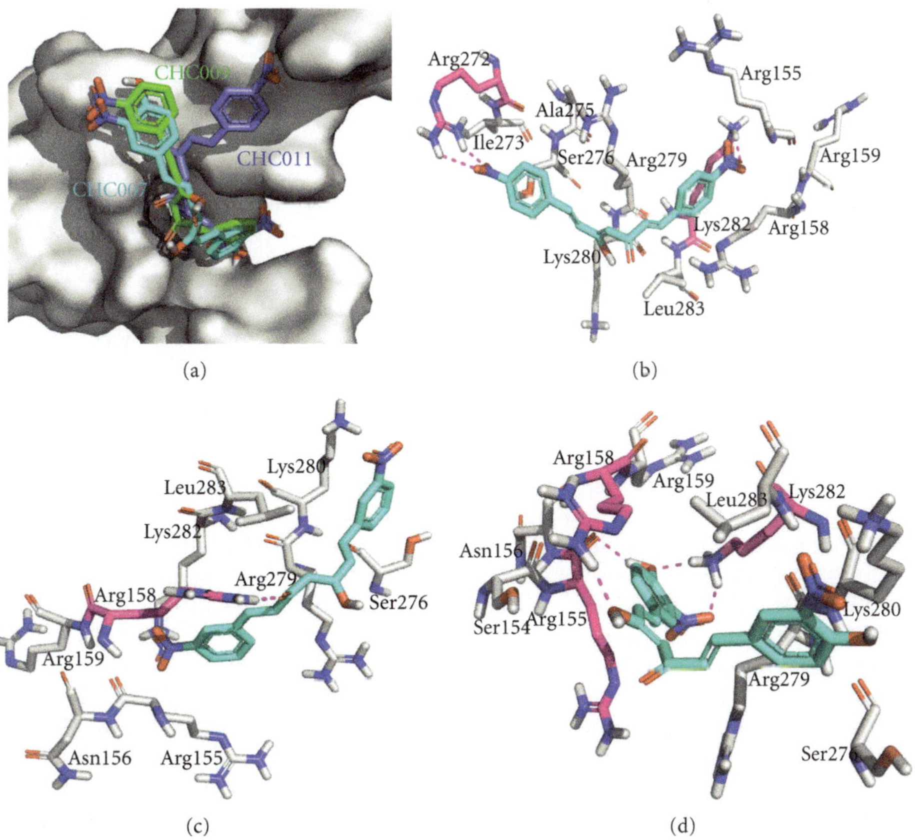

Figure 5: Binding modes of synthetic curcumin-based inhibitors (a) CHC011 (blue), CHC009 (green), and CHC007 (cyan) docked to DBR of Jun-Fos complex; (b) CHC011 (cyan) showing polar contacts with Arg272 and Lys282 (magenta); (c) CHC009 (cyan) showing polar contacts with Arg158 (magenta). (d) CHC007 (cyan) showing polar contacts with Arg155, Arg158, and Lys282 (magenta).

TABLE 2: Free energy of binding (ΔG) and predicted inhibition constant (KI) estimated with AutoDock 4.2.1 and interaction of inhibitors with Jun-Fos complex.

Compounds	ΔG (kcal/mol)	KI	Putative polar interactions	Hydrophobic residues in 5 Å region
T5224$^{\Psi}$	−9.96	49.64 nM	Arg158, Asn271, Ser278, Arg279, Lys282	Ala274, Ala275, Leu283
CHC011*	−9.59	−93.25 nM	Arg272, Lys282	Ile273, Ala275, Leu283
CHC009*	−9.52	104.26 nM	Arg158	Leu283
CHC007*	−9.15	196.96 nM	Arg155, Arg158, Lys282	Leu283
BJC004*	−9.12	207.86 nM	Lys153	Ala150, Ala151
BJC005*	−8.94	277.86 nM	Arg155, Arg158, Lys280, Lys282	Ala275, Leu283, Ile286
Curcumin sulphate	−8.20	976.64 nM	Arg158, Lys280, Lys282	Leu283
CHC010*	−6.73	11.59 μM	Ser278, Arg279	Ala274, Ala275
CHC008*	−5.86	50.65 μM	Arg155, Arg158, Ser-276, Lys282	Leu283
Cyclocurcumin	−5.75	61.42 μM	Arg155, Arg158	Ala151, Leu283
CHC003*	−5.73	62.98 μM	Arg158, Arg279, Lys280, Lys282	Ala275, Leu283
Demethoxycurcumin	−5.72	63.86 μM	Arg155, Arg158, Ser276, Arg279	Ala275, Leu283
BJC003*	−5.69	67.22 μM	Arg158, Arg279	Leu283
CHC004*	−5.66	71.36 μM	Arg158, Lys280, Lys282	Leu283
CHC006*	−5.45	101.79 μM	Arg158, Arg279	Leu283
CHC002*	−5.32	125.57 μM	Arg155, Arg158, Arg279, Lys280, Lys282	Ala275, Leu283
Bisdemethoxycurcumin	−5.30	130.44 μM	Arg158, Ser276, Arg279	Leu283
Curcumin (keto)	−5.27	136.46 μM	Arg158, Asn271, Arg279, Lys282	Ala274, Ala275, Leu283
Curcumin (enol)	−5.25	141.66 μM	Arg158, Ser276, Lys282	Ala275, Leu283
CHC005*	−5.24	144.93 μM	Arg158	Leu283
CHC001*	−5.19	156.49 μM	Arg158, Lys282	Ala275, Leu283, Ile286
α-Turmerone	−5.13	172.61 μM	Lys282	Ala150, Ala151, Leu283
β-Turmerone	−5.05	197.55 μM	Lys282	Ala275, Leu283
Tetrahydrocurcumin	−5.05	199.62 μM	Arg155, Arg158, Ser276, Arg279	Leu283
Curcumin glucuronide	−4.61	418.23 μM	Arg155, Arg158, Arg279, Lys282	Ala151, Leu283, Ile286
Dihydroguaiaretic acid$^{\Psi}$	−4.43	569.58 μM	Ser278, Arg279	Ala151, Ala275, Leu283
Resveratrol$^{\Psi}$	−4.20	829.30 μM	Ser154, Lys282	Ala151, Leu283
Hexahydrocurcuminol	−4.08	1.02 mM	Arg155, Arg158, Ser276, Lys280, Lys282	Ala275, Leu283
Hexahydrocurcumin	−4.07	1.04 mM	Arg158, Ser276, Arg279	Ala275, Leu283, Ile286

*Synthetic curcumin-based inhibitors of Jun-Fos-DNA complex formation.
$^{\Psi}$Known inhibitors of Jun-Fos-DNA complex formation.

and 63.86 μM, respectively (Figure 4(a)). The binding mode of curcumin sulphate depicted that sulphate and nearby methoxy group present at one aromatic ring of the molecule were in polar contact range with Lys282; however methoxy group present at the other side formed polar contact with side chain of Lys280 (Figure 4(b)). Keto group present in the linker region was in polar contact range with side chain of Arg158. The binding mode of cyclocurcumin showed that hydroxyl group present at one aromatic ring of the molecule formed polar contact with side chain of Arg155; however at the other side it formed polar contacts with Arg158 (Figure 4(c)). When demethoxycurcumin docked to Jun-Fos complex, hydroxyl and neighboring methoxy group present at one aromatic ring formed polar contact with side chains of Arg155 and Arg279, respectively, while hydroxyl group present at other side of the molecule formed polar contact with side chain of Ser276 (Figure 4(d)). In the linker region of the molecule keto and hydroxyl groups were in polar contact range with Arg158 and Arg279, respectively.

Amongst the synthetic curcumin-based inhibitors CHC011 bound to Jun-Fos complex with ΔG of −9.59 kcal/mol and predicted KI of 93.25 nM followed by CHC009 and CHC007 which docked with ΔG of −9.52 and −9.15 kcal/mol and predicted KI of 104.26 nM and 196.96 nM, respectively (Figure 5(a)). Similar results were observed in the in vitro studies by Hahm et al. in 2002 [28]. The binding mode studies depicted that $-NO_2$ group present at one aromatic ring of the CHC011 molecule formed polar contact with side chain of Arg272 while at the other side of the molecule it interacted with Lys282 (Figure 5(b)). When CHC009 docked to Jun-Fos complex, keto group present in the linker region of the molecule formed polar contact

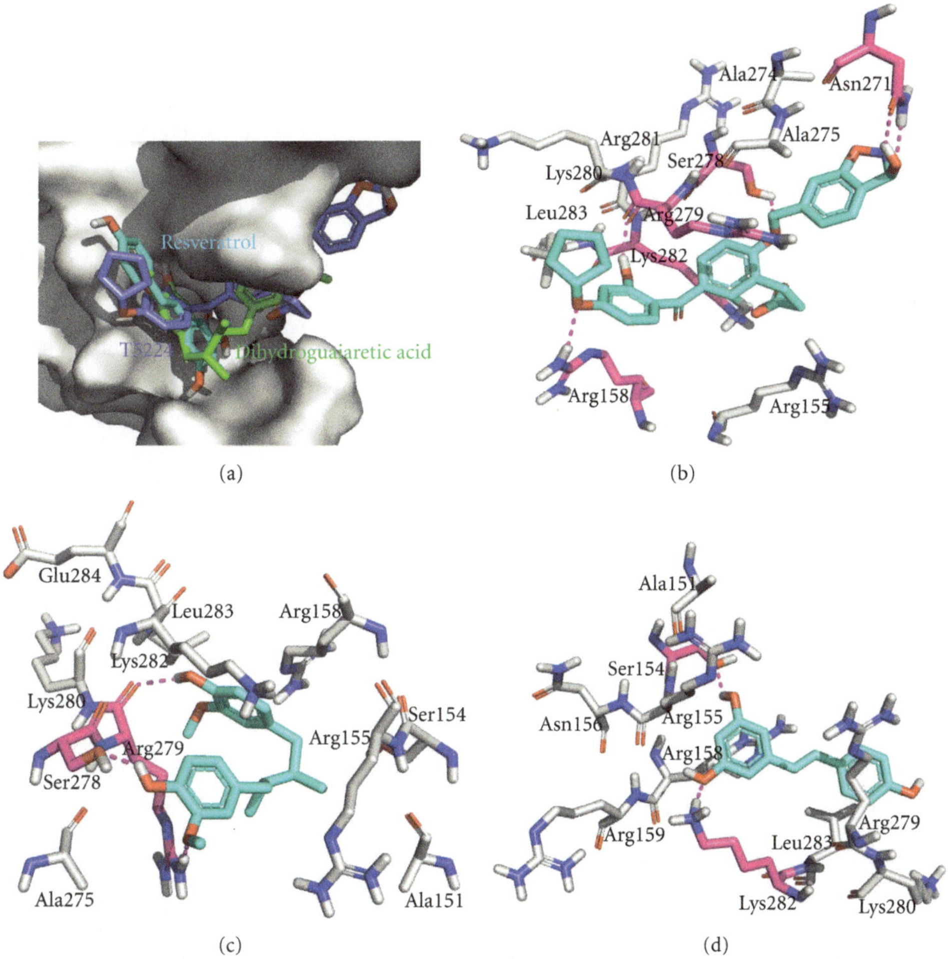

FIGURE 6: Binding modes of other known inhibitors. (a) T5224 (blue), dihydroguaiaretic acid (green), and resveratrol (cyan) docked to DBR of Jun-Fos complex (b) T5224 (cyan) showing polar contacts with Arg158, Asn271, Ser278, Arg279, and Lys282 (magenta); (c) Dihydroguaiaretic acid (cyan) showing polar contacts with Ser278 and Arg279 (magenta); (d) resveratrol (cyan) showing polar contacts with Ser154 and Lys282 (magenta).

with side chain of Arg158 (Figure 5(c)). Hydroxyl and $-NO_2$ group present at one aromatic ring of the CHC007 molecule formed polar contacts with backbone of Arg155 and side chain of Lys282, respectively, while the hydroxyl group present in the linker region of the molecule showed polar contact with side chain of Arg158 (Figure 5(d)).

Amongst the other known inhibitors T5224 [3-(5-(4-(cyclopentyloxy)-2-hydroxybenzoyl)-2-((3-hydroxybenzo [d]isoxazol-6-yl)methoxy)phenyl)propanoic acid] bound to Jun-Fos complex with ΔG of −9.96 kcal/mol and predicted KI of 49.64 nM followed by dihydroguaiaretic acid and resveratrol which docked with ΔG of −4.43 and −4.20 kcal/mol and predicted KI of 569.58 and 829.30 μM, respectively (Figure 6(a)). The binding mode studies of T5224 depicted that oxygen atom of cyclopentyloxy group formed polar contact with side chain of Arg158; however nearby hydroxyl group formed polar contact with Arg279. Hydroxyl group of 3-hydroxybenzo [d]isoxazol-6-yl)methoxy group formed polar contact with Asn271; however oxygen atom of its methoxy group formed polar contact with Ser278. Acid group of the T5224 molecule was in polar contact range with Lys282 (Figure 6(b)). When docked to Jun-Fos complex neighboring hydroxyl and methoxy groups present at one side of the dihydroguaiaretic acid molecule formed polar contacts with Ser278 and Arg279 respectively, whereas the hydroxyl group present at the other side of the molecule formed polar contact with backbone of Arg279 (Figure 6(c)). When docked to Jun-Fos complex neighboring hydroxyl groups attached to one of the aromatic ring of resveratrol molecule formed polar contacts with Ser 154 and side chain of Lys282, respectively (Figure 6(d)).

We observed that curcumin derivatives form polar contacts preferentially with residues like Arg155, Arg158, Lys276, Arg279, Lys280, and Lys282 when docked to DBR of Jun-Fos complex amongst which Arg155 and Arg158 are the key residues by which Jun-Fos complex binds to DNA. The results suggested that interaction of curcumin derivatives with residues like Arg155 and Arg158 could be the possible mechanism by which curcumin derivatives inhibit Jun-Fos-DNA complex formation. Ala151, Ala275, Leu283, and Ile286 were the hydrophobic residues present at binding site contributing to hydrophobic contacts with inhibitor molecules.

4. Conclusions

The present molecular docking study provides insights into the inhibition of Jun-Fos-DNA complex formation by curcumin derivatives. The involvement of residues like Arg155, Arg158, Lys276, Lys280, and Lys282 seems to play a key role in binding of curcumin derivatives to Jun-Fos complex through polar contacts which prevents its binding to DNA (AP1 site). Ala151, Ala275, Leu283 and Ile286 were the important hydrophobic residues present at binding site. Most of the curcumin derivatives were predicted to be more potent than inhibitors like resveratrol and dihydroguaiaretic acid. Curcumin sulphate was predicted to be the most potent inhibitor amongst all the natural curcumin derivatives docked.

Acknowledgments

A. Kumar thanks IITG and MHRD (Government of India) for the fellowship to pursue Ph.D. The authors thank developers of the free softwares used for this study and DBT, Government of India, for providing funds for establishment of Institutional Biotech Hub (BT/04/NE/2009) through which this work was carried out.

References

[1] R. Eferl and E. F. Wagner, "AP-1: a double-edged sword in tumorigenesis," *Nature Reviews Cancer*, vol. 3, no. 11, pp. 859–868, 2003.
[2] T. Curran and B. R. Franza, "Fos and jun: the AP-1 connection," *Cell*, vol. 55, no. 3, pp. 395–397, 1988.
[3] P. F. Johnson and S. L. McKnight, "Eukaryotic transcriptional regulatory proteins," *Annual Review of Biochemistry*, vol. 58, pp. 799–839, 1989.
[4] P. J. Mitchell and R. Tjian, "Transcriptional regulation in mammalian cells by sequence-specific DNA binding proteins," *Science*, vol. 245, no. 4916, pp. 371–378, 1989.
[5] W. Lee, P. Mitchell, and R. Tjian, "Purified transcription factor AP-1 interacts with TPA-inducible enhancer elements," *Cell*, vol. 49, no. 6, pp. 741–752, 1987.
[6] P. Angel, M. Imagawa, R. Chiu et al., "Phorbol ester-inducible genes contain a common *cis* element recognized by a TPA-modulated trans-acting factor," *Cell*, vol. 49, no. 6, pp. 729–739, 1987.
[7] P. Angel, K. Hattori, T. Smeal, and M. Karin, "The *jun* proto-oncogene is positively autoregulated by its product, Jun/AP-1," *Cell*, vol. 55, no. 5, pp. 875–885, 1988.
[8] P. Angel and M. Karin, "The role of Jun, Fos and the AP-1 complex in cell-proliferation and transformation," *Biochimica et Biophysica Acta*, vol. 1072, no. 2-3, pp. 129–157, 1991.
[9] P. Sassone-Corsi and I. M. Verma, "Modulation of c-fos gene transcription by negative and positive cellular factors," *Nature*, vol. 326, no. 6112, pp. 507–510, 1987.
[10] E. Shaulian and M. Karin, "AP-1 as a regulator of cell life and death," *Nature Cell Biology*, vol. 4, no. 5, pp. E131–E136, 2002.
[11] H. P. T. Ammon and M. A. Wahl, "Pharmacology of Curcuma longa," *Planta Medica*, vol. 57, no. 1, pp. 1–7, 1991.
[12] R. C. Srimal and B. N. Dhawan, "Pharmacology of diferuloyl methane (curcumin), a non steroidal anti inflammatory agent," *Journal of Pharmacy and Pharmacology*, vol. 25, no. 6, pp. 447–452, 1973.
[13] O. P. Sharma, "Antioxidant activity of curcumin and related compounds," *Biochemical Pharmacology*, vol. 25, no. 15, pp. 1811–1812, 1976.
[14] S. Toda, T. Miyase, and H. Arichi, "Natural antioxidants. III. Antioxidative components isolated from rhizome of Curcuma longa L," *Chemical and Pharmaceutical Bulletin*, vol. 33, no. 4, pp. 1725–1728, 1985.
[15] R. R. Satoskar, S. J. Shah, and S. G. Shenoy, "Evaluation of anti-inflammatory property of curcumin (diferuloyl methane) in patients with postoperative inflammation," *International Journal of Clinical Pharmacology Therapy and Toxicology*, vol. 24, no. 12, pp. 651–654, 1986.
[16] A. H. Conney, T. Lysz, T. Ferraro et al., "Inhibitory effect of curcumin and some related dietary compounds on tumor promotion and arachidonic acid metabolism in mouse skin," *Advances in Enzyme Regulation*, vol. 31, pp. 385–396, 1991.
[17] M. T. Huang, E. E. Deschner, H. L. Newmark, Z. Y. Wang, T. A. Ferraro, and A. H. Conney, "Effect of dietary curcumin and ascorbyl palmitate on azoxymethanol-induced colonic epithelial cell proliferation and focal areas of dysplasia," *Cancer Letters*, vol. 64, no. 2, pp. 117–121, 1992.
[18] M. T. Huang, Y. R. Lou, W. Ma, H. L. Newmark, K. R. Reuhl, and A. H. Conney, "Inhibitory effects of dietary curcumin on forestomach, duodenal, and colon carcinogenesis in mice," *Cancer Research*, vol. 54, no. 22, pp. 5841–5847, 1994.
[19] C. V. Rao, A. Rivenson, B. Simi, and B. S. Reddy, "Chemoprevention of colon carcinogenesis by dietary curcumin, a naturally occurring plant phenolic compound," *Cancer Research*, vol. 55, no. 2, pp. 259–266, 1995.
[20] R. Mohan, J. Sivak, P. Ashton et al., "Curcuminoids inhibit the angiogenic response stimulated by fibroblast growth factor-2, including expression of matrix metalloproteinase gelatinase B," *Journal of Biological Chemistry*, vol. 275, no. 14, pp. 10405–10412, 2000.
[21] T. S. Huang, S. C. Lee, and J. K. Lin, "Suppression of c-Jun/AP-1 activation by an inhibitor of tumor promotion in mouse fibroblast cells," *Proceedings of the National Academy of Sciences of the United States of America*, vol. 88, no. 12, pp. 5292–5296, 1991.
[22] S. S. Han, Y. S. Keum, H. J. Seo, and Y. J. Surh, "Curcumin suppresses activation of NF-kappaB and AP-1 induced by phorbol ester in cultured human promyelocytic leukemia cells," *Journal of Biochemistry and Molecular Biology*, vol. 35, no. 3, pp. 337–342, 2002.
[23] M. Hergenhahn, U. Soto, A. Weninger et al., "The chemopreventive compound curcumin is an efficient inhibitor of Epstein-Barr virus BZLF1 transcription in Raji DR-LUC cells," *Molecular Carcinogenesis*, vol. 33, no. 3, pp. 137–145, 2002.
[24] A. Mukhopadhyay, C. Bueso-Ramos, D. Chatterjee, P. Pantazis, and B. B. Aggarwal, "Curcumin downregulates cell survival mechanisms in human prostate cancer cell lines," *Oncogene*, vol. 20, no. 52, pp. 7597–7609, 2001.
[25] K. Nakamura, Y. Yasunaga, T. Segawa et al., "Curcumin down-regulates AR gene expression and activation in prostate cancer cell lines," *International Journal of Oncology*, vol. 21, no. 4, pp. 825–830, 2002.

[26] G. Kang, P. J. Kong, Y. J. Yuh et al., "Curcumin suppresses lipopolysaccharide-induced cyclooxygenase-2 expression by inhibiting activator protein 1 and nuclear factor kappab bindings in BV2 microglial cells," *Journal of Pharmacological Sciences*, vol. 94, no. 3, pp. 325–328, 2004.

[27] S. Park, D. K. Lee, and C. H. Yang, "Inhibition of fos-jun-DNA complex formation by dihydroguaiaretic acid and in vitro cytotoxic effects on cancer cells," *Cancer Letters*, vol. 127, no. 1-2, pp. 23–28, 1998.

[28] E. R. Hahm, G. Cheon, J. Lee, B. Kim, C. Park, and C. H. Yang, "New and known symmetrical curcumin derivatives inhibit the formation of Fos-Jun-DNA complex," *Cancer Letters*, vol. 184, no. 1, pp. 89–96, 2002.

[29] C. H. Park, J. H. Lee, and C. H. Yang, "Curcumin derivatives inhibit the formation of Jun-Fos-DNA complex independently of their conserved cysteine residues," *Journal of Biochemistry and Molecular Biology*, vol. 38, no. 4, pp. 474–480, 2005.

[30] H. K. Kim and C. H. Yang, "Synthetic curcumin derivatives inhibit Jun-Fos-DNA complex formation," *Bulletin of the Korean Chemical Society*, vol. 25, no. 12, pp. 1769–1774, 2004.

[31] J. L. Yap, J. Chauhan, K.-Y. Jung, L. Chen, E. V. Prochownik, and S. Fletcher, "Small-molecule inhibitors of dimeric transcription factors: antagonism of protein-protein and protein-DNA interactions," *Medicinal Chemistry Communications*, vol. 3, no. 5, pp. 541–551, 2012.

[32] M. F. Sanner, "Python: a programming language for software integration and development," *Journal of Molecular Graphics and Modelling*, vol. 17, no. 1, pp. 57–61, 1999.

[33] G. M. Morris, D. S. Goodsell, R. S. Halliday et al., "Automated docking using a Lamarckian genetic algorithm and an empirical binding free energy function," *Journal of Computational Chemistry*, vol. 19, no. 14, pp. 1639–1662, 1998.

MDAN-21: A Bivalent Opioid Ligand Containing mu-Agonist and Delta-Antagonist Pharmacophores and Its Effects in Rhesus Monkeys

Mario D. Aceto,[1] Louis S. Harris,[1] S. Stevens Negus,[1] Matthew L. Banks,[1] Larry D. Hughes,[1] Eyup Akgün,[2] and Philip S. Portoghese[2]

[1] *Department of Pharmacology and Toxicology, School of Medicine, Virginia Commonwealth University, Richmond, VA 23298, USA*
[2] *Department of Medicinal Chemistry, College of Pharmacy, University of Minnesota, Minneapolis, MN 55455, USA*

Correspondence should be addressed to Mario D. Aceto, mdaceto@vcu.edu

Academic Editor: Yoshio Okada

MDAN-21, 7′-{2-[(7-{2-[({(5α,6α)-4,5-Epoxy-3,14-dihydroxy-17-methylmorphin-6-yl}-aminocarbonyl)metoxy]-acetylamino}-heptylaminocarbonyl)-methoxy]-acetylamino}-naltrindole, a bivalent opioid ligand containing a mu-opioid receptor agonist (derived from oxymorphone) linked to the delta-opioid receptor antagonist (related to naltrindole) by a spacer of 21 atoms, was reported to have potent analgesic properties in mice. Tolerance, physical dependence, and conditioned place preference were not evident in that species. The finding that bivalent ligands in this series, with spacers 19 atoms or greater, were devoid of tolerance and dependence led to the proposal that MDAN-21 targets heteromeric mu-delta-opioid receptors. The present study focused on its effects in nonhuman primates (*Macaca mulatta*), a species with a physiology and behavioral repertoire not unlike humans. With regard to opioids, this species usually better predicts clinical outcomes. MDAN-21 substituted for morphine in morphine-dependent monkeys in the remarkably low dose range 0.006–0.032 mg/kg, subcutaneously. Although MDAN-21 failed to produce reliable thermal analgesia in the dose range 0.0032–0.032 mg/kg, intramuscularly, it was active in the same dose range and by the same route of administration, in the capsaicin-induced thermal allodynia assay. The results suggest that MDAN-21 may be useful in the treatment of opioid dependence and allodynia. The data provide additional evidence that opioid withdrawal is associated with sensitized pain.

1. Introduction

The alkaloid morphine has been used in the treatment of pain, cough, and diarrhea. Unfortunately, unpleasant and/or potentially dangerous side effects such as respiratory depression, nausea, vomiting, and constipation can accompany its use. Psychological and physiological processes such as abuse, tolerance, and physical dependence have been associated with chronic use and limit the utility of morphine and other mu-opioid agonists in the treatment of chronic pain.

In a continuing search for potent analgesics free of these undesirable side effects, many analogues of morphine and numerous semisynthetic and synthetic derivatives have been introduced. The evolution of this search was summarized in a succinct review of the exciting but vain quest for the Holy Grail of opioid research [1]. Results of an investigation of the effects of leucine and methionine enkephalin on morphine-induced analgesia suggested an interaction between mu- and delta-opioid receptors [2]. Over a decade later, investigators found that naltrindole, a selective delta-opioid receptor antagonist, blocked the morphine tolerance without diminishing its antinociceptive potency [3]. These reports and the finding that G-protein-coupled opioid receptors existed as heterodimers: mu-kappa [4], mu-delta [5, 6], and kappa-delta [7] led to a hypothesis-driven synthesis [8] of a series of conjugates containing the mu-opioid receptor agonist (derived from oxymorphone) linked with delta-opioid receptor antagonist (related to naltrindole) with spacers ranging from 16 to 21 atoms [9]. Studies in

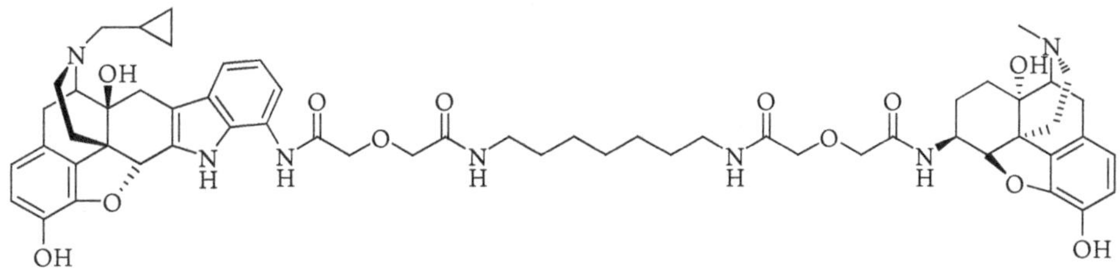

FIGURE 1: Chemical structure of MDAN-21.

mice indicated that 7′-{2-[(7-{2-[({(5α,6α)-4,5-Epoxy-3,14-dihydroxy-17-methylmorphin-6-yl}-aminocarbonyl)-metoxy]-acetylamino}-heptylaminocarbonyl)-methoxy]-acetylamino}-naltrindole or MDAN-21, (see Figure 1), with a 25.4 Å spacer, (21-atoms), was the most potent analgesic of the series. Significantly, naloxone, an opioid receptor antagonist, failed to precipitate withdrawal signs after chronic administration. It was subsequently shown that MDAN-21-treated mice failed to develop conditioned place preference [10]. These properties suggested that MDAN-21 may, to a significant extent, fulfill the long quest for a strong opioid analgesic with greatly reduced side effects. We decided to determine MDAN-21's effects on morphine-induced physical dependence and explore its analgesic profile in nonhuman primates. The rhesus monkey study models were selected because of their excellent relationship to the behavioral expressions of opioid effects in humans and their predictive value regarding pharmacotherapy [11] especially regarding opioids.

2. Material and Methods

2.1. Subjects. Male and female rhesus monkeys (*Macaca mulatta*) served as subjects. All monkeys had prior exposure to drugs and to the behavioral procedures in which they were tested. The subjects in the substitution of morphine assay were housed in groups, whereas those in the thermal nociception and capsaicin-induced thermal allodynia assays were individually housed. Food and water were freely available. Their diet consisted of Purina Lab Diet Fiber-Plus Monkey Biscuits no. 5049 (PMI Feeds, Inc., St. Louis, MO) supplemented with fresh fruit twice weekly. A 12 h light/12 h dark cycle was in effect (lights on from 7AM–7PM).

Animal maintenance and research were conducted in accordance with the guidelines provided by the NIH Committee on Laboratory Animal Resources. The facility was licensed by the United States Department of Agriculture and accredited by the Association for the Assessment and Accreditation of Laboratory Animal Care. The protocols were approved by the Institutional Animal Care and Use Committee. The health of the monkeys was monitored daily by technical and veterinary staff. Monkeys had visual, auditory and olfactory contact with other monkeys throughout the study. Monkeys also had access to puzzle feeders, mirrors, and chew toys to provide environmental enrichment. Nature videotapes or music were played at least twice a week in all housing rooms.

2.2. Substitution for Morphine Assay. Twenty-four monkeys in the weight range 3.5–7.5 kg comprised the group tested. The assay was based on that originally described by Deneau [12]. Modifications follow. Morphine was given subcutaneously (s.c.) daily at 6 AM, 12 noon, and 6 PM. All the animals had received morphine for at least 3 months and were maximally dependent on morphine. At least 4 monkeys/treatment were used and a minimal two-week wash-out period was allowed between testing. The assay was modified as indicated below [13, 14]. It was initiated by the injection (s.c.) of the test drug or control substances (morphine and vehicle) into animals in a group that had not received morphine for 14-15 hr and showed definite signs of withdrawal. Each animal was randomly chosen to receive one of the following treatments: (a) a dose MDAN-21; (b) morphine sulfate control (4.0 mg/kg); (c) vehicle control (1.0 mL/kg). Withdrawal signs were scored, absent, or present, once during each of five consecutive 30-min observation periods. Withdrawal signs included slowing of movement, drowsiness (sitting with eyed closed and lethargic or being indifferent to surroundings), fighting, vocalizing, rigidity of abdominal muscles, vocalization during palpation of abdominal muscles, restlessness (pacing), tremors, coughing, retching, vomiting, wet-dog shakes, and masturbation. The observer was "blind" regarding the assignment of treatments. The cumulative number of withdrawal signs was analyzed using the Kruskal-Wallis Analysis of Variance (ANOVA) and post hoc Mann-Whitney *U* Tests. The Stat View statistical package (Brainpower, Agoura Hills, CA) was utilized for these analyses. In all cases, significance (*P*) was set at 0.05.

2.3. Assay of Thermal Nociception. Three adult rhesus monkeys were studied in this assay as described previously [15]. They were seated in acrylic restraint chairs so that their tails hung down freely. During tail-withdrawal measurements, the bottom 15 cm of each monkey's shaved tail was immersed in a thermal container of warm water. If the subject did not withdraw its tail within 20 s, the tail was removed from the water by the experimenter, and a latency of 20 s was assigned to that measurement. Experiments were conducted no more than once a week. A stopwatch was used to measure and record time intervals. Each test session consisted of multiple cycles. Before MDAN-21 administration, baseline tail-withdrawal latencies from 38 and 50°C water were determined. Testing continued only if tail withdrawal from 38°C water did not occur before the 20 s cutoff, and if tail withdrawal occurred in ≤2 s from 50°C water. All monkeys met this criterion before every test session. The effects of MDAN-21 intramuscularly (i.m.) were evaluated using a time-course procedure, in which a single dose of MDAN-21 (0.0032–0.32 mg/kg) was administered at the start of the session, and tail withdrawal latencies from 50°C water were redetermined 3, 6, 10, 18, 32, 56, and 100 min after the injection.

MDAN-21 was completely ineffective up to the highest doses tested in two monkeys. Accordingly, a follow-up experiment was conducted in these two monkeys to determine whether MDAN-21 might function as an antagonist of the high-efficacy mu-agonist methadone. For these studies, monkeys were pretreated with vehicle or 0.32 mg/kg MDAN-21 15 min before treatment with 0.32 mg/kg methadone. Tail-withdrawal latencies were then determined 10, 20, and 30 min after methadone administration. Drug effects were expressed as Percent Maximum Possible Effect (%MPE) using the following equation: "%MPE = (Test Latency − Baseline Latency)/(20 − Baseline Latency)*100", where "Test Latency" was the tail-withdrawal latency from 50°C water obtained during each cycle after drug administration, "Baseline Latency" was the latency from 50°C water during the baseline determinations before drug injection, and "20" was the cutoff latency in s.

2.4. Assay of Capsaicin-Induced Thermal Allodynia. Three different male rhesus monkeys 5–12 kg were studied. Monkeys were seated in acrylic restraint chairs as described above. To determine tail-withdrawal latencies, the lower 15 cm of each monkeys shaved tail was immersed into a thermal container of warm water heated to the designated temperature (see below for temperatures). The latency in seconds for the monkey to remove its tail from the water was measured using a hand-held stopwatch. If the subject did not withdraw its tail within 20 s, the tail was removed from the water by the experimenter, and a latency of 20 s was assigned to that measurement. Experimental sessions were conducted once per week. At the beginning of each session, tail withdrawal latencies were determined for each monkey from water heated to 38, 42, 46, and 50°C, and the order of temperature presentations was randomized across sessions. By this procedure, baseline temperature-effect curves were determined in each monkey at the beginning of each session, and the highest temperature to produce a tail withdrawal latency >15 s was identified. Water heated to this "threshold" temperature then served as the thermal stimulus for subsequent studies of allodynia during that session. The threshold stimulus intensity was 42°C for two monkeys and 46°C for the third monkey throughout the study.

Allodynia was elicited by topical application of capsaicin as described previously [15, 16]. Following baseline tail withdrawal latency determinations, a topical patch treated with capsaicin solution or vehicle was prepared as described below (see Drugs), and the patch was applied to a region approximately 7 cm from the bottom of the tail for 5 min. After 5 min, the patch was removed. Tail withdrawal latencies were then redetermined 15, 30, 45, and 60 min after patch removal using the thermal stimulus identified from the baseline temperature-effect curve in each monkey (i.e., 42°C in two monkeys, 46°C in the third monkey). MDAN-21 (0.0032–0.32 mg/kg, i.m.) was administered 15 min before application of the capsaicin patch. Morphine (1.0 mg/kg, i.m., 15 min pretreatment) was tested for comparison.

Raw tail withdrawal latencies obtained 15, 30, 45, and 60 min after removal of the capsaicin patch were converted to Percent Maximum Possible Effect (%MPE) using the equation %MPE = [(Test Latency − Capsaicin Alone Latency)/(20 − Capsaicin Alone Latency)*100)], where "test latency" was the tail withdrawal latency obtained at each time point after drug pretreatment + capsaicin patch treatment, and "Capsaicin Alone Latency" was the latency obtained at the corresponding time point after treatment with the capsaicin patch alone.

2.5. Drugs. MDAN-21 was synthesized by Dr. Eyup Akgun (Medicinal Chemistry Laboratory, Dr. Philip Portoghese, Director). Morphine sulfate was purchased from Mallinckrodt, Inc., Hazelwood, MO. Methadone HCl was provided by the NIDA Drug Supply Program. All drugs were dissolved in sterile water for injection (Hospira, Inc., Forest Hills, IL). Capsaicin (Sigma Chemical Co., St. Louis, MO) was dissolved in vehicle composed of 70% alcohol and 30% sterile water and was delivered transdermally (topical patch) as described previously [15, 16]. The concentration of capsaicin in the solution was individually determined for each monkey as the lowest concentration to produce sustained decreases in tail-withdrawal latencies from the threshold temperature to ≤5 s throughout the 1 hr testing period (0.61 mg/mL (2 mM)) for all three monkeys in the study. Within 30 s of preparing the capsaicin patch, it was secured onto the monkey's tail with elastic tape and left on for 5 min.

3. Results

3.1. Substitution for Morphine in Morphine-Dependent Rhesus Monkeys Assay. Results of this assay are illustrated in Figure 2. It is evident that MDAN-21 had a short onset and long duration of action, that is, it effectively suppressed withdrawal as soon as 30 min and was still effective at 150 min. MDAN-21 and morphine suppressed the full spectrum of withdrawal signs exhibited by the vehicle controls. The effective dose range was 0.006–0.03 mg/kg. The Kruskal-Wallis ANOVA *P* values are as follows: 30 min-0.01; 60 min-0.0002; 90 min-0.0002; 120 min-0.0001; 150 min-0.0001. Post hoc Mann-Whitney comparisons with probability values of 0.05 or less are indicated in figure legends and with appropriate superscript letters in Figure 2.

3.2. Assay of Thermal Nociception. The mean ± S.E.M baseline tail withdrawal latency from 50°C was 0.75 ± 0.18 s. Figure 3 shows the time course of antinociceptive effects produced by doses of 0.0032–0.32 mg/kg MDAN-21 in three individual monkeys. In monkey M1470 (one of the two males), MDAN-21 doses of 0.0032 and 0.032 mg/kg produced a dose-dependent increase in tail withdrawal latencies that peaked after 18 and 32 min and then dissipated after 56 min. However, in the other two monkeys, MDAN-21 doses of 0.032 and 0.32 mg/kg were largely ineffective,

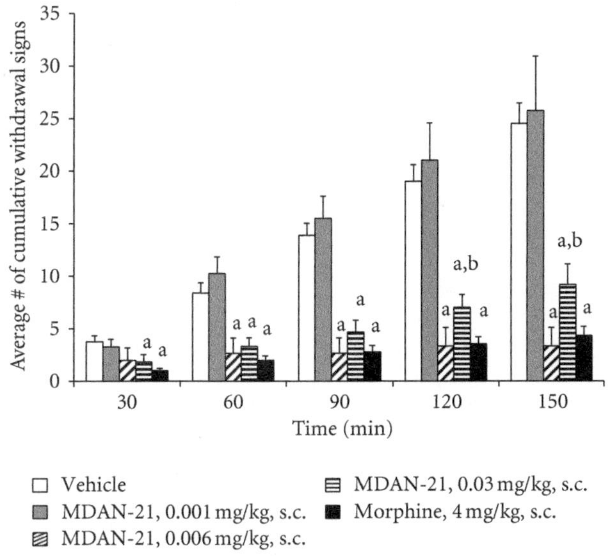

FIGURE 2: Effects of MDAN-21, morphine, and vehicle in morphine-dependent monkeys in spontaneous withdrawal. [a]Significantly less ($P < 0.05$) than vehicle and 0.001 mg/kg of MDAN-21. [b]Significantly less ($P < 0.05$) than MDAN-21. Each point shows mean ± S.E.M. from at least 4 monkeys.

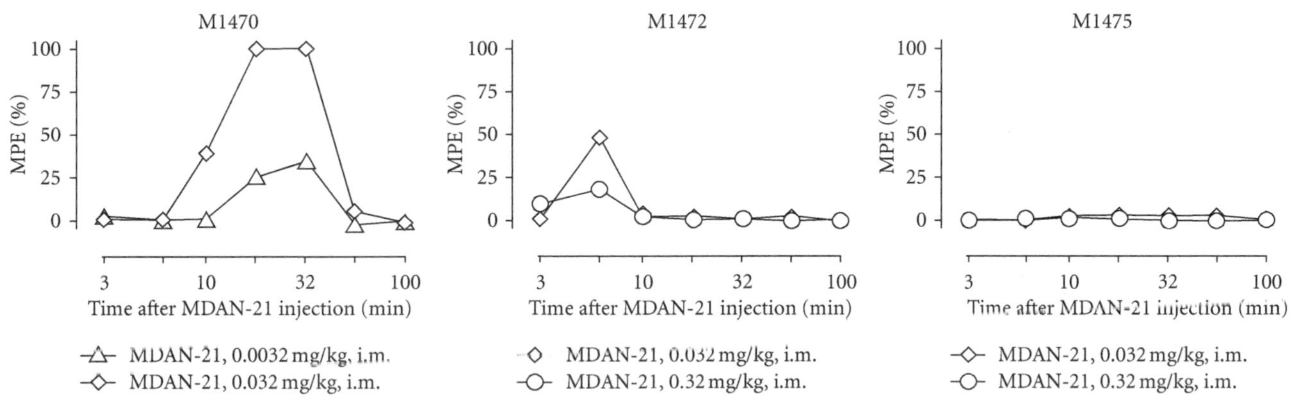

FIGURE 3: Effects of MDAN-21 in individual monkeys tested in a warm-water tail-withdrawal assay of thermal nociception. Abscissae: minutes after administration of MDAN-21 (log scale). Ordinates: percent maximum possible effect (%MPE). The identification number of the monkey is shown in the upper left of each panel.

producing only small increases in tail-withdrawal latencies after 6 min in monkey M1472. Higher doses were also tested in M1472 (2.6 mg/kg) and M1475 (1.0 mg/kg) and these doses were also ineffective (maximum effect of 15% MPE at any time in either monkey; data not shown).

Figure 4 shows the effects of pretreatment with vehicle or MDAN-21 on the antinociceptive effects of 0.32 mg/kg methadone in monkeys M1472 and M1475. After vehicle pretreatment, methadone produced a time-dependent antinociception that peaked after 20–30 min. Pretreatment with 0.32 mg/kg MDAN-21 had no effect on methadone antinociception in either monkey.

3.3. Capsaicin-Induced Thermal Allodynia Assay. The mean ± S.E.M baseline tail withdrawal latency at the threshold temperature was 19.59 ± 0.41 s, and treatment with capsaicin alone reduced tail withdrawal latencies at the threshold temperature to mean ± S.E.M. values of 1.71 ± 0.42, 1. 66 ± 0.23, 1.65 ± 0.41, and 1.49 ± 0.10 s at times 15, 30, 45, and 60 min after capsaicin patch removal, respectively. Figure 5 shows that pretreatment with MDAN-21 (0.0032–0.32 mg/kg) produced a dose- and time-dependent prevention of capsaicin-induced allodynia. The highest dose of 0.32 mg/kg MDAN-21 produced peak antiallodynic effects after 15 min, and these effects dissipated after 45 to 60 min. For comparison, the antiallodynic effect of 1.0 mg/kg morphine peaked after 30 min and remained at levels greater than 50% MPE in all three monkeys at 45 min and in two of three monkeys at 60 min.

4. Discussion

MDAN-21 was very effective in suppressing withdrawal signs in morphine-dependent monkeys. Its action was prompt.

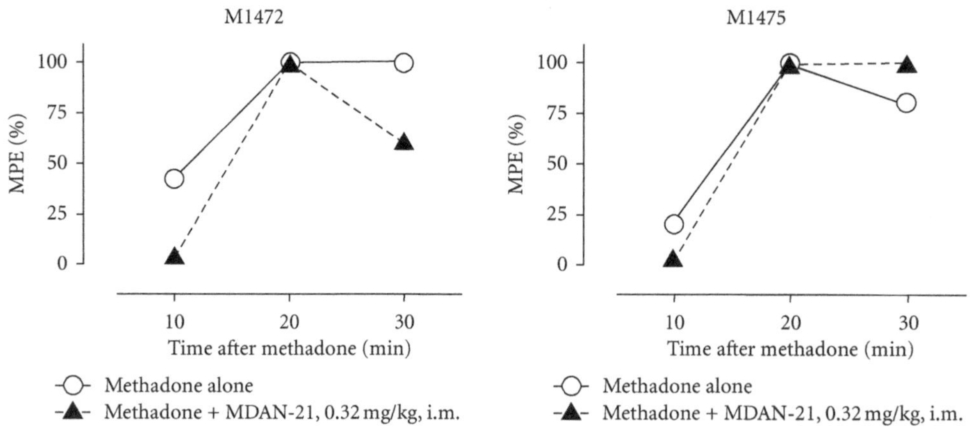

Figure 4: Antinociceptive effects of methadone administered alone or after pretreatment with MDAN-21 in monkey M1472 (left panel) and M1475 (right panel). Abscissae: minutes after methadone administration. Ordinates: percent maximum possible effect. Each point shows the results from a single determination in each monkey.

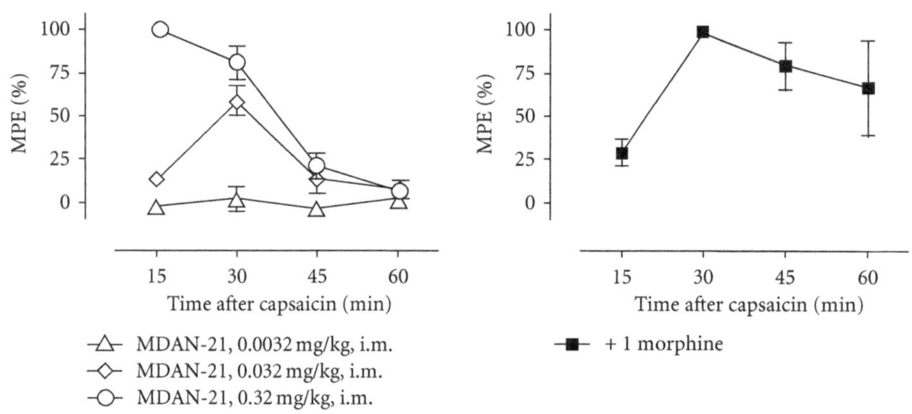

Figure 5: Effects of MDAN-21 (left panel) and morphine (right panel) in an assay of capsaicin-induced thermal allodynia. Abscissae: minutes after removal of the capsaicin patch. Ordinates: percent maximum possible effect. Each point shows mean ± S.E.M. from three monkeys.

It had a remarkably long duration of action compared to that in the capsaicin-thermal assay in nondependent monkeys. Perhaps, MDAN-21 is much more effective in situations involving physical dependence. Increased pain sensitivity in chronic pain patients and opioid abusers has been reported [17–19]. MDAN-21 failed to produce a reliable antinociceptive effect in the assay of thermal antinociception in nondependent monkeys. In one monkey, a dose of 0.032, MDAN-21 was fully effective, but neither this dose nor a 10-fold higher dose of 0.32 mg/kg was active in the other two monkeys. Conversely, MDAN-21 was as effective as morphine in all monkeys tested in the assay of capsaicin-induced thermal allodynia.

This profile of results is consistent with the possibility that MDAN-21 does not readily cross the blood-brain barrier in rhesus monkeys [15]. On the other hand, suppression of withdrawal signs in morphine-dependent monkeys by MDAN-21 suggests that it crosses the blood-brain barrier. It is known that quaternary compounds, do not readily cross the blood-brain barrier. Indeed, quaternary morphine compounds were reported virtually inactive in morphine-dependent rhesus monkeys and practically devoid of antinociceptive activity in mice [20]. The variability in effects produced by MDAN-21 in the assay of thermal nociception did not appear to result from low efficacy of MDAN-21 at mu-opioid receptors. If MDAN-21 had failed to produce antinociception in 2 of 3 monkeys due to low efficacy, then it would be expected to antagonize the effects of a higher efficacy mu-agonist in these two monkeys. However, MDAN-21 failed to antagonize the antinociceptive effects of methadone in either monkey. Nevertheless, it is possible that higher doses might have achieved a robust antinociceptive effect. Finally, species differences in mu-delta interactions on neural substrates that mediate thermal antinociception may be involved [21–23].

Nevertheless, MDAN-21was potently active in the capsaicin-induced thermal allodynia assay in the dose range of 0.032–0.32 mg/kg. It also attenuated withdrawal signs in morphine-dependent monkeys in spontaneous withdrawal for a longer time period. The results are in accord with the findings of other investigators that pain sensitivity is increased in opioid-treated animals, chronic pain patients and opioid abusers (see references above). The results also suggest that MDAN-21 may be useful in the pharmacotherapy of chronic pain and chronic opioid use.

Acknowledgments

This work was supported by NIDA DA 01533, DA 7–8859, RO1 DA 011460, and T32 DA007027. The authors thank Barbara R. Kipps for conducting the statistics and preparing the graphs in the physical dependence model. Thay also thank Crystal Reyns for excellent technical assistance in studies antinociception.

References

[1] A. D. Corbett, G. Henderson, A. T. McKnight, and S. J. Paterson, "75 Years of opioid research: the exciting but vain quest for the Holy Grail," *British Journal of Pharmacology*, vol. 147, supplement 1, pp. S153–S162, 2006.

[2] J. L. Vaught and A. E. Takemori, "A further characterization of the differential effects of leucine enkephalin, methionine enkephalin and their analogs on morphine-induced analgesia," *Journal of Pharmacology and Experimental Therapeutics*, vol. 211, no. 2, pp. 280–283, 1979.

[3] E. E. Abdelhamid, M. Sultana, P. S. Portoghese, and A. E. Takemori, "Selective blockage of delta opioid receptors prevents the development of morphine tolerance and dependence in mice," *Journal of Pharmacology and Experimental Therapeutics*, vol. 258, no. 1, pp. 299–303, 1991.

[4] D. Wang, X. Sun, L. M. Bohn, and W. Sadee, "Opioid receptor homo- and heterodimerization in living cells by quantitative bioluminescence resonance energy transfer," *Molecular Pharmacology*, vol. 67, no. 6, pp. 2173–2184, 2005.

[5] I. Gomes, A. Gupta, J. Filipovska, H. H. Szeto, J. E. Pintar, and L. A. Devi, "A role for heterodimerization of μ and δ opiate receptors in enhancing morphine analgesia," *Proceedings of the National Academy of Sciences of the United States of America*, vol. 101, no. 14, pp. 5135–5139, 2004.

[6] I. Gomes, B. A. Jordan, A. Gupta, N. Trapaidze, V. Nagy, and L. A. Devi, "Heterodimerization of mu and delta opioid receptors: a role in opiate synergy," *The Journal of Neuroscience*, vol. 20, no. 22, Article ID RC 110, 2000.

[7] B. A. Jordan and L. A. Devi, "G-protein-coupled receptor heterodimerization modulates receptor function," *Nature*, vol. 399, no. 6737, pp. 697–700, 1999.

[8] P. S. Portoghese, "From models to molecules: opioid receptor dimers, bivalent ligands, and selective opioid receptor probes," *Journal of Medicinal Chemistry*, vol. 44, no. 14, pp. 2259–2269, 2001.

[9] D. J. Daniels, N. R. Lenard, C. L. Etienne, P. Y. Law, S. C. Roerig, and P. S. Portoghese, "Opioid-induced tolerance and dependence in mice is modulated by the distance between pharmacophores in a bivalent ligand series," *Proceedings of the National Academy of Sciences of the United States of America*, vol. 102, no. 52, pp. 19208–19213, 2005.

[10] N. R. Lenard, D. J. Daniels, P. S. Portoghese, and S. C. Roerig, "Absence of conditioned place preference or reinstatement with bivalent ligands containing mu-opioid receptor agonist and delta-opioid receptor antagonist pharmacophores," *European Journal of Pharmacology*, vol. 566, no. 1–3, pp. 75–82, 2007.

[11] M. D. Aceto, "Assessment of physical dependence techniques for the evaluation of abused drugs," in *Modern Methods in Pharmacology*, M. W. Adler and A. Cowan, Eds., vol. 6 of *Testing and Evaluation of Drugs of Abuse*, pp. 67–79, Wiley-Lis, New York, NY, USA, 1990.

[12] G. A. Deneau, *An analysis of factors influencing the development of physical dependence of narcotic analgesics in rhesus monkeys with methods of predicting physical dependence liability in man*, thesis, University of Michigan, Ann Arbor, Mich, USA, 1956.

[13] M. D. Aceto, R. E. Flora, and L. S. Harris, "The effects of naloxone and nalorphine during the development of morphine dependence in rhesus monkeys," *Pharmacology*, vol. 15, no. 1, pp. 1–9, 1977.

[14] M. D. Aceto, L. S. Harris, and E. L. May, "Dependence studies of new compounds in the rhesus monkey, rat, and mouse," *National Institute on Drug Abuse Research Monograph Series*, vol. 67, pp. 399–452, 1986.

[15] G. P. Do Carmo, R. Polt, E. J. Bilsky, K. C. Rice, and S. S. Negus, "Behavioral pharmacology of the μ/δ opioid glycopeptide MMP2200 in rhesus monkeys," *Journal of Pharmacology and Experimental Therapeutics*, vol. 326, no. 3, pp. 939–948, 2008.

[16] E. R. Butelman, J. W. Ball, T. J. Harris, and M. J. Kreek, "Topical capsaicin-induced allodynia in unanesthetized primates: pharmacological modulation," *Journal of Pharmacology and Experimental Therapeutics*, vol. 306, no. 3, pp. 1106–1114, 2003.

[17] P. Compton, P. Athanasos, and D. Elashoff, "Withdrawal hyperalgesia after acute opioid physical dependence in non-addicted humans: a preliminary study," *Journal of Pain*, vol. 4, no. 9, pp. 511–519, 2003.

[18] M. Doverty, J. M. White, A. A. Somogyi, F. Bochner, R. Ali, and W. Ling, "Hyperalgesic responses in methadone maintenance patients," *Pain*, vol. 90, no. 1-2, pp. 91–96, 2001.

[19] J. J. Lipman and B. Blumenkopf, "Comparison of subjective and objective analgesic effects of intravenous and intrathecal morphine in chronic pain patients by heat beam dolorimetry," *Pain*, vol. 39, no. 3, pp. 249–256, 1989.

[20] A .E. Jacobson, "Biological evaluation of compounds for their physical dependence potential and abuse liability.V11. Drug testing program of the committee on drug dependence, Inc, (1983)," *National Institute on Drug Abuse Research Monograph Series*, vol. 49, pp. 352–360, 1984.

[21] A. Mansour, H. Khachaturian, M. E. Lewis, H. Akil, and S. J. Watson, "Autoradiographic differentiation of mu, delta, and kappa opioid receptors in the rat forebrain and midbrain," *Journal of Neuroscience*, vol. 7, no. 8, pp. 2445–2464, 1987.

[22] J. B. Daunais, S. R. Letchworth, L. J. Sim-Selley, H. R. Smith, S. R. Childers, and L. J. Porrino, "Functional and anatomical localization of mu opioid receptors in the striatum, amygdala, and extended amygdala of the nonhuman primate," *Journal of Comparative Neurology*, vol. 433, no. 4, pp. 471–485, 2001.

[23] D. Peckys and G. B. Landwehrmeyer, "Expression of mu, kappa, and delta opioid receptor messenger RNA in the human CNS: a ^{33}P *in situ* hybridization study," *Neuroscience*, vol. 88, no. 4, pp. 1093–1135, 1999.

12

Synthetic Methods, Chemistry, and the Anticonvulsant Activity of Thiadiazoles

Bhawna Sharma,[1] Amita Verma,[2] Sunil Prajapati,[1] and Upendra Kumar Sharma[1]

[1] *Institute of Pharmacy, Bundelkhand University, Jhansi 284128, India*
[2] *Department of Pharmaceutical Sciences, Faculty of Health Sciences, Sam Higginbottom Institute of Agriculture, Technology and Sciences-Deemed University, Allahabad 211007, India*

Correspondence should be addressed to Bhawna Sharma; bms1928@yahoo.com

Academic Editor: Hussein El-Subbagh

The chemistry of heterocyclic compounds has been an interesting field of study for a long time. Heterocyclic nucleus 1,3,4-thiadiazole constitutes an important class of compounds for new drug development. The synthesis of novel thiadiazole derivatives and investigation of their chemical and biological behavior have gained more importance in recent decades. The search for antiepileptic compounds with more selective activity and lower toxicity continues to be an active area of intensive investigation in medicinal chemistry. During the recent years, there has been intense investigation of different classes of thiadiazole compounds, many of which possess extensive pharmacological activities, namely, antimicrobial activity, anticonvulsant, antifungal antidiabetic, anti-inflammatory, antioxidant, and antituberculosis activities, and so forth. The resistance towards available drugs is rapidly becoming a major worldwide problem. The need to design new compounds to deal with this resistance has become one of the most important areas of research today. Thiadiazole is a versatile moiety that exhibits a wide variety of biological activities. Thiadiazole moiety acts as "hydrogen binding domain" and "two-electron donor system." It also acts as a constrained pharmacophore. On the basis of the reported literature, we study here thiadiazole compounds and their synthetic methods chemistry and anticonvulsant activity.

1. Introduction

Epilepsy is the name of a brain disorder characterized predominantly by recurrent and unpredictable interruptions of normal brain function, called epileptic seizures [1, 2]. The current therapy of epilepsy with antiepileptic drugs is associated with side effects, dose-related and chronic toxicity, and teratogenic effects [3, 4]. Epilepsy is not a singular disease entity but a variety of disorders reflecting underlying brain dysfunction that may result from many different causes. Therefore, there is continuing demand for new anticonvulsant agents. So, there is an urgent requirement for the dieovery and development of some novel anticonvulsant agents with more selective activity and lower toxicity for the effective treatment of epilepsy. Several five-membered aromatic systems having three heteroatoms at symmetrical positions such as thiadiazoles have been studied extensively owing to their interesting pharmacological activities. There is a broad variety of heterocyclic compounds which are having medicinal importance, and recently, much attention has been focused on thiadiazole derivatives in view of their broad spectrum activities. Thiadiazole is one such heterocyclic nucleus. There are several isomers of thiadiazole, that is 1,2,3-thiadiazole, 1,2,4-thiadiazole, 1,2,5-thiadiazole, and 1,3,4-thiadiazole [5], 1,3,4-Thiadiazole is the main isomer of thiadiazole series having versatile pharmacological activities.

2. Thiadiazole

Thiadiazole is a heterocyclic organic compound that has a five-member ring having one sulphur and two nitrogen atoms [6]. 1,3,4-Thiadiazoles represent one of the most biologically active classes of compounds, possessing a wide spectrum of activities. Thiadiazoles have become very important

TABLE 1: Physical properties of 1,3,4-thiadiazole.

Bond length		Bond angle	
Type	Å	Type	(°)
a	1.371	α	112.2
b	1.302	β	114.6
c	1.721	γ	86.4
d	1.721	δ	114.6
e	1.302	ε	112.2

compounds in medicine, agriculture, and many fields of technology. A large number of 1,3,4-thiadiazoles have been patented in the agricultural field as herbicides and bactericides [7].

X-ray analysis shows the following structure parameter for 1,3,4-thiadiazole ring (see Table 1 and Structure **(1)**):

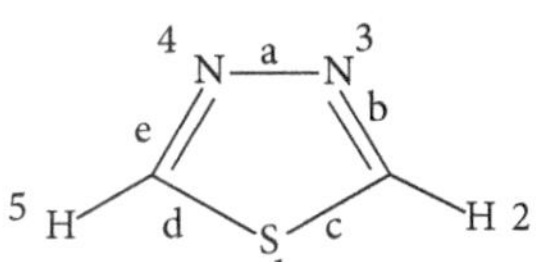

(1)

3. Chemistry of Thiadiazole

A recent literature survey revealed that the 1,3,4-thiadiazole moiety has been widely used by the medicinal chemist in the past to explore its biological activities. The development of 1,3,4-thiadiazole chemistry is linked to the discovery of phenylhydrazines and hydrazine in the late nineteenth century. The first 1,3,4-thiadiazole was described by Fischer in 1882, but the true nature of the ring system was demonstrated first in 1890 by Freund and Kuh.

Thiadiazole is a five-member heterocyclic compound having one sulphur and two nitrogen atoms. There are several isomers of thiadiazole, that is, 1,2,3-thiadiazole **(2)**, 1,2,5-thiadiazole **(4)**, 1,2,4-thiadiazole **(3)**, and 1,3,4-thiadiazole **(5)**. There are four possible systems in thiadiazole.

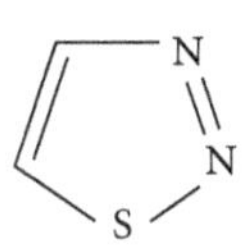

1,2,3-thiadiazole

(2)

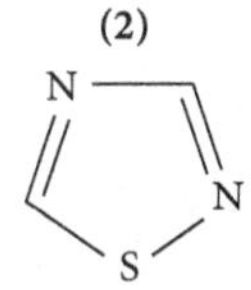

1,2,4-thiadiazole

(3)

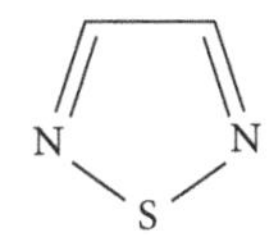

1,2,5-thiadiazole

(4)

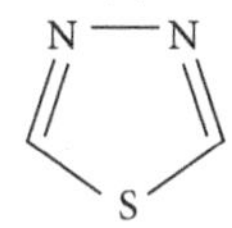

1,3,4-thiadiazole

(5)

4. 1,3,4-Thiadiazoles

1,3,4-Thiadiazole was first described in 1882 by Fischer and further developed by Bush and his coworkers, but true nature of the ring system was demonstrated first in 1956 by Goerdler et al. [8]. The advent of sulphur drugs and the later discovery of mesoionic compound greatly accelerated the rate of progress in the field of thiadiazole. Thiadiazole carrying mercapto, hydroxyl, and amino substituents can exist in many tautomeric forms. The 1,3,4-thiadiazoles are conveniently divided into three subclasses:

(a) aromatic systems which include the neutral thiadiazoles and constitute a major part of this paper;

(b) mesoionic systems which are defined as five-membered heterocycles which are not covalent or polar and possess a sextet of electrons in association with the five atoms comprising the ring;

(c) nonaromatic systems such as the 1,3,4-thiadiazoles and the tetrahydro 1,3,4-thiadiazoles. In the partially reduced systems, the position of the double bond is denoted by the prefix Δ, with being a Δ^2-1,3,4-thiadiazole (Structures **(6)**, **(7)**, and **(8)**).

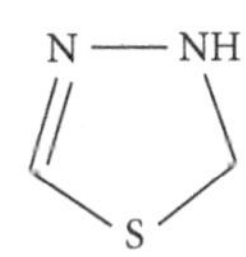

(6)

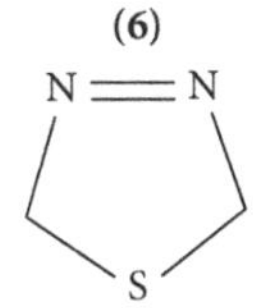

(7)

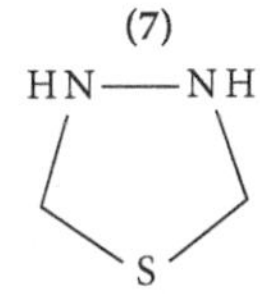

(8)

5. Synthetic Procedures of 1,3,4-Thiadiazoles

(a) Formation of One Bond. The most common procedure for the synthesis of 5-substituted 2-amino-thiadiazole

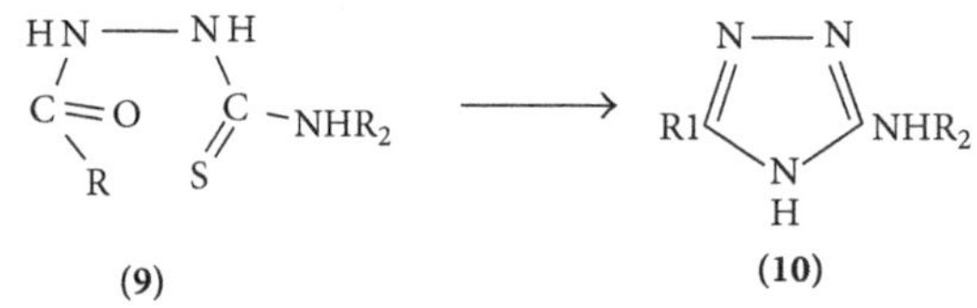

Scheme 1

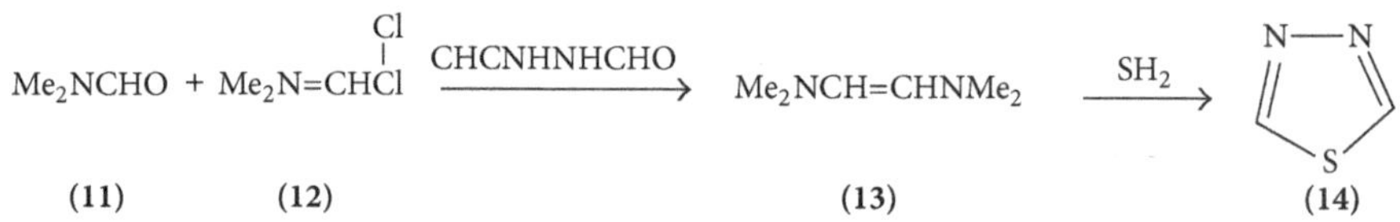

Scheme 2

is the acylation of a thiosemicarbazide followed by dehydration. Sulphuric acid, polyphosphoric acid, and phosphorous halides are some of the reagents used. The most recent procedure utilizes 1.5 moles of methane sulphonic as a dehydrating agent and the thiadiazoleare obtained in the high yield and good purity. 5-Alkyl-2-methyl amino-1,3,4-thiadiazoles are prepared from a suitable carboxylic acid and methyl thio-semicarbazide in the presence of three parts of polyphosphoric acid and one part of concentreted sulphuric acid. 2-alkylamino-1,3,4-thiadiazole *n* substituted in the 5th position can be prepared in high yields by the reaction of 4-alkylthiosemicarbazides with orthoformate esters in the presence of small amount of concentreted hydrochloric acid [9] (Scheme 1).

(b) Formation of Two Bonds. This is the most widely used procedure for the synthesis of thiadiazoles, thiazolidines, and mesoionic thiadiazoles.

(1) Cyclization. The parent molecule 1,3,4-thiadiazole was synthesized in 1956 by a four-step reaction sequence start utilizing hydrazine and from thiosemicarbazide. A second procedure utilizes hydrazine and potassium dithioformate. Dehydration of DMF with thionyl chloride or phosgene gives the formamide chloride which on treatment with N,N′Diformylhydrazine gives the dihydrochloride of the free base which is liberated with sodium ethoxide, which then cyclizes to thiadiazole in the presence of hydrogen sulphide in an overall yield of 65%. 2-Amino 1,3,4-thiadiazole is also prepared from thiosemicarbazide and a mixture of formic and hydrochloric acid in a tedious procedure with an overall yield of 65% (Scheme 2).

(2) Dipolar Cycloadditions. This procedure has been widely used during the last decade for both synthetic and mechanistic reasons.

From Diazo Compounds. The reaction of aryl sulphonyl-substituted sulfines with diazomethane gives Δ^3-1,3,4-thiadiazole 1-oxide which, however, is unstable and rearranged via an isomerization of the Δ^3 to the Δ^2-thiadiazole oxide. This is followed by an elimination and readdition of sulfonic acid and loss of water in a pummerer-type aromatization to give the rearranged thiadiazole (Scheme 3).

5.1. Method of Synthesis of 1,3,4-Thiadiazole

(a) From Hydrazine. 3-Thiocarbamoyl thione methyl carbonate on oxidation with H_2O_2 gave alkoxythiadiazole [10] (Scheme 4).

(I) Thiobenzoyl hydrazine cyclized to produce 1,3,4-thiadiazole-thiobenzoyl hydrazine [11] (Scheme 5).

(II) Thione carbazate are cyclized by cyanogens chloride or bromide to give 1,3-thiadiazole [12] (Scheme 6).

(b) From Semicarbazide. When 4-phenyl-1-(thiobenzole) semicarbazide reacts in the presence of concentreted HCl giving 2-hydroxyl-5-phenyl-1,3,4-thiadiazole [13] (Scheme 7).

(c) From Thiosemicarbazide. Many syntheses of the 1,3,4-thiadiazole proceed from thiosemicarbazide or substituted thiosemicarbazide.

Method 1. Gupta et al. have shown that thiosemicarbazide cyclizes directly to 2-amino-5-diazole with acetyl chloride. This simple route to 2-amino 5-substituted-1,3,4-thiadiazole seems to be quite general. In the example shown, R may be methyl17, nor hydnocarpyl, benzyl, cyclopropyl [14], and many others (Scheme 8).

(d) From Thiosemicarbazone. 2-Amino-5-substituted thiadiazoles are prepared by oxidative cyclization of thiosemicarbazones with ferric chloride found 1,3,4-thiadiazole [15] (Scheme 9).

(e) From Resin. Resin with TMSCl, MCPBA, and R^2 R^3 NH gave 1,3,4-S thiadiazole [16] (Scheme 10).

(f) From Bithioureas. Bithiourea and substituted bithiourease have been converted to 1,3,4-thiadiazole by several methods.

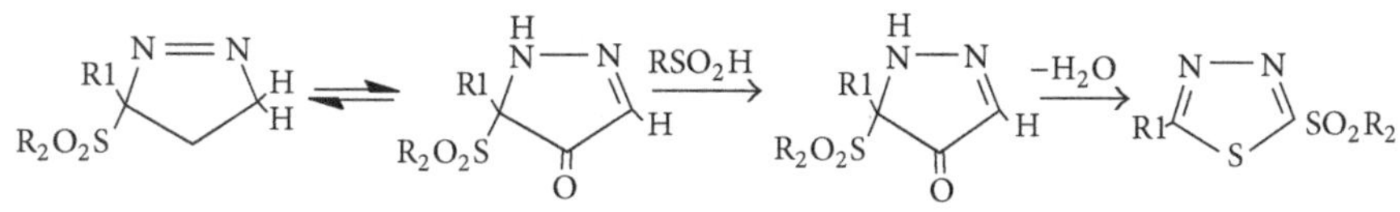

(16) (17) (18) (19)

Scheme 3

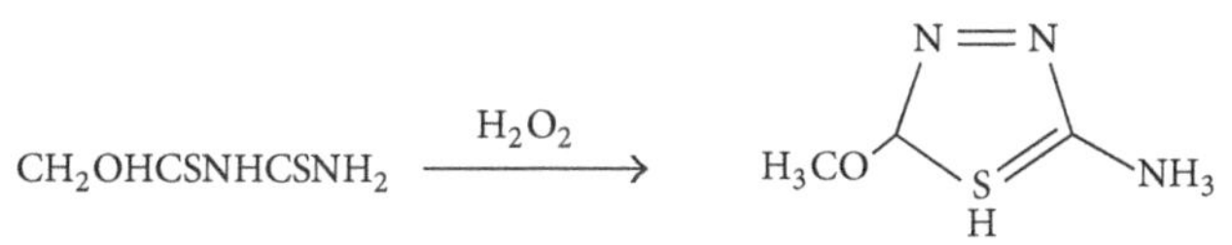

(20) (21)

Scheme 4

Method 1. Bithiourea when treated with 3% hydrogen peroxide is cyclized to 2,5-diamino-1,3,4-thiadiazole (Scheme 11).

Method 2. Acetic anhydride acts on bithiourea to form a diacetyl derivative of 2,5-diamino-1,3,4-thiadiazole. The acetyl group I is easily removed by hydrolysis to give the parent thiazole [17].

5.2. Chemical Properties

5.2.1. Reactivity. Some of the characteristic reactions of the 1,3,4-thiadiazole nucleus are ring opening by strong base ease of nucleophilic attack and the formation of mesoionic compounds by quaternization. The substituents in the 2 and 5 positions have a large effect in determining the reactivity of the molecule as a whole. Thus, the ambient nucleophilicity of 2-aminothiazoles gives rise to electrophilic attack on both the amino group and the nuclear nitrogen atom. Ring formation between these two nitrogen atoms is also a common reaction. 2-Mercaptothiazoles react similarly to arenethiols while a methyl group on the thiadiazole ring has reactivity similar to that in a picoline. Nucleophiles easily displace halogen atom from the thiadiazole nucleus. This is due to the electronegativity of the two nuclear nitrogen atoms which impart a low electron density to the carbon atom of the nucleus.

5.2.2. Thermal and Photochemical Reactions. Thermal and photochemical fragmentations of 1,3,4-thiadiazole often follow the fragmentation pattern observed in the mass spectrometer. The cis-2,5-di-t-butyl-2,5-dihydro-1,3,4-thiadiazole gives while photolysis yields the katazine. The cis 2,5-dihydro-1,3,4-thiadiazole 1,1-dioxide (cis-4) undergoes thermolysis at 50°C to give the azine. The trans isomer (trans-4), however, undergoes thermolysis only above 145°C to give the alkylidine hydrazide plus sulphur monoxide which disproportionate into sulphur dioxide. Compounds were independently prepared from dimethylpropanyl, and the hydrazole thermolysis of the isomeric 2,3-dihydro-1,3,4-thiazole 1,1-thiadiazole 1,1 dioxide also gives rise. The concerted [4 + 1] cycloelimination of sulphur dioxidefrom the trans isomer cannot take place due to steric hindrance (Scheme 12).

(1) Reaction with Electrophile. Because unsaturation rings have excessive πe^{-}, the two nitrogen atoms e^{-} are pulled towards them and ring carbon atoms remain with low e^{-} density and consequently no electrophilic substitution possible in unsubstituted 1,3,4-thiadiazole ring.

(a) Attack at Ring Nitrogen (Quaternization). Ratio of 3 or 4 substituted product depends on ring substitution, in which nitrogen has high e^{-} density (Scheme 13).

(b) Electrophilic Attack at Carbon. Due to the electron density at the carbon atoms in 1,3,4-thiadiazole, such reaction as nitration, acetylation, sulphonation, halogenations. However, 2-amino-substituted 1,3,4-thiadiazoles (1a–i) heat with bromine in acetic acid to give 5-bromo derivatives (2a–i). Similarly, the thiadiazoles (3b) yield (3d–f) the corresponding 5-bromo derivatives (4b) and (4d–f) (Scheme 14).

(c) Electrophilic Attack at Sulphur. Although direct oxidation at sulphur atom in 1,3,4-thiadiazoles has not been reported, Δ^3 1,3,4-thiadiazole 1-oxides and 1,1 dioxides can be indirect mean (Scheme 15).

(d) Electrophilic Substitution at Ring. Strong e^{-} donar substituted at ring position 2 activates towards electrophilic agents under drastic condition (conc. H_2SO_4 + fuming HNO_3).

Nitration of 2 ammonia 1,3,4-thiadiazoles with fuming nitric acid at 40°C gave 2-amino-1,3,4-thiadiazoles (21A) (Scheme 16).

5.3. Amination. Direct nuclear amination of certain thiadiazoles is possible, illustrating the case of nucleophilic attack. (R=H) react with hydroxylamine in the presence of base to give (R=NH_2) presumably via the imine (Scheme 17).

(III) *Nucleophilic Attack.* 1,3,4-Thiadiazole ring reacts with strong nucleophiles, treatment of the parent compound with base leading to ring fission. 2-Amino and 2-methylamino 1, 3, 4-thiadiazole rearrange in the presence of methylamine to the triazolinethione (Scheme 18).

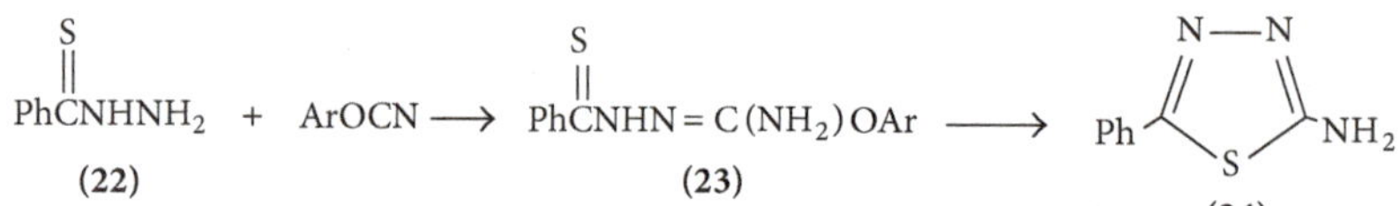

(22) (23) (24)

Scheme 5

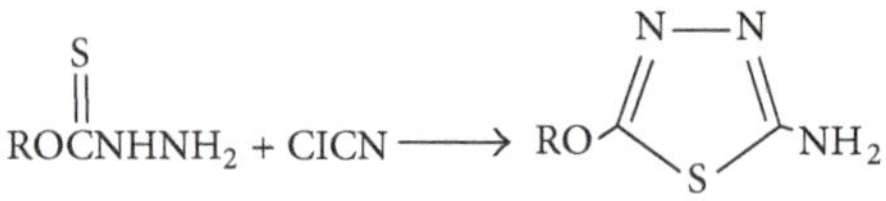

(25) (26)

Scheme 6

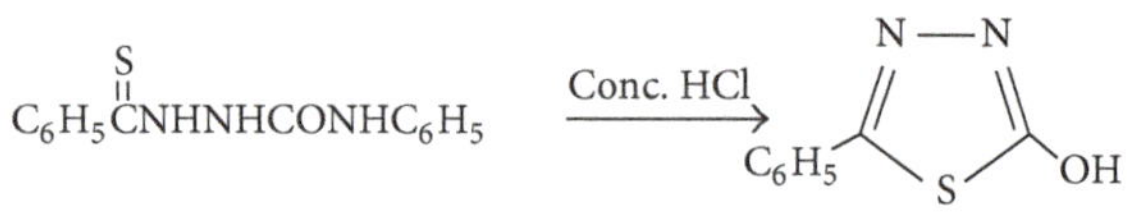

(27) (28)

Scheme 7

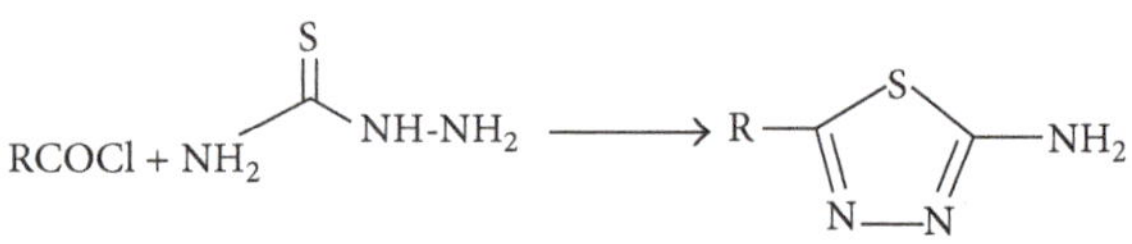

(29) (30) (31)

Scheme 8

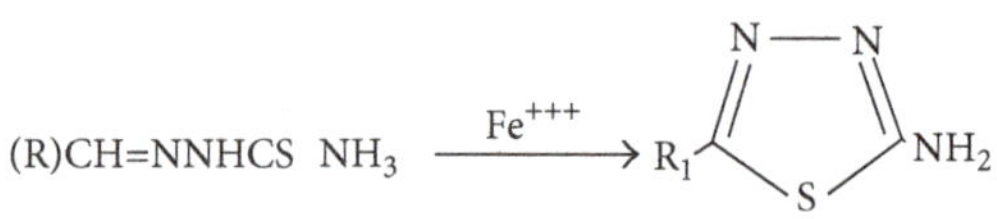

(32) (33)

R = Ar, ArCO, CH_3CO, EtCO

Scheme 9

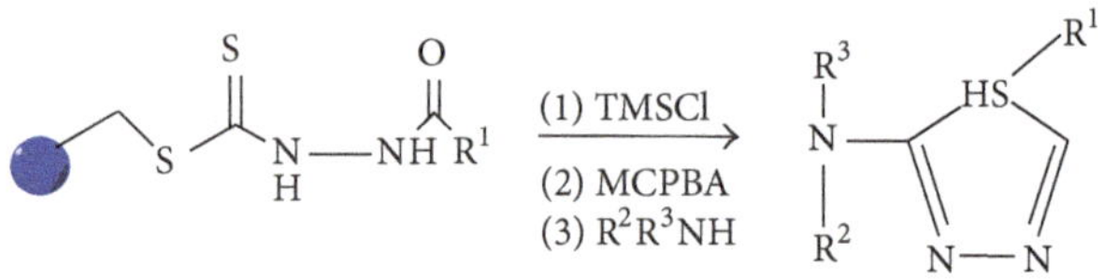

(34) (35)

Scheme 10

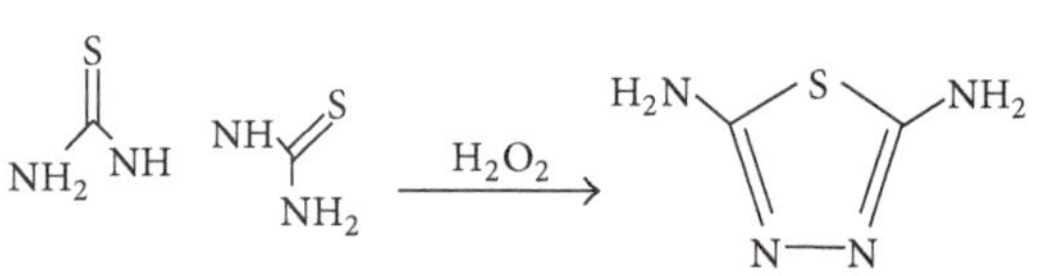

(36) (37)

Scheme 11

(2) *Nucleophilic Substitution.* Due to the electronegativity of the two nitrogen atoms in the ring, the carbon atom has low electron density and gives many possible nucleophilic reactions [18]. Halogenate 1,3,4-thiadiazoles are important intermediates in which the halogen atom is readily by nucleophiles. 5-Chloro-thiadiazole reacts readily with nucleophiles to give a series of 5 substituted-2-aryl-1,3,4-thiadiazoles (Scheme 19).

Amino thiadiazoles are rather weak bases; they are nucleophilic enough to be readily acetylated by acid chlorides or acetic acids (Scheme 20).

Aminothiadiazole forms diazonium salts and shows the coupling activity [19] (Scheme 21).

2-Amino-1,3,4-thiadiazoles and aromatic aldehydes form shiff bases [20] (Scheme 22).

5.4. Miscellaneous Reaction. Direct introduction of mercapto group on 2-phenyl-1, 3, 4-thiadiazoles is possible by the use of phosphorous pentasulphide [21] (Scheme 23).

5.4.1. Reduction of Thiadiazole Ring. 2-Amino-1,3,4-thiadiazoles reduces to benzaldehyde thio semicarbazone [22] (Scheme 24).

5.4.2. Reaction Involving Formation. On heating, 2-amino thiadiazole reacts with many dicarbonyl group containing compounds and forms various types of rings attached to thiadiazole (Scheme 25).

5.4.3. Rearrangement Reaction. Various 2-chloro and 2-amino substituted thiadiazoles undergo rearrangement by ring opening intermediate formation and again ring closure to form 4N substituted 1,3,4-thiadiazole 5(1)-terione [23, 24] (Scheme 26).

(a) Dimorph Rearrangement. See Scheme 27.

(b) Under Same Condition. See Scheme 28.

(c) Acid Catalysed. See Scheme 29.

(d) Cycloaddition Reaction. Aminobenzothiadiazoles are produced when 1,3,4-thiadiazoleinethiones (R^1=H, Me, Ac, R^2=Me, Ph) are reacted with benzyne, most likely by a 1,3-dipolar addition mechanism followed by elimination of RSCN (Scheme 30).

5.4.4. Reaction of Constituents on Carbon. Treatment of 2-methyl 1,5 diphenyl-1-1,3,4-thiadiazole (R=H) with *n*-butyllithium at −78°C gives the lithio derivative (R=Li)

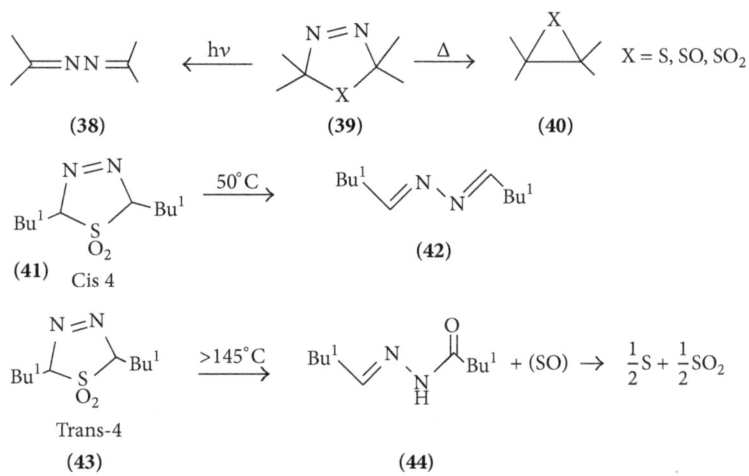

Scheme 12

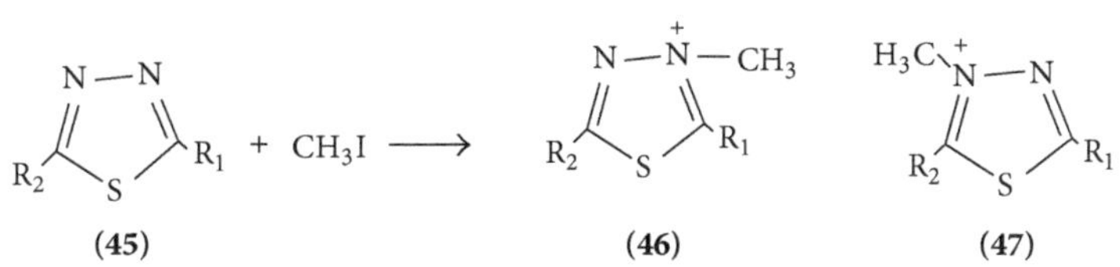

Scheme 13

which, on treatment with methyl iodide, gives the expected 2-ethyl homologue (R=Me); if (R=Li) is allowed to warm from −78°C to 25°C the dimer is formed, which on heating above 150°C converts into starting material. The dimerization proceeds via intermediates and not the ketone amine (Scheme 31).

5.4.5. Sulphur Substituents. The mesoionic 4,5-diphenyl-1-1,3,4-thiadiazole-2-thiolate reacts with methyl azodicarboxylate to yield the azothiadiazole and not the tetrazine betaine as previously claimed from 2-amino-5-phenyl-1,3,4-thiadiazole and nitrosobenzene (Scheme 32).

5.4.6. Amino Substituents. The reaction of unsubstituted 2-amino-1,3, and 4-thiadiazole with 1,3-dicarbonyl compounds is independent of the nature of the dicarbonyl compounds. The reaction of pentane-2-4-dione gives the 4,6-dimethyl 1,2-thiocyanatepyrimidine. The formation may proceed via the cation. With ethylacetoacetate, however, if a mixture formed also being converted in to on heating.

Biological Activities of 1,3,4-Thiadiazole Derivatives. During the recent years, there has been intense investigation of different classes of thiadiazole compounds, many of which possess extensive pharmacological activities. Among of these compounds having 1,3,4-thiadiazole nucleus are known to exhibit unique antimicrobial activity [25–38]. Antifungal [39–41], antidiabetic [42], anti-inflammatory [38, 43–54], antileishmanial activity [55–58], antituberculosis activity, [59–61], anticancer activity, [34, 62–68] anti-HIV activity [69, 70], antioxidant/radioprotective Activity [71–73], Carbonic anhydrase inhibitors [74, 75]. Anti-*Helicobacter pylori* activity [76, 77]. This paper focuses on the therapeutic importance of novel thiadiazole derivatives as anticonvulsant agents for future.

Anticonvulsant Activity of Thiadiazole Derivatives. A series of 3-aryl amino/amino-4-aryl-5-imino-D2-1,2,4-thiadiazole has been synthesized and screened for anticonvulsant activity. All the synthesized compounds were evaluated against maximal electroshock-induced seizure (MES) and subcutaneous-pentylenetetrazole- (ScPTZ-) induced seizure models in mice. Among the compounds tested, all showed protection from MES seizures, whereas only **(104) (105)** was found to be active in the ScPTZ test. The present results revealed that a number of 3-aryl amino/amino-4-aryl-5-imino-D2-1,2,4-thiadiazoles exhibit a range of activities in anticonvulsant screen [78].

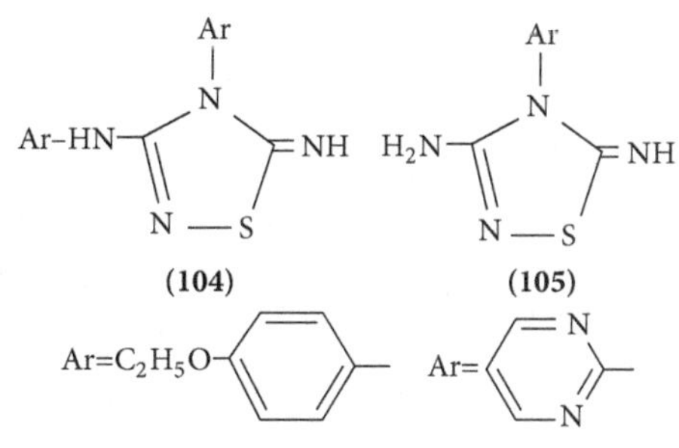

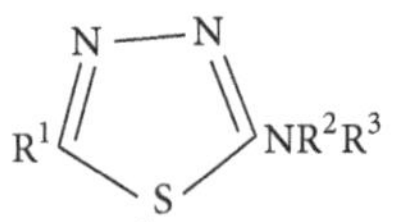

(1) $R^1 = H$
(2) $R^1 = Br$

(48)

(a) $R^2 = R^3 = H$
(b) $R^2 = H$; $R^3 = Me$
(c) $R^2 = R^3 = NO$
(d) $R^2 = H$; $R^3 = COMe$
(e) $R^2 = H$; $R^3 = COPh$
(f) $R^2 = R^3 = Me$
(g) $R^2 = Me$; $R^3 = NO$
(h) $R^2 = Me$; $R^3 = COMe$
(i) $R^2 = Me$; $R^3 = COPh$

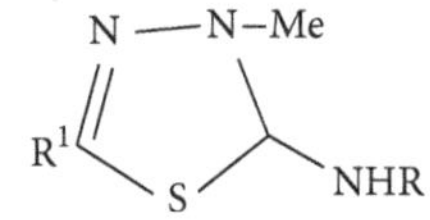

(3) $R^1 = H$
(4) $R^1 = Br$

(49)

(a) $R^2 = H$
(b) $R^2 = Me$
(c) $R^2 = NO$
(d) $R^2 = COMe$
(e) $R^2 = COPh$
(f) $R^2 = Br$

Scheme 14

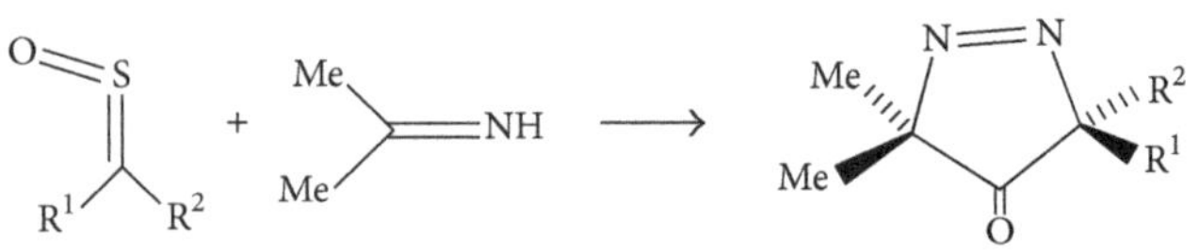

(50) **(51)** **(52)**

Scheme 15

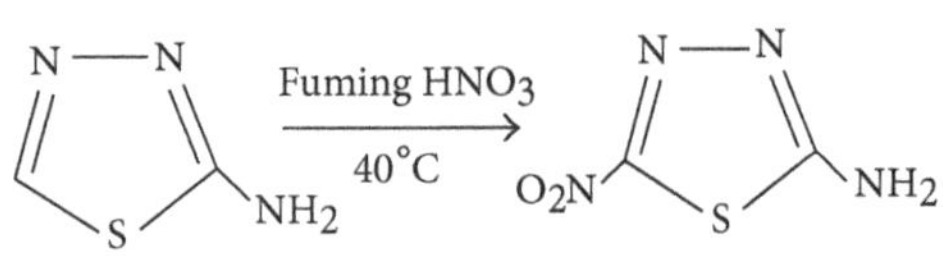

(53) **(54)**

Scheme 16

A series of new substituted 1,2,4-thiadiazoles were synthesized by appropriate route and screened for anticonvulsant, neurotoxic, and sedative-hypnotic activities. The structures of the synthesized compounds were confirmed by IR spectroscopy, C-13 NMR, and elemental (nitrogen and sulphur) analysis. After i.p. injection of the compounds to mice or rate at doses of 30, 100, and 300 mg/kg, body weights were examined in the maximal electroshock-induced seizure (MES) and subcutaneous-pentylenetetrazole- (scPTZ-) induced seizure models after 0.5 and 4 h. All the compounds showed protection against MES screen after 0.5 h. Compounds were active at the doses of 100 mg/kg and 300 mg/kg dose i.p. It may be concluded that the synthesized compounds were potent against MES-induced seizures than ScPTZ-induced seizures [79] (Structure **(106)**).

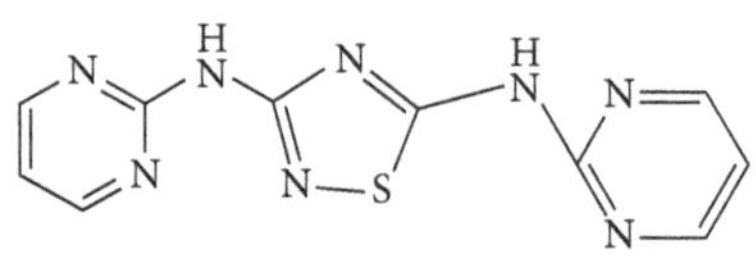

(106)

A series of five-membered heterocyclics were synthesized by the reaction between isoniazid and various substituted isothiocyanates and tested for their anticonvulsant activity by determining their ability to provide protection against convulsions induced by electroconvulsiometer. Among the synthesized compounds, (**107f**) 2-(4-chlorophenyl) amino-5-(4-pyridyl)-1,3,4-thiadiazole and (**108f**) 2-(4-chlorophenyl)amino-5-(4-pyridyl)-1,3,4-oxadiazole were found promising compounds of the series [80].

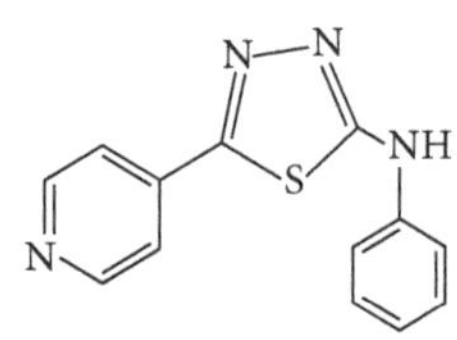

(107a–f)

Compound	R
107a, 108a	-H
107b, 108b	O-CH_3
107c, 108c	p-CH_3
107d, 108d	O-OCH_3
107e, 108e	P-OCH_3
107f, 108f	P-Cl

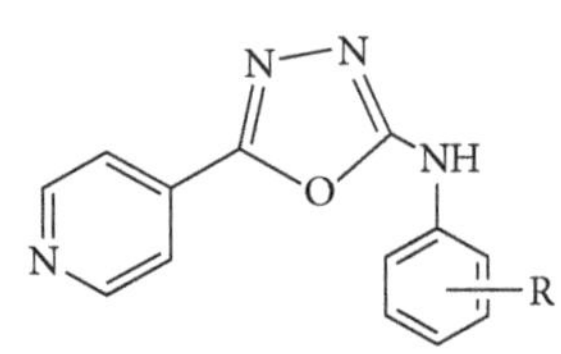

(108a–f)

A series of aromatic aldehyde imine derivative of 2-thiobezyl-1,3,4-thiadiazole were synthesized. These derivatives **109(a–e)**, **110(ia–ie)**, **110(iia–iie)** show good anticonvulsant activity. Among these compounds, chlorobenzyl substituted compound shows the potent anticonvulsant activity against MES method [81].

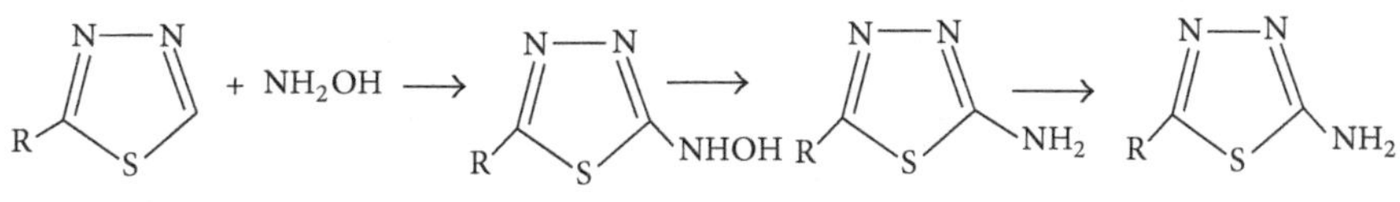

(55) (56) (57) (58)

Scheme 17

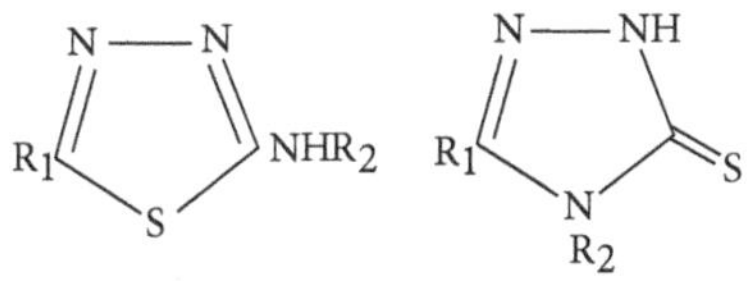

(59) (60)

Scheme 18

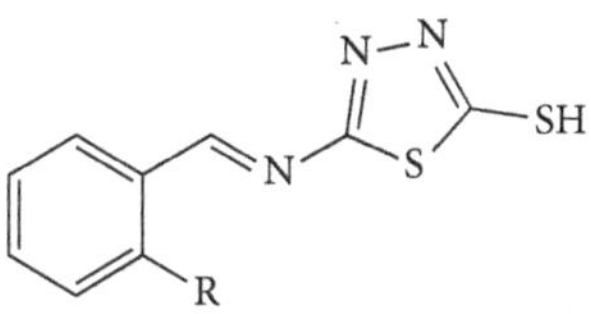

(109a–e)

Compounds	R
109a, 110ia, 110iia	2-Cl
109b, 110ib, 110iib	4-Cl
109c, 110ic, 110iic	2-NO_2
109d, 110id, 110iid	4-NO_2
109e, 110ie, 110iie	4-F

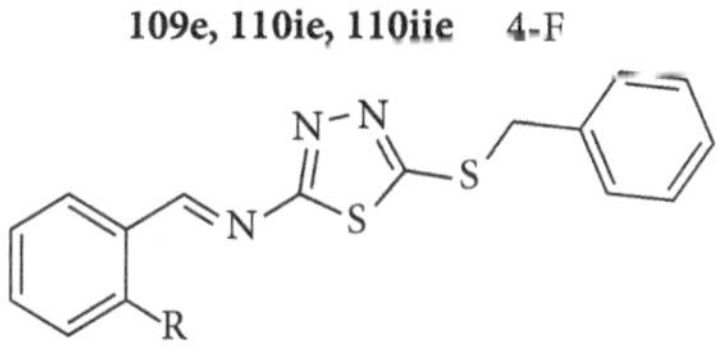

(110ia–8ie)

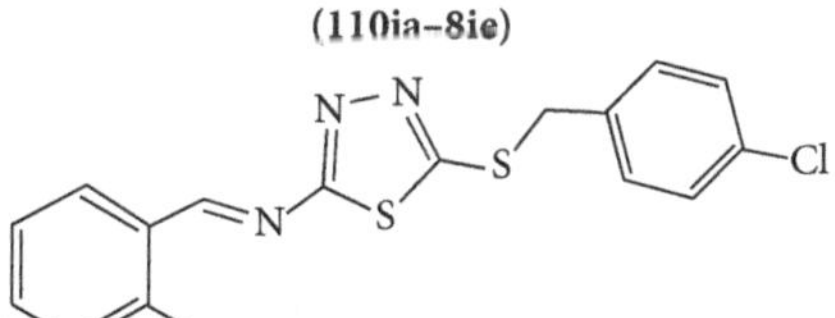

(110iia–15iie)

A variety of new 3-[5-substituted phenyl-1,3,4-thiadiazol-2-yl]-2-styryl quinazoline-4(3*H*)-ones were synthesized and evaluated for anticonvulsant activity b. Compounds were examined in the maximal-electroshock- (MES-) induced seizures and subcutaneous-pentylenetetrazole- (scPTZ-) induced seizure models. Compounds **(111a)**, **(111b)**, and **(111c)** showed good anticonvulsant activity in the test models [82].

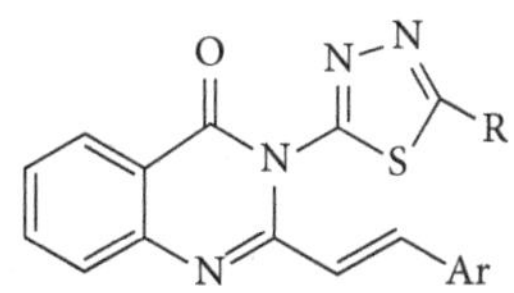

111a = R = C_6H_5; Ar = 4-ClC_6H_4

111b = R = 3-ClC_6H_4; Ar = 4-ClC_6H_4

111c = R = 4-ClC_6H_4; Ar = pyridyl

Two new series of 2,5-disubstituted-1,3,4-thiadiazoles were synthesized for their possible anticonvulsant, antibacterial, and antifungal activities. The degree of protection afforded by these compounds at a dose of 100 mg/kg i.p. against pentylenetetrazole-induced convulsions in mice ranged from 0% to 90%. Among these compounds, **112a** (90%) and **112g** (70%) showed maximum protection [83].

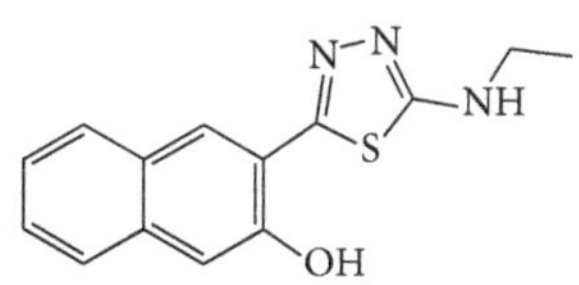

(112a)

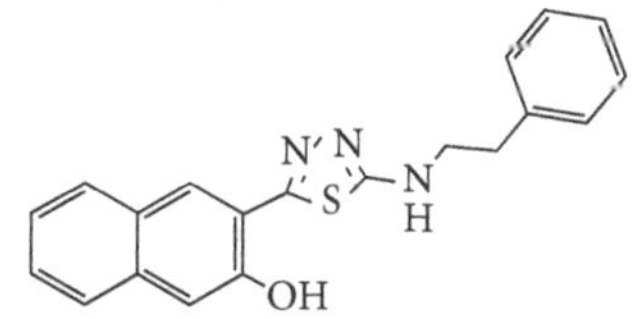

(112b)

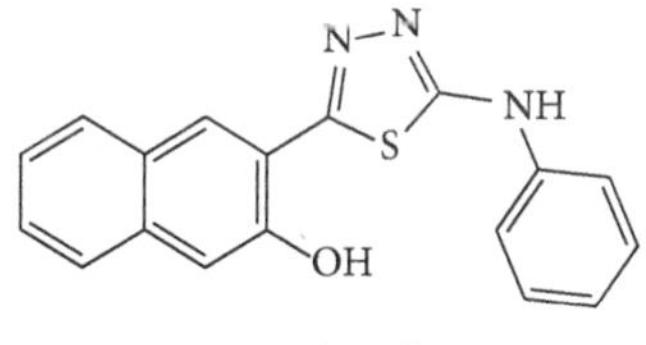

(112c)

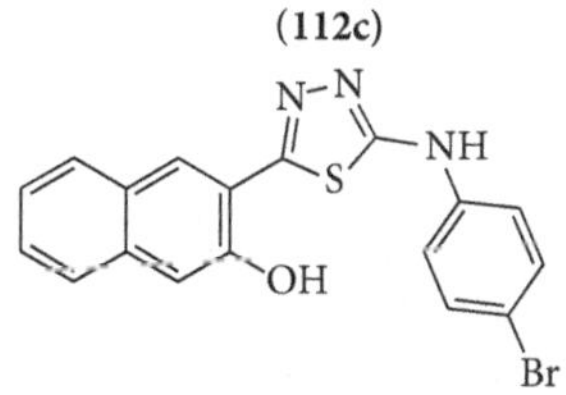

(112d)

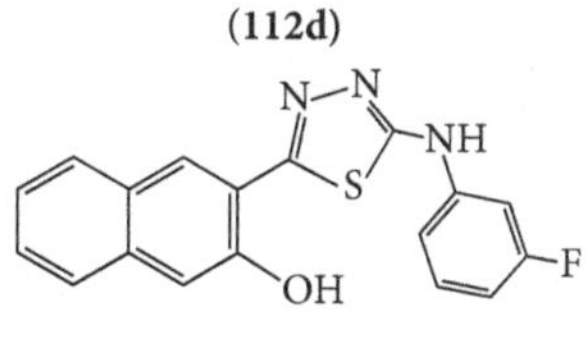

(112e)

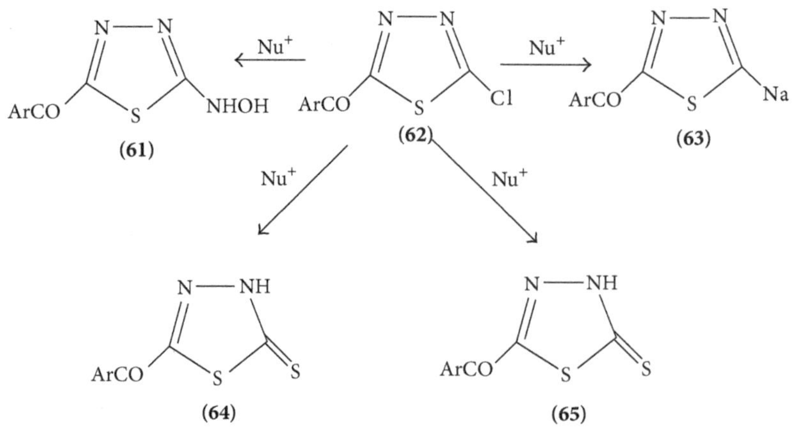

Scheme 19

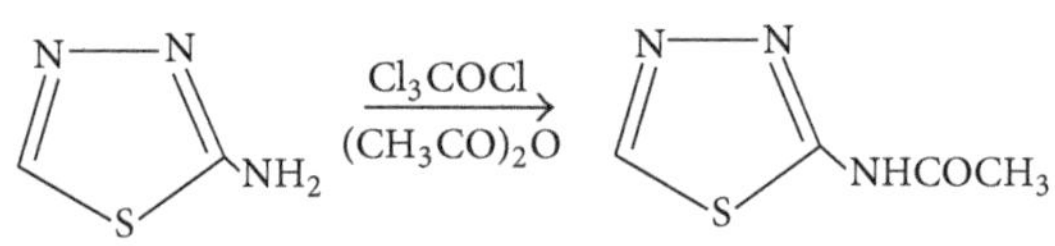

Scheme 20

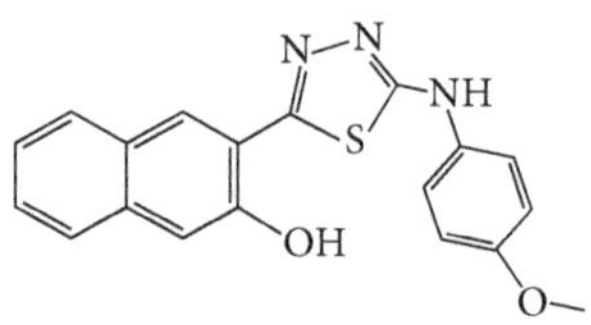

(112f)

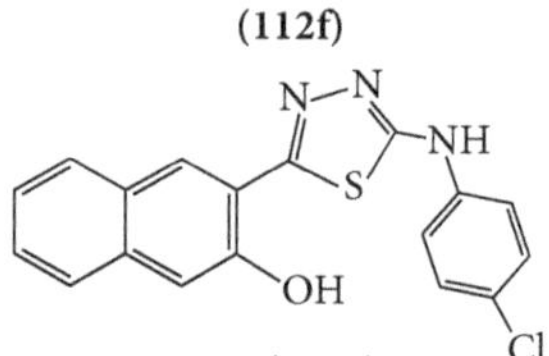

(112g)

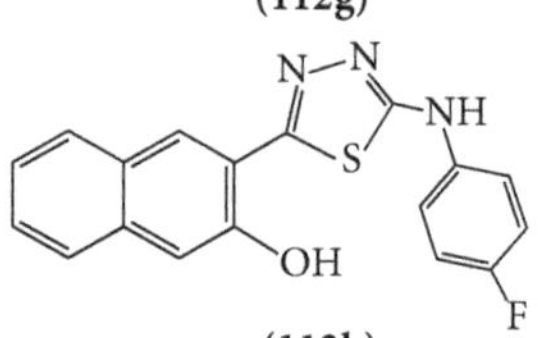

(112h)

3-[(2-Methyl-1*H*-3-indolyl)methyl]-4-aryl-4, 5-dihydro-1*H*-1,2,4-triazole-5-thiones and their respective *N*-{5-[(2-methyl-1*H*-3-indolyl)methyl]-1,3,4-thiadiazol-2-yl}-*N*-arylamines have been prepared. Behavioral effects, induced by the members of both series, in conjunction with their activity in some specific tests (forced swim, pentetrazol convulsions) on mice, show that these derivatives cross the blood-brain barrier and could develop an antidepressant activity comparable to that of imipramine. Blood-brain barrier penetration is also supported by the lipophilicity data obtained for all analogs [84]. (Structure **(113)**).

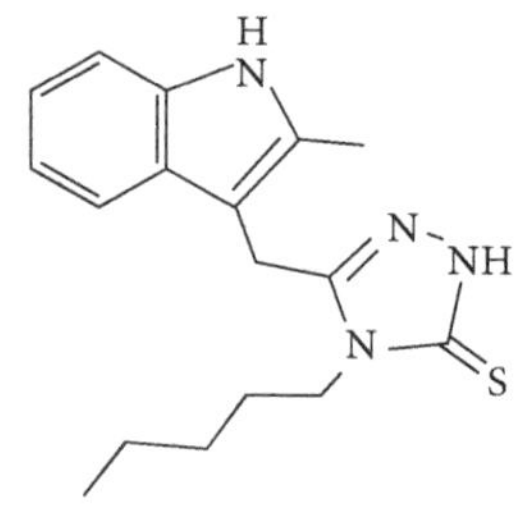

(113)

Several novel 2-amino-5-[4-chloro-2-(2-chlorophenoxy)phenyl]-1,3,4-thiadiazole derivatives 4a–d were synthesized, and their anticonvulsant activity was determined by evaluation of the ability of these compounds to protect mice against convulsion induced by lethal doses of pentylenetetrazole (PTZ) and maximal electroshock (MES). The result of anticonvulsant data shows that among the synthesized compounds, 5-[4-chloro-2-(2-chlorophenoxy)phenyl]-N-ethyl-1,3,4-thiadiazol-2-amine (**114, 115, 116**) was the most active compound in both MES and PTZ tests with an ED50 of 20.11 and 35.33 mg/kg, respectively [85].

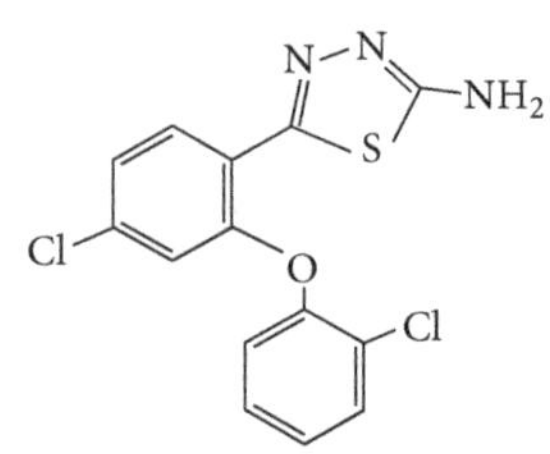

(114)

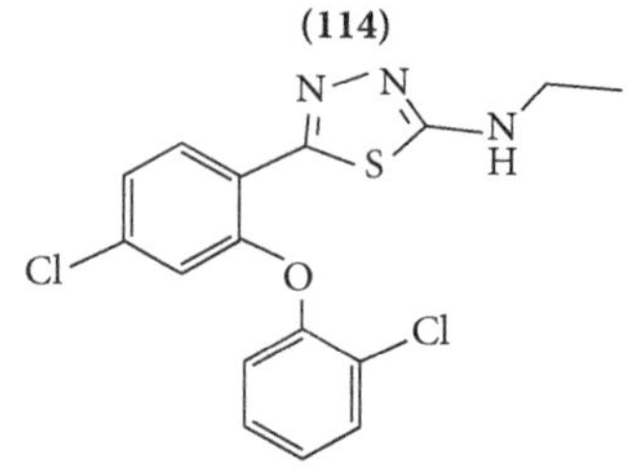

(115)

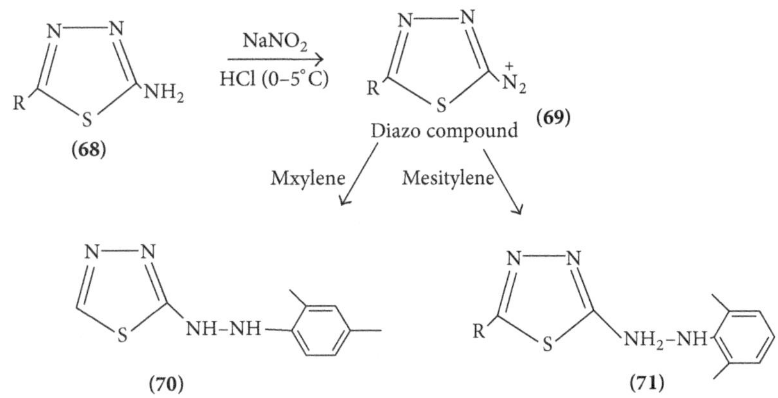

Scheme 21

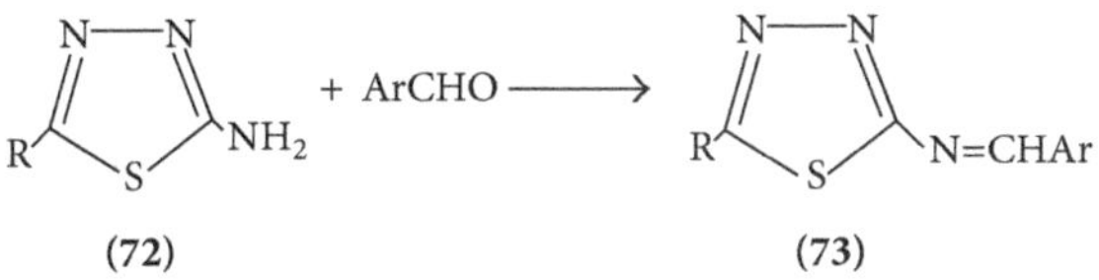

Scheme 22

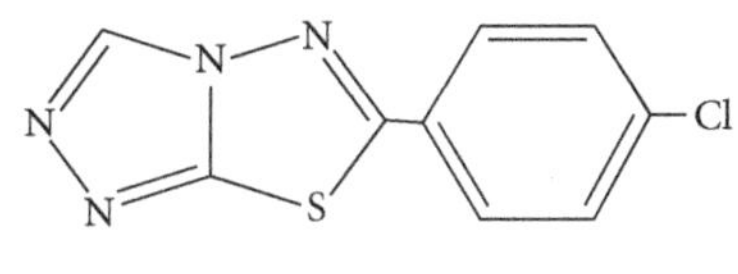

(117)

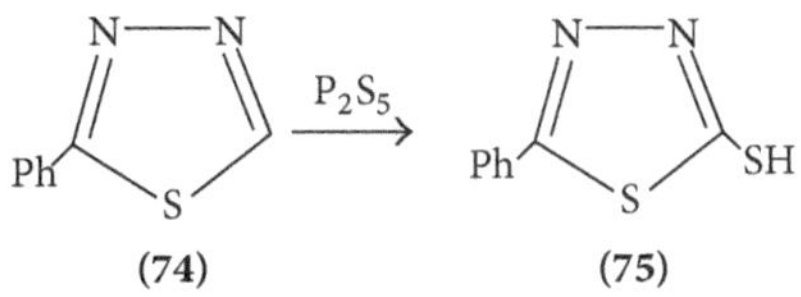

Scheme 23

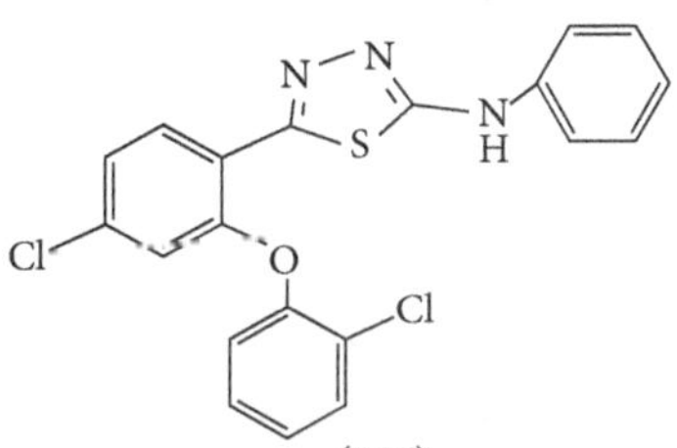

(116)

The present study describes the synthesis and anticonvulsant activity evaluation of 6-substituted-[1, 2, 4]triazolo[3, 4-b][1, 3, 4]thiadiazole derivatives and their partially dehydrogenated products 5,6-dihydro-6-substituted-[1, 2, 4]triazolo[3, 4-b][1, 3, 4]thiadiazole derivatives. The bioevaluation demonstrated that most compounds in the series exhibited potent anticonvulsant activity in the maximal electroshock test. Among which, 6-(4-chlorophenyl)-[1, 2, 4]triazolo[3,4-b][1, 3, 4]thiadiazole **(117)** emerged as the most promising candidate on the basis of its favorable ED50 value of 23.7 mg/kg and PI value of 10.8. In addition, the potency of compound **117** against seizures induced by pentylenetetrazole, 3-mercaptopropionic acid, and bicuculline in the chemical-induced seizure tests suggested that compound **117** displayed broad-spectrum activity in several models, and it may exert its anticonvulsant activity through affecting the GABAergic system [86].

Synthesis and pharmacological evaluation of a number of substituted 1,3,4-thiadiazole the first member of the series 2-(aminomethyl)-5-(2-biphenyl)-1,3,4-thiadiazole **(118)** was found to possess potent anticonvulsant property in rats and mice and compared favourable with the standard anticonvulsant drug phenytoin, Phenobarbital and carbamazepine in a number of test situations. The potency of compound was maintained on alkylation of the side chain nitrogen atom; however, aryl substitution on chain lengthening caused a drop in potency replacement of the two biphenyl group by phenyl or benzyl also lead to inactive compound [87].

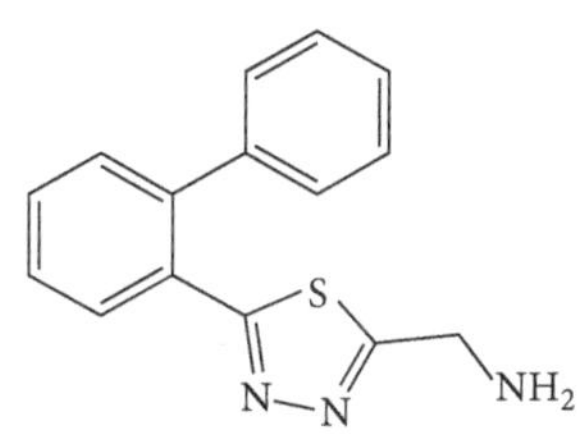

(118)

Various N-(5-chloro-6-substituted-benzothiazol-2-yl)-N′-(substituted phenyl)-[1, 3, 4]thiadiazole-2,5-diamines were designed and synthesized starting from substituted acetophenones. Structures of all the compounds were confirmed on the basis of spectral and elemental analyses. All the newly synthesized compounds were screened for their anticonvulsant activity and were compared with the standard drug phenytoin sodium. Interestingly, all the compounds showed protections against seizures in the range 50%–100% indicative of the promising nature of the compounds against seizure spread. Compound **119** showed complete protection against MES-induced seizures [88].

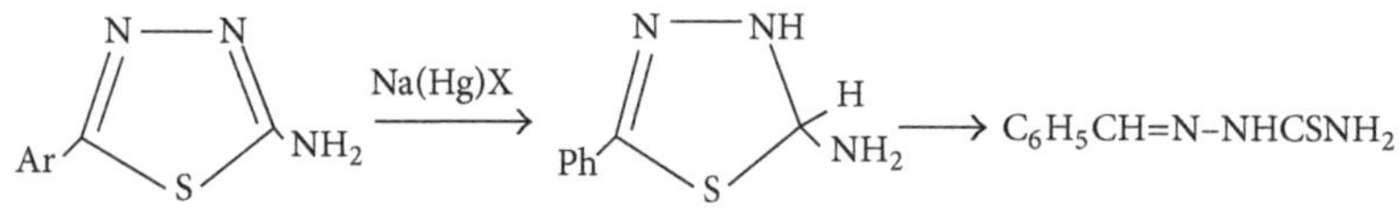

(76) **(77)** **(78)**

Scheme 24

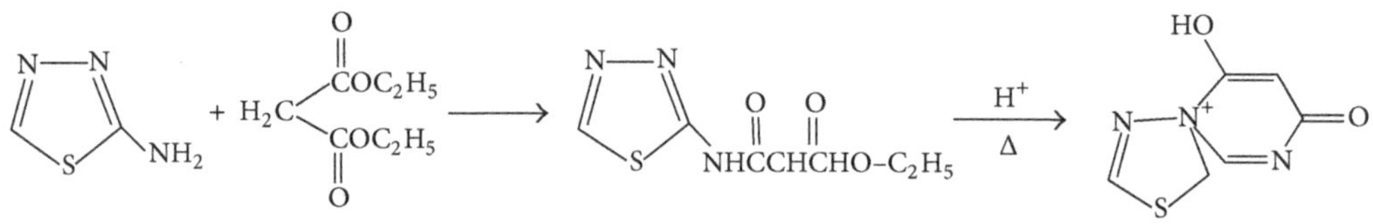

(79) **(80)** **(81)**

Scheme 25

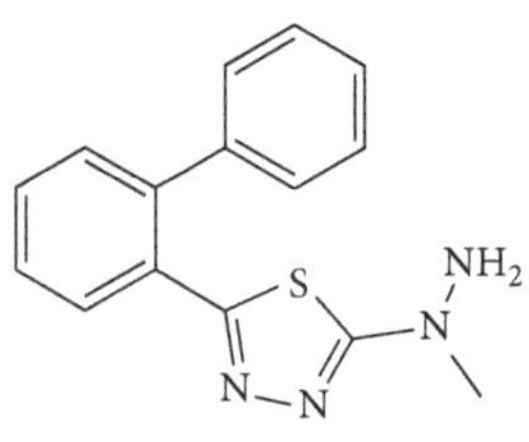

(119)

A series of 1,2,4-thiadiazoles **(120a–e)** were prepared and evaluated for anticonvulsant activity by Siddiqui et al. The compound with para-chloro substitution **(120c)** showed maximal activity in MES test and blocked strychnine seizures to some extent whereas other compounds of the series were less active [89].

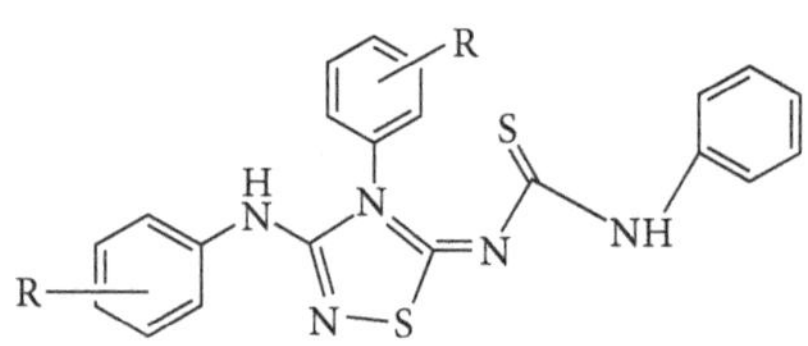

(120a–e)

R = H, 2-Cl, 4-Cl, 4-NO_2, 2-CH_3

A series of 1-(substituted phenyl)-3-[(5-substituted phenyl)-1,3,4-thiadiazol-2-yl]-2-thioxodihydropyrimidine-4,6 (1*H*,5*H*)-diones **121(a–w)** were designed, synthesized in good yields. The compounds were evaluated for anticonvulsant activity. The compounds were potent in MES test and were less neurotoxic as compared to standard drug phenytoin [90].

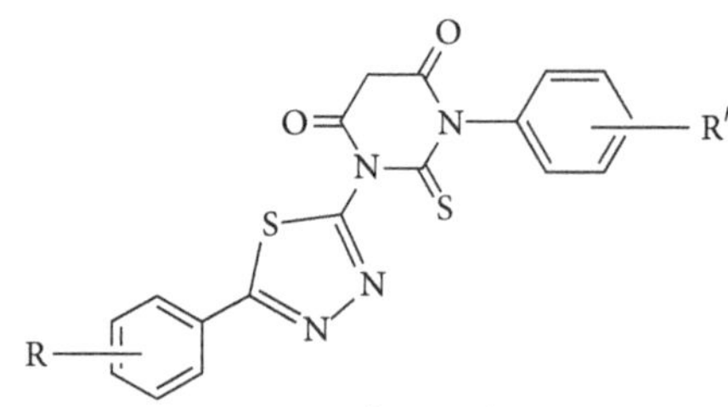

(121a–w)

R = 3-NO_2, 4-F, 2-Cl

R′ = H, 2-CH_3, 3-CH_3, 4-CH_3, 2-Cl, 3-Cl, 4-Cl

A series of thiadiazole derivatives were synthesized with differently substituted benzoic acids which were cyclized to give differently substituted thiazolidin-4-one. Elemental analysis, IR, HNMR,C NMR, and mass spectral data confirmed the structure of the synthesized compounds. The derivatives of these moieties were evaluated for anticonvulsant activity by MES model and neurotoxicity by rotarod method. The synthesized compounds showed good potential for anticonvulsant activity besides this, and the compounds also showed neurotoxic effect. It was observed that compounds with OCH3 at 3, 4 position of phenyl ring showed less protection against convulsions as compared to compounds having unsubstituted phenyl ring [91].

The synthesis and anticonvulsant activity of a series of 2-aryl-5-hydrazino-1,3,4-thiadiazoles are described. The combination of preferred aromatic substituents in the 2-position coupled with alkyl substitution on the hydrazine moiety led to a number of potent compounds lacking sedation, ataxia, or lethality. 5-(2-Biphenylyl)-2-(1-methylhydrazino)-1,3,4-thiadiazole (4m) represents a new class of anticonvulsant agent and compares favorably with the standard drugs phenytoin, phenobarbital, and carbamazepine [92] (Structure **(122)**).

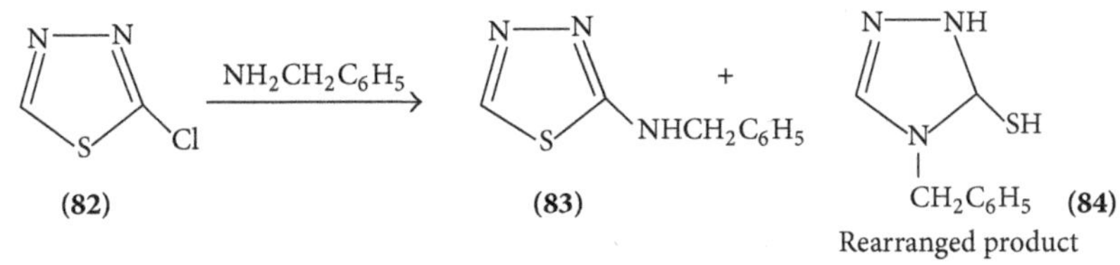

Scheme 26

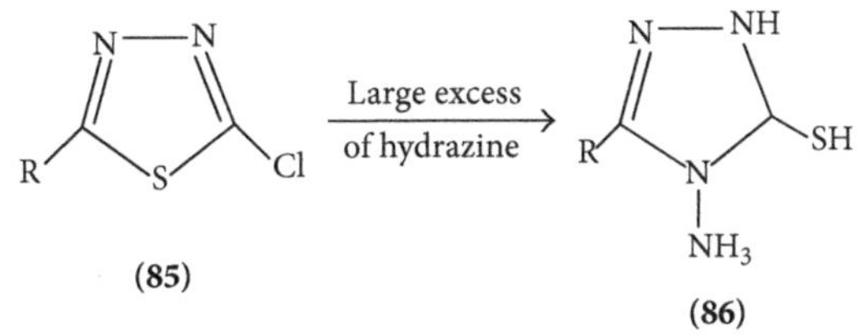

Scheme 27

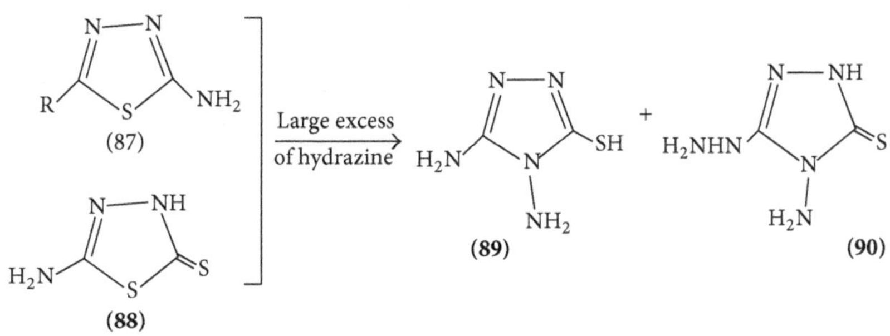

Scheme 28

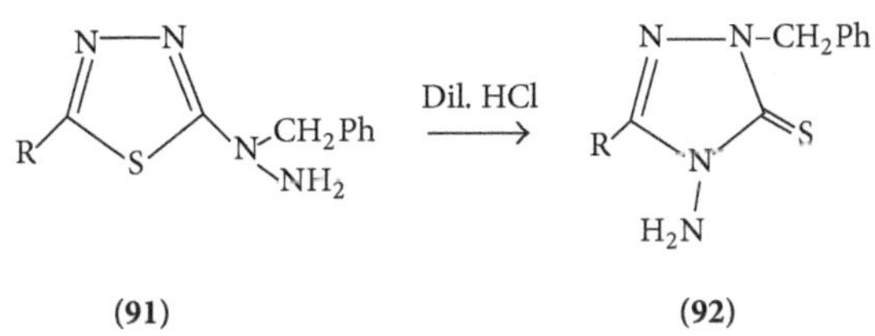

Scheme 29

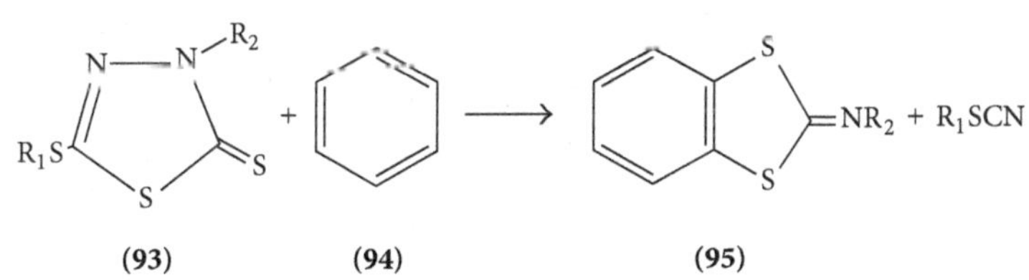

Scheme 30

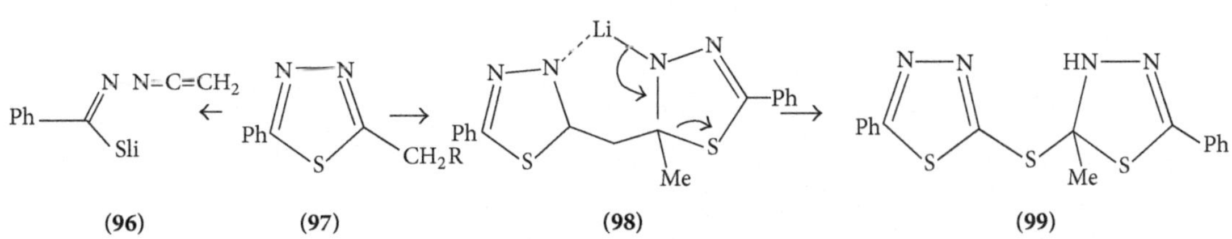

Scheme 31

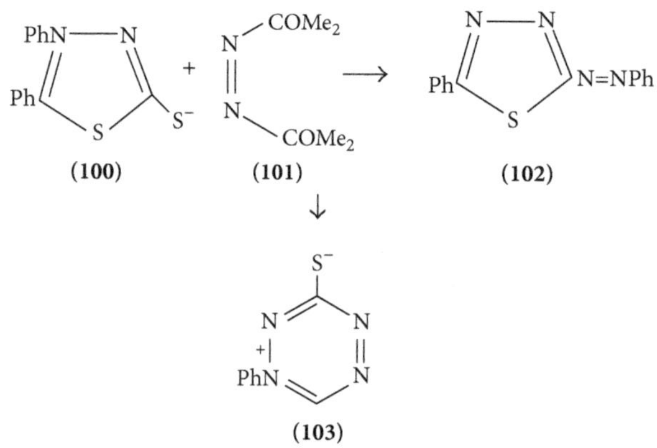

SCHEME 32

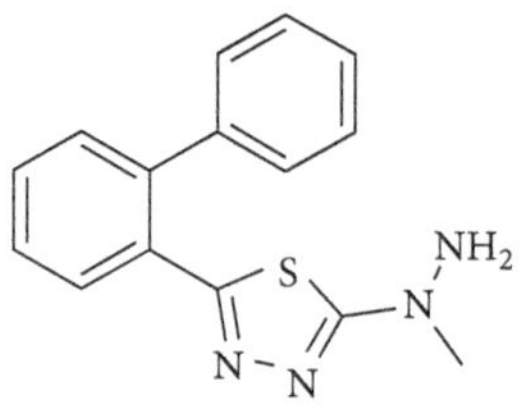

(122)

A novel series of NN′-{5-[(1H-indol-3-ylmethyl)-1,3,4-thiadiazol]-2-yl}-N4-(4-substituted benzaldehyde) semicarbazones, N1-{5-[(1H-indol-3-ylmethyl)-1,3,4-thiadiazol-2-yl}-N4-[1-(4-substituted phenyl)ethanone]-semicarbazones and N1-{5-[(1H-indol-3-ylmethyl)-1,3,4-thiadiazol-2-yl}-N4-[1-(4-substituted phenyl) (phenyl) methanone]-semicarbazones were synthesized and evaluated for their anticonvulsant potential using maximal electroshock seizure (MES) and subcutaneous pentylenetrtrazole (scPFZ) models. The minimal motor impairment (neurotoxicity) was determined by rotarod test. The results of the present study confirmed the requirements of various structural features of four binding site pharmacophore model for anticonvulsant activity [93].

Treatment of 2-bromoacetylbenzofuran with 1*H*-benzotriazole afforded 1-(benzofuran-2-yl)-2-(benzotriazol-1-yl) ethanone which reacted with phenylisothiocyanate to give the corresponding thioacetanilide derivatives. Treatment of the latter ethanone and thioacetanilide derivatives with hydrazonoyl chlorides afforded the corresponding pyrazole and 1,3,4-thiadiazolederivatives. The thioacetanilide derivative reacted with α-haloketones and α-halodiketones to afford thiophene and thiazole derivatives, respectively. The newly synthesized compounds were found to possess anticonvulsant and anti-inflammatory activities with the same mechanism of action of selective COX-2 inhibitors [94].

References

[1] W. Loscher, "New visions in the pharmacology of anticonvulsion," *European Journal of Pharmaceutical Sciences*, vol. 342, pp. 1–13, 1998.

[2] A. Gringauz John Wiley & Sons, New York, NY, USA, 1st edition, 1997.

[3] M. J. Brodie, "Monotherapy trials: prerequisite data," *Epilepsy Research*, vol. 45, no. 1–3, pp. 61–64, 2001.

[4] R. H. Mattson, "Efficacy and adverse effects of established and new antiepileptic drugs," *Epilepsia*, vol. 36, supplement 2, pp. S13–S26, 1995.

[5] A. Demirbas, D. Sahin, N. Demirbas, and S. A. Karaoglu, "Synthesis of some new 1,3,4-thiadiazol-2-ylmethyl-1,2,4-triazole derivatives and investigation of their antimicrobial activities," *European Journal of Medicinal Chemistry*, vol. 44, no. 7, pp. 2896–2903, 2009.

[6] R. T. Morrison and R. N. Boyd, *Organic Chemistry*, Prentice Hall, New Delhi, India, 6th edition, 2005.

[7] T.-B. Wei, H. Liu, J.-H. Hu et al., "Microwave promoted efficient synthesis of 2,5-disubstituted 1,3,4-thiadiazole," *Indian Journal of Chemistry B*, vol. 45, pp. 2754–2756, 2006.

[8] J. Goerdler, J. Ohm, and O. Tegmeyer, "Darstellung und Eigenschaften des 1.2.4- und des 1.3.4-Thiodiazols," *Chemische Berichte*, vol. 89, pp. 1534–1543, 1956.

[9] A. R. Katritzky and R. Charlews, in *Comprehensive Heterocyclic Chemistry*, vol. 6, pp. 568–575, Ist edition, 1984.

[10] E. Formm, E. Layer, and K. Nerz, *Advances in Heterocyclic Chemistry*, vol. 9, p. 188, 1968.

[11] H. B. Koing and H. A. Ofle, *Chemical Abstracts*, vol. 53, p. 10253, 1957.

[12] British Patent, Chemical abstract, 59, 1650, 916061, 1963.

[13] F. Kurzer, "Thiadiazoles. Part XI. Synthesis and cyclisation of N-(thiobenzamido)guanidines and related compounds," *Journal of the Chemical Society*, vol. 1961, pp. 1617–1625, 1961.

[14] R. R. Gupta, M. Kumar, and V. Gupta, *Heterocyclic Chemistry Springen*, vol. 2, pp. 566–573, 2005.

[15] G. Young and W. Eyne, "III.—Oxidation of benzalthiosemicarbazone," *Journal of the Chemical Society, Transactions*, vol. 79, pp. 54–60, 1901.

[16] J. Y. Hweng, H. Choi, D. H. Lee, and Y. D. Gong, "Solid-phase synthesis of 1,3,4-oxadiazole and 1,3,4-thiadiazole derivatives via selective, reagent-based cyclization of acyldithiocarbazate resins," *Journal of Combinatorial Chemistry*, vol. 7, pp. 816–819, 2005.

[17] P. C. Guha, "Ring closure of hydrazodithio- and -monothiodicarbonamides with acetic anhydride," *Journal of the American Chemical Society*, vol. 45, pp. 1036–1042, 1923.

[18] T. A. Bachetti and B. J. Banieli, *Advances in Heterocyclic Chemistry*, vol. 9, p. 1711, 1968.

[19] J. Goerdeler and H. Hanbrich, "Über die Kupplungsaktivität einiger heterocyclischer Diazoverbindungen," *Chemische Berichte*, vol. 93, pp. 397–405, 1960.

[20] M. Otha, H. Oya, and A. J. Miture, *Charm Soc Japan Chemical Abstract*, vol. 48, p. 10006, 1953.

[21] C. J. Anisworth, "The investigation of some substituted 1,3,4-thiadiazoles," *Journal of the American Chemical Society*, vol. 80, no. 19, p. 5201, 1958.

[22] F. F. Medovschikowa and I. Y. Postovskii, *Journal of Heterocyclic Chemistry*, vol. 9, p. 197, 1968.

[23] H. Sai Kachi and M. Kanaka, *Yakugakuzasthi Chemi Abstract*, vol. 58, p. 4543, 1963.

[24] H. Sai Kachi and M. Kanaka, *Yakugakuzasthi*, vol. 58, p. 7304, 1962.

[25] S. N. Swamy, Basappa, B. S. Priya et al., "Synthesis of pharmaceutically important condensed heterocyclic 4,6-disubstituted-1,2,4-triazolo-1,3,4-thiadiazole derivatives as antimicrobials," *European Journal of Medicinal Chemistry*, vol. 41, no. 4, pp. 531–538, 2006.

[26] S. Hussain, J. Sharma, and M. Amir, "Synthesis and antimicrobial activities of 1,2,4-triazole and 1,3,4-thiadiazole derivatives of 5-amino-2-hydroxybenzoic acid," *E-Journal of Chemistry*, vol. 5, no. 4, pp. 963–968, 2008.

[27] A. A. Aly and R. E. Sayed, "Synthesis and biological activity of new 1,3,4-thiadiazole derivatives," *Chemical Papers*, vol. 60, no. 1, pp. 56–60, 2006.

[28] R. Dua and S. K. Srivastava, "Synthesis, characterization and antimicrobial activity," *International Journal of Pharma and Bio Sciences*, vol. 1, no. 2, 2010.

[29] F. Hadizadeh and R. Vosooghi, "Synthesis of α-[5-(5-amino-1,3,4-thiadiazol-2-yl)-2-imidazolylthio] acetic acids," *Journal of Heterocyclic Chemistry*, vol. 45, no. 5, pp. 1477–1479, 2008.

[30] J. Salimon, N. Salih, A. Hameed, H. Ibraheem, and E. Yousif, "Synthesis and antibacterial activity of some new 1,3,4-oxadiazole and 1,3,4-thiadiazole derivatives," *Journal of Applied Sciences Research*, vol. 6, no. 7, pp. 866–870, 2010.

[31] A. M. Qandil, H. N. Tumah, and M. A. Hassan, "Synthesis and antibacterial activity of N4-benzoyl-N 1-dihydroxybenzoylthiosemicarbazides and their cyclic 1,3,4-thiadiazole derivatives," *Acta Pharmaceutica Sciencia*, vol. 48, no. 2, pp. 95–107, 2006.

[32] R. S. Lamani, N. S. Shetty, R. R. Kambl, and A. M. Khazi, "Synthesis and antimicrobial studies of novel methylene bridged benzisoxazolyl imidazo[2,1-*b*][1,3,4]thiadiazole derivatives," *European Journal of Medicinal Chemistry*, vol. 44, pp. 2828–2833, 2008.

[33] A. Foroumadi, S. Emami, A. Hassanzadeh et al., "Synthesis and antibacterial activity of N-(5-benzylthio-1,3,4-thiadiazol-2-yl) and N-(5-benzylsulfonyl-1,3,4-thiadiazol-2-yl)piperazinyl quinolone derivatives," *Bioorganic and Medicinal Chemistry Letters*, vol. 15, no. 20, pp. 4488–4492, 2005.

[34] T. Onkol, D. S. Doruer, L. Uzun, S. Adak, S. Ozkan, and M. F. Ahin, "Synthesis and antimicrobial activity of new 1,2,4-triazole and 1,3,4-thiadiazole derivatives," *Journal of Enzyme Inhibition and Medicinal Chemistry*, vol. 23, no. 2, pp. 277–284, 2008.

[35] N. Siddiqui and M. S. Alam, "5-(1H-indol-3-ylmethyl)-N-(substituted phenyl)-1, 2, 4-thiadiazol-2-amine derivatives: synthesis and biological screening," *Biosciences Biotechnology Research Asia*, vol. 6, no. 1, pp. 261–264, 2009.

[36] P. Karegoudar, D. J. Prasad, M. Ashok, M. Mahalinga, B. Poojary, and B. S. Holla, "Synthesis, antimicrobial and anti-inflammatory activities of some 1,2,4-triazolo[3,4-b][1,3,4] thiadiazoles and 1,2,4-triazolo[3,4-b][1,3,4]thiadiazines bearing trichlorophenyl moiety," *European Journal of Medicinal Chemistry*, vol. 43, no. 4, pp. 808–815, 2008.

[37] T. Karabasanagouda, A. V. Adhikari, and N. S. Shetty, "Synthesis and antimicrobial activities of some novel 1,2,4-triazolo[3,4-*b*]-1,3,4-thiadiazoles and 1,2,4-triazolo[3,4-*b*]-1,3,4-thiadiazines carrying thioalkyl and sulphonyl phenoxy moieties," *European Journal of Medicinal Chemistry*, vol. 42, no. 4, pp. 521–529, 2007.

[38] V. Mathew, J. Keshavayya, V. P. Vaidya, and D. Giles, "Studies on synthesis and pharmacological activities of 3,6-disubstituted-1,2,4-triazolo[3,4-*b*]-1,3,4-thiadiazoles and their dihydro analogues," *European Journal of Medicinal Chemistry*, vol. 42, no. 6, pp. 823–840, 2007.

[39] C. J. Chen, B. A. Song, S. Yang et al., "Synthesis and antifungal activities of 5-(3,4,5-trimethoxyphenyl)-2-sulfonyl-1,3,4-thiadiazole and 5-(3,4,5-trimethoxyphenyl)-2-sulfonyl-1,3,4-oxadiazole derivatives," *Bioorganic and Medicinal Chemistry*, vol. 15, pp. 3981–3989, 2007.

[40] H. S. Chen, H. S. LI, H. S. Z, Y. F. Han, and Z. W. Wang, "New fungicidally active pyrazolyl-substituted 1,3,4-thiadiazole compounds and their preparation," *Chinese Chemical Letters*, vol. 10, no. 5, pp. 365–366, 1999.

[41] M. Barboiu, M. Cimpoesu, C. Guran, and C. T. Supumn, "1, 3, 4-thiadiazole derivatives. Part 9[1]. Synthesis and biological activity of metal complexes of 5-(2-aminoethyl)-2-amino-1,3,4-thiadiazole," *Metal Based Drugs*, vol. 3, no. 5, pp. 227–232, 1996.

[42] S. R. Pattan, P. kekare, N. S. Dighe et al., "Synthesis and biological evaluation of some 1,3,4-thiadiazoles," *Journal of Chemical and Pharmaceutical Research*, vol. 1, no. 1, pp. 191–198, 2009.

[43] M. Amir, H. Kumar, and S. A. Javed, "Condensed bridgehead nitrogen heterocyclic system: Synthesis and pharmacological activities of 1,2,4-triazolo-[3,4-*b*]-1,3,4-thiadiazole derivatives of ibuprofen and biphenyl-4-yloxy acetic acid," *European Journal of Medicinal Chemistry*, vol. 43, no. 10, pp. 2056–2066, 2008.

[44] M. Amir, H. Kumar, and S. A. Javed, "Synthesis and pharmacological evaluation of condensed heterocyclic 6-substituted-1,2,4-triazolo[3,4-*b*]-1,3,4-thiadiazole derivatives of naproxen," *Bioorganic and Medicinal Chemistry Letters*, vol. 17, pp. 4504–4508, 2007.

[45] L. S. Varandas, C. A. M. Fraga, A. L. P. Miranda, and E. J. Barreiro, "Design, synthesis and pharmacological evaluation of new nonsteroidal antiinflammatory 1,3,4-thiadiazole derivatives," *Letters in Drug Design and Discovery*, vol. 2, no. 1, pp. 62–67, 2005.

[46] M. Moise, V. Sunel, L. Profire, M. Popa, J. Desbrieres, and C. Peptu, "Synthesis and biological activity of some new 1,3,4-thiadiazole and 1,2,4-triazole compounds containing a phenylalanine moiety," *Molecules*, vol. 14, no. 7, pp. 2621–2631, 2009.

[47] H. Kumar, S. A. Javed, S. A. Khan, and M. Amir, "1,3,4-Oxadiazole/thiadiazole and 1,2,4-triazole derivatives of biphenyl-4-yloxy acetic acid: Synthesis and preliminary evaluation of biological properties ," *European Journal of Medicinal Chemistry*, vol. 43, no. 12, pp. 2688–2698, 2008.

[48] S. Schenone, C. Brullo, O. Bruno et al., "New 1,3,4-thiadiazole derivatives endowed with analgesic and anti-inflammatory activities," *Bioorganic and Medicinal Chemistry*, vol. 14, no. 6, pp. 1698–1705, 2006.

[49] S. G. Kucukguzel, I. Kucukguze, I. E. Tatar et al., "Synthesis of some novel heterocyclic compounds derived from diflunisal hydrazide as potential anti-infective and anti-inflammatory agents," *European Journal of Medicinal Chemistry*, vol. 42, no. 7, pp. 893–901, 2007.

[50] M. A. Hilfiker, N. Wang, X. Hou et al., "Discovery of novel aminothiadiazole amides as selective EP3 receptor antagonists," *Bioorganic and Medicinal Chemistry Letters*, vol. 19, no. 15, pp. 4292–4295, 2009.

[51] A. K. Gadad, M. B. Palkar, K. Anand, M. N. Noolvi, T. S. Boreddy, and J. Wagwade, "Synthesis and biological evaluation of 2-trifluoromethyl/sulfonamido-5,6-diaryl substituted imidazo[2,1-b]-1,3,4-thiadiazoles: a novel class of cyclooxygenase-2 inhibitors," *Bioorganic and Medicinal Chemistry*, vol. 16, no. 1, pp. 276–283, 2008.

[52] U. S. Goksen, N. G. Kelekci, O. Goktas et al., "1-Acylthi-osemicarbazides, 1,2,4-triazole-5(4*H*)-thiones, 1,3,4-thiadiazoles and hydrazones containing 5-methyl-2-benzoxazolinones: synthesis, analgesic-anti-inflammatory and antimicrobial activities," *Bioorganic and Medicinal Chemistry*, vol. 15, no. 17, pp. 5738–5751, 2007.

[53] A. S. Kalgutar, A. B. Marnett, B. C. Crews, R. P. Remmel, and L. J. Marnett, "Ester and Amide derivatives of the nonsteroidal anti-inflammatory drug, indomethacin, as selective cyclooxygenase-2 inhibitors," *Journal of Medicinal Chemistry*, vol. 43, no. 15, pp. 2860–2870, 2000.

[54] M. Duflos, M. R. Nourrisson, J. Brelet et al., "N-Pyridinyl-indole-3-(alkyl)carboxamides and derivatives as potential systemic and topical inflammation inhibitors," *European Journal of Medicinal Chemistry*, vol. 36, no. 6, pp. 545–553, 2001.

[55] V. J. Ram, A. Goel, and M. Kandpal, "Tetraazaacenaphthene, tetraazaphenalene and 1,3,4-thiadiazole derivatives as potential leishmanicides," *Bioorganic and Medicinal Chemistry Letters*, vol. 7, no. 6, pp. 651–656, 1997.

[56] E. F. da Silva, M. M. Canto-Cavalheiro, V. R. Braz, L. Cysne-Finkelstein, L. L. Leon, and A. Echevarria, "Synthesis, and biological evaluation of new 1,3,4-thiadiazolium-2-phenylamine derivatives against Leishmania amazonensis promastigotes and amastigotes," *European Journal of Medicinal Chemistry*, vol. 37, no. 12, pp. 979–984, 2002.

[57] M. Behrouzi-Fardmoghadam, F. Poorrajab, S. K. Ardestani, S. Emami, A. Shafiee, and A. Foroumadi, "Synthesis and in vitro anti-leishmanial activity of 1-[5-(5-nitrofuran-2-yl)-1,3,4-thiadiazol-2-yl]- and 1-[5-(5-nitrothiophen-2-yl)-1,3,4-thiadiazol-2-yl]-4-aroylpiperazines," *Bioorganic and Medicinal Chemistry*, vol. 16, no. 8, pp. 4509–4515, 2008.

[58] A. Foroumadi, S. Emami, S. Pournourmohammadi, A. Kharazmi, and A. Shafiee, "Synthesis and in vitro leishmanicidal activity of 2-(1-methyl-5-nitro-1H- imidazol-2-yl)-5-substituted-1,3,4-thiadiazole derivatives," *European Journal of Medicinal Chemistry*, vol. 40, no. 12, pp. 1346–1350, 2005.

[59] S. Karakus and S. Rollas, "Synthesis and antituberculosis activity of new N-phenyl-N′-[4-(5-alkyl/arylamino-1,3,4-thiadiazole-2-yl)phenyl]thioureas," *Il Farmaco*, vol. 57, no. 7, pp. 577–581, 2002.

[60] E. E. Oruc, S. Rollas, F. Kandermirli, N. Shvets, and A. S. Dimoglo, "1,3,4-thiadiazole derivatives. Synthesis, structure elucidation, and structure–antituberculosis activity relationship investigation," *Journal of Medicinal Chemistry*, vol. 47, pp. 6760–6676, 2004.

[61] A. Foroumadi, Z. Kargar, A. Sakhteman et al., "Synthesis and antimycobacterial activity of some alkyl [5-(nitroaryl)-1,3,4-thiadiazol-2-ylthio]propionates," *Bioorganic and Medicinal Chemistry Letters*, vol. 16, no. 5, pp. 1164–1167, 2006.

[62] S. Karakus, U. Coruh, B. Barlas-Durgun et al., "Synthesis and cytotoxic activity of some 1,2,4-triazoline-3-thione and 2,5-disubstituted-1,3,4-thiadiazole derivatives," *Marmara Pharmaceutical Journal*, vol. 14, no. 2, pp. 84–90, 2010.

[63] J. Matysiak and A. Opolski, "Synthesis and antiproliferative activity of N-substituted 2-amino-5-(2,4-dihydroxyphenyl)-1,3,4-thiadiazoles," *Bioorganic and Medicinal Chemistry*, vol. 14, no. 13, pp. 4483–4489, 2006.

[64] V. Padmavathi, G. Sudhakar Reddy, A. Padmaja, P. Kondaiah, and Ali-Shazia, "Synthesis, antimicrobial and cytotoxic activities of 1,3,4-oxadiazoles, 1,3,4-thiadiazoles and 1,2,4-triazoles," *European Journal of Medicinal Chemistry*, vol. 44, no. 5, pp. 2106–2112, 2009.

[65] M. X. Wei, L. Feng, X. Q. Li, X. Z. Zhou, Z.-H. Shao, and H. Shao, "Synthesis of new chiral 2,5-disubstituted 1,3,4-thiadiazoles possessing γ-butenolide moiety and preliminary evaluation of in vitro anticancer activity," *European Journal of Medicinal Chemistry*, vol. 44, no. 8, pp. 3340–3344, 2009.

[66] D. A. Ibrahim, "Synthesis and biological evaluation of 3,6-disubstituted [1,2,4]triazolo[3,4-b][1,3,4]thiadiazole derivatives as a novel class of potential anti-tumor agents," *European Journal of Medicinal Chemistry*, vol. 44, pp. 2776–2781, 2009.

[67] J. Matysiak, A. Nasulewicz, M. Pelczynska, M. Switalska, I. Jaroszewicz, and A. Opolski, "Synthesis and antiproliferative activity of some 5-substituted 2-(2,4-dihydroxyphenyl)-1,3,4-thiadiazoles ," *European Journal of Medicinal Chemistry*, vol. 41, pp. 475–482, 2006.

[68] A. T. Mavrova, D. Wesselinova, Y. A. Tsenov, and P. Denkova, "Synthesis, cytotoxicity and effects of some 1,2,4-triazole and 1,3,4-thiadiazole derivatives on immunocompetent cells," *European Journal of Medicinal Chemistry*, vol. 44, no. 1, pp. 63–69, 2009.

[69] P. Zhan, X. Liu, Z. Li et al., "Novel 1,2,3-thiadiazole derivatives as HIV-1 NNRTIs with improved potency: synthesis and preliminary SAR studies," *Bioorganic and Medicinal Chemistry*, vol. 17, pp. 5920–5927, 2009.

[70] P. Zhan, X. Liu, Z. Fang, Z. Li, C. Pannecouque, and E. D. Clercq, "Synthesis and anti-HIV activity evaluation of 2-(4-(naphthalen-2-yl)-1,2,3-thiadiazol-5-ylthio)-*N*-acetamides as novel non-nucleoside HIV-1 reverse transcriptase inhibitors," *European Journal of Medicinal Chemistry*, vol. 44, pp. 4648–4653, 2009.

[71] C. Kus, G. A. Kilcigil, S. Ozbey et al., "Synthesis and antioxidant properties of novel *N*-methyl-1,3,4-thiadiazol-2-amine and 4-methyl-2*H*-1,2,4-triazole-3(4*H*)-thione derivatives of benzimidazole class," *Bioorganic and Medicinal Chemistry*, vol. 16, pp. 4294–4303, 2008.

[72] S. Dhanya, A. M. Isloor, P. Shetty, K. Satyamoorthy, and A. S. Bharath Prasad, *Arabian Journal of Chemistry*. In press.

[73] D. Cressier, C. Prouillac, P. Hernandez et al., "Synthesis, antioxidant properties and radioprotective effects of new benzothiazoles and thiadiazoles," *Bioorganic and Medicinal Chemistry*, vol. 17, no. 14, pp. 5275–5284, 2009.

[74] V. Dudutien, L. Baranauskien, and D. Matulis, "Benzimidazo[1,2-*c*][1,2,3]thiadiazole-7-sulfonamides as inhibitors of carbonic anhydrase," *Bioorganic and Medicinal Chemistry Letters*, vol. 17, pp. 3335–3338, 2007.

[75] G. L. Almajan, A. Innocenti, L. Puccetti et al., "Carbonic anhydrase inhibitors. Inhibition of the cytosolic and tumor-associated carbonic anhydrase isozymes I, II, and IX with

a series of 1,3,4-thiadiazole- and 1,2,4-triazole-thiols," *Bioorganic and Medicinal Chemistry Letters*, vol. 15, no. 9, pp. 2347–2352, 2005.

[76] J. Mirzaei, F. Siavoshi, S. Emami et al., "Synthesis and in vitro anti-Helicobacter pylori activity of N-[5-(5-nitro-2-heteroaryl)-1,3,4-thiadiazol-2-yl]thiomorpholines and related compounds," *European Journal of Medicinal Chemistry*, vol. 43, no. 8, pp. 1575–1580, 2008.

[77] A. Foroumadi, A. Rineh, S. Emami et al., "Synthesis and anti-Helicobacter pylori activity of 5-(nitroaryl)-1,3,4-thiadiazoles with certain sulfur containing alkyl side chain," *Bioorganic and Medicinal Chemistry Letters*, vol. 18, no. 11, pp. 3315–3320, 2008.

[78] A. Gupta, P. Mishra, S. K. Kashaw, V. Jatav, and J. P. Stables, "Synthesis and anticonvulsant activity of some novel 3-aryl amino/amino-4-aryl-5-imino-Δ2-1,2,4-thiadiazoline," *European Journal of Medicinal Chemistry*, vol. 43, no. 4, pp. 749–754, 2008.

[79] A. Gupta, P. Mishra, S. N. Pandeya, S. K. Kashaw, V. Kashaw, and J. P. Stables, "Synthesis and anticonvulsant activity of some substituted 1,2,4-thiadiazoles," *European Journal of Medicinal Chemistry*, vol. 44, no. 3, pp. 1100–1105, 2009.

[80] M. S. Yar and M. W. Akhter, "Synthesis and anticonvulsant activity of substituted oxadiazole and thiadiazole derivatives," *Acta Poloniae Pharmaceutica*, vol. 66, no. 4, pp. 393–397, 2009.

[81] B. Ahamad and M. Yusuf, "synthesis of aromatic aldehyde imine derivative of 2-thiobenzyl-1,3,4-thiadiazole and evaluation of their anticonvulsant activity," *Indian Journal of Chemistry B*, vol. 49, pp. 241–246, 2010.

[82] V. Jatav, P. Mishra, S. Kashaw, and J. P. Stables, "CNS depressant and anticonvulsant activities of some novel 3-[5-substituted 1,3,4-thiadiazole-2-yl]-2-styryl quinazoline-4(3*H*)-ones," *European Journal of Medicinal Chemistry*, vol. 43, pp. 1945–1954, 2008.

[83] H. N. Dogan, A. Duran, S. Rollas, G. Sener, M. K. Uysal, and D. Gulen, "Synthesis of new 2, 5-disubstituted-1, 3, 4-thiadiazoles and preliminary evaluation of anticonvulsant and antimicrobial activities," *Bioorganic & Medicinal Chemistry*, vol. 10, no. 9, pp. 2893–2898, 2002.

[84] A. Varvaresou, T. Siatra-Papastaikoudi, A. Tsotinis, A. T. Kakoulidou, and A. Vamvakides, "Synthesis, lipophilicity and biological evaluation of indole-containing derivatives of 1,3,4-thiadiazole and 1,2,4-triazole," *Il Farmaco*, vol. 53, no. 5, pp. 320–326, 1998.

[85] A. Foroumadi, V. Sheibani, A. Sakhteman et al., "Synthesis and anticonvulsant activity of novel 2-amino-5-[4-chloro-2-(2- chlorophenoxy) phenyl]-1,3,4-thiadiazole derivatives," *Daru Journal of Pharmaceutical Sciences* , vol. 15, no. 2, pp. 89–93, 2007.

[86] X. Q. Deng, Z. Q. Dong, M. X. Song, B. Shu, S. B. Wang, and Z. S. Quan, "Synthesis and anticonvulsant activities of some triazolothiadiazole derivatives," *Archiv der Pharmazie*, vol. 345, no. 7, pp. 565–573, 2012.

[87] M. R. Stillings, P. W. Anthony, and S. W. Donald, "Substituted 1,3,4-thiadiazoles with anticonvulsant activity. 2. Aminoalkyl derivatives," *Journal of Medicinal Chemistry*, vol. 29, no. 11, pp. 2280–2284, 1986.

[88] N. Siddiqui, A. Rana, S. Khan et al., "Synthesis and preliminary screening of benzothiazol-2-yl thiadiazole derivatives for anticonvulsant activity," *Acta Pharmaceutica*, vol. 59, no. 4, pp. 441–451, 2009.

[89] N. Siddiqui, S. Ali, S. A. Khan, S. Drabu, A. Rana, and M. Alam, "Synthesis of 3-arylamino-4-aryl-5-(N-arylthiocarbonylimino)-4, 5-dihydro-1,2,4-thiadiazoles as anticonvulsant agents," *Indian Journal of Heterocyclic Chemistry*, vol. 14, no. 2, pp. 159–160, 2004.

[90] N. Siddiqui, M. F. Arshad, S. A. Khan, and W. Ahsan, "Synthesis, anticonvulsant and neurotoxicity screening of 1-(substituted phenyl)-3-[(5-substituted phenyl)-1,3,4-thiadiazole-2-yl]-2-thioxodihydro pyrimidine-4,6 (1H, 5H)-diones," *Journal of Pharmaceutical Research*, vol. 7, no. 2, pp. 122–125, 2008.

[91] P. Mullick, S. A. Khan, S. Verma, and O. Alam, "Thiadiazole derivatives as potential anticonvulsant agents," *Bulletin of the Korean Chemical Society*, vol. 32, no. 3, pp. 1011–1016, 2011.

[92] C. B. Chapleo, P. L. Mayers, J. F. Saville et al., "Substituted 1,3,4-thiadiazoles with anticonvulsant activity. 1. Hydrazines," *Journal of Medicinal Chemistry*, vol. 29, no. 11, pp. 2273–2280, 1986.

[93] H. Rajak, C. K. Behera, R. S. Pawar, P. K. Singour, and M. D. Kharya, "Synthesis and anticonvulsant evaluation of some novel 2,5-disubstituted 1,3,4-thiadiazoles: pharmacophore model studies," *Acta Poloniae Pharmaceutica*, vol. 67, no. 5, pp. 503–510, 2010.

[94] M. D. Kamal, H. A. Gawad, E. A. Rageb, M. Ellithey, and H. A. Mohamed, "Synthesis, anticonvulsant, and anti-inflammatory evaluation of some new benzotriazole and benzofuran-based heterocycles," *Bioorganic and Medicinal Chemistry*, vol. 14, no. 11, pp. 3672–3680, 2006.

13

Room Temperature Synthesis and Antibacterial Activity of New Sulfonamides Containing *N,N*-Diethyl-Substituted Amido Moieties

Olayinka O. Ajani,[1] Oluwole B. Familoni,[2] Feipeng Wu,[3] Johnbull O. Echeme,[1] and Zheng Sujiang[4]

[1] *Department of Chemistry, Covenant University, Canaanland, P.M.B. 1023, Ogun State, Ota, Nigeria*
[2] *Department of Chemistry, University of Lagos, Lagos State, Akoka 100001, Nigeria*
[3] *New Functional Polymeric Material Group, Technical Institute of Physics and Chemistry, Chinese Academy of Sciences (CAS), Beijing 100190, China*
[4] *Test Center of Antimicrobial Materials, Technical Institute of Physics and Chemistry, Chinese Academy of Sciences (CAS), Beijing 100190, China*

Correspondence should be addressed to Olayinka O. Ajani, ola.ajani@covenantuniversity.edu.ng

Academic Editor: Patrick Bednarski

Sulfonamide drugs which have brought about an antibiotic revolution in medicine are associated with a wide range of biological activities. We have synthesized a series of α-tolylsulfonamide, **1–11** and their substituted *N,N*-diethyl-2-(phenylmethylsulfonamido) alkanamide derivatives, **12–22** in improved and excellent yields in aqueous medium at room temperature through highly economical synthetic routes. The chemical structures of the synthesized compounds **1–22** were confirmed by analytical and spectral data such as IR, ^{1}H- and ^{13}C-NMR, and mass spectra. The *in vitro* antibacterial activity of these compounds along with standard clinical reference, streptomycin, was investigated on two key targeted organisms. It was observed that 1-(benzylsulfonyl)pyrrolidine-2-carboxylic acid, **2** emerged as the most active compound against *Staphylococcus aureus* at MIC value of 1.8 μg/mL while 4-(3-(diethylamino)-3-oxo-2-(phenylmethylsulfonamido) propyl)phenyl phenylmethanesulfonate, **22** was the most active sulfonamide scaffold on *Escherichia coli* at MIC value of 12.5 μg/mL.

1. Introduction

The development of sulfonamides is a fascinating and informative area in medicinal chemistry [1–3]. Its functional group has a long and rich history in organic chemistry and drug discovery [4, 5]. The *p*-toluenesulfonamide and benzenesulfonamides have been widely explored in synthetic chemistry [4, 5]; however, few work has been done on the α-tolylsulfonamide. For instance, synthesis of benzenesulfonamide derivative of pipecolic acid [6] and that of glycine [7] had been reported. In addition, benzenesulfonamide of alanine was achieved as a result of synthetic usage of such amino acid as a linker to 6*H*-1,3,4-thiadiazine scaffold [8] while naphthylsulfonamide was prepared as antagonist of chemokine receptor [9]. Domagk's discovery of antibacterial activity for the azo dye prontosil led to the first effective chemotherapeutic agent, sulfanilamide [10]. A retrospective look at sulfonamides leaves no doubt that besides providing the first effective treatment of bacterial infections [10, 11], they also unleashed an antibiotic revolution in medicine [12–15] to rationally design new therapeutic agents [16, 17]. These compounds provided an excellent lead for structural modification and ushered in the modern era of chemotherapy and drug design. Sulfonamides inhibit the multiplication of bacteria by acting as competitive inhibitors of *p*-aminobenzoic acid (PABA) in the folic acid metabolism cycle [18, 19]. In fact, the discovery that sulfonamides act through folate inhibition resulted in the development of dihydrofolate reductase inhibitors such as trimethoprim [20, 21].

Furthermore, sulfonamide moiety has a crucial functionality because of its wide variety of reported biological [22–25] and pharmacological activities such as anticancer [26–28], carbonic anhydrase inhibitory [29–31], antibacterial [32–34], antimalarial [35, 36], antitumor [37, 38], antihypertensive [39], anti-inflammatory [40–42], and antiprotozoal activities [43]. Some sulfonamides have been established as potent drugs in treatment of insomnia and other sleepless challenges in man by antagonizing orexin neural activity [44–46]. Sulfonamidophenyl porphyrins have a great potential utility as model for activatable photosensitizers [47]. Sulfonamide has also been reported to possess good herbicidal [48] and corrosion inhibitory properties [49, 50]. Multidrug resistance is one of the major immediate threats to human health today [51]. For instance, methicillin is a good antibacterial agent, yet, methicillin resistance among *Staphylococcus aureus* [52] and *Staphylococcus epidermidis* [53, 54] as well as other drug resistance in *Escherichia coli* [55] had been identified to be of great concern in public health. Epidemiological studies have also revealed that the emergent of new diseases is on the increase and quite alarming [56, 57].

In addition, the choice of *S. aureus* and *E. coli* as the targeted organisms in this study was due to wide range of infectious diseases and life threatening conditions associated with such isolates. *S. aureus,* which produces heat stable toxin, is among the invasive gram positive known as pyogenic cocci implicated in several diseases of human [58, 59]. From the literatures, *S. aureus* had shown to be very resistant to a wide variety of antibiotics [60]. Infections caused by methicillin-resistant *S. aureus* (MRSA) and vancomycin-resistant *S. aureus* are associated with high morbidity and mortality, high treatment cost, and long stays in hospitals [60]. *E. coli,* a facultative anaerobe of wide distribution in the environment, has been implicated in the cause of urinary tract infections, meningitis, sepsis, wound infections, nosocomial pneumonia, and arthritis. A subgroup enterohemorrhagic *E. coli* (EHEC) can cause severe potentially fatal illness known as hemorrhagic colitis with symptoms of blood diarrhea and severe abdominal pain [61].

Based on the numerous applications of sulfonamides and challenges associated with drug usage and multidrug resistance microorganisms aforementioned among others, there is a continuous need for the synthesis of new organic compounds as potential antimicrobial agents for the replacement of the old existing ones currently available in the market or to enhance the potency of the former ones. Thus, it is conceivable to develop a series of functionalized sulfonamides in order to investigate the *in vitro* antibacterial activities of such scaffolds on the targeted organisms, namely, *S. aureus* and *E. coli.*

2. Results and Discussion

2.1. Chemistry. In the continuation of our effort on the discovery of rationally designed therapeutic agents [62, 63], we have herein reported the synthesis of a series of new alkylarylsulfonamide with potent antibacterial activity. We started with the reaction of equimolar proportion of readily available amino acid, L-pipecolic acid with α-toluene sulfonylchloride (α-TsCl) via a continuous magnetic stirring at room temperature in the presence of aqueous sodium carbonate for 48 h according to a known procedure [64]. After the completion of the reaction which was TLC monitored, the excess unreacted α-TsCl was recovered with dichloromethane (DCM) using separatory funnel while the aqueous layer was worked up by acidifying to a pH of 2.2 to get clear solution which upon freeze drying and column purification afforded 1-(benzylsulfonyl)piperidine-2-carboxylic acid **1** in excellent yield (98%). Being motivated by this encouraging discovery, we then proceeded on the coupling of nine other amino acids with α-TsCl under similar condition to obtain α-tolylsulfonamides **2–10** in good to excellent yields (Scheme 1).

In contrast to monosulfonylation that occurred in formation of **1–10**, the synthesis of **11** was achieved in higher yield only when disulfonylation of the required amino acid (tyrosine) was utilized. This was noticed when monosulfonylation of tyrosine afforded **11** in lower yield (45%) when compared with other sulfonamide products **1–10** with their yields ranging from 87.5% to 98.8%. This was because under the monosulfonylation of tyrosine, the reaction terminated when the α-TsCl, being the limiting reagent, has been exhausted living excess of the tyrosine in the aqueous medium. This was as a result of two nucleophilic attacking sites available on the L-tyrosine moiety. This means that the phenolate anion was competing with the amino group in terms of the nucleophilic potential on the α-TsCl precursor, thereby resulting in an unusually lower yield (45%). Based on this development, the stoichiometric ratio was revisited in the production of **11** to give allowance for selective disulfonylation of tyrosine. Hence, the reaction for the production of **11** was reconducted in such a way that the molar equivalent of α-TsCl was doubled that of tyrosine in order to accommodate nucleophilicity from both NH_2 and PhO^- of the latter, then the yield for **11** increased drastically to 89.6% (Scheme 2). Structures of all eleven α-tolylsulfonamides were confirmed by spectroscopic means which include IR, mass spectra, ^1H, and ^{13}C NMR as well as the elemental analytical data. Thus, the detail structural elucidation was here-in confirmed using compound **1** as the representative of the α-tolylsulfonamide templates. The infrared spectrum of compound **1** gave rise to the absorption bands of OH and carbonyl of carboxylic acid at 3422 and 1736 cm^{-1}, respectively. Also, the CH stretching of both aliphatic and aromatic occurred at the vibrational frequency of 2974 and 2822 cm^{-1}, respectively. The two IR bands for SO_2 unit were observed at 1159 and 1238 cm^{-1} as expected, whereas the band accountable for C=C of aromatic was observed at 1603 cm^{-1}. The only easily accountable band resulting from a bending vibrational mode at the finger print region was Ar-H which appeared at 700 cm^{-1}.

In the ^1H-NMR of compound **1**, all the five aromatic protons resonated downfield as a singlet at δ 7.52 ppm while the CH_2 of benzylic group appeared as a two-proton singlet at δ 4.28 ppm. Considering the upfield region, it was discovered that CH proton adjacent to COOH resonated

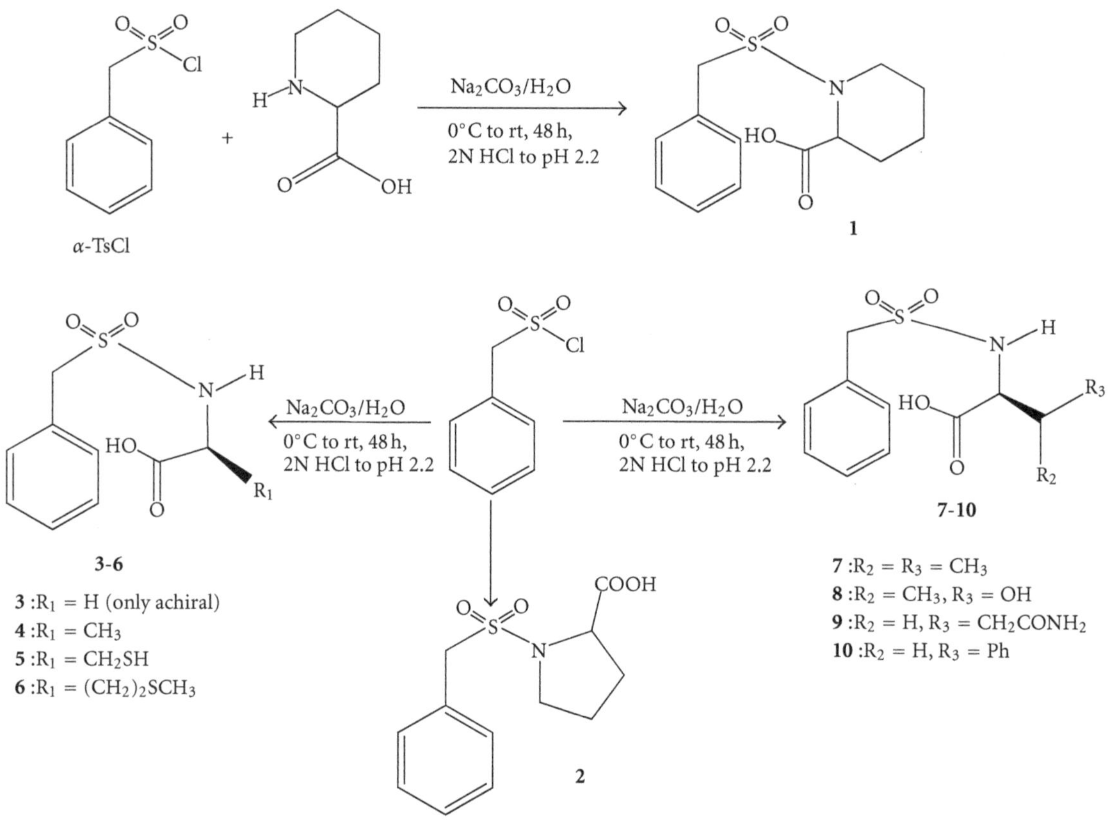

SCHEME 1: Synthesis of α-tolylsulfonamide derivatives **1–10**.

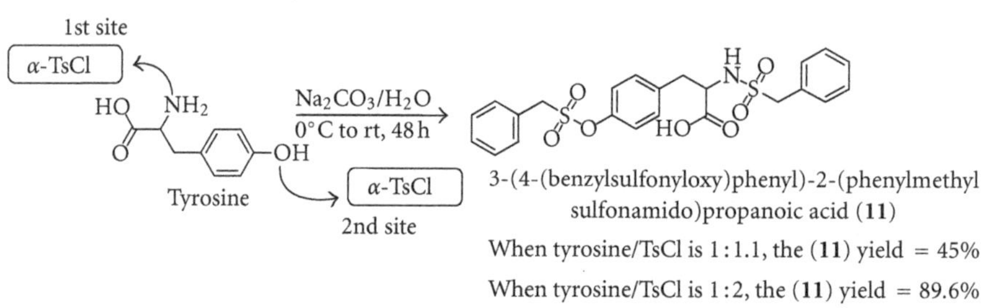

SCHEME 2: Synthesis of sulfonamide **11** by selective disulfonylation of tyrosine.

as a doublet of doublet at δ 4.08–4.02 ppm with coupling constants of 3.44 Hz and 15.12 Hz. This was as a result of the fact that its CH_2 neighbouring protons were not totally chemically equivalent even though they were attached to the same carbon atom. All other remaining pipecolic protons were well accounted for upfield between δ 3.57–3.53 and 1.81–1.68 ppm. The justification for their appearance near TMS scale lies in the fact that they were attached to sp^3 hybridized centre. Although, the molecular ion peak was not observed, the base peak, however, occurred at *m/z* 178.1 which was as a result of the loss of benzylic radical. Other prominent peaks which were formed as a result of some fragmentation patterns appeared at *m/z* 180.1, 179.1, 165.1, 121.0, 77.0, and 64.0 with the intensities of 55%, 65%, 30%, 42%, 13%, and 31.7%, respectively.

The second stage of this experiment involved the chemical transformation of the carboxylic acid side chain moieties of the sulfonamides **1–11** into *N,N*-diethyl alkanamides **12–22** using highly efficient nonconventional method in its slightly modified version [65]. The reaction optimization for this aspect was carried out using amination of **1** with diethyl amine under triethyl amine basified condition as a representative. Thus, having established the *C*-amidoalkylation of the sulfonamide **1** to achieve the alkanamide **12** in higher yield as a bountiful success, we were motivated to extend our investigation to the reaction of other α-tolylsulfonamides **2–11** with diethylamine under the same condition (Scheme 3). These attempts also afforded the *N,N*-diethyl-substituted products **13–22** in good to excellent yields as envisaged. The spectroscopic assignment was consistent with that of proposed structures for the obtained *N,N*-diethyl alkanamides **12–22**. So, it is necessary to consider the spectral data of **12** as a typical representative of the *N,N*-diethyl alkanamides in order to authenticate such compounds. The IR spectrum of **12** had no absorption band above 3028 cm^{-1} indicating the absence of –OH bond. This in turn confirmed the effective amidation of COOH of compound **1**. The bands at 3028 and 2951 cm^{-1} depicted CH of aromatic and aliphatic, respectively. Also, the C=O and C=C frequency appeared at 1720 and 1593 cm^{-1}, respectively, while the two bands of SO_2 were observed at 1188 and 1148 cm^{-1}. Furthermore, the chemical shifts and multiplicity patterns of ^1H- and ^{13}C-NMR correlated well with that of the proposed alkylated sulfonamides **12–22**. The ^1H-NMR spectrum of **12** in D_2O showed five aromatic protons as singlet at δ 7.54 and CH_2 of benzylic group as singlet at δ 4.30. One triplet at δ 1.40–1.37 (6H) was assigned to two methyl of diethyl group that were chemically equivalent while one quartet at δ 3.21–3.16 (4H) was assigned to two methylene protons adjacent to the methyl protons with a coupling constant of 7.20 Hz. The ^{13}C-NMR of **12** in D_2O in the presence of a drop of dioxane revealed the presence of seventeen different carbon atoms with C=O having the highest signal at δ 173.5 ppm while the CH_3 carbon atoms appeared to have the least signals at δ 11.4 ppm.

2.2. Antibacterial Activity. The general antibacterial sensitivity testing (inhibition zone, mm) of all the series of twenty-two synthesized sulfonamides alongside with that of streptomycin clinical standard was assayed on test organisms (*E. coli* and *S. aureus*) using agar diffusion technique [66]. The choice of *E. coli* as the gram −ve organism is because it is easily transmissible through food, water, soil, animal, and man [60]. *E. coli* is a normal flora of human body which causes a lot of vancomycin-resistant *Enterococci* and methicillin-resistant *Staphylococcus aureus* (MRSA) [51]. Based on our previous report [62], the choice of streptomycin as clinical standards is due to the fact that, at low concentrations, streptomycin only inhibits the growth of the bacteria through induction of prokaryotic ribosomes to misread *m*RNA [67] and it also possesses broad spectrum of antibacterial activity. There were reported cases of *E. coli* and *Staphylococcus aureus* being susceptible to streptomycin [4, 68]. The biological relevance of the synthesized sulfonamides here-in was authenticated by screening them *in vitro* against *Staphylococcus aureus* ATCC 6538 (*S. aureus*) and *Escherichia coli* ATCC 25922 (*E. coli*) with the reported selectivity index (SI) duly calculated from zones of inhibition (ZOI). Hence, it should be noted afterwards that the abbreviated forms were given in the bracket for the sake of brevity and conciseness.

From the result of sensitivity testing, probable activities of α-tolylsulfonamide family on the test organisms were categorized based on the size of zone of inhibition (Table 1). Interestingly, it was observed that some of the compounds exhibited probable significant activities based on the large zone of inhibition reported. For instance, compounds **6, 7, 12, 14,** and **22** were highly active on *E. coli* while compounds **1, 2, 4, 5, 10, 11, 17,** and **20** exhibited moderate activities on the same organism. All other compounds showed low activities on *E. coli* except **3** and **16** which showed no activity at all on the *E. coli* even at 1000 μg/mL. The scenario of comparative study of effect of the sulfonamides and streptomycin on *E. coli* could be vividly understood by observing the selectivity index (SI). All the sulfonamides have selectivity indices ranging from 0.29 for compound **18** to 0.96 for compound **22** (i.e., less than 1). This implies that streptomycin (SI = 1) was probably more active than any of the sulfonamide scaffolds as regarding the inhibition of *E. coli* growth. In the same vein, looking through the effect on *S. aureus*, compounds **2, 5, 17, 21,** and **22** were highly active; **4, 6, 8, 10, 11, 12, 14,** and **19** were moderately active; **1, 3, 7, 9, 13, 15, 18,** and **20** exhibited low activity while **16** showed no activity on *S. aureus* (Table 1). The comparative study of α-toluenesulfonamides to streptomycin on *S. aureus* growth inhibition is worthy, of commendation. From the SI values, compounds **9, 12, 18,** and **20** competed favourably with streptomycin while **2, 4, 5, 6, 8, 13, 14, 17, 19, 21,** and **22** (SI = 1.08–2.31) showed even a better activity than streptomycin on *S. aureus*. All other compounds exhibited lesser activity than streptomycin (SI = 0.46–0.92) on *S. aureus* except **16** which showed no activity.

Due to high zones of inhibition obtained during general sensitivity testing, the minimum inhibitory concentration (MIC) was conducted, first at 100 μg/mL using Russell and Furr method [69]. However, those compounds that could not affect the inhibition of microbial growth at this concentration were further repeated for MIC test at 1000 μg/mL. The result of the MIC of this class of compounds on *E. coli* and *S. aureus* was as shown in Table 2. Interestingly, all the sulfonamides tested showed a concentration-dependent inhibitory effect on the *in vitro* microbial growth assays [70]. Considering the MIC testing on the gram negative organism (*E. coli*), it was observed that compounds **1, 6, 7, 11, 12, 14,** and **22** inhibited the microbial growth at varying values less than or equal to 100 μg/mL, whereas, all other compounds were active on *E. coli* at higher concentration (between 125 and 1000 μg/mL) except **3** and **16** which had no activity even at 1000 μg/mL. Specifically, MIC values of the synthesized compounds on *E. coli* were reported to be 50 μg/mL for **12** and **14**; 100 μg/mL for **1** and **11**; 125 μg/mL for **2, 4, 5, 10, 17,** and **20**; 250 μg/mL for **8, 9, 13, 15, 19,** and **21**.

Although, the most active sulfonamides on *E. coli* were **6, 7,** and **22** with MIC values of 25, 25, and 12.5 μg/mL, respectively, none of them could compete with streptomycin (MIC value of 6.25 μg/mL) in terms of activity. The two rings presence in **22** and their π character might be responsible for it being the most active as deduced from the finding of Aissaoui and coworkers [71]. In addition, eight sulfonamides (**2, 5, 6, 8, 14, 17, 21,** and **22**) inhibited the *S. aureus* growth at concentration ranging from 1.8 to 100 μg/mL. All other compounds were able to affect the expected inhibition

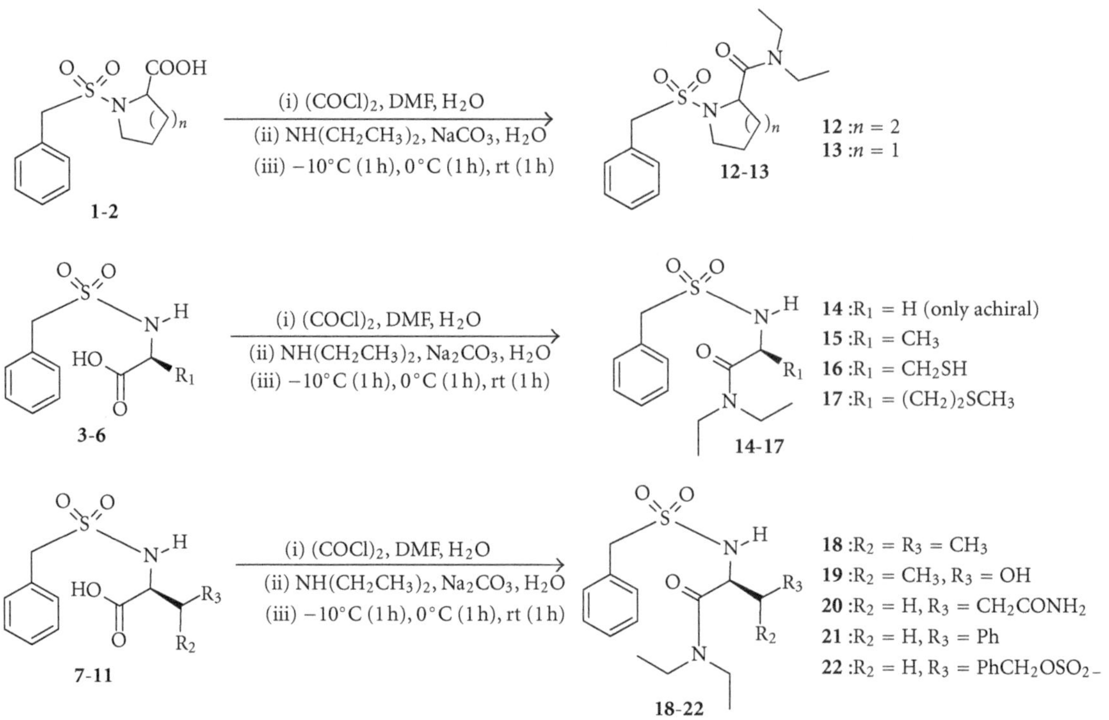

SCHEME 3: Synthesis of *N,N*-diethylamide substituted α-tolylsulfonamides, 12–22.

from 125 μg/mL to 1000 μg/mL except **16** which had no activity even at 1000 μg/mL. The significant antibacterial activity of the synthesized compounds may be explained by the ability of its sulfonamide binding site to mimic *p*-aminobenzoic acid (PABA) which is an essential growth factor in the targeted organisms as earlier documented in the literatures [14–16]. The explanation for this encouraging activity of the synthesized sulfonamides could be traceable to the mode of action of sulfonamide drugs. This is based on the inhibition of DNA synthesis [72] by interfering with *para*-aminobenzoic acid (PABA) in biosynthesis of folic acid [73].

Furthermore, from the structure activity relationship (SAR) study, it was observed that the nature of side chains (R_1 and R_2-CH-R_3) of the α-tolylsulfonamides and the presence of *N,N*-diethylated amido moieties {$(CH_3CH_2)_2N$–C=O} of the amide contributed immensely toward synergistic or antagonistic effect on the reported *in vitro* antibacterial activity. For instance, the antibacterial activity of structurally related α-tolylsulfonamides containing $R_1 = (CH_2)_2SCH_3$, **6**; $R_2 = R_3 = CH_3$, **7**, on *E. coli* were very high with MIC value of 25 ppm whereas the activity reduced to moderate when $R_1 = CH_3$, **4**; $R_1 = CH_2SH$, **5** with MIC value of 125 ppm. The synergistic effect due to the presence of *N, N*-diethylamido moieties was noticed in some synthetic conversions such as **1** → **12**, **3** → **14** and **11** → **22**. Thus, α-tolylsulfonamides **1** and **11** had moderate activity on *E. coli* while their corresponding diethylamido **12** and **22** had better and higher activity on *E. coli*. In fact, α-tolylsulfonamide **3** had no activity on *E. coli* whereas its *N, N*-diethyl amidated counterpart, **14** had very high activity with MIC value of 50 ppm. The synergistic effect noticed herein might be as a result of electron donating nature of the R_1 side chain. On the contrary, antagonistic effect was observed in the synthetic modification of **5** → **16** and **6** → **17** as far as presence of *N, N*-diethyl-substituted amido moieties was concerned. This reverse order of activity on *E. coli* might be as a result of the presence of sulfur atom in **5** and **6** which might have altered the electron donating prowess. Considering the structural relationship effect on the activity of the synthesized compounds against *S. aureus*, it was discovered that *N,N*-diethyl-substituted amides **14**, **17**, **21**, and **22** had better activity than their α-tolylsulfonamide precursors **3**, **6**, **10**, and **11**.

3. Conclusion

It was discovered that aqueous medium approach at ambient temperature used here-in was a highly efficient procedure for the preparation of various α-tolylsulfonamide derivatives **1–11** in good to excellent yield. The synthetic modification of these derivatives as precursors to furnish disubstituted amide bearing sulfonamides **12–22** was also successfully achieved. The antibacterial evaluation study showed 1-(benzylsulfonyl) pyrrolidine-2-carboxylic acid, **2** to be the most a ctive sulfonamide on *S. aureus* while 4-(3-(diethylamino)-3-oxo-2-(phenylmethylsulfonamido)propyl) phenylphenylmethane sulfonate, **22** emerged as the most active on the growth inhibition of *E. coli*. Thus, this work will be very useful for further studies in terms of toxicity effect and structural activity relationship (SAR) study by using various types of

Table 1: Result of general sensitivity testing with zones of inhibitory in (mm) and selectivity index.

Compd. no.	*In vitro* antibacterial activity Organisms			
	E. coli ATCC 25922		*S. aureus* ATCC 6538	
	ZOI (mm)	SI	ZOI (mm)	SI
1	++	0.61	+	0.92
2	++	0.54	+ + +	2.31
3	–	—	+	0.69
4	++	0.57	++	1.08
5	++	0.57	+ + +	2.15
6	+ + +	0.86	++	1.15
7	+ + +	0.89	+	0.46
8	+	0.46	++	1.15
9	+	0.36	+	1.00
10	++	0.57	++	0.92
11	++	0.75	++	0.92
12	+ + +	0.84	++	1.00
13	+	0.43	+	1.08
14	+ + +	0.82	++	1.38
15	+	0.36	+	0.92
16	–	—	–	—
17	++	0.57	+ + +	2.23
18	+	0.29	+	1.00
19	+	0.43	++	1.08
20	++	0.54	+	1.00
21	+	0.39	+ + +	2.00
22	+ + +	0.96	+ + +	2.00
Str.	+ + +	1.00	++	1.00

+: Less active 5–12 mm; ++: moderately active 13–19 mm; +++: highly active 20–31 mm; –: resistance; str.: streptomycin clinical reference; Z.O.I.: zone of inhibition; S.I.: selective index obtained by comparing inhibition zone of compound to that of streptomycin standard; *E. coli*: *Escherichia coli* (ATCC 25922)$^{G-}$; *S. aureus*: *Staphylococcus aureus* (ATCC 6538)$^{G+}$; G–: gram negative; G+: gram positive.

dialkylated substituents to monitor the trend of improvement on their biological and pharmacological properties.

4. Experimental

4.1. General Conditions. The ^{1}H-NMR spectra were recorded in D_2O on NMR Bruker DPX 400 spectrometer operating at 400 MHz. TMS was used as an internal standard with the deuterium signal of the solvent as the lock and chemical shifts δ recorded in ppm. The ^{13}C-NMR spectra were run in addition of a few drops of Dioxane but at 100 MHz frequency. The melting points were determined on XT-4 Digital Binocular Microscope melting point apparatus manufactured by Beijing Technical Instrument Co. Ltd. and were uncorrected. IR spectra were run on Varian Excalibur HE 3100 FT-IR Spectrometer while the mass spectra were obtained using Waters GCT Premier Spectrometer. The elemental analyses (C, H, and N) of the compounds were performed using Flash EA 1112 Elemental Analyzer. Lyophilization was carried out where necessary by using FD-1 Freeze Drier while concentration and removal of solvents were achieved with RE-2000B Rotary Evaporator.

In addition, the pH was monitored and confirmed during acidification by using Portable pH Meter Model PHB4. All drying was conducted at a reduced pressure with DHG-9023A Vacuum Oven. The reaction progress was monitored with TLC using $CHCl_3/CH_3OH$ solvent system and the developed plates were visualized under UV lamp and/or in iodine tank where necessary. Column chromatographic purifications were carried out on Merck silica gel F (Mesh 200–300). Organic solutions were dried over anhydrous Na_2SO_4 and concentrated with a Buchi Rotary Evaporator at reduced pressure. At all stages of the experiments, the synthetic protocols were effected in bone dried solvents under nitrogen atmosphere in dried glassware which were wiped with stream flow of nitrogen gas prior to use and $SOCl_2$ was freshly distilled prior to use. Other reagents were used directly after ascertaining the purity condition.

4.2. Synthesis

*4.2.1. General Procedure of α-Tolylsulfonamide Derivatives **1**–**11**.* To a solution of L-amino acid (5 mmol) in H_2O (6 mL) was added Na_2CO_3 (1.113 g, 10.5 mmol) with a continuous stirring until all the solutes had dissolved. The clear solution was cooled to $-10°C$ and α-toluenesulfonyl chloride, α-TsCl (1.144 g, 6 mmol) was added in three batches over a period of 1 h. It was warmed up to 0°C and stirred there for 1 h. Finally, the reacting mixture was then warmed up to room temperature and allowed to stir there for 48 h. The reaction was quenched by the addition of DCM (10 mL) and transferred into separatory funnel where the excess of α-TsCl was removed by extraction. The aqueous layer was then worked up to give a clear solution by the addition of 2N HCl until the pH 2.2 was attained. The clear liquid was then lyophilized at $-52°C$ under reduced pressure for 12 h to obtain the crude solid product which was purified by column chromatography ($CHCl_3/CH_3OH$, 3 : 1) to afford α-toluene sulfonamides **1**–**11** in excellent yields.

*(1) 1-(Benzylsulfonyl)piperidine-2-carboxylic Acid, **1**.* Reagents: L-pipecolic acid, yield 98%, mp 248°C (dec), $R_f = 0.87$. ^{1}H-NMR (D_2O, 400 MHz) δ: 7.52 (s, 5H, Ar-H), 4.28 (s, 2H, CH_2-SO_2), 4.08–4.02 (dd, $J_1 = 3.44$ Hz, $J_2 = 15.12$ Hz, 1H, HOOC-CH-CH_2), 3.57–3.53 (m, 1H, CHa of CH_2-N), 3.16–3.10 (m, 1H, CHb of CH_2-N), 2.40–2.36 (m, 1H, CH), 2.01–1.92 (m, 2H, 2 × CH), 1.81–1.68 (m, 3H, CH & CH_2) ppm. ^{13}C-NMR (Dioxane, 100 MHz) δ: 172.5 (CO), 132.5, 131.2 (2CH aromatic), 129.5 (2CH aromatic), 128.9, 57.7, 57.6, 44.8, 26.5, 22.1 (2CH_2) ppm. IR (KBr) cm^{-1}: 3422 (OH of acid), 2974 (CH aromatic), 2822 (CH aliphatic), 1736 (C=O of COOH), 1603 (C=C), 1238, 1159 (SO_2 two bands), 700 (Ar-H). MS: in m/z [rel. %]: 269.1 [M^+-CH_2, 3.2%], 180.1 [55%], 179.1 [65%], 178.1 [M^+-$PhCH_2$, 100%], 165.1 [30%], 121.0 [42%], 77.0 [Ph^+, 13%], 64.0 [SO_2^+, 31.7%]. Anal. Calcd. for 283.35 $C_{13}H_{17}NO_4S$: C, 55.11; H, 6.05; N, 4.94. Found: C, 55.29; H, 5.94; N, 4.86.

Table 2: Result of MIC test of α-tolylsulfonamide on targeted organisms (μg/mL).

Compd. no.	Minimum inhibitory concentration (μg/mL) Organisms			
	E. coli ATCC 25922		*S. aureus* ATCC 6538	
	at 100 μg/mL	at 1000 μg/mL	at 100 μg/mL	at 1000 μg/mL
1	100	<1000	>100	250
2	>100	125	1.8	<1000
3	>100	—	>100	250
4	>100	125	>100	125
5	>100	125	50	<1000
6	25	<1000	100	<1000
7	25	<1000	>100	500
8	>100	250	100	<1000
9	>100	250	>100	250
10	>100	125	>100	125
11	100	<1000	>100	125
12	50	<1000	>100	125
13	>100	250	>100	250
14	50	<1000	62.5	<1000
15	>100	250	>100	1000
16	>100	—	>100	—
17	>100	125	25	<1000
18	>100	500	>100	1000
19	>100	250	>100	125
20	>100	125	>100	250
21	>100	250	25	<1000
22	12.5	<1000	25	<1000
Str.	6.25	<1000	>100	125

>100 means that if there was no growth inhibition at 100 μg/mL, it was repeated at 1000 μg/mL, <1000 μg/mL means that growth inhibition has already been experienced at lower concentration less than or equal to 100 μg/mL; hence, there is no need to repeat the test at 1000 μg/mL. — means no activity was observed even at 1000 μg/mL. Str. means streptomycin clinical reference.

(2) 1-(Benzylsulfonyl)pyrrolidine-2-carboxylic Acid, **2**. L-amino acid is L-proline; yield 1.24 g (92%); mp 108–110°C; R_f = 0.84. ^{1}H-NMR (D_2O, 400 MHz) δ: 7.47 (s, 5H, Ar-H), 4.46–4.43 (dd, J_1 = 7.2 Hz, J_2 = 15.76 Hz, 1H, HOOC-CH-CH_2(a,b)), 4.24 (s, 2H, CH_2-SO_2), 3.48–3.45 (t, J = 7.28 Hz, 2H, N-CH_2-CH_2), 2.46-2.45 (m, 1H, CHa of CH_2), 2.22-2.21 (m, 1H, CHb of CH_2), 2.15–2.09 (quintet, J = 6.8 Hz, 2H, CH_2-CH_2-CH_2(a,b)) ppm. ^{13}C-NMR (Dioxane, 100 MHz) δ: 173.1 (CO), 132.6, 131.2 (2CH aromatic), 129.5 (2CH aromatic), 128.9, 60.6, 57.7, 47.1, 29.2, 24.3 ppm. IR (KBr) cm^{-1}: 3441 (OH of acid), 2980 (CH aromatic), 2828 (CH aliphatic), 1728 (C=O of COOH), 1620 (C=C), 1219, 1151 (SO_2 two bands), 700 (Ar-H). MS: in *m/z* [rel. %]: 270.1 [MH^+, 6.5%], 269.1 [M^+, 9%], 179.1 [18.4%], 178.1 [M^+-$PhCH_2$, 100%], 176.1 [32.4%], 122.0 [49%], 105.0 [32%]. Anal. Calcd. for 269.32 $C_{12}H_{15}NO_4S$: C, 53.52; H, 5.61; N, 5.20. Found: C, 53.77; H, 5.49; N, 5.34.

(3) 2-(Phenylmethylsulfonamido)acetic Acid, **3**. Amino acid is glycine; yield 1.08 g (94%); mp 150-151°C; R_f = 0.51. ^{1}H-NMR (D_2O, 400 MHz) δ: 7.45 (s, 5H, Ar-H), 4.19 (s, 2H, CH_2-SO_2), 3.76 (s, 2H, CH_2-COOH) ppm. ^{13}C-NMR (Dioxane, 100 MHz) δ: 170.6 (CO), 132.7, 131.2 (2CH aromatic), 129.5 (2CH aromatic), 128.9, 57.7, 40.8 ppm. IR (KBr) cm^{-1}: 3433 (OH of acid), 3030 (N–H), 2990 (CH aromatic), 2832 (CH aliphatic), 1736 (C=O of COOH), 1616 (C=C), 1215, 1171 (SO_2 two bands), 702 (Ar-H). MS: in *m/z* [rel. %]: 212.1 [M^+-OH, 7.9%], 180.1 [73%], 179.1 [88%], 178.1 [M^+ - $PhCH_2$, 100%], 91.1 [$PhCH_2^+$, 48%], 64 [26%] 45 [$^+$COOH, 2.4%]. Anal. Calcd. for 229.26 $C_9H_{11}NO_4S$: C, 47.15; H, 4.84; N, 6.11. Found: C, 46.90; H, 5.01; N, 5.97.

(4) 2-(Phenylmethylsulfonamido)propanoic Acid, **4**. L-amino acid is L-alanine; yield 1.18 g (97%); mp 126–128°C; R_f = 0.81. ^{1}H-NMR (D_2O, 400 MHz) δ: 7.47 (s, 5H, Ar-H), 4.22 (s, 2H, CH_2-SO_2), 4.14–4.08 (q, J = 7.28 Hz, 1H, CH-CH_3), 1.60-1.58 (d, J = 7.28 Hz, 3H, CH_3-CH) ppm. ^{13}C-NMR (Dioxane, 100 MHz) δ: 176.8 (CO), 131.0, 130.6 (2CH aromatic), 129.3 (2CH aromatic), 129.1, 60.4, 52.1, 19.6 ppm. IR (KBr) cm^{-1}: 3424 (OH of acid), 2974 (CH aromatic), 2822 (CH aliphatic), 1751 (C=O of COOH), 1599 (C=C), 1213, 1169 (SO_2 two bands), 698 (Ar-H). MS: in *m/z* [rel. %]: 212.1 [22%], 180.1 [81.5%], 179.1 [91%], 178.1 [85%], 165.1 [M-$PhCH_3$, 55%], 122.0 [80%], 121.0 [100%],

77.0 [Ph^+, 71.4%], 64.0 [SO_2^+, 54.6%], 51.0 [28%]. Anal. Calcd. for 243.28 $C_{10}H_{13}NO_4S$: C, 49.37; H, 5.39; N, 5.76. Found: C, 49.29; H, 5.28; N, 5.94.

(5) 3-Mercapto-2-(phenylmethylsulfonamido)propanoic Acid, **5**. L-amino acid is cysteine; yield 1.22 g (89%); mp 171–173°C; R_f = 0.49. ^{1}H-NMR (D_2O, 400 MHz) δ: 7.49 (s, 5H, Ar-H), 4.50-4.47 (dd, J_1 = 4.24 Hz, J_2 = 7.92 Hz, 1H, CH_2-CH-COOH), 4.24 (s, 2H, CH_2-SO_2), 3.55–3.50 (dd, J_1 = 4.24 Hz, J_2 = 20 Hz, 1H, CHa of CH_2-CH), 3.39–3.33 (dd, J_1 = 7.92 Hz, J_2 = 20 Hz, 1H, CHb of CH_2-CH) ppm. ^{13}C-NMR (Dioxane, 100 MHz) δ: 171.5 (CO), 132.6, 131.3 (2CH aromatic), 129.5 (2CH aromatic), 128.7, 57.8, 52.7, 37.2 ppm. IR (KBr) cm^{-1}: 3439 (OH of acid), 2978 (CH aromatic), 2832 (CH aliphatic), 1728 (C=O of COOH), 1618 (C=C), 1207, 1159 (SO_2 two bands), 810 (Ar-H). MS: in *m/z* [rel. %]: 214.1 [31.7%], 123.0 [100%], 122.0 [90%], 92.1 [$PhCH_3^+$, 33%], 91.0 [$PhCH_2^+$, 88%], 77.0 [Ph^+, 8%], 65.0 [HSO_2^+, 34%], 45.0 [$^+$COOH, 28%], 36.0 [34%]. Anal. Calcd. for 275.35 $C_{10}H_{13}NO_4S_2$: C, 43.62; H, 4.76; N, 5.09. Found: C, 43.45; H, 4.94; N, 5.07.

(6) 4-(Methylthio)-2-(phenylmethylsulfonamido)butanoic Acid, **6**. L-amino acid is L-methionine; yield 1.33 g (87.7%); mp 89-90°C; R_f = 0.68. ^{1}H-NMR (D_2O, 400 MHz) δ: 7.47 (s, 5H, Ar-H), 4.23 (s, 2H, CH_2-SO_2), 4.01–3.97 (t, J = 8.84 Hz, 1H, CH-CH_2), 2.76–2.72 (t, J = 7.40 Hz, 2H, CH_2-CH_2-S), 2.33–2.29 (m, 1H, CH), 2.28–2.21 (m, 1H, CH), 2.19 (s, 3H, CH_3-S) ppm. ^{13}C-NMR (Dioxane, 100 MHz) δ: 172.9 (CO), 132.6, 131.2 (2CH aromatic), 129.5 (2CH aromatic), 128.7, 57.9, 52.8, 29.8, 29.4, 14.7 ppm. IR (KBr) cm^{-1}: 3442 (OH of acid), 2974 (CH aromatic), 2833 (CH aliphatic), 2774 (CH aliphatic), 1742 (C=O of COOH), 1590 (C=C), 1211, 1161 (SO_2 two bands), 698 (Ar-H). Anal. Calcd. for 303.40 $C_{12}H_{17}NO_4S_2$: C, 47.51; H, 5.65; N, 4.62. Found: C, 47.49; H, 5.64; N, 4.66.

(7) 3-Methyl-2-(phenylmethylsulfonamido)butanoic Acid, **7**. L-amino acid is L-valine; yield 1.34 g (98.7%); mp 137-138°C; R_f = 0.83. ^{1}H-NMR (D_2O, 400 MHz) δ: 7.49 (s, 5H, Ar-H), 4.25 (s, 2H, CH_2-SO_2), 4.04-4.03 (d, J = 4.4 Hz, 1H, CH-CH-COOH), 2.47–2.38 (m, 1H, CH-CH-$(CH_3)_2$), 1.15–1.13 (d, J = 7.04 Hz, 3H, CH_3-CH), 1.13–1.11 (d, J = 7.08 Hz, 3H, CH_3-CH) ppm. ^{13}C-NMR (Dioxane, 100 MHz) δ: 172.4 (CO), 132.5, 131.2 (2CH aromatic), 129.5 (2CH aromatic), 128.9, 59.2, 57.7, 29.8, 18.1, 17.8 ppm. IR (KBr) cm^{-1}: 3447 (OH of acid), 2974 (CH aromatic), 2833 (CH aliphatic), 2783 (CH aliphatic), 1730 (C=O of COOH), 1618 (C=C), 1225, 1165 (SO_2 two bands), 700 (Ar-H). MS: in *m/z* [rel. %]: 271.1 [M^+, 14%], 91.0 [$PhCH_2^+$, 30%], 75.0 [65%], 72.1 [100%], 55.0 [79%], 29.0 [50%]. Anal. Calcd. for 271.34 $C_{12}H_{17}NO_4S$: C, 53.12; H, 6.32; N, 5.16. Found: C, 53.31; H, 6.50; N, 5.20.

(8) 3-Hydroxy-2-(phenylmethylsulfonamido)butanoic Acid, **8**. L-amino acid is L-threonine; yield 1.25 g (91.5%); mp 194-195°C; R_f = 0.48. ^{1}H-NMR (D_2O, 400 MHz) δ: 7.49 (s, 5H, Ar-H), 4.48–4.46 (m, 1H, CH), 4.24 (s, 2H, CH_2-SO_2), 4.05-4.04 (d, J = 4.4 Hz, 1H, CH-CH-COOH), 1.42–1.41 (d, J = 6.64 Hz, 3H, CH_3-CH) ppm. ^{13}C-NMR (Dioxane, 100 MHz) δ: 171.5 (CO), 132.5, 131.2 (2CH aromatic), 129.5 (2CH aromatic), 128.9, 66.1, 59.4, 57.7, 19.8 ppm. IR (KBr) cm^{-1}: 3404 (OH of acid), 2976 (CH aromatic), 2824 (CH aliphatic), 1740 (C=O of COOH), 1601 (C=C), 1219, 1157 (SO_2 two bands), 700 (Ar-H). Anal. Calcd. for 273.31 $C_{11}H_{15}NO_5S$: C, 48.34; H, 5.53; N, 5.12. Found: C, 48.29; H, 5.61; N, 4.98.

(9) 5-Amino-5-oxo-2-(phenylmethylsulfonamido)pentanoic Acid, **9**. L-amino acid is L-glutamine; yield 1.41 g (94.1%); mp 211–214°C; R_f = 0.38. ^{1}H-NMR (D_2O, 400 MHz) δ: 7.48 (s, 5H, Ar-H), 4.47–4.43 (dd, J_1 = 5.04 Hz, J_2 = 14.32 Hz, 1H, HOOC-CH-CH_2), 4.23 (s, 2H, CH_2-SO_2), 2.62–2.55 (m, 1H, CH), 2.49–2.44 (dd, J_1 = 9.2 Hz, J_2 = 18.72 Hz, 2H, CH_2-CH_2-CON), 2.27–2.20 (m, 1H, CH) ppm. ^{13}C-NMR (Dioxane, 100 MHz) δ: 181.1 (CO of acid), 163.2 (CO of amide), 132.5, 131.2 (2CH aromatic), 129.5 (2CH aromatic), 128.9, 59.2, 57.7, 32.3, 26.2 ppm. IR (KBr) cm^{-1}: 3246 (OH of acid), 3075, 3053 (NH two bands), 2983 (CH aromatic), 2951 (CH aliphatic), 1703 (C=O of COOH), 1659 (C=O amide), 1412 (OH bending in-plane), 1221, 1193 (SO_2 two bands), 696 (Ar-H), 631 (N-H bending with wagging). Anal. Calcd. for 300.34 $C_{12}H_{16}N_2O_5S$: C, 47.99; H, 5.37; N, 9.33. Found: C, 48.03; H, 5.56; N, 9.41.

(10) 3-Phenyl-2-(phenylmethylsulfonamido)propanoic Acid, **10**. L-amino acid is L-phenyl alanine; yield = 1.58 g (98.8%); mp 158-159°C (dec); R_f = 0.84. ^{1}H-NMR (D_2O, 400 MHz) δ: 7.46 (s, 10H, 2 × Ar-H), 4.34–4.31 (dd, J_1 = 5.60 Hz, J_2 = 13.28 Hz, 1H, $PhCH_2$-CH-COOH), 4.22 (s, 2H, CH_2-SO_2), 3.41–3.36 (dd, J_1 = 5.60 Hz, J_2 = 20.00 Hz, CHa of CH_2-Ph), 3.28–3.22 (dd, J_1 = 7.60 Hz, J_2 = 20 Hz, 1H, CHb of CH_2-Ph) ppm. ^{13}C-NMR (Dioxane, 100 MHz) δ: 172.3 (CO of acid), 134.8, 132.6, 131.2 (2 × CH aromatic), 130.3 (2 × CH aromatic), 130.1 (2 × CH aromatic), 129.5 (2 × CH aromatic), 128.9 (2 × CH aromatic), 57.7, 55.0, 36.4 ppm. IR (KBr) cm^{-1}: 3441 (OH of acid), 2974 (NH), 2822 (CH aromatic), 2776 (CH aliphatic), 1740 (C=O of COOH), 1609 (C=C), 1221, 1171 (SO_2 two bands), 702 (Ar-H), 623 (N-H bending). MS: in *m/z* [rel. %]: 270.1 [4%], 212.1 [12%], 180.1 [90%], 179.1 [95%], 178.1 [100%], 165.1 [60%], 122.0 [34%], 121.0 [67%], 64 [SO_2^+, 70%]. Anal. Calcd. for 319.38 $C_{16}H_{17}NO_4S$: C, 60.17; H, 5.37; N, 4.39. Found: C, 59.98; H, 5.21; N, 4.42.

(11) 3-(4-(Benzylsulfonyloxy)phenyl)-2-(phenylmethylsulfonamido)propanoic Acid, **11**. L amino acid is tyrosine; yield = 2.19 g (89.6%); mp 195-196°C; R_f = 0.82. ^{1}H-NMR (D_2O, 400 MHz) δ: 7.43 (s, 10H, 2 × Ar-H), 7.22–7.20 (d, J = 8.4 Hz, 2H, OTs-H), 6.93–6.91 (d, J = 8.4 Hz, 2H, OTs-H), 4.31–4.28 (dd, J_1 = 5.60 Hz, J_2 = 13.08 Hz, 1H, $PhCH_2$-CH-COOH), 4.19 (s, 4H, 2 × CH_2-SO_2), 3.31-3.26 (dd, J_1 = 5.60 Hz, J_2 = 20.00 Hz, CHa of CH_2-Ph), 3.19–3.14 (dd, J_1 = 7.52 Hz, J_2 = 20 Hz, 1H, CHb of CH_2-Ph) ppm. ^{13}C-NMR (Dioxane, 100 MHz) δ: 172.8 (CO of acid), 156.5, 140.9, 132.0, 131.8 (2CH aromatic), 131.3 (2 × CH aromatic),

129.6 (2 × CH aromatic), 129.0 (2 × CH aromatic), 126.6, 123.6 (2 × CH aromatic), 117.1 (2 × CH aromatic), 111.8 (2 × CH aromatic), 58.4, 55.4, 36.1 ppm. IR (KBr) cm^{-1}: 3435 (OH of acid), 3167 (NH), 3028 (CH aromatic), 2949 (CH aliphatic), 1724 (C=O of COOH), 1597 (C=C), 1194, 1148 (SO_2 two bands), 789 (S-OR ester), 694 (Ar-H), 631 (N-H bending). MS: in *m/z* [rel. %]: 180.1 [74%], 179.1 [90%], 178.1 [100%], 165.1 [30%], 122.0 [10%], 91.1 [$PhCH_2^+$, 47%], 64 [SO_2^+, 24%]. Anal. Calcd. for 489.57 $C_{23}H_{23}NO_7S_2$: C, 56.43; H, 4.76; N, 2.86. Found: C, 56.38; H, 4.79; N, 2.69.

4.2.2. General Procedure of N,N-Diethyl-2-(phenylmethysulfonamido)propanamide, ***12–22***. To a solution of α-tolylsulfonamide derivatives **1–11** (2.96 mmol) in H_2O (10 mL) in a streaming flow of nitrogen gas was added oxalyl chloride (0.34 mL, 3.85 mmol, 1.30 eq.) via dropping pipette followed by carefully controlled addition of 1 drop of DMF. The resulting mixture was stirred at room temperature for 2 h to get crude acid chloride which was kept air-tighted prior to use. In a separate 250 mL three-necked round bottom flask, equipped with a magnetic stirring bar, was added Na_2CO_3 (0.628 g, 5.92 mmol, 2 equiv.) to H_2O (10 mL) followed by diethyl amine DEA (0.4 mL, 3.85 mmol, 1.3 equiv.) in a continuous stirring and cooled to −15°C. Then, earlier kept acid chloride was added in such a way to maintain the internal temperature of the reaction mixture at around −10°C. The reacting mixture was then stirred at −10°C for 1 h; at 0°C for 1 h; finally at room temperature for 1 h. The reaction was terminated, worked up by acidifying with 2 N HCl, and concentrated in a rotary evaporator. The clear solution obtained was freeze-dried to get crude solid product which was purified by column chromatography ($CHCl_3/CH_3OH$, 3 : 1) to afford *N,N*-dimethylacetamide of α-tolylsulfonamide derivatives **12–22**.

(1) 1-(Benzylsulfonyl)-N,N-diethylpiperidine-2-carboxamide, ***12***. Yield 0.99 g (99.0%); mp 210-211°C; R_f = 0.72. ^{1}H-NMR (D_2O, 400 MHz) δ: 7.54 (s, 5H, Ar-H), 4.30 (s, 2H, CH_2-SO_2), 4.08–4.04 (dd, J_1 = 3.52 Hz, J_2 = 20.00 Hz, 1H, CHa of CH_2-N), 3.60–3.57 (m, 1H, CH-CON), 3.21–3.16 (q, J = 7.20 Hz, 4H, 2 × CH_2-CH_3), 2.42–2.38 (dd, J_1 = 3.32 Hz, J_2 = 20.00 Hz, 1H, CHb of CH_2-N), 2.03–1.99 (m, 2H, CH_2), 1.86–1.73 (m, 3H, CH & CH_2), 1.40–1.37 (t, J = 7.20 Hz, 6H, 2 × CH_3-CH_2) ppm. ^{13}C-NMR (Dioxane, 100 MHz) δ: 173.5 (C=O), 132.9, 131.1 (2 × CH aromatic), 129.8 (2 × CH aromatic), 128.6, 60.6, 57.8, 47.4, 43.4 (CH_2), 29.4, 24.2, 18.1, 11.4 (CH_3) ppm. IR (KBr) cm^{-1}: 3028 (CH aromatic), 2951 (CH aliphatic), 1720 (C=O), 1593 (C=C), 1188, 1148 (SO_2 two bands), 696 (Ar-H). Anal. Calcd. for $C_{17}H_{26}N_2O_3S$ (338.47): C, 60.33; H, 7.74; N, 8.28. Found: C, 60.29; H, 6.94; N, 7.98.

(2) 1-(Benzylsulfonyl)-N,N-diethylpyrrolidine-2-carboxamide, ***13***. Yield 0.94 g (97.9%); mp 185–187°C; R_f = 0.71. ^{1}H-NMR (D_2O, 400 MHz) δ: 7.53 (s, 5H, Ar-H), 4.54–4.50 (m, 1H, CH-CON), 4.28 (s, 2H, CH_2-SO_2), 3.56–3.54 (m, 2H, CH_2-N), 3.19–3.14 (q, J = 7.28 Hz, 4H, 2 × CH_2-CH_3), 2.54–2.51 (m, 1H, CH), 2.28–2.24 (m, 1H, CH), 2.16–2.14 (m, 2H, CH_2), 1.39–1.35 (t, J = 7.28 Hz, 6H, 2 × CH_3-CH_2) ppm. ^{13}C-NMR (Dioxane, 100 MHz) δ: 173.2 (C=O), 132.6, 131.1 (2 × CH aromatic), 129.4 (2 × CH aromatic), 128.8, 60.8, 57.7, 47.1, 43.3 (2 × CH_2), 29.2, 24.2, 11.4 (2 × CH_3) ppm. Anal. Calcd. for $C_{16}H_{24}N_2O_3S$ (324.45): C, 59.23; H, 7.46; N, 8.63. Found: C, 59.09; H, 7.46; N, 8.48.

(3) N,N-Diethyl-2-(phenylmethylsulfonamido)acetamide, ***14***. Yield 0.78 g (92.6%); mp 213–215°C; R_f = 0.51. ^{1}H-NMR (D_2O, 400 MHz) δ: 7.53 (s, 5H, Ar-H), 4.29 (s, 2H, CH_2-SO_2), 4.01 (s, 2H, CH_2-CON), 3.20-3.14 (q, J = 7.30 Hz, 4H, 2 × CH_2-CH_3), 1.39–1.36 (t, J = 7.30 Hz, 6H, 2 × CH_3-CH_2) ppm. ^{13}C-NMR (Dioxane, 100 MHz) δ: 174.1 (C=O), 133.4, 131.6 (2 × CH aromatic), 129.5 (2 × CH aromatic), 128.6, 57.7, 49.8, 43.1 (CH_2), 11.4 (CH_3) ppm. IR (KBr) cm^{-1}: 3217 (N-H), 3036 (CH aromatic), 2947 (CH aliphatic), 1712 (C=O), 1601 (C=C), 1219, 1194, (SO_2 two bands), 694 (Ar-H). Anal. Calcd. for $C_{13}H_{20}N_2O_3S$ (284.38): C, 54.91; H, 7.09; N, 9.85. Found: C, 55.13; H, 6.94; N, 10.08.

(4) N,N-Diethyl-2-(phenylmethylsulfonamido)propanamide, ***15***. Yield 0.87 g (98.5%); mp 238–240°C; R_f = 0.56. ^{1}H-NMR (D_2O, 400 MHz) δ: 7.50 (s, 5H, Ar-H), 4.26 (s, 2H, CH_2-SO_2), 4.22–4.16 (q, J = 7.28 Hz, 1H, CH-CH_3), 3.18–3.12 (q, J = 7.32 Hz, 4H, 2 × CH_2-CH_3), 1.65–1.63 (d, J = 7.28 Hz, 3H, CH_3-CH), 1.37–1.33 (t, J = 7.32 Hz, 6H, 2 × CH_3-CH_2) ppm. ^{13}C-NMR (Dioxane, 100 MHz) δ: 173.8 (C=O), 132.7, 131.2 (2 × CH aromatic), 129.5 (2 × CH aromatic), 128.9, 57.7, 49.8, 43.1 (CH_2), 16.2, 11.4 (CH_3) ppm. IR (KBr) cm^{-1}: 3058 (N-H), 3036 (CH aromatic), 2951 (CH aliphatic), 1719 (C=O), 1601 (C=C), 1219, 1196, 1148 (SO_2 two bands), 696 (Ar-H). Anal. Calcd. for $C_{14}H_{22}N_2O_3S$ (298.41): C, 56.35; H, 7.43; N, 9.39. Found: C, 56.11; H, 7.33; N, 9.28.

(5) N,N-Diethyl-3-mercapto-2-(phenylmethylsulfonamido)propanamide, ***16***. Yield 0.87 g (89.0%); mp 198–200°C; R_f = 0.71. ^{1}H-NMR (D_2O, 400 MHz) δ: 7.46 (s, 5H, Ar-H), 4.54–4.51 (dd, J_1 = 4.24 Hz, J_2 = 7.92 Hz, 1H, CH_2-CH-COOH), 4.21 (s, 2H, CH_2-SO_2), 3.57–3.52 (dd, J_1 = 4.24 Hz, J_2 = 20.00 Hz, 1H, CHa of CH_2-CH), 3.41–3.36 (dd, J_1 = 7.92 Hz, J_2 = 20.00 Hz, 1H, CHb of CH_2-CH), 3.18–3.12 (q, J = 7.35 Hz, 4H, 2 × CH_2-CH_3), 1.37–1.33 (t, J = 7.35 Hz, 6H, 2 × CH_3-CH_2) ppm. ^{13}C-NMR (Dioxane, 100 MHz) δ: 173.5 (C=O), 132.7, 131.2 (2 × CH aromatic), 129.5 (2 × CH aromatic), 128.9, 57.4, 49.5, 43.1 (CH_2), 35.5, 11.4 (CH_3) ppm. IR (KBr) cm^{-1}: 2997 (CH aromatic), 2911 (CH aliphatic), 1719 (C=O), 1591 (C=C), 1194, 1144 (SO_2 two bands), 696 (Ar-H). Anal. Calcd. for $C_{14}H_{22}N_2O_3S_2$ (330.47): C, 50.88; H, 6.71; N, 8.48. Found: C, 50.71; H, 6.99; N, 7.97.

(6) N,N-Diethyl-4-(methylthio)-2-(phenylmethylsulfonamido)butanamide, ***17***. Yield 0.96 g (90.6%); mp 170–172°C; R_f = 0.65. ^{1}H-NMR (D_2O, 400 MHz) δ: 7.49 (s, 5H, Ar-H), 4.24 (s, 2H, CH_2-SO_2), 4.23–4.20 (t, J = 6.72 Hz, 1H, CH-CH_2), 3.16–3.10 (q, J = 7.32 Hz, 4H, 2 × CH_2-CH_3), 2.77–2.73 (t,

$J = 7.40$ Hz, 2H, S-CH_2-CH_2), 2.34–2.24 (m, 2H, CH-CH_2-CH_2-S), 2.19 (s, 3H, CH_3-S), 1.35–1.31 (t, $J = 7.32$ Hz, 6H, 2 × CH_3-CH_2) ppm. ^{13}C-NMR (Dioxane, 100 MHz) δ: 173.8, 132.7, 131.2 (2 × CH aromatic), 129.5, (2 × CH aromatic), 57.7, 49.8, 43.1 (CH_2), 30.8, 29.5, 16.2, 11.4 (CH_3) ppm. IR (KBr) cm^{-1}: 3028 (CH aromatic), 2945 (CH aliphatic), 1722 (C=O), 1620 (C=C), 1200, 1126 (SO_2 two bands) 698 (Ar-H). Anal. Calcd. for $C_{16}H_{26}N_2O_3S_2$ (358.53): C, 53.60; H, 7.31; N, 7.81. Found: C, 53.55; H, 7.22; N, 7.69.

(7) N,N-Diethyl-3-methyl-2-(phenylmethylsulfonamido)butanamide, ***18***. Yield 0.94 g (97.3%); mp 226–230°C; R_f = 0.69. ^{1}H-NMR (D_2O, 400 MHz) δ: 7.50 (s, 5H, Ar-H), 4.25 (s, 2H, CH_2-SO_2), 4.00-3.97 (d, $J = 2.84$ Hz, 1H, CH-CH-CON), 3.17–3.11 (q, $J = 7.32$ Hz, 4H, 2 × CH_2-CH_3), 2.45–2.38 (m, 1H, CH), 1.36–1.32 (t, $J = 7.32$ Hz, 6H, 2 × CH_3-CH_2), 1.15–1.13 (d, $J = 7.00$ Hz, 3H, CH_3-CH), 1.12–1.10 (d, $J = 7.00$ Hz, 3H, CH_3-CH) ppm. ^{13}C-NMR (Dioxane, 100 MHz) δ: 173.8 (C=O), 132.7, 131.2 (2 × CH aromatic), 129.5 (2 × CH aromatic), 128.9, 60.8, 57.7, 49.8, 43.1 (CH_2), 16.2, 15.5, 11.4 (CH_3) ppm. IR (KBr) cm^{-1}: 3053 (CH aromatic), 2945 (CH aliphatic), 1718 (C=O), 1611 (C=C), 1219, 1194 (SO_2 two bands), 696 (Ar-H). Anal. Calcd. for $C_{16}H_{26}N_2O_3S$ (326.46): C, 58.87; H, 8.03; N, 8.58. Found: C, 59.01; H, 7.96; N, 8.61.

(8) N,N-Diethyl-3-hydroxy-2-(phenylmethylsulfonamido) butanamide, ***19***. Yield 0.78 g (80.2%); mp 245°C (dec); R_f = 0.53. ^{1}H-NMR (D_2O, 400 MHz) δ: 7.55 (s, 5H, Ar-H), 4.54–4.52 (m, 1H, CH-CH-CH_3), 4.31 (s, 2H, CH_2-SO_2), 4.09–4.08 (d, $J = 3.96$ Hz, 1H, CH-CH-CON), 3.22–3.17 (q, $J = 7.32$ Hz, 4H, 2 × CH_2-CH_3), 2.85 (s, 1H, OH), 1.49–1.47 (d, $J = 6.60$ Hz, 3H, CH_3-CH), 1.41–1.37 (t, $J = 7.32$ Hz, 6H, 2 × CH_3-CH_2) ppm. ^{13}C-NMR (Dioxane, 100 MHz) δ: 173.5 (C=O), 132.9, 131.3 (2 × CH aromatic), 129.5 (2 × CH aromatic), 128.9, 61.2, 58.1, 49.5, 42.5 (CH_2), 16.6, 11.1 (CH_3) ppm. IR (KBr) cm^{-1}: 3396 (OH), 3030 (CH aromatic), 2945 (CH aliphatic), 1720 (C=O), 1601 (C=C), 1221, 1194 (SO_2 two bands), 694 (Ar-H). Anal. Calcd. for $C_{15}H_{24}N_2O_4S$ (328.43): C, 54.86; H, 7.37; N, 8.53. Found: C, 54.71; H, 7.26; N, 8.65.

(9) N^1,N^1-Diethyl-2-(phenylmethylsulfonamido)pentanediamide, ***20***. Yield 0.98 g (93.2%); mp 251–253°C; R_f = 0.58. ^{1}H-NMR (D_2O, 400 MHz) δ: 7.51 (s, 5H, Ar-H), 4.52–4.48 (dd, J_1 = 5 Hz, J_2 = 14.32 Hz, 1H, NOC-CH-$CH_{2a,b}$), 4.26 (s, 2H, CH_2-SO_2), 3.17–3.12 (q, $J = 7.32$ Hz, 4H, 2 × CH_2-CH_3), 2.67–2.60 (m, 1H, CH_a of CH_2), 2.53–2.48 (t, J = 8 Hz, 2H, CO-CH_2-CH_2), 2.30–2.24 (m, 1H, CH_b of CH_2), 1.37–1.33 (t, $J = 7.32$ Hz, 6H, 2 × CH_3-CH_2) ppm. ^{13}C-NMR (Dioxane, 100 MHz) δ: 182.7 (C=O), 177.1 (C=O), 132.6, 131.2 (2 × CH aromatic), 129.5 (2 × CH aromatic), 128.9, 57.7, 53.3, 43.1 (CH_2), 30.1, 25.2, 11.4 (CH_3) ppm. IR (KBr) cm^{-1}: 3075 (NH), 3053 (CH aromatic), 2951 (CH aliphatic), 1703 (C=O), 1659 (C=O of CON), 1601 (C=C), 1221, 1193 (SO_2 two bands), 696 (Ar-H). Anal. Calcd. for $C_{16}H_{25}N_3O_4S$ (355.46): C, 54.06; H, 7.09; N, 11.82. Found: C, 53.95; H, 6.88; N, 12.01.

(10) N,N-Diethyl-3-phenyl-2-(phenylmethylsulfonamido) propanamide, ***21***. Yield 1.00 g (90.2%); mp 227–229°C; R_f = 0.70. ^{1}H-NMR (D_2O, 400 MHz) δ: 7.49 (s, 10H, 2 × Ar-H), 4.39–4.36 (dd, J_1 = 5.60 Hz, J_2 = 7.60 Hz, 1H, NOC-CH-$CH_{2a,b}$), 4.24 (s, 2H, CH_2-SO_2), 3.44–3.39 (dd, J_1 = 5.60 Hz, J_2 = 20.00 Hz, 1H, CH_a of $CH_{2a,b}$), 3.31–3.26 (dd, J_1 = 7.60 Hz, J_2 = 20.00 Hz, 1H, CH_b of $CH_{2a,b}$), 3.16–3.10 (q, $J = 7.32$ Hz, 4H, 2 × CH_2-CH_3), 2.78 (s, 2H, CH_2), 1.35–1.31 (t, $J = 7.32$ Hz, 6H, 2 × CH_3-CH_2) ppm. ^{13}C-NMR (Dioxane, 100 MHz) δ: 172.4 (C=O), 135.1, 132.7, 131.2 (2 × CH aromatic), 130.3 (2 × CH aromatic), 130.1 (2 × CH aromatic), 129.5 (2 × CH aromatic), 128.9 (2 × CH aromatic), 57.8, 55.2, 43.1 (CH_2), 36.5, 11.5 (CH_3) ppm. IR (KBr) cm^{-1}: 2976 (NH), 2828 (CH aromatic), 2774 (CH aliphatic), 1736 (C=O), 1620 (C=C), 1206, 1153, 1148 (SO_2 two bands), 698 (Ar-H). Anal. Calcd. for $C_{20}H_{26}N_2O_3S$ (374.51): C, 64.14; H, 7.00; N, 7.48. Found: C, 64.00; H, 6.84; N, 7.29.

(11) 4-(3-(Diethylamino)-3-oxo-2-(phenylmethylsulfonamido) propyl)phenylphenylmethanesulfonate, ***22***. Yield 1.44 g (89.3%); mp 265°C (dec); R_f = 0.69. ^{1}H-NMR (D_2O, 400 MHz) δ: 7.50 (s, 10H, 2 × Ar-H), 7.29–7.27 (d, $J = 8.00$ Hz, 2H, Ar-H), 6.99–6.97 (d, $J = 8.00$ Hz, 2H, Ar-H), 4.37–4.33 (dd, J_1 = 5.60 Hz, J_2 = 7.60 Hz, 1H, NOC-CH-$CH_{2a,b}$), 4.26 (s, 2H, CH_2-SO_2), 3.37–3.32 (dd, J_1 = 5.60 Hz, J_2 = 20.00 Hz, 1H, CH_a of $CH_{2a,b}$), 3.26–3.21 (dd, J_1 = 7.60 Hz, J_2 = 20.00 Hz, 1H, CH_b of $CH_{2a,b}$), 3.17–3.12 (q, $J = 7.32$ Hz, 4H, 2 × CH_2-CH_3), 1.37–1.33 (t, $J = 7.32$ Hz, 6H, 2 × CH_3-CH_2) ppm. ^{13}C-NMR (Dioxane, 100 MHz) δ: 172.4 (C=O), 155.9, 141.3, 131.8 (2 × CH aromatic), 131.3 (2 × CH aromatic), 129.6 (2 × CH aromatic), 129.0 (2 × CH aromatic), 126.6, 123.0 (2 × CH aromatic), 116.9 (2 × CH aromatic), 111.6 (2 × CH aromatic), 57.8, 55.2, 43.2 (CH_2), 35.6, 11.6 (CH_3) ppm. IR (KBr) cm^{-1}: 3058 (NH), 3036 (CH aromatic), 2951 (CH aliphatic), 1719 (C=O), 1601 (C=C), 1219, 1196, 1148 (SO_2 two bands), 696 (Ar-H). Anal. Calcd. for $C_{27}H_{32}N_2O_6S_2$ (544.69): C, 59.54; H, 5.92; N, 5.14. Found: C, 59.43; H, 5.99; N, 4.98.

4.3. Antibacterial Activity Assays. The antimicrobial properties of the sulfonamides were investigated in the form of the general sensitivity testing and minimum inhibitory concentration (MIC) with respect to freshly cultured targeted organisms. The two organisms of interest in this present study are one gram positive (*Staphylococcus aureus* ATCC 6538) and one gram negative (*Escherichia coli* ATCC 25922) organisms which are associated with the gastrointestinal tract damage in man and animal.

4.3.1. Preparation of the Inoculum. The standard strains of *S. aureus* and *E. coli* used were obtained from Test Center of Antimicrobial Materials, TIPC, Beijing. No clinically isolated organism was used based on inavailability of such as at the time of this study. The strains were propagated on nutrient agar plates and maintained on the plate at 4°C. The isolates were subcultured in nutrient broth at 37°C for 8 h prior to antibacterial testing.

4.3.2. Antibacterial Sensitivity Testing of the Synthesized Compounds. Agar well diffusion technique as described by Adeniyi and coworkers was used to determine the antibacterial activity of the synthesized compounds [66]. Sensitivity test agar plates were seeded with 0.1 mL of an overnight culture of each bacterial strain (equivalent to 10^7–10^8 CFU mL^{-1}). The seeded plates were allowed to set and a standard cork borer of 8 mm diameter was used to cut uniform wells on the surface of the agar. The wells were then filled with 0.3 mL of each sulfonamide solution in an appropriate solvent at a concentration of 1000 μg/mL (0.02 g of sulfonamide dissolved in 20 mL distilled water). All the plates were incubated at 37°C for 24 h. The assay was conducted at regular intervals of 24 h until a marked decline in the potency of the sulfonamide solution to inhibit the growth of the test organisms was noticed. Zones of clearance round each well means inhibition and the diameter of such zones were measured. The procedure was repeated for the streptomycin (standard).

4.3.3. Determination of Minimum Inhibitory Concentration (MIC). Agar well dilution method as described by Russell and Furr was used to determine the minimum inhibitory concentration (MIC) of the sulfonamides and streptomycin [69]. Different dilutions of the sulfonamides were prepared first at $\leq$100 μg/mL to give final concentrations in the range of 100, 50, 25, 12.5, 6.25, and 1.8 μg/mL. The different dilutions of sulfonamide derivatives that could not inhibit the microbial growth at $\leq$100 μg/mL were later prepared at $\leq$1000 μg/mL to give final concentrations in the range of 1000, 500, 250, 125, and 62.5 μg/mL. Two milliliter (2 mL) of each dilution was mixed with 18 mL of Mueller Hinton agar (MHA, Difco, France) and poured into Petri dishes and allowed to set. The agar was streaked with an overnight broth culture of the bacterial strains and incubated overnight. The plates were then examined for the presence or absence of growth. The minimum concentration that completely inhibited macroscopic growth was regarded as the minimum inhibitory concentration of the respective sulfonamide. The procedure was repeated for streptomycin (standard). Selectivity index (SI) is the ratio of the zone of the inhibition of compound to that of the streptomycin.

Conflict of Interests

The authors hereby declare that there is no conflict of interest as regard this paper.

Acknowledgments

O. O. Ajani is grateful to the Academy of Science for the Developing World (TWAS) and Chinese Academy of Sciences (CAS) for the sponsorship of this work. This present research work was birthed from the 2009 TWAS-CAS PG Fellowship Award which was jointly funded by both TWAS and CAS.

References

[1] M. A. Bhat, M. Imran, S. A. Khan, and N. Siddiqui, "Biological activities of sulfonamides," *Indian Journal of Pharmaceutical Sciences*, vol. 67, no. 2, pp. 151–159, 2005.

[2] C. T. Supuran, A. Casini, A. Mastrolorenzo, and A. Scozzafava, "COX-2 selective inhibitors, carbonic anhydrase inhibition and anticancer properties of sulfonamides belonging to this class of pharmacological agents," *Mini-Reviews in Medicinal Chemistry*, vol. 4, no. 6, pp. 625–632, 2004.

[3] A. K. Gadad, C. S. Mahajanshetti, S. Nimbalkar, and A. Raichurkar, "Synthesis and antibacterial activity of some 5-guanylhydrazone/thiocyanato-6-arylimidazo[2,1-*b*]-1,3,4-thiadiazole-2-sulfon amide derivatives," *European Journal of Medicinal Chemistry*, vol. 35, no. 9, pp. 853–857, 2000.

[4] H. Eshghi, M. Rahimizadeh, M. Zokaei et al., "Synthesis and antimicrobial activity of some new macrocyclic *bis*-sulfonamide and disulphides," *European Journal of Chemistry*, vol. 2, no. 1, pp. 47–50, 2011.

[5] D. W. Hopper, M. D. Vera, D. How et al., "Synthesis and biological evaluation of ((4-keto)-phenoxy)methyl biphenyl-4-sulfonamides: a class of potent aggrecanase-1 inhibitors," *Bioorganic and Medicinal Chemistry Letters*, vol. 19, no. 9, pp. 2487–2491, 2009.

[6] C. Juli, M. Sippel, J. Jäger et al., "Pipecolic acid derivatives as small-molecule inhibitors of the legionella MIP protein," *Journal of Medicinal Chemistry*, vol. 54, no. 1, pp. 277–283, 2011.

[7] A. Scozzafava and C. T. Supuran, "Protease inhibitors—part 5. Alkyl/arylsulfonyl- and arylsulfonylureido-/arylureido-glycine hydroxamate inhibitors of *Clostridium histolyticum* collagenase," *European Journal of Medicinal Chemistry*, vol. 35, no. 3, pp. 299–307, 2000.

[8] J. Schröder, A. Henke, H. Wenzel et al., "Structure-based design and synthesis of potent matrix metalloproteinase inhibitors derived from a 6*H*-1,3,4-thiadiazine scaffold," *Journal of Medicinal Chemistry*, vol. 44, no. 20, pp. 3231–3243, 2001.

[9] D. F. Burdi, S. Chi, K. Mattia et al., "Small molecule antagonists of the CC chemokine receptor 4 (CCR4)," *Bioorganic and Medicinal Chemistry Letters*, vol. 17, no. 11, pp. 3141–3145, 2007.

[10] G. A. Domagk, "A contribution to chemotherapy of bacterial infections," *Clinical Infectious Diseases*, vol. 8, no. 1, pp. 163–166, 1986.

[11] D. Vicente and E. Pérez-Trallero, "Tetracyclines, sulfonamides, and metronidazole," *Enfermedades Infecciosas y Microbiología Clínica*, vol. 28, no. 2, pp. 122–130, 2010.

[12] Z. Chen, W. Xu, K. Liu et al., "Synthesis and antiviral activity of 5-(4-chlorophenyl)-1,3,4-thiadiazole sulfonamides," *Molecules*, vol. 15, no. 12, pp. 9046–9056, 2010.

[13] A. J. McCarroll, T. D. Bradshaw, A. D. Westwell, C. S. Matthews, and M. F. G. Stevens, "Quinols as novel therapeutic agents. 7. [1] Synthesis of antitumor 4-[1-(arylsulfonyl-1*H*-indol-2-yl)]-4-hydroxycyclohexa-2,5-dien-1-ones by Sonogashira reactions," *Journal of Medicinal Chemistry*, vol. 50, no. 7, pp. 1707–1710, 2007.

[14] B. L. Wilkinson, L. F. Bornaghi, T. A. Houston et al., "Carbonic anhydrase inhibitors: Inhibition of isozymes I, II, and IX with triazole-linked *O*-glycosides of benzene sulfonamides," *Journal of Medicinal Chemistry*, vol. 50, no. 7, pp. 1651–1657, 2007.

[15] A. Ali, G. S. K. K. Reddy, H. Cao et al., "Discovery of HIV-1 protease inhibitors with picomolar affinities incorporating *N*-aryl-oxazolidinone-5-carboxamides as novel P2 ligands,"

Journal of Medicinal Chemistry, vol. 49, no. 25, pp. 7342–7356, 2006.

[16] H. Kourlas and S. Morey, "Sulfonamide allergies and possible cross-reactivity," *Journal of Pharmacy Practice*, vol. 20, no. 5, pp. 399–402, 2007.

[17] M. Gurrath, "Peptide-binding G protein-coupled receptors: new opportunities for drug design," *Current Medicinal Chemistry*, vol. 8, no. 13, pp. 1605–1648, 2001.

[18] H. T. Lai and J. H. Hou, "Light and microbial effects on the transformation of four sulfonamides in eel pond water and sediment," *Aquaculture*, vol. 283, no. 1–4, pp. 50–55, 2008.

[19] M. J. O. 'Neil, P. E. Heckelman, C. B. Koch, and K. J. Roman, Eds., *The Merck Index*, Merck, Whitehouse Station, NJ, USA, 14th edition, 2006.

[20] N. Anand and W. A. Remers, *Burger's Medicinal Chemistry, Drug Discovery and Development*, John Wiley & Sons, New York, NY, USA, 2010.

[21] D. E. Golan, "Dihydrofolate reductase inhibitors," in *Principle of Pharmacology: The Pathophysiologic Basis of Drug Therapy*, D. E. Golan, A. H. Tashjian, E. J. Armstrong, and A. W. Armstrong, Eds., pp. 577–580, Lippincott Williams & Wilkins, Philadelphia, Pa, USA, 2nd edition, 2008.

[22] P. Purushottamachar, A. Khandelwal, T. S. Vasaitis, R. D. Bruno, L. K. Gediya, and V. C. O. Njar, "Potent anti-prostate cancer agents derived from a novel androgen receptor down-regulating agent," *Bioorganic and Medicinal Chemistry*, vol. 16, no. 7, pp. 3519–3529, 2008.

[23] B. R. Stranix, J. F. Lavallée, G. Sévigny et al., "Lysine sulfonamides as novel HIV-protease inhibitors: Nε-acyl aromatic α-amino acids," *Bioorganic and Medicinal Chemistry Letters*, vol. 16, no. 13, pp. 3459–3462, 2006.

[24] W. G. Harter, H. Albrect, K. Brady et al., "The design and synthesis of sulfonamides as caspase-1 inhibitors," *Bioorganic and Medicinal Chemistry Letters*, vol. 14, no. 3, pp. 809–812, 2004.

[25] N. S. Reddy, M. R. Mallireddigari, S. Cosenza et al., "Synthesis of new coumarin 3-(*N*-aryl) sulfonamides and their anticancer activity," *Bioorganic and Medicinal Chemistry Letters*, vol. 14, no. 15, pp. 4093–4097, 2004.

[26] F. Abbate, A. Casini, T. Owa, A. Scozzafava, and C. T. Supuran, "Carbonic anhydrase inhibitors: E7070, a sulfonamide anticancer agent, potently inhibits cytosolic isozymes I and II, and transmembrane, tumor-associated isozyme IX," *Bioorganic and Medicinal Chemistry Letters*, vol. 14, no. 1, pp. 217–223, 2004.

[27] K. Na and Y. H. Bae, "Self-assembled hydrogel nanoparticles responsive to tumor extracellular pH from pullulan derivative/sulfonamide conjugate: characterization, aggregation, and adriamycin release *in vitro*," *Pharmaceutical Research*, vol. 19, no. 5, pp. 681–688, 2002.

[28] K. Fukuoka, J. Usuda, Y. Iwamoto et al., "Mechanisms of action of the novel sulfonamide anticancer agent E7070 on cell cycle progression in human non-small cell lung cancer cells," *Investigational New Drugs*, vol. 19, no. 3, pp. 219–227, 2001.

[29] A. Weber, A. Casini, A. Heine et al., "Unexpected nanomolar inhibition of carbonic anhydrase by COX-2-selective celecoxib: new pharmacological opportunities due to related binding site recognition," *Journal of Medicinal Chemistry*, vol. 47, no. 3, pp. 550–557, 2004.

[30] A. G. Habeeb, P. N. P. Rao, and E. E. Knaus, "Design and synthesis of celecoxib and rofecoxib analogues as selective cyclooxygenase-2 (COX-2) inhibitors: replacement of sulfonamide and methylsulfonyl pharmacophores by an azido bioisostere," *Journal of Medicinal Chemistry*, vol. 44, no. 18, pp. 3039–3042, 2001.

[31] A. L. Blobaum and L. J. Marnett, "Structural and functional basis of cyclooxygenase inhibition," *Journal of Medicinal Chemistry*, vol. 50, no. 7, pp. 1425–1441, 2007.

[32] I. R. Ezabadi, C. Camoutsis, P. Zoumpoulakis et al., "Sulfonamide-1,2,4-triazole derivatives as antifungal and antibacterial agents: synthesis, biological evaluation, lipophilicity, and conformational studies," *Bioorganic and Medicinal Chemistry*, vol. 16, no. 3, pp. 1150–1161, 2008.

[33] P. E. Sum, A. T. Ross, P. J. Petersen, and R. T. Testa, "Synthesis and antibacterial activity of 9-substituted minocycline derivatives," *Bioorganic and Medicinal Chemistry Letters*, vol. 16, no. 2, pp. 400–403, 2006.

[34] S. Joshi, N. Khosla, D. Khare, and R. Sharda, "Synthesis and *in vitro* study of novel Mannich bases as antibacterial agents," *Bioorganic and Medicinal Chemistry Letters*, vol. 15, no. 1, pp. 221–226, 2005.

[35] M. Padmanilayam, B. Scorneaux, Y. Dong et al., "Antimalarial activity of N-alkyl amine, carboxamide, sulfonamide, and urea derivatives of a dispiro-1,2,4-trioxolane piperidine," *Bioorganic and Medicinal Chemistry Letters*, vol. 16, no. 21, pp. 5542–5545, 2006.

[36] J. N. Domínguez, C. León, J. Rodrigues, N. G. De Domínguez, J. Gut, and P. J. Rosenthal, "Synthesis and antimalarial activity of sulfonamide chalcone derivatives," *Farmaco*, vol. 60, no. 4, pp. 307–311, 2005.

[37] G. Bouchain, S. Leit, S. Frechette et al., "Development of potential antitumor agents. Synthesis and biological evaluation of a new set of sulfonamide derivatives as histone deacetylase inhibitors," *Journal of Medicinal Chemistry*, vol. 46, no. 5, pp. 820–830, 2003.

[38] A. Yokoi, J. Kuromitsu, T. Kawai et al., "Profiling novel sulfonamide antitumor agents with cell-based phenotypic screens and array-based gene expression analysis," *Molecular Cancer Therapeutics*, vol. 1, no. 4, pp. 275–286, 2002.

[39] M. Banerjee, A. Poddar, G. Mitra, A. Surolia, T. Owa, and B. Bhattacharyya, "Sulfonamide drugs binding to the colchicine site of tubulin: thermodynamic analysis of the drug-tubulin interactions by isothermal titration calorimetry," *Journal of Medicinal Chemistry*, vol. 48, no. 2, pp. 547–555, 2005.

[40] T. Inaba, K. Tanaka, R. Takeno, H. Nagaki, C. Yoshida, and S. Takano, "Synthesis and antiinflammatory activity of 7-methanesulfonylamino-6-phenoxychromones. Antiarthritic effect of the 3-formylamino compound (T-614) in chronic inflammatory disease models," *Chemical and Pharmaceutical Bulletin*, vol. 48, no. 1, pp. 131–139, 2000.

[41] J. J. Talley, S. R. Bertenshaw, D. L. Brown et al., "*N*-[[(5-methyl-3-phenylisoxazol-4-yl)phenyl]sulfonyl]propanamide, sodium salt, parecoxib sodium: a potent and selective inhibitor of COX-2 for parenteral administration," *Journal of Medicinal Chemistry*, vol. 43, no. 9, pp. 1661–1663, 2000.

[42] S. M. Sondhi, M. Johar, N. Singhal, S. G. Dastidar, R. Shukla, and R. Raghubir, "Synthesis and anticancer, antiinflammatory, and analgesic activity evaluation of some sulfa drug and acridine derivatives," *Monatshefte fur Chemie*, vol. 131, no. 5, pp. 511–520, 2000.

[43] K. Chibale, H. Haupt, H. Kendrick et al., "Antiprotozoal and cytotoxicity evaluation of sulfonamide and urea analogues of quinacrine," *Bioorganic and Medicinal Chemistry Letters*, vol. 11, no. 19, pp. 2655–2657, 2001.

[44] C. Brisbare-Roch, J. Dingemanse, R. Koberstein et al., "Promotion of sleep by targeting orexin system in rats, dogs and humans," *Nature Medicine*, vol. 13, no. 2, pp. 150–155, 2007.

[45] M. G. Lee, O. K. Hassani, and B. E. Jones, "Discharge of identified orexin/hypocretin neurons across the sleep-waking cycle," *Journal of Neuroscience*, vol. 25, no. 28, pp. 6716–6720, 2005.
[46] I. V. Estabrooke, M. T. McCarthy, E. Ko et al., "Fos expression in orexin neurons varies with behavioral state," *Journal of Neuroscience*, vol. 21, no. 5, pp. 1656–1662, 2001.
[47] J. Bhaumik, R. Weissleder, and J. R. McCarthy, "Synthesis and photophysical properties of sulfonamidophenyl porphyrins as models for activatable photosensitizers," *Journal of Organic Chemistry*, vol. 74, no. 16, pp. 5894–5901, 2009.
[48] M. A. Gonzalez, D. B. Gorman, C. T. Hamilton, and G. A. Roth, "Process development herbicide pyroxsulam," *Organic Process Research and Development*, vol. 12, no. 2, pp. 301–303, 2008.
[49] T. Arslan, F. Kandemirli, E. E. Ebenso, I. Love, and H. Alemu, "Quantum chemical studies on the corrosion inhibition of some sulphonamides on mild steel in acidic medium," *Corrosion Science*, vol. 51, no. 1, pp. 35–47, 2009.
[50] M. M. El-Naggar, "Corrosion inhibition of mild steel in acidic medium by some sulfa drugs compounds," *Corrosion Science*, vol. 49, no. 5, pp. 2226–2236, 2007.
[51] N. B. Dyatkina, C. D. Roberts, J. D. Keicher et al., "Minor groove DNA binders as antimicrobial agents. 1. Pyrrole tetraamides are potent antibacterials against vancomycin resistant *Enteroccoci* and methicillin resistant *Staphylococcus aureus*," *Journal of Medicinal Chemistry*, vol. 45, no. 4, pp. 805–817, 2002.
[52] G. W. Kaatz, F. McAleese, and S. M. Seo, "Multidrug resistance in *Staphylococcus aureus* due to overexpression of a novel multidrug and toxin extrusion (MATE) transport protein," *Antimicrobial Agents and Chemotherapy*, vol. 49, no. 5, pp. 1857–1864, 2005.
[53] N. Nayak, T. C. Nag, G. Satpathy, and S. B. Ray, "Ultrastructural analysis of slime positive & slime negative *Staphylococcus epidermidis* isolates in infectious keratitis," *Indian Journal of Medical Research*, vol. 125, no. 6, pp. 767–771, 2007.
[54] A. Masunari and L. C. Tavares, "A new class of nifuroxazide analogues: synthesis of 5-nitrothiophene derivatives with antimicrobial activity against multidrug-resistant *Staphylococcus aureus*," *Bioorganic and Medicinal Chemistry*, vol. 15, no. 12, pp. 4229–4236, 2007.
[55] K. Nishino, S. Yamasaki, M. Hayashi-Nishino, and A. Yamaguchi, "Effect of NlpE overproduction on multidrug resistance in *Escherichia coli*," *Antimicrobial Agents and Chemotherapy*, vol. 54, no. 5, pp. 2239–2243, 2010.
[56] K. E. Jones, N. G. Patel, M. A. Levy et al., "Global trends in emerging infectious diseases," *Nature*, vol. 451, no. 7181, pp. 990–993, 2008.
[57] P. K. Anderson, A. A. Cunningham, N. G. Patel, F. J. Morales, P. R. Epstein, and P. Daszak, "Emerging infectious diseases of plants: pathogen pollution, climate change and agrotechnology drivers," *Trends in Ecology and Evolution*, vol. 19, no. 10, pp. 535–544, 2004.
[58] O. C. Nwinyi, S. N. Chinedu, O. O. Ajani, C. O. Ikpo, and K. O. Ogunniran, "Antibacterial effects of extracts of *Ocimum gratissimum* and piper guineense on *Escherichia coli* and *Staphylococcus aureus*," *African Journal of Food Science*, vol. 3, no. 3, pp. 77–81, 2009.
[59] D. A. A. Mossel and P. van Netten, "*Staphylococcus aureus* and related *Staphylococci* in foods: ecology, proliferation, toxinogenesis, control and monitoring," *Journal of Applied Bacteriology Symposium Supplement*, vol. 69, no. 19, pp. 123–146, 1990.
[60] O. C. Nwinyi, N. S. Chinedu, and O. O. Ajani, "Evaluation of antibacterial activity of *Pisidium guajava* and *Gongronema latifolium*," *Journal of Medicinal Plant Research*, vol. 2, no. 8, pp. 189–192, 2008.
[61] G. E. Dolores and J. G. Doyle, "*Escherichia coli* in diarrhea disease," *Annals of Internal Medicine*, vol. 247, no. 1, pp. 81–90, 2001.
[62] O. O. Ajani, C. A. Obafemi, O. C. Nwinyi, and D. A. Akinpelu, "Microwave assisted synthesis and antimicrobial activity of 2-quinoxalinone-3-hydrazone derivatives," *Bioorganic and Medicinal Chemistry*, vol. 18, no. 1, pp. 214–221, 2010.
[63] O. O. Ajani and O. C. Nwinyi, "Microwave-assisted synthesis and evaluation of antimicrobial activity of 3-{3-(*s*-aryl and *s*-heteroaromatic)acryloyl}-2*H*-chromen-2-one derivatives," *Journal of Heterocyclic Chemistry*, vol. 47, no. 1, pp. 179–187, 2010.
[64] W. Zhang, Y. Qin, S. Zhang, and M. Luo, "Synthesis of novel *N*-heterocyclic carbene-Rh complexes derived from L-proline and their catalysis in the addition of arylboronic acids to aldehydes," *Arkivoc*, vol. 2005, no. 14, pp. 39–48, 2005.
[65] J. T. Kuethe and G. L. Beutner, "Synthesis of 2-arylindole-4-carboxylic amides: [2(4-fluorophenyl)-1*H*-indol-4-yl]-1-pyrrolidinylmethanone," *Organic Syntheses*, vol. 86, no. 1, pp. 92–104, 2009.
[66] B. A. Adeniyi, H. A. Odelola, and B. A. Oso, "Antimicrobial potentials of *Diospyros mespiliformis* (Ebenaceae)," *African Journal of Medical and Medicinal Science*, vol. 25, no. 3, pp. 221–224, 1996.
[67] D. Voet and J. D. Voet, "The genetic code," in *Biochemistry*, pp. 1341–1342, John Wiley & Sons, New York, NY, USA, 3rd edition, 2004.
[68] O. O. Ajani and O. C. Nwinyi, "Synthesis and evaluation of antimicrobial activity of phenyl and furan-2-yl[1,2,4]triazolo[4,3-a]quinoxalin-4(5*H*)-one and their hydrazone precursors," *Canadian Journal of Pure and Applied Sciences*, vol. 3, no. 3, pp. 983–992, 2009.
[69] A. D. Russell and J. R. Furr, "The antibacterial activity of a new chloroxylenol preparation containing ethylenediamine tetraacetic acid," *Journal of Applied Bacteriology*, vol. 43, no. 2, pp. 253–260, 1977.
[70] C. R. Andrighetti-Fröhner, K. N. De Oliveira, D. Gaspar-Silva et al., "Synthesis, biological evaluation and SAR of sulfonamide 4-methoxychalcone derivatives with potential antileishmanial activity," *European Journal of Medicinal Chemistry*, vol. 44, no. 2, pp. 755–763, 2009.
[71] H. Aissaoui, R. Koberstein, C. Zumbrunn et al., "*N*-glycine-sulfonamides as potent dual orexin 1/orexin 2 receptor antagonists," *Bioorganic and Medicinal Chemistry Letters*, vol. 18, no. 21, pp. 5729–5733, 2008.
[72] E. Pérez-Trallero and L. Iglesias, "Tetracyclines, sulfonamides and metronidazole," *Enfermedades Infecciosas y Microbiología Clínica*, vol. 21, no. 9, pp. 520–533, 2003.
[73] I. Levin, M. Mevarech, and B. A. Palfey, "Characterization of a novel bifunctional dihydropteroate synthase/dihydropteroate reductase enzyme from *Helicobacter pylori*," *Journal of Bacteriology*, vol. 189, no. 11, pp. 4062–4069, 2007.

14

Identification of a Novel Series of Potent TrkA Receptor Tyrosine Kinase Inhibitors

Stéphane L. Raeppel,[1] Frédéric Gaudette,[1] Hannah Nguyen,[2] Normand Beaulieu,[3] James Wang,[4] Christiane Maroun,[3] Jeffrey M. Besterman,[3] and Arkadii Vaisburg[1]

[1] *Department of Medicinal Chemistry, MethylGene Inc., 7150 Rue Frederick-Banting, Suite 200, Montréal, QC, Canada H4S 2A1*
[2] *Department of Cell Biology and Pharmacology, MethylGene Inc., 7150 Rue Frederick-Banting, Suite 200, Montréal, QC, Canada H4S 2A1*
[3] *Department of Cell Biology and Pharmacology, MethylGene Inc., 7150 Rue Frederick-Banting, Suite 200, Montréal, QC, Canada H4S 2A1*
[4] *Department of PK/Analytical Chemistry, MethylGene Inc., 7150 Rue Frederick-Banting, Suite 200, Montréal, QC, Canada H4S 2A1*

Correspondence should be addressed to Stéphane L. Raeppel, raeppels@chemrf.com

Academic Editor: O. Bruno

A novel series of *N*-(3-(6-substituted-aminopyridin-3-yloxy)phenyl)-2-oxo-3-phenylimidazolidine-1-carboxamides targeting TrkA receptor tyrosine kinase was identified. SAR study of the series allowed us to design and synthesize compounds possessing inhibitory activity of TrkA kinase enzyme in the low nanomolar range with low residual activity against c-Met and with no significant activity against VEGFR2.

Tropomyosin-related kinases (Trks) are receptor tyrosine kinases normally expressed in neuronal tissue where they play important role in both development and function of the nervous system [1]. The Trk receptor family is composed of three members (A, B, and C) activated by specific ligands called neurotrophins (NT). Each Trk receptor contains an extracellular domain (ligand binding), a transmembrane region, and an intracellular domain (including kinase domain) which upon binding of their respective ligand (nerve growth factor (NGF) for TrkA, brain-derived growth factor (BDNF) and NT-4/5 for TrkB, and NT3 for TrkC) triggers oligomerization of the receptors, phosphorylation of specific tyrosine residues in the kinase domain, and downstream signal transduction pathways, including survival, proliferation, and differentiation in normal and neoplastic neuronal cells [2–4]. While Trks are expressed at low levels outside the nervous system in the adult, deregulation of TrkA and TrkB and their cognate ligands has been described in numerous types of cancers including prostate, breast, colorectal, ovarian, lung, pancreas, melanoma, thyroid, and neuroblastoma and occurs mainly through wild type receptor overexpression, activation, amplification, and/or mutation [5, 6]. Importantly, increased Trks activation in tumor tissues correlates with an aggressive phenotype and poor clinical outcome [6].

There are a limited number of reported selective TrkA receptor tyrosine kinase inhibitors in the literature (Figure 1) [7–9]. Developed by Cephalon (now a group member of Teva Pharmaceutical Industries Ltd.), lestaurtinib (**CEP-701**) is a potent multitargeted tyrosine kinases inhibitor targeting mainly TrkA, Flt3, and JAK2 and is in clinical trials for the treatment of myeloproliferative disorders [10, 11]. More recently, Cephalon disclosed indenopyrrolocarbazole **12a** as a potent and selective TrkA inhibitor displaying antitumor properties [12]. Moreover, organoruthenium-pyridocarbazole ($\mathbf{R_{Ru}}$)-**21** from Pagano et al. [13], oxindole **3** from GSK [14], isothiazole **5n** from Pfizer [15], thiazole **20h** from BMS [16], and **AZ-23** from AstraZeneca were reported as potent and selective TrkA inhibitors as well [17, 18].

We have previously reported on four novel series of kinase inhibitors with a 2-oxo-3-phenylimidazolidine-1-carboxamide head group based on thieno[3,2-*b*]pyridine

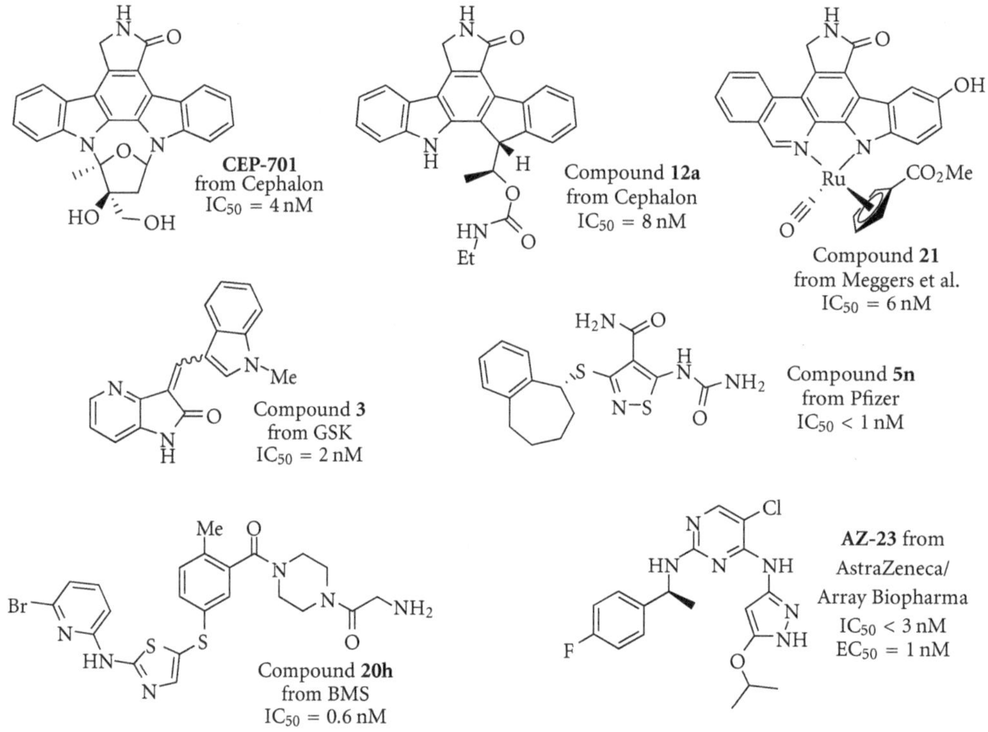

FIGURE 1: TrkA inhibitors disclosed in the literature and their reported IC_{50} and/or EC_{50}.

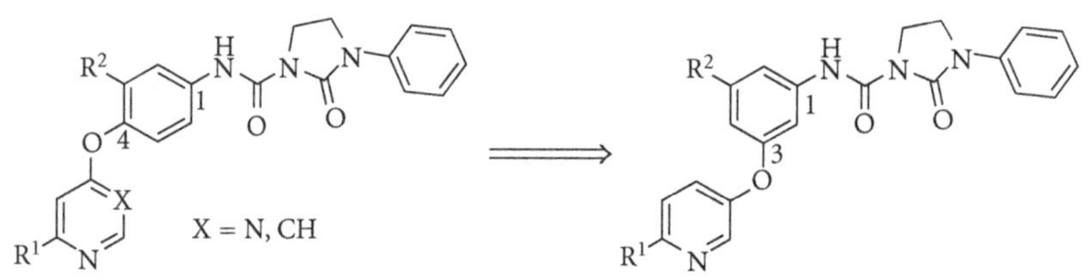

FIGURE 2: Transposition from 1,4- into 1,3-substitution pattern for the central phenyl ring.

[19], quinoline [20], pyrimidine, and pyridine scaffolds [21], exemplified by structures **1–4**, respectively (Table 1). Bi-cyclic RTK inhibitors **1** and **2** strongly inhibited both c-Met and VEGFR2 enzymes while compounds **3** and **4** showed some preference for the c-Met kinase enzyme. In our efforts to identify new kinase inhibitors with a different kinase inhibitory profile from the ones of compounds **1–4** (and their analogues), we viewed compound **4** as a valuable starting point for further exploration. As the first exercise, we changed the substitution pattern between the central phenyl ring and the pyridine core of compound **4** as illustrated in Figure 2 (transposition from 1,4- to 1,3-substitution pattern like in isothiazole **5n** from Pfizer, thiazole **20h** from BMS, and pyrimidine **AZ-23** from AstraZeneca), thus synthesizing compound **5**. This compound was roughly ten times weaker than the parent compound **4** against both c-Met and VEGFR2 enzymes, but surprisingly showed potent inhibition of TrkA when tested at 0.1 μM. As a consequence, we initiated the SAR study around compound **5** as a prototype of selective TrkA inhibitors. The IC_{50} of compound **5** was determined to be 4 nM against the TrkA human enzyme.

A straightforward synthesis of the target molecule **5** was developed and is illustrated in Scheme 1. (The details for the synthesis and the characterization of all the new compounds are well described in [21].) 5-Bromo-2-nitropyridine was reacted with 3-aminophenol in the presence of cesium carbonate in acetonitrile at room temperature. The choice of the base, solvent, and temperature is critical in order to obtain the intermediate **A** in fairly good yield; otherwise multiple products are formed. Intermediate **A** was subsequently reacted with 2-oxo-3-phenylimidazolidine-1-carbonyl chloride generated *in situ* from the reaction between 1-phenylimidazolidin-2-one and triphosgene to afford intermediate **B**. After reduction of the nitro group and subsequent derivatization of 6-aminopyridine intermediate **C**, compound **5** was obtained in good yield.

To continue with the SAR study around compound **5**, we explored the replacement of the oxygen linker by keeping unchanged the 2-oxo-3-phenylimidazolidine-1-carboxamide head group, which is presumed to occupy the hydrophobic back pocket of the kinase catalytic domain, and 6-acetamidopyridine scaffold, which is supposed to

Table 1: 2-Oxo-3-phenylimidazolidine-1-carboxamide kinase inhibitors based on different scaffolds.

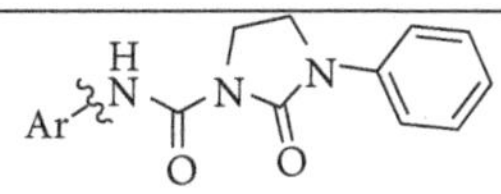

Cpd	Ar	c-Met[a] IC_{50} (μM)	VEGFR2[a] IC_{50} (μM)	TrkA[a] % inh. at 0.1 μM
1	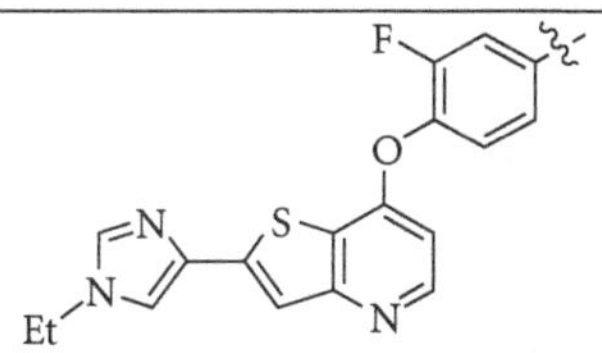	0.02	0.01	82
2	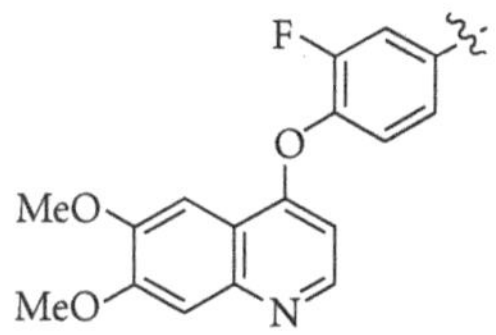	0.03	0.01	94
3	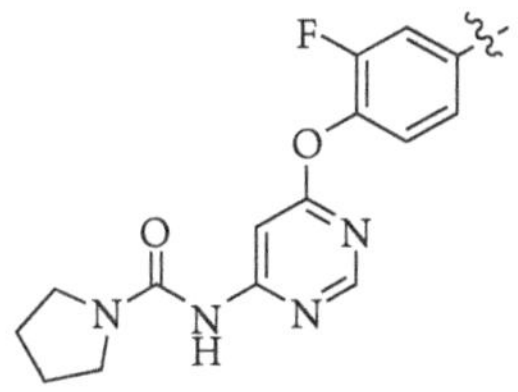	0.05	0.15	16
4	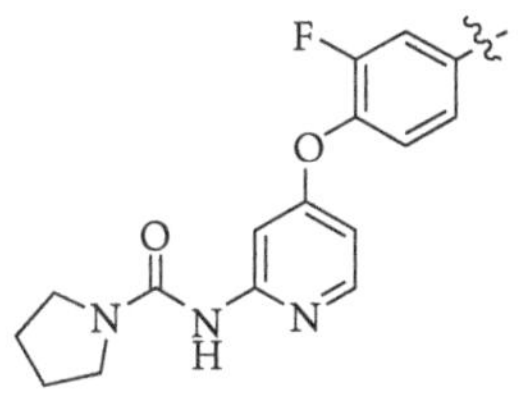	0.03	0.23	49
5	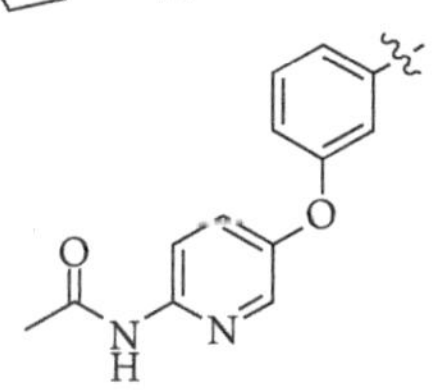	0.31	2.50	98

[a] IC_{50} and percentage of inhibition values are reported as the average of ≥3 experiments.

interact with the ATP catalytic site in the hinge region of the target kinase enzyme (Table 2). Not surprisingly, the replacement of the oxygen linker in compound **5** by an ureido, ethynyl, sulfamoyl, or carbamoyl spacer (compounds **6–9**) completely abolished the activity against TrkA kinase enzyme.

We next investigated the presence of a small substituent located on the central phenyl ring (positions 2 and 4) of compound **5** (Table 3). In contrast with either thiazole **20h** from Bristol-Myers Squib [16] or **AZ 23** from AstraZeneca [17, 18] (Figure 1), in which the presence of a small substituent on the central phenyl or pyridinyl ring improves the activity and the selectivity against the TrkA kinase enzyme, in our series of compounds that trend was not observed. Thus, compounds **10–12** turned out to be inactive against TrkA.

We finally examined the influence of each extremity of compound **5** where R^1 is presumed to extend into the solvent exposed area and R^2 is thought to occupy the hydrophobic back pocket of the kinase catalytic domain—both sites potentially capable of modulating physicochemical and pharmacokinetic properties of the inhibitors (compounds **13–17**, Table 4). Removal of the acetyl group from compound **5** afforded compound **13** with a lower potency against TrkA, probably due to the absence of a strong hydrogen bond donor (NH amide) and/or a hydrogen bond acceptor (carbonyl amide) which is presumed to interact with the hinge region of the kinase binding domain. Introduction of relatively small electronegative fluorine in the *para* position of the phenyl ring is well tolerated (compound **14**). Finally, replacement of the acetamido group by a functionalized ureido moiety is also accepted, giving rise to potent TrkA inhibitors (compounds **15–17**) and allowing us to incorporate water solubilizing groups.

Compound **5** was profiled and tested at 100 nM against 24 other kinase enzymes using Millipore's Kinaseprofiler

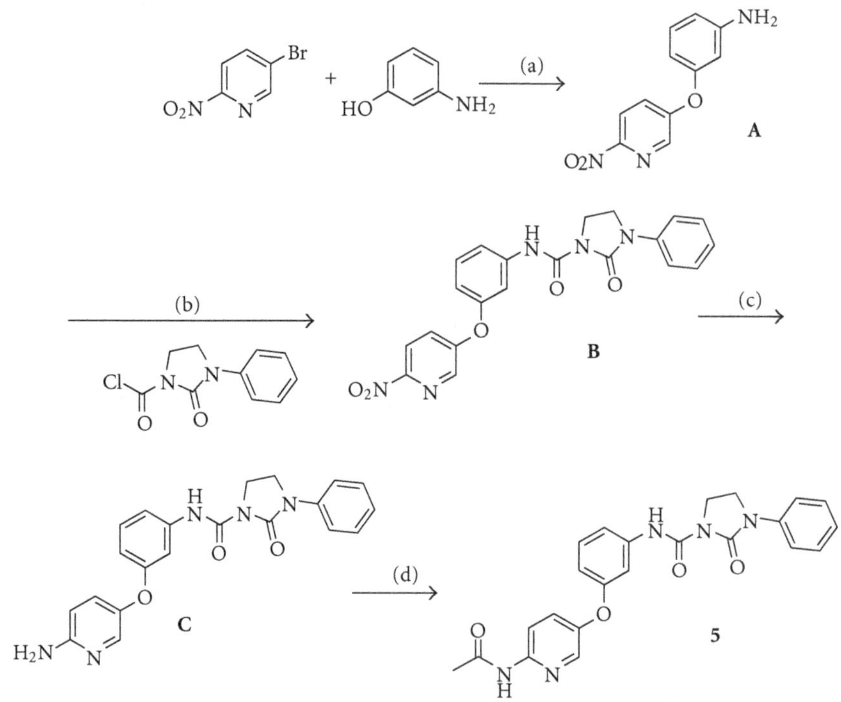

Scheme 1: Synthesis of compound **5**: (a) Cs_2CO_3, MeCN, rt; (b) DIPEA, DCM, rt; (c) iron, NH_4Cl, MeOH/water, reflux; (d) Ac_2O, rt.

Table 2: Influence of the linker in the 6-acetamidopyridine series on the TrkA potency.

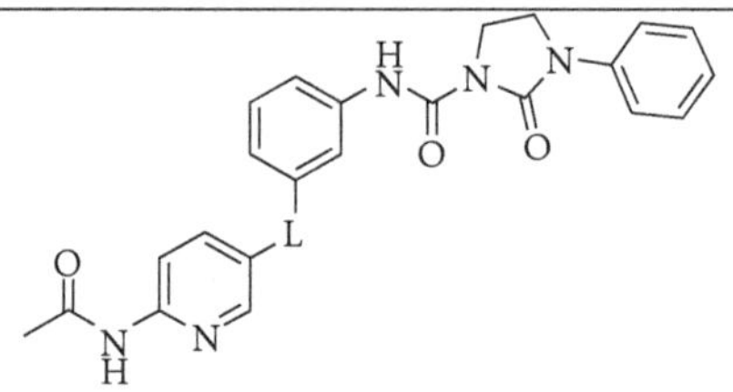

Cpd	L	TrkA[a] % inh. at 30 nM
5	O	96
6	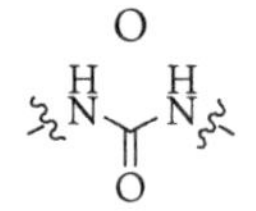	5
7		16
8	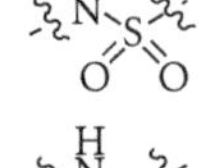	6
9	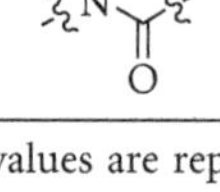	2

[a] percentage of inhibition values are reported as the average of ≥3 experiments.

Table 3: Influence of the central phenyl ring substistution in the 6-acetamidopyridine series on the TrkA potency.

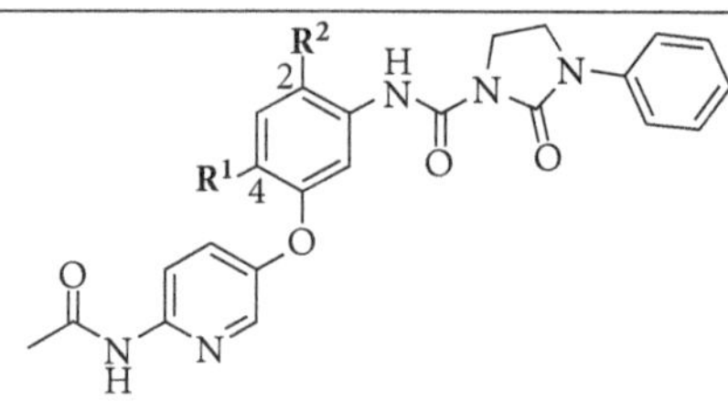

Cpd	R^1	R^2	TrkA[a] % inh. at 30 nM
5	H	H	96
10	Me	H	6
11	OMe	H	15
12	H	Cl	0

[a] percentage of inhibition values are reported as the average of ≥3 experiments.

assay services. In addition to targeting TrkA receptor tyrosine kinase (98% inhibition), it was also moderately active against Flt3 (83% inhibition) but, importantly, did not show significant inhibitory activity against the other representative human kinases (<40% inhibition): ALK, Bmx, CHK1, cKit, c-Raf, EphB4, FAK, GSK3β, Haspin, IKKβ, JAK2, JAK3, LIMK1, MEK1, c-Met, PDK1, PI3-Kinaseβ, Pim-1, PKBα, Ret, Ron, and Tie2. When tested at 10 nM against TrkB, compound **5** showed 40% inhibition.

Compound **5** was further evaluated for its pharmacokinetic properties in the rat (Table 5). Compound **5** showed a 1.5 h half-life, a high rate of clearance, an appreciable steady-state volume of distribution, and excellent oral bioavailability.

In conclusion, a novel series of *N*-(3-(6-substituted-aminopyridin-3-yloxy)phenyl)-2-oxo-3-phenylimidazolidine-1-carboxamides was designed and synthesized. The most attractive compounds of the series—**5** and **14–17**—are selective for the TrkA kinase enzyme in the low nanomolar

Table 4: Influence of R^1 and R^2 substituents on the TrkA potency.

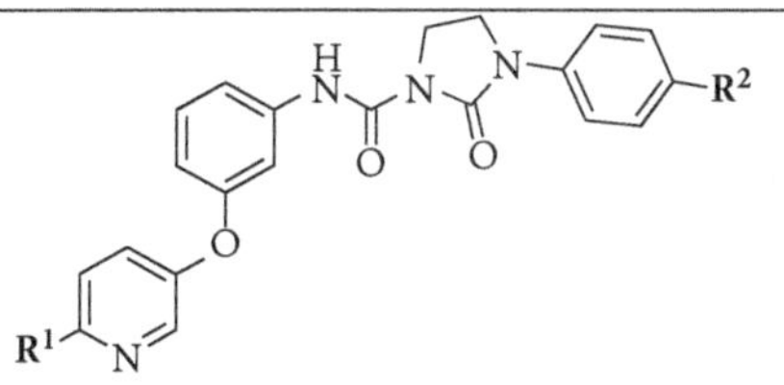

Cpd	R^1	R^2	TrkA IC_{50} (nM)
5	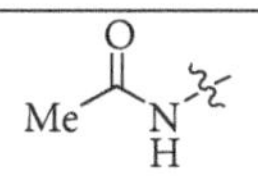	H	4
13	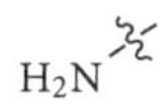	H	20
14	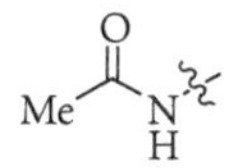	F	3
15	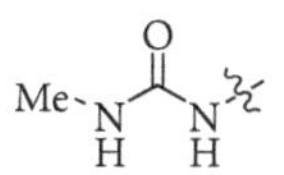	H	5
16	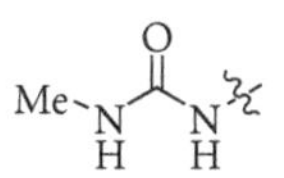	F	2
17	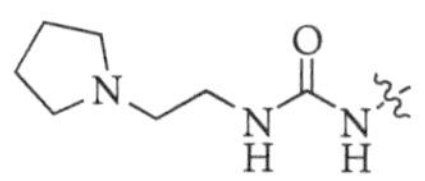	F	4

[a] IC_{50} values are reported as the average of ≥2 experiments.

Table 5: Rat pharmacokinetic profile of compound **5**.

PK properties[a]	Cpd 5
$T_{1/2}$ (h), i.v.	1.5
Cl (L/h/kg) i.v.	3.3
Vss (L/kg) i.v.	3.7
t_{max} (h), p.o.	4.0
$N.\ C_{max}$ [μM/(mg/kg)], p.o.	0.08
$N.$ AUC [μM.h/(mg/kg)], p.o.	0.83
F (%)	99

[a] i.v. dose: 2.58 mg/kg; p.o. dose: 5.25 mg/kg.

range (comparable with the known TrkA inhibitors) with only residual activity against other RTKs. These novel chemical entities represent a valuable starting point in design and synthesis of TrkA selective inhibitors both as tools to further validate TrkA as a promising biological target and for the development of potential new anticancer therapies.

Acknowledgment

The authors thank Dr. Ljubomir Isakovic for the synthesis of compound **4**.

References

[1] A. Patapoutian and L. F. Reichardt, "Trk receptors: mediators of neurotrophin action," *Current Opinion in Neurobiology*, vol. 11, no. 3, pp. 272–280, 2001.

[2] E. J. Huang and L. F. Reichardt, "Trk receptors: roles in neuronal signal transduction," *Annual Review of Biochemistry*, vol. 72, pp. 609–642, 2003.

[3] L. F. Reichardt, "Neurotrophin-regulated signalling pathways," *Philosophical Transactions of the Royal Society B*, vol. 361, no. 1473, pp. 1545–1564, 2006.

[4] S. D. Skaper, "The biology of neurotrophins, signalling pathways, and functional peptide mimetics of neurotrophins and their receptors," *CNS and Neurological Disorders - Drug Targets*, vol. 7, no. 1, pp. 46–62, 2008.

[5] G. M. Brodeur, J. E. Minturn, R. Ho et al., "Trk receptor expression and inhibition in neuroblastomas," *Clinical Cancer Research*, vol. 15, no. 10, pp. 3244–3250, 2009.

[6] C. J. Thiele, Z. Li, and A. E. McKee, "On Trk—the TrkB signal transduction pathway is an increasingly important target in cancer biology," *Clinical Cancer Research*, vol. 15, no. 19, pp. 5962–5967, 2009.

[7] R. L. Hudkins and K. A. Josef, "Pyrazole trk kinase inhibitors for the treatment of cancer," *Expert Opinion on Therapeutic Patents*, vol. 17, no. 3, pp. 351–356, 2007.

[8] T. Wang, D. Yu, and M. L. Lamb, "Trk kinase inhibitors as new treatments for cancer and pain," *Expert Opinion on Therapeutic Patents*, vol. 19, no. 3, pp. 305–319, 2009.

[9] K. J. Martin, N. Shpiro, R. Traynor, M. Elliott, and J. S. C. Arthur, "Comparison of the specificity of Trk inhibitors in recombinant and neuronal assays," *Neuropharmacology*, vol. 61, no. 1-2, pp. 148–155, 2011.

[10] S. J. Miknyoczki, H. Chang, A. Klein-Szanto, C. A. Dionne, and B. A. Ruggeri, "The Trk tyrosine kinase inhibitor CEP-701 (KT-5555) exhibits significant antitumor efficacy in preclinical xenograft models of human pancreatic ductal adenocarcinoma," *Clinical Cancer Research*, vol. 5, no. 8, pp. 2205–2212, 1999.

[11] D. E. Gingrich, S. X. Yang, G. W. Gessner, T. S. Angeles, and R. L. Hudkins, "Synthesis, modeling, and in vitro activity of (3′S)-epi-K-252a analogues. Elucidating the stereochemical requirements of the 3′-sugar alcohol on trkA tyrosine kinase activity," *Journal of Medicinal Chemistry*, vol. 48, no. 11, pp. 3776–3783, 2005.

[12] R. Tripathy, T. S. Angeles, S. X. Yang, and J. P. Mallamo, "TrkA kinase inhibitors from a library of modified and isosteric Staurosporine aglycone," *Bioorganic and Medicinal Chemistry Letters*, vol. 18, no. 12, pp. 3551–3555, 2008.

[13] N. Pagano, E. Y. Wong, T. Breiding et al., "From imide to lactam metallo-pyridocarbazoles: distinct scaffolds for the design of selective protein kinase inhibitors," *Journal of Organic Chemistry*, vol. 74, no. 23, pp. 8997–9009, 2009.

[14] E. R. Wood, L. Kuyper, K. G. Petrov, R. N. Hunter, P. A. Harris, and K. Lackey, "Discovery and in vitro evaluation of potent TrkA kinase inhibitors: oxindole and aza-oxindoles," *Bioorganic and Medicinal Chemistry Letters*, vol. 14, no. 4, pp. 953–957, 2004.

[15] B. Lippa, J. Morris, M. Corbett et al., "Discovery of novel isothiazole inhibitors of the TrkA kinase: structure-activity relationship, computer modeling, optimization, and identification of highly potent antagonists," *Bioorganic and Medicinal Chemistry Letters*, vol. 16, no. 13, pp. 3444–3448, 2006.

[16] S. H. Kim, J. S. Tokarski, K. J. Leavitt et al., "Identification of 2-amino-5-(thioaryl)thiazoles as inhibitors of nerve growth

factor receptor TrkA," *Bioorganic and Medicinal Chemistry Letters*, vol. 18, no. 2, pp. 634–639, 2008.

[17] T. Wang, M. L. Lamb, D. A. Scott et al., "Identification of 4-aminopyrazolylpyrimidines as potent inhibitors of Trk kinases," *Journal of Medicinal Chemistry*, vol. 51, no. 15, pp. 4672–4684, 2008.

[18] K. Thress, T. MacIntyre, H. Wang et al., "Identification and preclinical characterization of AZ-23, a novel, selective, and orally bioavailable inhibitor of the Trk kinase pathway," *Molecular Cancer Therapeutics*, vol. 8, no. 7, pp. 1818–1827, 2009.

[19] S. Raeppel, S. Claridge, O. Saavedra et al., "N-(3-fluoro-4-(2-arylthieno[3,2-b]pyridin-7-yloxy)phenyl)-2-oxo-3-phenylimidazolidine-1-carboxamides: a novel series of dual c-Met/VEGFR2 receptor tyrosine kinase inhibitors," *Bioorganic and Medicinal Chemistry Letters*, vol. 19, no. 5, pp. 1323–1328, 2009.

[20] M. Mannion, S. Raeppel, S. Claridge et al., "N-(4-(6,7-Disubstituted-quinolin-4-yloxy)-3-fluorophenyl)-2-oxo-3-phenylimidazolidine-1-carboxamides: A novel series of dual c-Met/VEGFR2 receptor tyrosine kinase inhibitors," *Bio-organic and Medicinal Chemistry Letters*, vol. 19, no. 23, pp. 6552–6556, 2009.

[21] S. Raeppel, O. Saavedra, S. Claridge et al., "Kinase inhibitors and uses thereof," Patent application WO 2008/046216.

15

Pharmacophore Modelling and 3D-QSAR Studies on N^3-Phenylpyrazinones as Corticotropin-Releasing Factor 1 Receptor Antagonists

Paramjit Kaur, Vikas Sharma, and Vipin Kumar

Institute of Pharmaceutical Sciences, Kurukshetra University, Haryana, Kurukshetra 136119, India

Correspondence should be addressed to Vipin Kumar, vipbhardwaj@rediffmail.com

Academic Editor: Armando Rossello

Pharmacophore modelling-based virtual screening of compound is a ligand-based approach and is useful when the 3D structure of target is not available but a few known active compounds are known. Pharmacophore mapping studies were undertaken for a set of 50 N^3-phenylpyrazinones possessing Corticotropin-releasing Factor 1 (CRF 1) antagonistic activity. Six point pharmacophores with two hydrogen bond acceptors, one hydrogen bond donor, two hydrophobic regions, and one aromatic ring as pharmacophoric features were developed. Amongst them the pharmacophore hypothesis AADHHR.47 yielded a statistically significant 3D-QSAR model with 0.803 as R^2 value and was considered to be the best pharmacophore hypothesis. The developed pharmacophore model was externally validated by predicting the activity of test set molecules. The squared predictive correlation coefficient of 0.91 was observed between experimental and predicted activity values of test set molecules. The geometry and features of pharmacophore were expected to be useful for the design of selective CRF 1 receptor antagonists.

1. Introduction

Anxiety and depression are among the most common disorders seen in medical practice. The coexistence of anxiety and depression with medical illness is a topic of considerable clinical and research interest [1]. Depression is a serious mental health problem, with significant consequences in terms of human suffering, lost productivity, and even loss of life [2]. Corticotropin-releasing factor 1 (CRF 1) receptor antagonists have been sought since the stress-secreted peptide (adrenocorticotropin-releasing hypothalamic peptide) was isolated in 1981. Although evidence is mixed concerning the efficacy of CRF 1 receptor antagonist as antidepressants, CRF 1 receptor antagonist might be novel pharmacotherapies for anxiety and addiction [3]. Two well-characterized receptor subtypes, CRF 1 and CRF 2, have been identified. These G-protein-coupled receptors are widely distributed throughout the central and peripheral nervous systems [4]. Clinical evidence supports the hypothesis that overproduction of CRF 1 may underlie the pathology of depression, anxiety, and stress-related disorders and suggests that CRF 1 receptor antagonists could be useful for the treatment of these conditions [5].

To reduce the overall cost associated with the discovery and development of a new drug, the computer-aided molecular design methods have been identified as the most promising candidates to focus on the experimental efforts in modern medicinal chemistry. Pharmacophore mapping is one of the major elements of drug design in the absence of structural data of the target receptor. With the aim of providing useful pharmacophoric information for the future efforts in the development of more potent molecules in the series of phenylpyrazinones and to get insight into the structural and molecular properties, the ligand-based 3D-QSAR study was performed using pharmacophore techniques with PHASE module from Schrodinger, New York [6, 7].

2. Experimental

2.1. Dataset. The *in vitro* biological data of a series of 50 phenylpyrazinones as CRF 1 receptor antagonists were used

TABLE 1: Various substituents attached to basic structure of N[3]-phenylpyrazinones.

Comp. No.	Series	R1	Ar	X	Y	n	IC_{50}	pIC_{50}	Set
1.	a	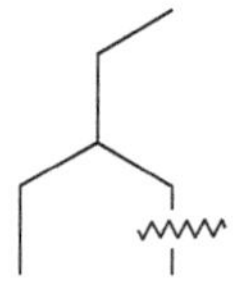	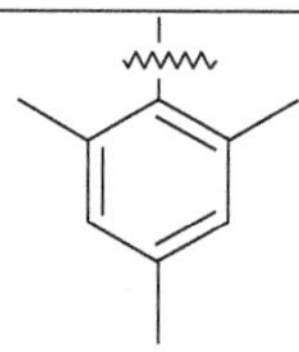	Cl	H	—	39.0	7.409	Training
2.	a	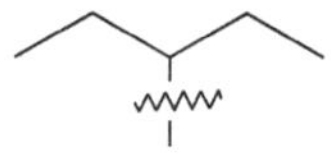	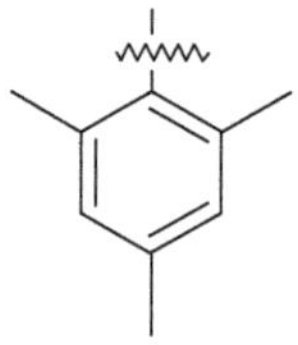	Cl	H	—	2.9	8.685	Training
3.	a		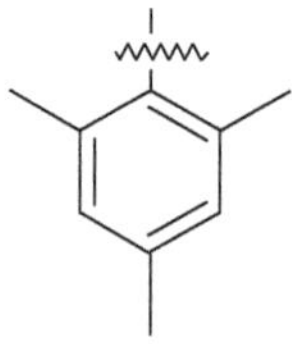	Cl	H	—	2.4	8.620	Test
4.	a	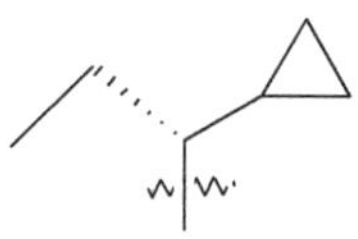	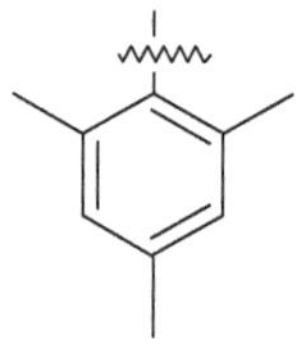	Cl	H	—	0.63	9.2	Test
5.	a	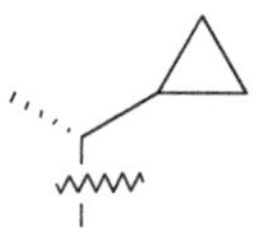	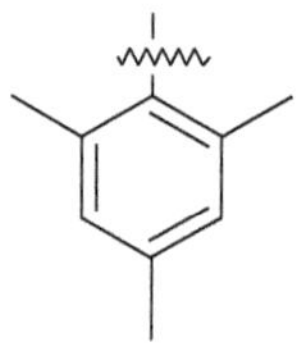	Cl	H	—	1.7	8.856	Training
6.	a	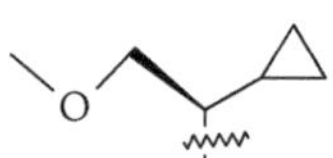	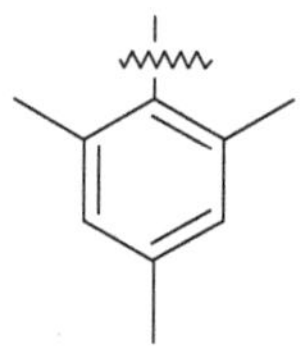	Cl	H	—	1.3	8.98	Training
7.	a	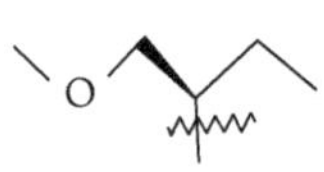	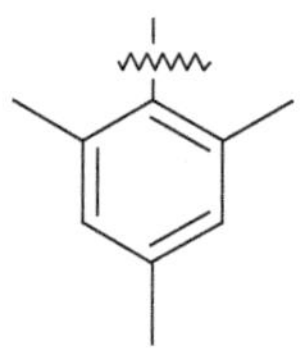	Cl	H	—	1.4	8.85	Training
8.	a	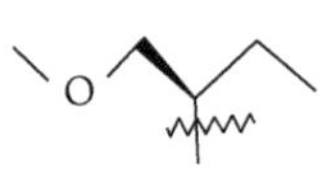	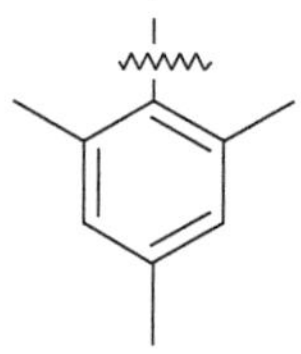	Cl	H	—	12	7.921	Test
9.	a	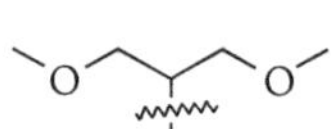	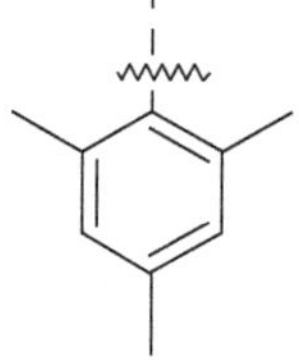	Cl	H	—	16	7.796	Training

Table 1: Continued.

Comp. No.	Series	R1	Ar	X	Y	n	IC_{50}	pIC_{50}	Set
10.	a	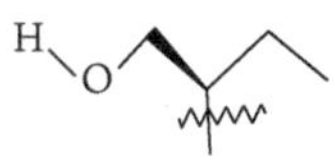	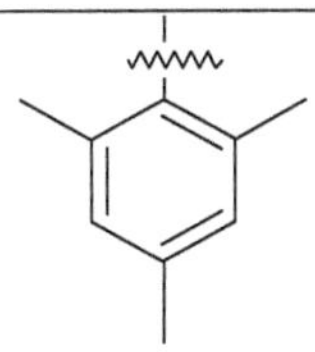	Cl	H	—	35	7.585	Training
11.	a	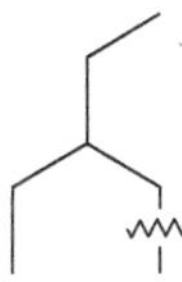	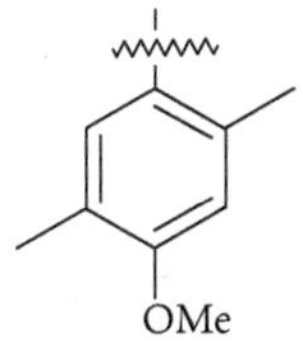	Cl	H	—	76	6.041	Training
12.	a	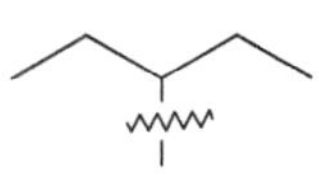	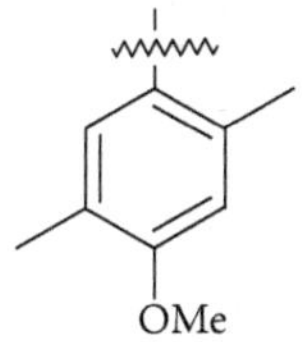	Cl	H	—	0.80	9.096	Test
13.	a		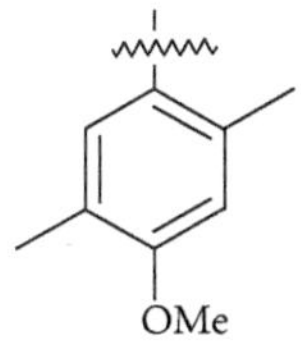	Cl	H	—	0.62	8.89	Training
14.	a	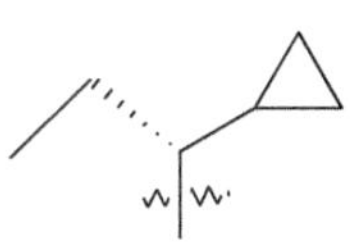	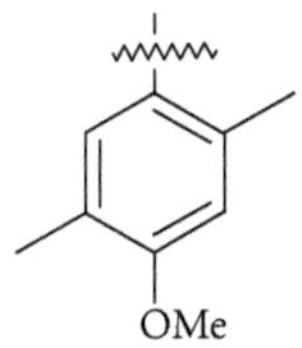	Cl	H	—	0.27	9.568	Test
15.	a		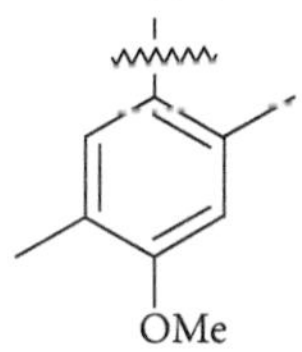	Cl	H	—	0.42	9.376	Training
16.	a	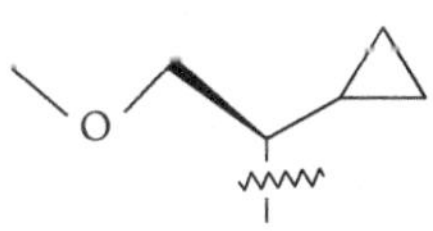	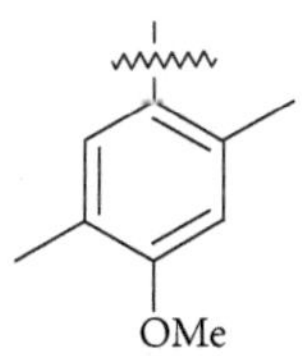	Cl	H	—	0.26	9.585	Training
17.	a	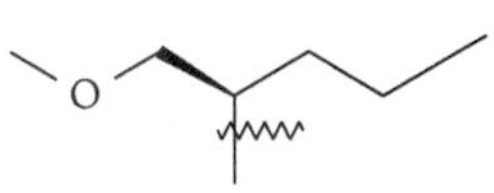	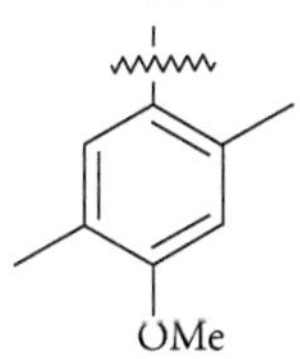	Cl	H	—	0.59	9.285	Training
18.	a	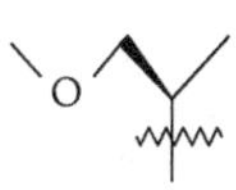	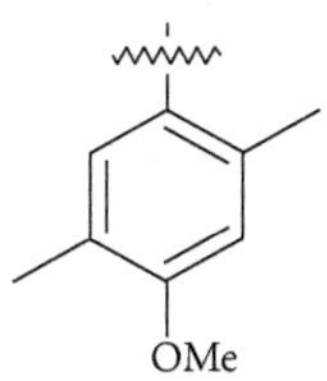	Cl	H	—	1.7	8.88	Training

Table 1: Continued.

Comp. No.	Series	R1	Ar	X	Y	n	IC_{50}	pIC_{50}	Set
19.	a	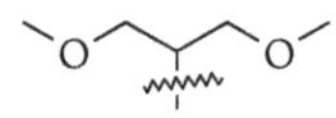	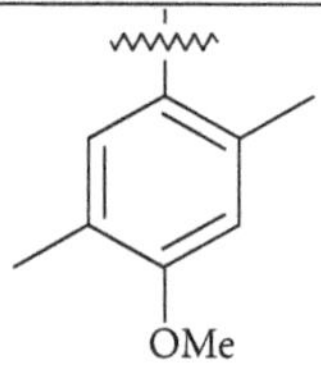	Cl	H	—	2.1	8.74	Training
20.	a	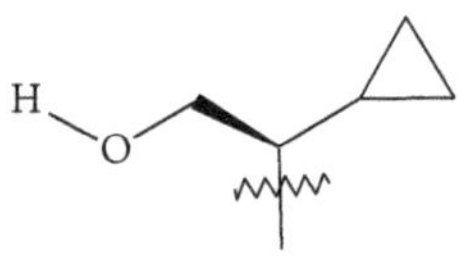	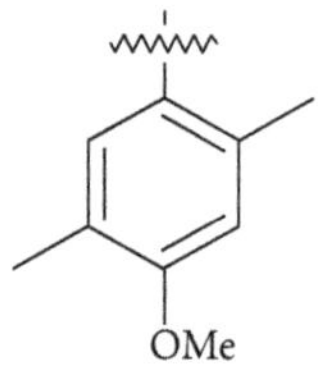	Cl	H	—	10	8.77	Training
21.	a	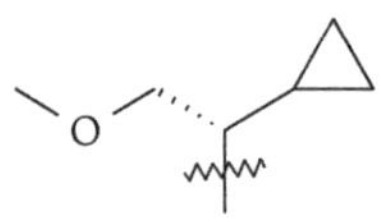	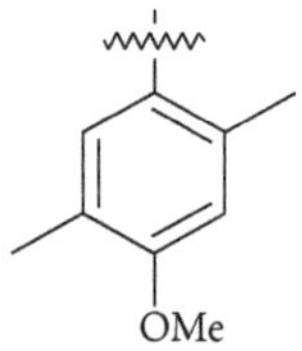	Cl	H	—	0.29	9.568	Training
22.	a	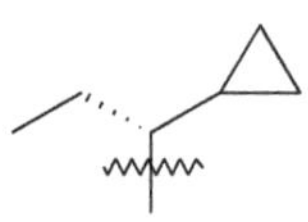	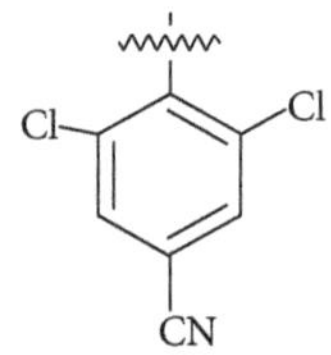	Cl	H	—	1.2	9.212	Training
23.	a	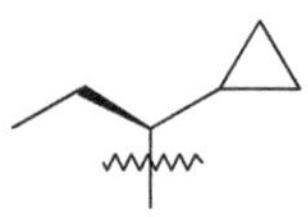	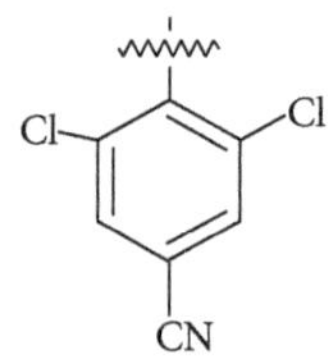	Cl	H	—	2.8	8.42	Training
24.	a		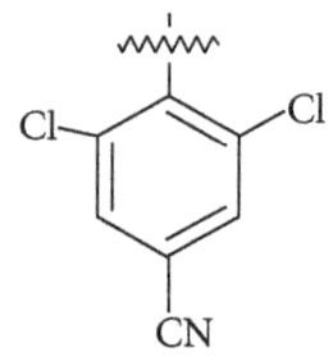	Cl	H	—	2.4	9.376	Training
25.	a	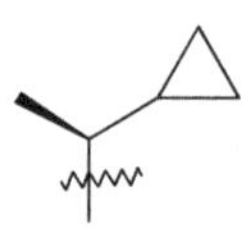	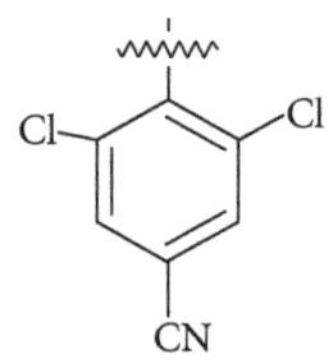	Cl	H	—	6.3	8.89	Training
26.	a	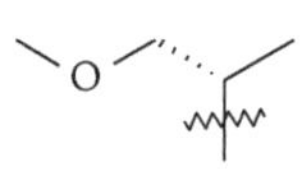	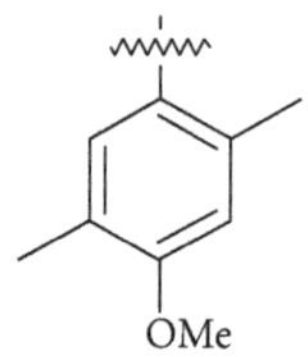	Cl	H	—	5.8	8.236	Test
27.	a		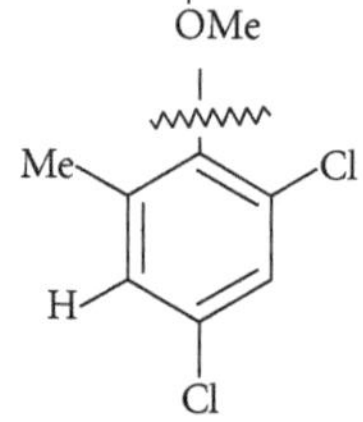	Cl	H	—	0.53	9.275	Training

Table 1: Continued.

Comp. No.	Series	R1	Ar	X	Y	n	IC_{50}	pIC_{50}	Set
28.	a		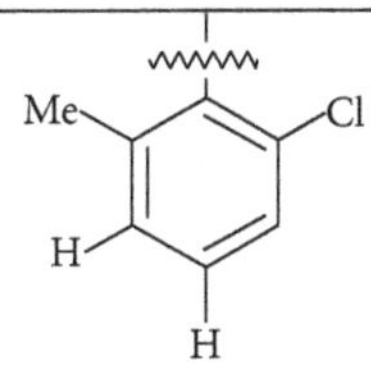	Cl	H	—	9.3	8.032	Training
29.	a		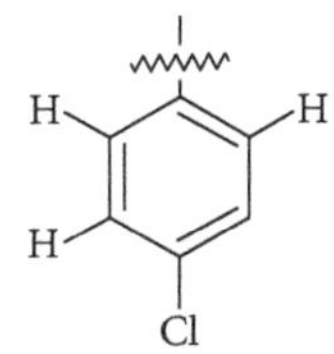	Cl	H	—	180	6.745	Training
30.	a		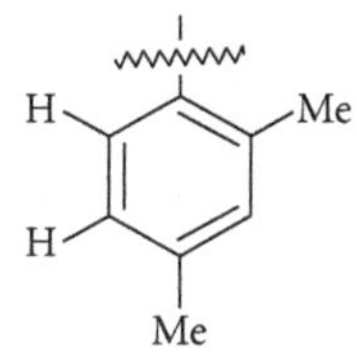	Cl	H	—	6.6	8.181	Test
31.	a	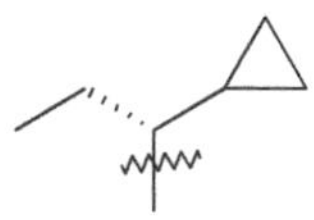	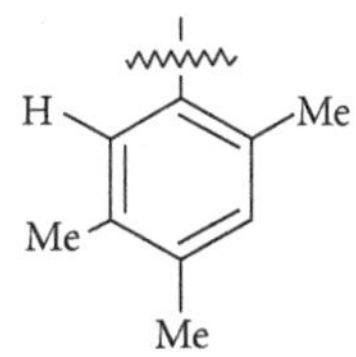	Cl	H	—	2.4	8.62	Training
32.	a	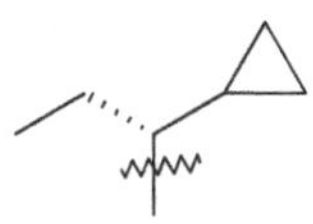	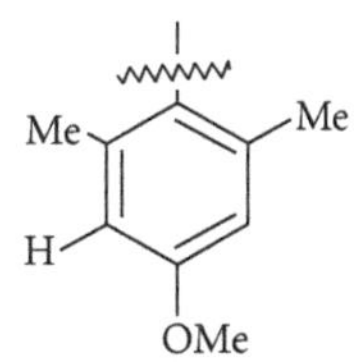	Cl	H	—	0.52	9.284	Training
33.	a	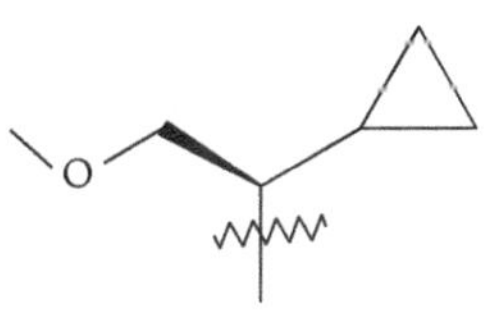	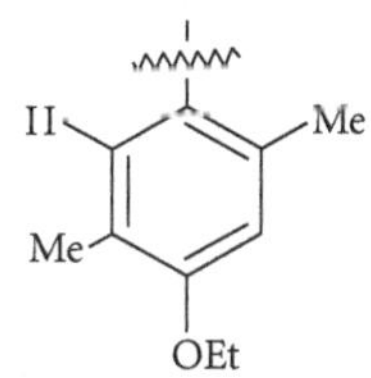	Cl	H	—	1.5	8.824	Test
34.	a	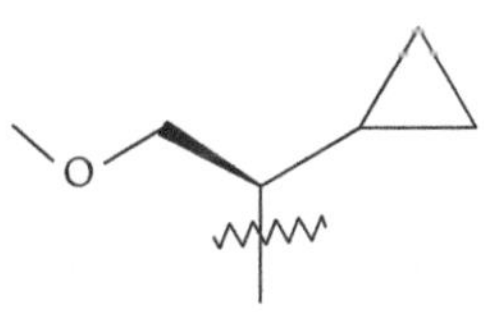	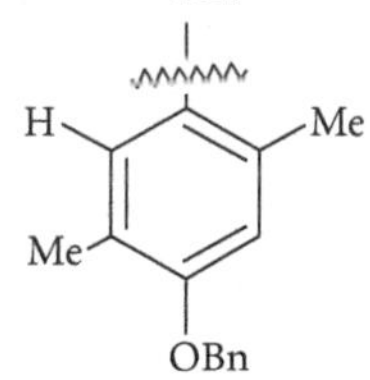	Cl	H	—	910	6.041	Training
35.	a	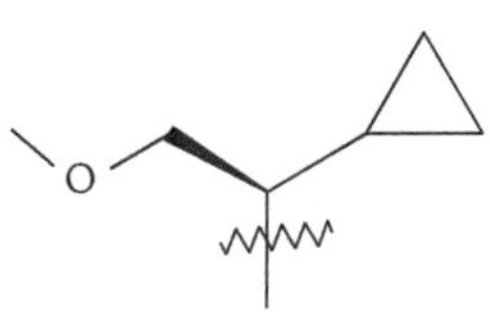	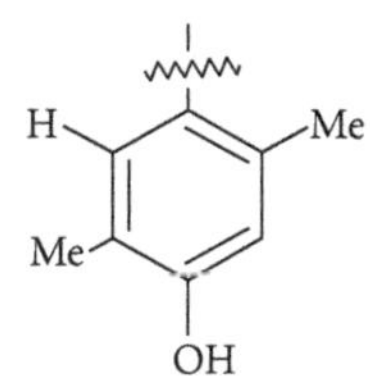	Cl	H	—	280	6.553	Training
36.	a		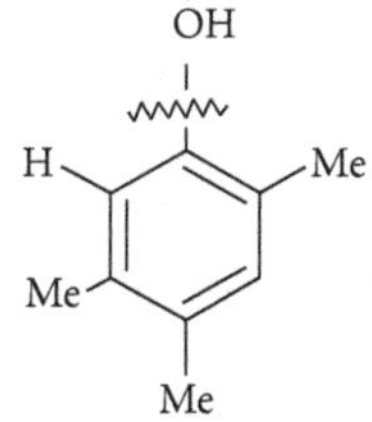	Cl	Et	—	1.5	8.824	Test

Table 1: Continued.

Comp. No.	Series	R1	Ar	X	Y	n	IC_{50}	pIC_{50}	Set
37.	a	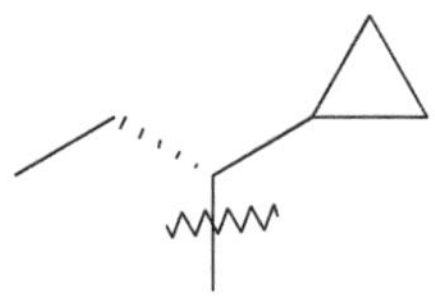	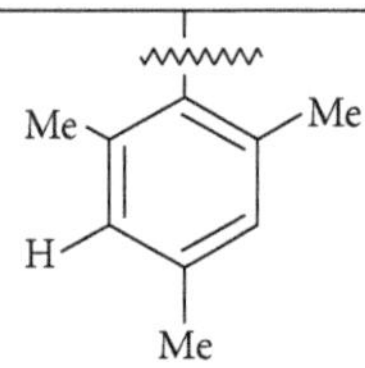	Cl	Et	—	4.6	8.62	Training
38.	a		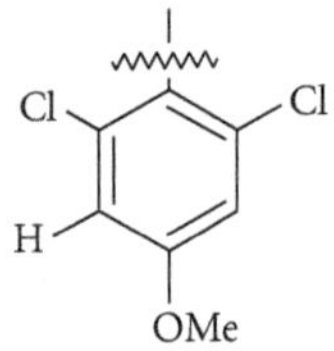	Cl	H	—	1.1	8.959	Training
39.	a	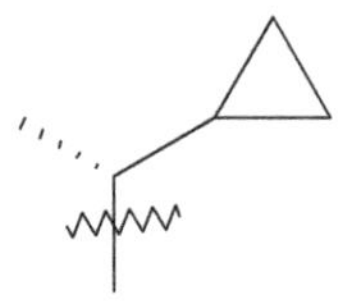	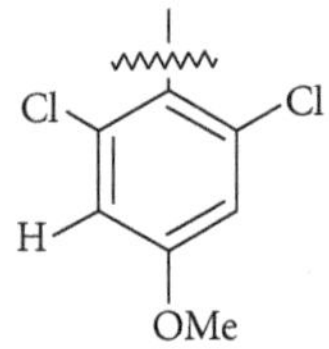	Cl	Et	—	3.4	7.469	Test
40.	b	Cl	—	CH2	—	1	0.62	9.207	Training
41.	b	Br	—	CH2	—	1	0.94	9.026	Test
42.	b	Br	—	O		2	19	7.722	Test
43.	a	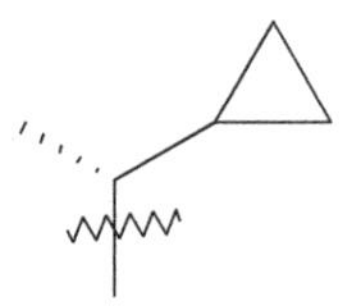	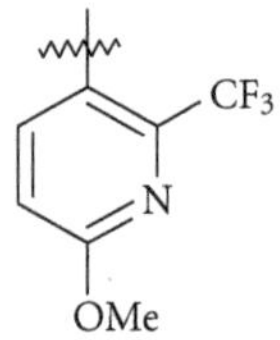	H	H	—	150	6.824	Test
44.	a	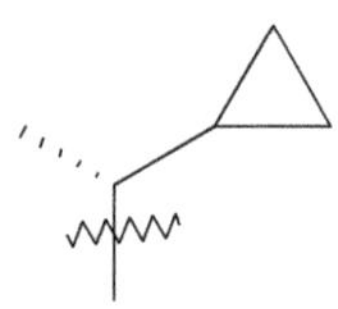	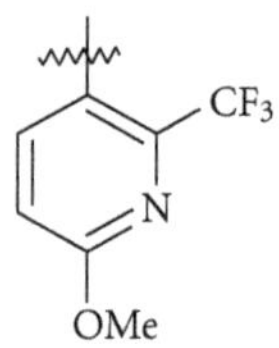	Cl	H	—	1.8	8.745	Training
45.	a	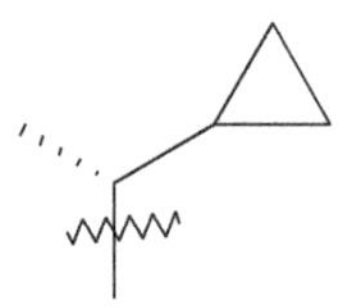	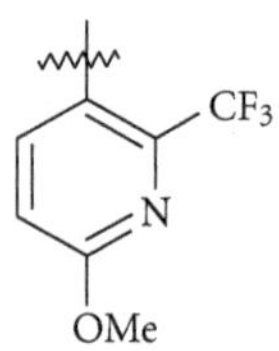	Br	H	—	1.2	8.921	Test
46.	a		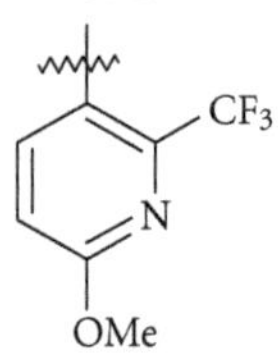	Me	H	—	3.8	8.44	Training
47.	a		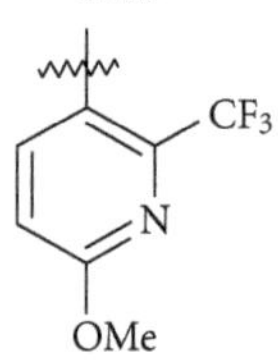	C≡CN	H	—	6.7	8.25	Training
48.	a		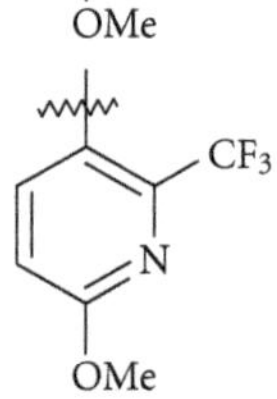	C≡CH	H	—	5.6	8.46	Training

Table 1: Continued.

Comp. No.	Series	R1	Ar	X	Y	n	IC_{50}	pIC_{50}	Set
49.	a		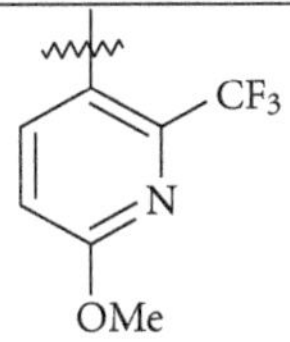	C_2H_5	H	—	180	6.744	Test
50.	a		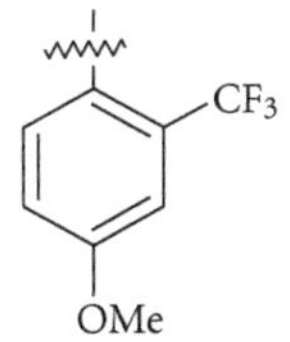	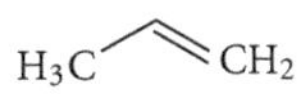	H	—	1200	5.921	Training

Table 2: Parameters of six featured pharmacophore hypothesis.

Serial number	Hypothesis	Survival score	R^2	F
1	AADHHR.47	2.980	0.803	138.5
2	ADHHHR.203	2.892	0.789	129.7
3	DHHHRR.611	2.863	0.7565	105.6
4	AAHHHR.19	2.700	0.7747	118.1

for the present studies [6]. The CRF 1 antagonistic activity was expressed as IC_{50}, that is, concentration in μm required for 50% inhibition of enzyme activity. The dataset was divided randomly into training set and test set by considering the 70% of the total molecules in the training set and 30% in the test set. The basic structures of the N^3-phenylpyrazinones are shown in Figure 1, and various substituents are listed in Table 1. Thirty-five molecules forming the training set were used to generate pharmacophore model and prediction of the activity of test set (15 compounds) molecules was used as a method to validate the proposed models.

2.2. PHASE Methodology. The 3D-QSAR studies were carried out using PHASE [8–10] version 3.0 implemented in the Maestro 8.5 molecular modeling package from Schrodinger, Molecular Modeling Interface Inc., LLC, New York, NY USA. Phase is a versatile product for pharmacophore perception, structure alignment, activity prediction, and 3D database searching. Phase provides support for lead discovery, SAR development, lead optimization, and lead expansion. Phase may also be used as a source of molecular alignments for third-party 3D-QSAR programs. Phase is well suited to drug discovery projects for which no receptor structure is available. Phase utilizes fine-grained conformational sampling and a range of scoring techniques to identify common pharmacophore hypotheses, which convey characteristics of 3D chemical structures that are purported to be critical for binding. A given hypothesis may be combined with known activity data to create 3D-QSAR models that identify overall aspects of molecular structure that govern activity. These models may be used in conjunction with the hypothesis to mine a 3D database for molecules that are most likely to exhibit strong activity toward the target [11].

2.3. Preparing Ligands. LigPrep [12] was used to attach hydrogen, converts 2D structures to 3D, generates stereoisomer, and, optionally, neutralizes charged structures or determines the most probable ionization state at user-defined pH. All the structures were ionized at neutral pH 7. Conformers for each ligand were generated using ConfGen by applying OPLS-2005 force field method [13, 14] with implicit GB/SA distance-dependent dielectric solvent model at cutoff root mean square deviation (RMSD) of 1 (MacroModel 9.6 2010) with 1,000 iterations using water as solvent.

2.4. Pharmacophore Hypothesis Generation. PHASE can identify the spatial arrangements of functional groups that are common and essential for the biological activity of the ligands under investigation [10, 15]. The most dominating features, hydrogen bond acceptor (A), hydrogen bond donor (D), hydrophobic group (H), and negatively charged group (N) were defined by a set of chemical structural patterns with the requirement that all five match. Pharmacophore-matching tolerance was set to 1 A°. Hypotheses were generated by a systematic variation of number of sites and the number of matching active compounds. Common pharmacophore hypotheses (CPH) were considered, which indicated at least five sites common to all molecules. Further, the best CPH was selected depending on the survival score, until at least one hypothesis was found and scored successfully. The hypotheses were scored using default parameters for site, vector, volume, selectivity, number of matches, and energy terms. The regression analysis was performed by constructing a series of models with an increasing number of PLS factors. Pharmacophore-based 3D-QSAR models were generated for the hypotheses using the 35 member training set with three PLS factors and a grid spacing of 1 A°. The

TABLE 3: Distances between different sites of model AADHHR.47.

Site-1	Site-2	Distances (A°)	Site-1	Site-2	Distances (A°)
A2	A4	3.572	A4	R12	3.474
A2	D6	2.520	D6	H10	5.541
A2	H10	3.291	D6	H11	6.688
A2	H11	4.947	D6	R12	3.243
A2	R12	5.054	H10	H11	5.984
A4	D6	3.850	H10	R12	8.343
A4	H10	6.431	H11	R12	7.161
A4	H11	3.802			

TABLE 4: Angles between different sites of model AADHHR.47.

Site-1	Site-2	Site-3	Angle (°)	Site-1	Site-2	Site-3	Angle (°)
A4	A2	D6	76.2	A2	H10	A4	21.3
A4	A2	H10	139.1	A2	H10	D6	15.3
A4	A2	H11	49.9	A2	H10	H11	55.8
A4	A2	R12	43.4	A2	H10	R12	1.5
D6	A2	H10	144.6	A4	H10	D6	36.6
D6	A2	H11	123.9	A4	H10	H11	35.4
D6	A2	R12	32.9	A4	H10	R12	22.8
H10	A2	H11	90.9	D6	H10	H11	70.8
H10	A2	R12	177.5	D6	H10	R12	13.8
H11	A2	R12	91.4	H11	H10	R12	57.2
A2	A2	D6	39.5	A2	H11	A4	45.9
A2	A4	H10	19.6	A2	H11	D6	18.2
A2	A4	H11	84.2	A2	H11	H10	33.4
A2	A4	R12	91.7	A2	H11	R12	44.9
D6	A4	H10	59.0	A4	H11	D6	29.3
D6	A4	H11	121.9	A4	H11	H10	78.7
D6	A4	R12	52.3	A4	H11	R12	9.7
H10	A4	H11	65.9	D6	H11	H10	51.5
H10	A4	R12	111.2	D6	H11	R12	26.8
H11	A4	R12	159.6	H10	H11	R12	78.2
A2	D6	A4	64.3	A2	R12	A4	44.9
A2	D6	H10	20.1	A2	R12	D6	25.0
A2	D6	H11	37.9	A2	R12	H10	1.0
A2	D6	R12	122.1	A2	R12	H11	43.7
A4	D6	H10	84.4	A4	R12	D6	69.8
A4	D6	H11	28.9	A4	R12	H10	45.9
A4	D6	R12	57.9	A4	R12	H11	10.7
H10	D6	H11	57.7	D6	R12	H10	24.0
H10	D6	R12	142.2	D6	R12	H11	68.4
H11	D6	R12	84.7	H10	R12	H11	44.6

evaluation of generated CPHs was performed by correlating the observed and the estimated activity for the training set of 35 molecules. PLS analyses were performed in which a series of models were constructed with an increasing number of PLS factors. Score hypotheses step was employed to align the actives to the hypotheses and calculate the score for the actives. CPHs of significant statistical values were selected for molecular alignments.

2.5. Validation of Pharmacophore Model. For accurate and reliable predictions of biological activities of new compounds, the main target was to develop QSAR models,

TABLE 5: Experimental and predicted IC_{50} values of training set molecules based on hypothesis AADHHR.47.

Comp. No.	Experimental activity (IC_{50})	Predicted activity (IC_{50})	Fitness score	Comp. No.	Experimental activity (IC_{50})	Predicted activity (IC_{50})	Fitness score
1.	7.409	8.2	1.96	24	9.376	8.88	2.37
2.	8.685	9.02	2.14	25	8.89	9.25	1.29
5.	8.856	9.12	2.26	27	9.275	9.45	1.31
6.	8.98	9.22	1.67	28	8.032	8.32	1.46
7.	8.85	9.13	1.90	29	6.745	7.156	1.33
9.	7.796	8.14	2.01	31	8.62	8.61	1.37
10.	7.585	8.22	1.56	32	9.284	9.05	1.43
11.	6.041	7.15	1.10	34	6.041	7.44	1.31
13.	8.89	9.18	1.89	35	6.553	8.68	1.55
15.	9.376	9.22	2.48	37	8.62	9.18	2.08
16.	9.585	9.74	1.66	38	8.959	8.57	2.00
17.	9.285	9.27	1.75	40	9.207	9.56	1.92
18.	8.88	9.01	2.03	44	8.745	9.18	2.68
19.	8.74	8.82	2.01	46	8.421	8.44	2.60
20.	8.77	8.89	1.79	47	8.174	8.25	2.59
21.	9.568	9.45	1.43	48	8.252	8.46	3.00
22.	9.212	9.22	2.35	50	5.921	7.32	1.08
23.	8.42	9.16	1.25				

TABLE 6: Experimental and predicted IC_{50} values of test set molecules based on hypothesis AADHHR.47.

Comp. No.	Experimental activity (IC_{50})	Predicted activity (IC_{50})	Fitness score	Comp. no.	Experimental activity (IC_{50})	Predicted activity (IC_{50})	Fitness score
3.	8.620	8.68	1.99	36	8.824	9.23	1.95
4.	9.200	9.10	2.23	39	7.469	7.42	1.89
8.	7.921	8.15	2.01	41	9.026	9.07	1.81
12.	9.096	9.02	2.27	42	7.722	8.41	1.94
14.	9.568	9.19	2.23	43	6.824	7.12	2.51
26.	8.236	8.42	2.21	45	8.921	9.11	0.87
30.	8.181	8.13	1.18	49	6.744	7.14	2.61
33.	8.824	9.20	1.75				

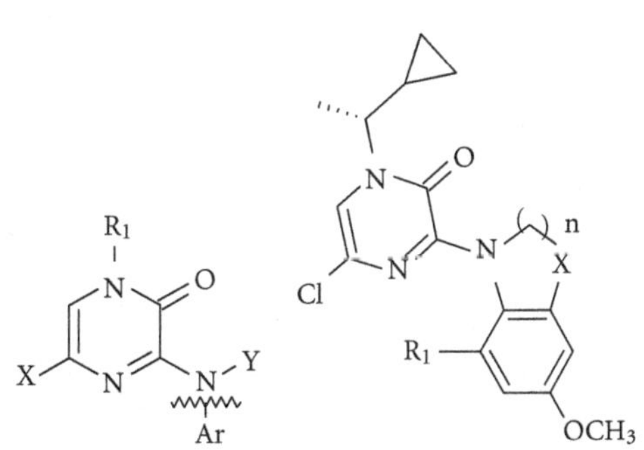

(a) (b)

FIGURE 1: Basic structures of N^3-phenylpyrazinones.

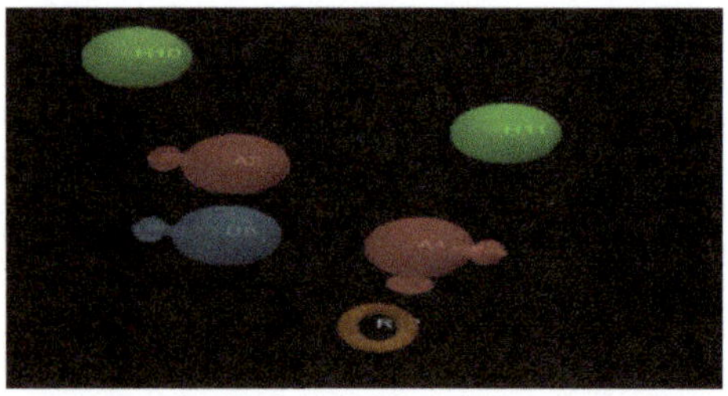

Figure 2: PHASE-generated pharmacophore model AADHHR.47 illustrating hydrogen bond acceptor (A2, A4; pink), hydrogen bond donor (D6; blue), hydrophobic region (H10, H11; green), and aromatic ring (R12; orange) features with distances (in Å) between different sites of AADHHR.47.

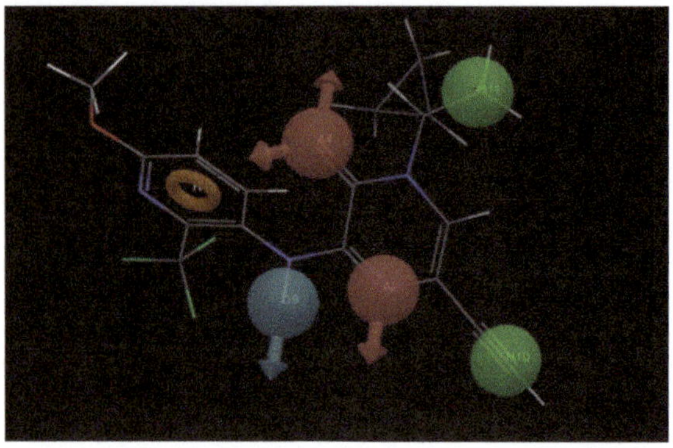

Figure 3: Best pharmacophore model AADHHR.47 aligned with molecule 48. Pharmacophore features are color coded: 2 hydrogen bond acceptors (A2, A4; pink), 1 hydrogen bond donor (D6; blue), 2 hydrophobic regions (H10, H11; green), and 1 aromatic ring (R12; orange).

which were statistically robust both internally as well as externally. The data set was divided into a training set and a test set as external validation is considered to be a conclusive proof for judging predictability of a model. The training set was used to generate pharmacophore model and prediction of the activity of test set was used as a method to validate the proposed models. The robustness of the developed pharmacophore hypotheses was internally validated by statistical parameters, that is, squared correlation coefficient (R^2) and variance ratio (F). Validation is a crucial aspect of pharmacophore design, particularly when the model is built for the purpose of predicting activities of molecules in external test series [16]. In the present case, the developed pharmacophore model was externally validated by predicting the activity of test set molecules. The correlation between the experimental and predicted activities of the test set molecules was determined.

3. Results and Discussion

CRF 1 receptor antagonists may offer therapeutic potential for the treatment of diseases resulting from elevated levels of CRF 1 such as anxiety and depression. Efforts to identify structurally diverse CRF 1 receptor antagonists led to the discovery of pyrazinone-based compounds and it was found that CRF 1 receptor binding affinity was improved when the branching point was on the carbon atom bonded directly to the pyrazinone ring [6]. Ligand-based drug design relies on knowledge of other molecules that bind to the biological target of interest. These molecules may be used to derive a pharmacophore which defines the minimum necessary structural characteristics a molecule must possess in order to bind to the target [7]. In other words, a model of the biological target is built based on the knowledge of what binds to it and this model in turn may be used to design new molecular entities that interact with the target.

Thirty-five molecules forming the training set were used to develop the pharmacophore models. The pharmacophoric features selected for creating sites were hydrogen bond acceptor (A), hydrogen bond donor (D), hydrophobic region (H), and aromatic ring (R). Pharmacophore models containing four to six features were generated. The four and five featured pharmacophore hypotheses were rejected due to low value of survival score (less than 2.5), as they were unable to define the complete binding space of the selected molecules. Six featured pharmacophore hypotheses were selected and subjected to stringent scoring function analysis.

The results of six featured pharmacophore hypotheses, labeled AADHHR.47, ADHHHR.203, DHHHRR.611, and AAHHHR.19, are presented in Table 2. The first hypothesis AADHHR.47 is the best hypothesis in this study, characterized by highest survival score (2.980) and best regression coefficient (0.803).

The pharmacophore hypothesis AADHHR.47 is presented in Figure 2. The features represented by this hypothesis are two hydrogen bond acceptors (A), one hydrogen bond donor (D), two hydrophobic regions (H), and one aromatic ring (R). The distances and angles between different sites of AADHHR.47 are given in Tables 3 and 4, respectively.

For each ligand, one aligned conformer based on the lowest RMSE of feature atom coordinates from those of the corresponding reference feature was superimposed on AADHHR.47. The fitness scores for all ligands were observed on the best scored pharmacophore model AADHHR.47. The greater the fitness score, the greater the activity prediction of the compound. The fit function does not only check if the feature is mapped or not but also contains a distance term, which measures the distance that separates the feature on the molecule from the centroid of the hypothesis feature.

Table 5 shows the fitness score for all the molecules of training set. Figure 3 shows the AADHHR.47 aligned with ligand having maximum fitness score, that is, molecule 48 of the training set. Beside this survival score analysis, another validation method to characterize the quality of AADHHR.47 is represented by its capacity for correct activity prediction of training set molecules.

The predicted CRF 1 antagonistic activity of training set molecule exhibited R^2 value of 0.803 (RMSD = 0.618) with experimental CRF 1 antagonistic activity using model AADHHR.47 (Figure 4). The validity and predictive character of AADHHR.47 was further assessed by using the test set prediction.

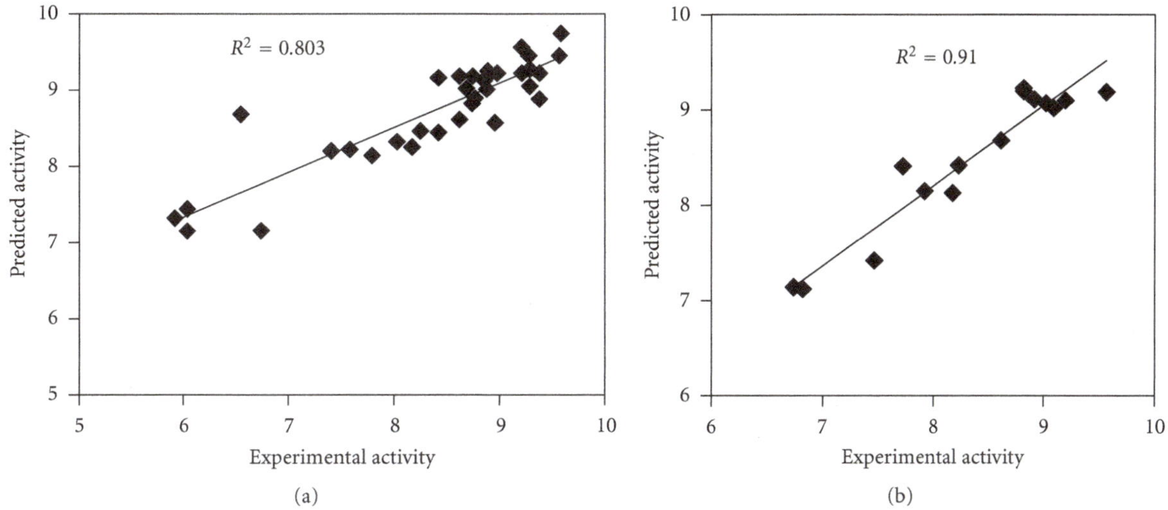

FIGURE 4: Relation between experimental and predicted CRF 1 antagonistic activity values of training set (a) and test set molecules (b) using model AADHHR.47.

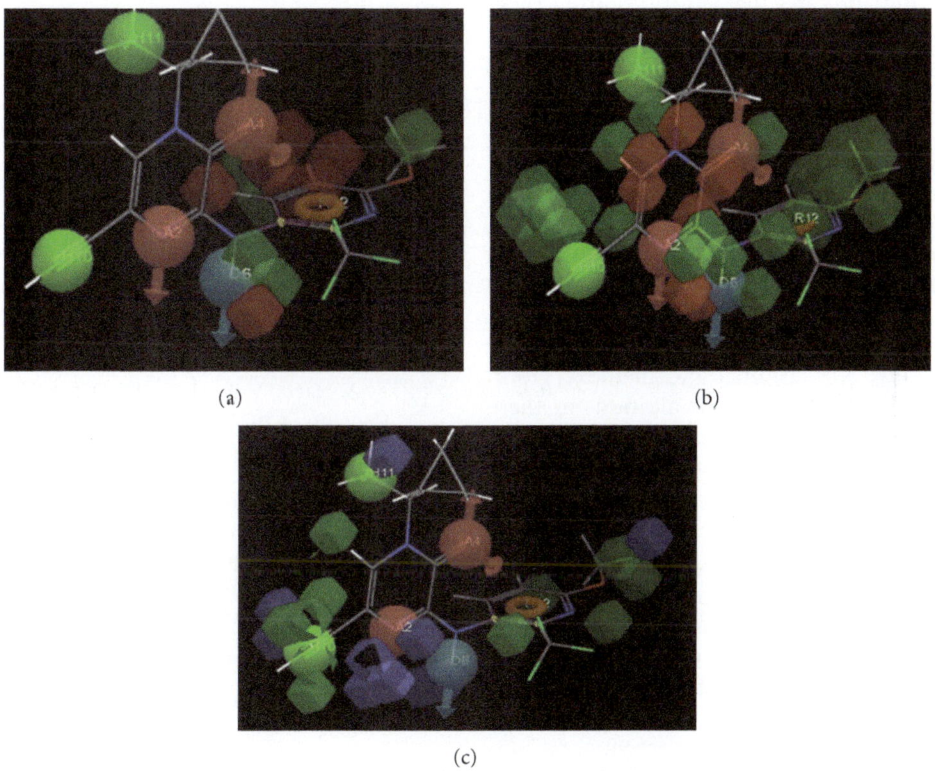

FIGURE 5: 3-D QSAR model based on molecule 48 of training set illustrating hydrogen bond donor feature (a), hydrogen bond acceptor feature (b) and hydrophobicity feature (c).

The test set having fifteen molecules was analyzed. All the test set molecules were built and minimized as well as used in conformational analysis like all training set molecules. Then the activities of test set molecules were predicted using AADHHR.47 and compared with the actual activity. Actual and predicted activity values of test set molecules are given in Table 6.

The predicted CRF 1 antagonistic activity of test molecule exhibited R^2 value of 0.91 (RMSD = 0.2961) with experimental CRF 1 antagonistic activity using model AADHHR.47 (Figure 4). For a reliable model, the squared predictive correlation coefficient should exceed 0.60 [17, 18]. The results of this study reveal that model AADHHR.47 can be used for the prediction of CRF 1 antagonistic activity.

3.1. 3D QSAR Analysis. Additional insight into the CRF 1 antagonistic activity can be gained by visualizing the 3D QSAR model in the context of one or more ligands in the series with varying activity. This information can then be used to design new or more active analogues. 3D QSAR model based on the molecules of training and test set using various features, that is, hydrogen bond acceptor (A), hydrogen bond donor (D), hydrophobic regions (H), and aromatic ring (R) has been studied.

3.1.1. Hydrogen Bond Donor Field Predictions. The 3D QSAR model based on molecule 48 of the training set using hydrogen bond donor feature is shown in Figure 5(a). Red region near and around the meta- and parahydrogen of pyridine ring substituted at position 3 through NH terminal indicates that the substitutions at these positions by groups having more hydrogen bond donor property favor the CRF 1 antagonistic activity. Green region around the double bond between C_5 and C_6 on pyridine ring, NH terminal connecting pyrazinone and pyridine ring, methoxy group at C_4 on pyridine ring, indicates that substitutions at these positions by groups having hydrogen bond donor property do not favor CRF 1 antagonistic activity.

3.1.2. Hydrogen Bond Acceptor Field Predictions. The 3D QSAR model based on molecule 48 of the training set using hydrogen bond acceptor feature is shown in Figure 5(b). Red region around N1, carbonyl group at C_2 and in between C_2 and C_3, double bond between C_5 and C_6 (all 3 on pyrazinone ring), indicates that the substitutions at these positions by groups having more hydrogen bond acceptor property favour the CRF 1 antagonistic activity. Green region around C_5 on pyrazinone ring, NH terminal and methoxy group on pyridine ring, indicates that the substitutions at these positions by groups having more hydrogen bond acceptor property do not favour the CRF 1 antagonistic activity.

3.1.3. Hydrophobicity Field Prediction. The 3D QSAR model based on molecule 48 of the training set using hydrophobicity feature is shown in Figure 5(c). Green region around ethynyl group substituted at position 5, at C_6 (both on pyrazinone ring), double bond between C_5 and C_6, methoxy group at C_4 (both on pyridine ring), indicates that the substitutions at these positions by groups having more hydrophobicity favour CRF 1 antagonistic activity and substitutions at these positions by more hydrophobic groups will result in increase in CRF 1 antagonistic activity. Blue region around methyl substituent of N^1, position 4 (both on pyrazinone ring), methoxy group at C_4 on pyridine ring, indicates that groups having more hydrophobic property do not favour CRF 1 antagonistic activity.

4. Conclusions

This study shows the generation of a pharmacophore model AADHHR.47 for N^3 phenylprazinones acting as CRF 1 antagonists. Pharmacophore modelling correlates activities with the spatial arrangement of various chemical features. The first hypothesis AADHHR.47 is the best hypothesis in this study, characterized by the best regression coefficient (0.803), degree of freedom (138.5), and highest survival score (2.980). Hypothesis AADHHR.47 represents the best pharmacophore model for determining CRF 1 antagonistic activity. AADHHR.47 consists of two hydrogen bond acceptors, one hydrogen bond donor, two hydrophobic regions, and one aromatic ring features. AADHHR.47 model had strong correlation between experimental and estimated activity of the training ($R^2 = 0.803$) and test ($R^2 = 0.91$) set molecules. Thus, AADHHR.47 pharmacophore model was able to accurately predict CRF 1 antagonistic activity, and the validation results also provide additional confidence in the proposed pharmacophore model. The obtained results suggested that the proposed 3D-QSAR model AADHHR.47 can be useful to rationally design new N^3-phenylprazinones molecules as CRF 1 antagonists and also to identify new promising molecules as CRF 1 antagonists in large 3D database of molecules.

References

[1] M. R. Dimatteo, H. S. Lepper, and T. W. Croghan, "Depression is a risk factor for noncompliance with medical treatment meta-analysis of the effects of anxiety and depression on patient adherence," *Archives of Internal Medicine*, vol. 160, no. 14, pp. 2101–2107, 2000.

[2] C. G. Beevers, "Cognitive vulnerability to depression: a dual process model," *Clinical Psychology Review*, vol. 25, no. 7, pp. 975–1002, 2005.

[3] E. P. Zorrilla and G. F. Koob, "Progress in corticotropin-releasing factor-1 antagonist development," *Drug Discovery Today*, vol. 15, no. 9-10, pp. 371–383, 2010.

[4] R. A. Hartz, V. T. Ahuja, W. D. Schmitz et al., "Synthesis and structure-activity relationships of N^3-pyridylpyrazinones as corticotropin-releasing factor-1 (CRF1) receptor antagonists," *Bioorganic and Medicinal Chemistry Letters*, vol. 20, no. 6, pp. 1890–1894, 2010.

[5] X. Han, S. S. Pin, K. Burris et al., "Synthesis and structure-activity relationship of imidazo[1,2-a] benzimidazoles as corticotropin-releasing factor 1 receptor antagonists," *Bioorganic & Medicinal Chemistry Letters*, vol. 15, no. 18, pp. 4029–4032, 2005.

[6] R. A. Hartz, V. T. Ahuja, A. G. Arvanitis et al., "Synthesis, structure-activity relationships, and in vivo evaluation of N3-phenylpyrazinones as novel corticotropin-releasing factor-1 (CRF1) receptor antagonists," *Journal of Medicinal Chemistry*, vol. 52, no. 14, pp. 4173–4191, 2009.

[7] O. F. Guner, *Pharmacophore Perception, Development and Use in Drug Design*, International University Line, San Diego, Calif, USA, 2000.

[8] PHASE, Version 3.0, Schrödinger, LLC, New York, NY, USA, 2008.

[9] S. L. Dixon, A. M. Smondyrev, E. H. Knoll, S. N. Rao, D. E. Shaw, and R. A. Friesner, "PHASE: a new engine for pharmacophore perception, 3D QSAR model development, and 3D database screening—1. Methodology and preliminary results," *Journal of Computer-Aided Molecular Design*, vol. 20, no. 10-11, pp. 647–671, 2006.

[10] S. L. Dixon, A. M. Smondyrev, and S. N. Rao, "PHASE: a novel approach to pharmacophore modeling and 3D database

searching," *Chemical Biology and Drug Design*, vol. 67, no. 5, pp. 370–372, 2006.

[11] Maestro, Version 8.5, Schrödinger, LLC, New York, NY, USA, 2008.

[12] Ligprep 2.0, Schrodinger, LLC, New York, NY, USA, 2010.

[13] G. A. Kaminski, R. A. Friesner, J. Tirado-Rives, and W. L. Jorgensen, "Evaluation and reparametrization of the OPLS-AA force field for proteins via comparison with accurate quantum chemical calculations on peptides," *Journal of Physical Chemistry B*, vol. 105, no. 28, pp. 6474–6487, 2001.

[14] N. D. Amnerkar and K. P. Bhusari, "Synthesis, anticonvulsant activity and 3D-QSAR study of some prop-2-eneamido and 1-acetyl-pyrazolin derivatives of aminobenzothiazole," *European Journal of Medicinal Chemistry*, vol. 45, no. 1, pp. 149–159, 2010.

[15] D. A. Evans, T. N. Doman, D. A. Thorner, and M. J. Bodkin, "3D QSAR methods: phase and catalyst compared," *Journal of Chemical Information and Modeling*, vol. 47, no. 3, pp. 1248–1257, 2007.

[16] B. D. Boyd, "Successes of computer-assisted molecular design," in *Reviews in Computational Chemistry*, vol. 4, pp. 355–371, VCH, New York, NY, USA, 1990.

[17] H. Dureja, V. Kumar, S. Gupta, and A. K. Madan, "Topochemical models for the prediction of lipophilicity of 1,3-disubstituted propan-2-one analogs," *Journal of Theoretical and Computational Chemistry*, vol. 6, no. 3, pp. 435–448, 2007.

[18] S. Wold, "Validation of QSAR's," *Quantitative Structure-Activity Relationships*, vol. 10, no. 3, pp. 191–193, 1991.

16

The Development of Models Based on Linear and Nonlinear Multivariate Methods to Predict ADME/PK Properties Using Physicochemical Properties of Kinase, Protease Inhibitors, and GPCR Antagonists

Deepu Bakasta[1] and M. G. Shambhu[2]

[1] *Department of Biotechnology, PES Institute of Technology, Bangalore 560068, India*
[2] *Department of Biotechnology, The Oxford College of Engineering, Bangalore 560068, India*

Correspondence should be addressed to M. G. Shambhu; shambumg13@gmail.com

Academic Editor: Armando Rossello

Oral bioavailability of a drug compound is the significant property for potential drug candidates. Measuring this property can be costly and time-consuming. Quantitative structure-property relationships (QSPRs) are used to estimate the percentage of oral bioavailability, and they are an attractive alternative to experimental measurements. A data set of 217 drug and drug-like compounds with measured values of the percentage of oral bioavailability taken from the small molecule ChemBioBase database was used to develop and test a QSPR model. Descriptors were calculated for the compounds using Codessa 2.1 tool. Nonlinear general regression neural network model was generated using the DTREG predictive modeling program software. The calculated percentage of oral bioavailability model performs well, with root-mean-square (rms) errors of 4.55% oral bioavailability units for the training set, 14.32% oral bioavailability units for the test set, and 19.12% oral bioavailability units for the external prediction set. Given the structural diversity and bias of the data set, this is a good first attempt at modeling oral bioavailability using QSPR methods. The model can be used as a potential virtual screen or property estimator. With a larger data supply less biased toward the high end values of the percentage of oral bioavailability, a more successful model could likely be developed.

1. Introduction

Rational drug discovery requires early methods of all factors influencing on the likely success of a drug candidate in the subsequent stages of the drug development. The study of absorption, distribution, metabolism, excretion, and pharmacokinetics (ADME/PK) has become a major discipline in drug discovery through the application of well-established in vitro and in vivo methodologies [1]. PK plays a crucial role in the pharmaceutical research and development and, because of the major role of drug metabolism, drug discovery research in this area is covered by groups coalesced around the name drug metabolism and pharmacokinetics (DMPKs) [2–6]. Elimination is the product of metabolism and excretion. Pharmacokinetics describe how the body reacts to a specific drug after administration. Pharmacokinetic properties of drugs may be affected by factors such as the administration site and the administered drug dose; these may influence the rate of absorption. One more process, liberation, plays an important role in pharmacokinetics: liberation means the release of drug from the formulation [7].

The primary goal of the drug discovery and development process is to find a molecule possessing both good pharmacodynamics and good pharmacokinetic properties [8]. Ideally, a new drug should be efficacious and selective, target tissue(s) specific, and orally absorbed, cause minimal or no adverse effects due to metabolite activity or toxicity, and be distributed/excreted in such a fashion as to permit dosage once a day. Successful optimization of all these properties is an extremely challenging task. QSPR methods have

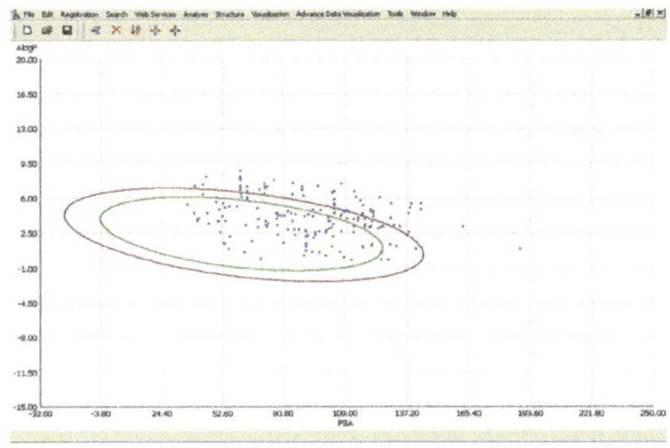

FIGURE 1: AlogP versus PSA.

TABLE 1: Statistics results of R^2, RMSE, MAE, and standard deviation for training set, test set, external prediction set.

	R^2	RMSE	MAE	Standard deviation
Training set	0.98	4.55	2.90	37.68
Test set	0.77	14.32	12.16	22.64
External prediction set	0.86	19.14	14.76	37.20

successfully been used to model physicochemical properties of organic compounds. Several computational models have also been reported for such biopharmaceutical properties as %HIA, blood-brain barrier, skin and ocular permeation, pharmacokinetics and metabolism. However, all these studies involved sets of closely related structural analogues, and models based on limited chemical space generally lack predictive value outside their structural classes. Broadly applicable QSPR models of biopharmaceutical properties must be built using compounds which cover both a wide range of the property being modeled as well as of the chemical structure space [9].

Bioavailability is a key pharmacokinetic parameter which expresses the proportion of a drug administered by any nonvascular route that gains access to the systemic circulation [10]. It was found that 30% of drugs fail during the drug discovery process. Therefore, there is a need of a robust and accurate computational model (QSAR/QSPR) which can predict the oral bioavailability of compounds without carrying out any experiments.

2. Materials and Methods

Development of model using linear and nonlinear multivariate statistics to predict the oral bioavailability of a compound.

2.1. Quantitative Structure Property Relationship. The QSPR methodology used in this project consists of three main parts: representation of molecular structure, feature selection, and mapping. The general assumption in QSPR modeling is that molecular structure causes the observed behavior of a compound, linking a series of chemical structures to properties of interest; in this case, oral bioavailability should provide a method for modeling the property. General regression neural network method present in DTREG was used to generate the oral bioavailability model. This neural network has 3 layers and 1 hidden layer, and it uses kernel function.

2.2. Preparation of Data Set. The set of 217 drug and drug-like compounds and their experimentally derived percentage of oral bioavailability values used in this project were gathered from the literature sources. The 217 compounds in the working data set were classified by using sphere exclusion algorithm [11] into a training set of 159 compounds, test set of 50 compounds, and an external prediction set of 8 compounds. The external prediction set was chosen in such a way as to cover the range of percentage of oral bioavailability values in the data set, and it spans the range of 0%–100%. The compounds in the external prediction set were never used during the model development process but were reserved to validate potential models. The structures of the 217 compounds were extracted from the small molecule ChemBioBase [12] database with ISIS which consists of kinase, protease inhibitors, and GPCR antagonists. Accurate geometries are necessary for the calculation of certain descriptors thought to be necessary for modeling physical and chemical properties. As a result, LigPrep [13] tool was used to generate accurate three-dimensional geometries, and conformational analysis was performed in order to get the local minima structure which is necessary for calculating geometric descriptors.

2.3. Descriptor Generation and Analysis. A total of 11 descriptors were generated for each of the 217 compounds using Codessa 2.1 tool [14]. The descriptors fall into three general categories: topological, electronic, and geometric. Topological descriptors like number of single bonds, number of aromatic bonds, number of oxygen atoms, nitrogen atoms, relative number of oxygen atoms, and relative number of nitrogen atoms are derived from the information of the two-dimensional structure of the molecule.

Electronic descriptors were calculated by MOPAC using the AM1 Hamiltonian electronic descriptors include maximum partial charge for hydrogen atoms. Geometric descriptors include cube root of gravitational index, SAAA (surface area of hydrogen bond acceptor atoms/number of hydrogen bond acceptor atoms), and normalized 2D projection on YZ plane. Accurate three-dimensional geometries of the molecules are necessary to calculate descriptors of this nature. A fourth class of descriptors can be derived by combining electronic and geometric information to form hybrid descriptors. By combining the molecular surface area with partial atomic charges, charged-partial surface area (CPSA) descriptors can be calculated. This includes HASA-2, area-weighted surface charge of hydrogen bonding acceptor atoms. Eleven descriptors span the following ranges: number of single bonds (26–90), number of aromatic bonds (0–12), YZ shadow (29.9–110.46), GRAV-3 (25.97–42.16), SAAA (28.38–32.14), number of oxygen atoms (0–7), number of nitrogen atoms (1–8), relative number of oxygen atoms (0–0.12), relative number of nitrogen atoms (0.02–0.17), maximum partial

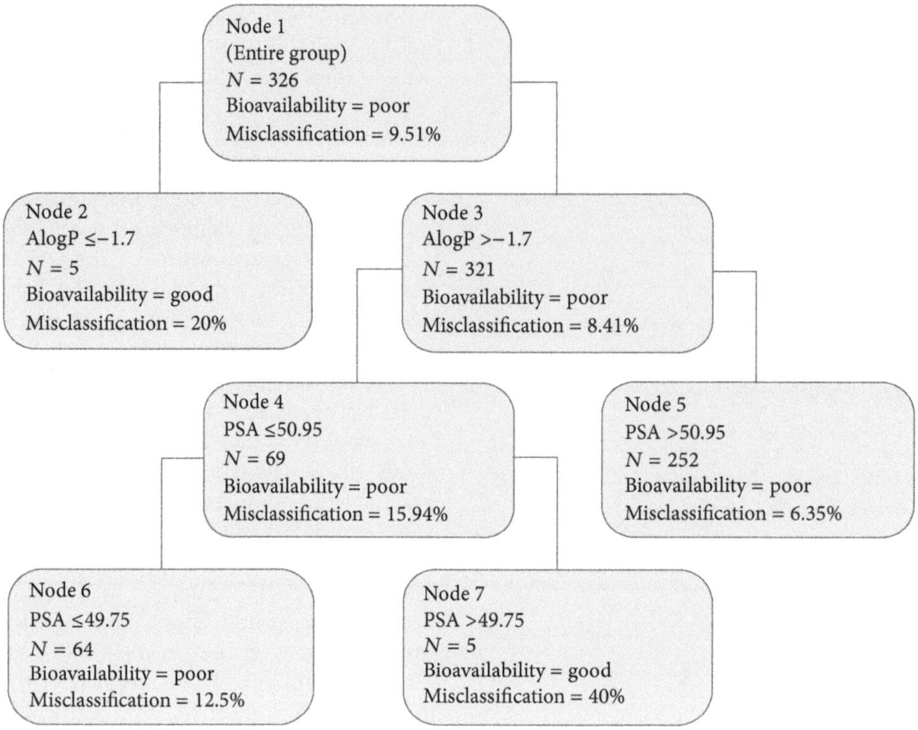

FIGURE 2: Single decision tree method.

TABLE 2: Correlation matrix for descriptors.

Variable	G. index	S. bonds	SAAA	YZ shadow	HASA-2	P. charge	O. atoms	R. O. atoms	N. atoms	R. N. atoms	A. bonds
G. index	1										
S. bonds	0.26798	1									
SAAA	0.41483	−0.0544	1								
YZ shadow	0.50115	0.31235	0.21755	1							
HASA-2	0.13998	−0.05302	−0.02074	0.18598	1						
P. charge	−0.03232	−0.1199	0.03092	0.03528	0.30933	1					
O. atoms	0.11838	0.09889	0.04789	−0.00181	0.12292	0.13269	1				
R. O. atoms	−0.02622	−0.31383	0.05442	−0.14418	0.11698	0.16337	0.58024	1			
N. atoms	0.21829	0.27275	0.02458	0.21158	0.50736	−0.10747	−0.11228	−0.29378	1		
R. N. atoms	0.02254	−0.32032	0.05143	−0.01294	0.43585	--0.10581	−0.24878	−0.1698	0.52116	1	
A. bonds	0.2846	0.2069	0.00138	0.33825	0.14602	0.159	0.01585	−0.1204	−0.02972	−0.24046	1

charge for a hydrogen atom (0.03–0.1), and HASA-2 (2.89–32.2). First two descriptors indicate amount of structural flexibility; next two descriptors encode molecular size, shape, and bulk properties. Remaining descriptors are all hydrogen bonding descriptors.

2.4. Descriptors

(1) Topological descriptors include number of single bonds, number of oxygen atoms, relative number of oxygen atoms, number of nitrogen atoms, relative number of nitrogen atoms, and number of aromatic bonds [15].

(2) Geometric descriptors include the following:

(a) GRAV-3: cube root of gravitational index [16],

(b) SAAA-2: surface area of hydrogen bond acceptor atoms/number of hydrogen bond acceptor atoms; these two are calculated using java program,

(c) SHDW-6: normalized 2D projection of molecule on YZ plane [16],

(3) maximum partial charge for hydrogen atoms [17],

(4) HASA-2: area-weighted surface charge of hydrogen bonding acceptor atoms [18].

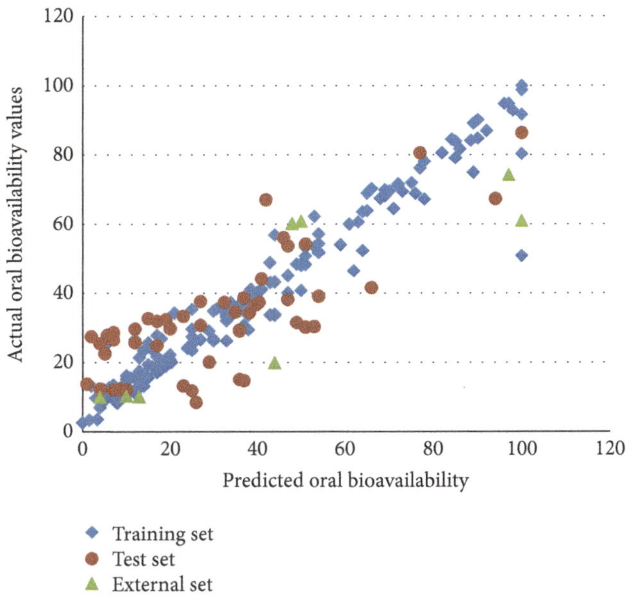

FIGURE 3: Actual percentage of oral bioavailability values versus predicted values.

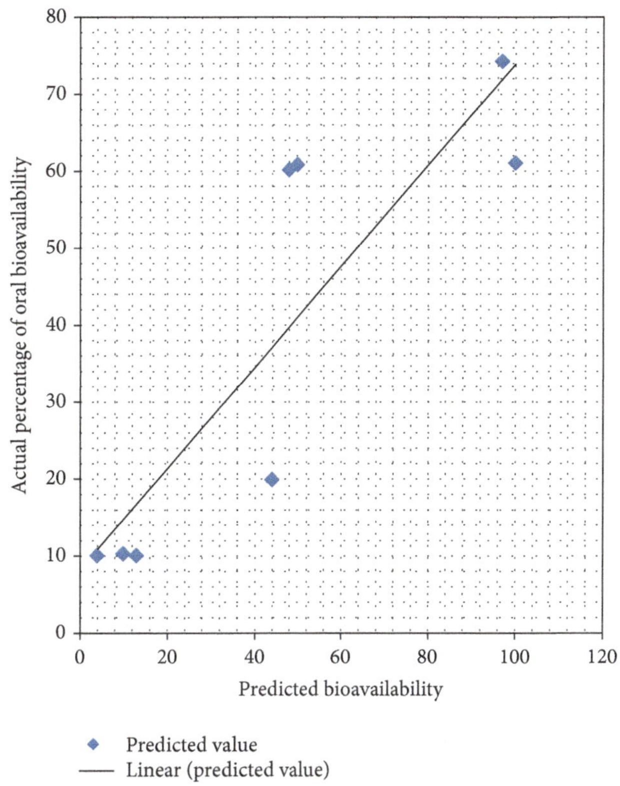

FIGURE 4: Actual percentage of oral bioavailability values versus predicted values (external prediction set).

The NSB descriptor is encoding single bonds, and this may be an indication of the amount of structural flexibility [19]. The SHDW-6 and GRAV-3 descriptors are encoding molecular size, shape, and bulk properties [17]. These size descriptors may be important with respect to the ability of the drug to penetrate cell membranes. The three remaining descriptors are all hydrogen bonding descriptors (Table 2).

3. Results and Discussion

3.1. Absorption Ellipse Model. In this model, Alogp and PSA descriptors were used as predictor variables. Well-absorbed and poor-absorbed compounds were plotted against AlogP versus PSA. A 95% confidence ellipse for the well absorbed dataset was also computed and plotted. The 95% confidence ellipse represents the region of chemical space where we can expect to find well-absorbed compounds (90%) 95 out of 100 times (Figure 1).

CIMPL 1.0 [19] version tool was used to generate 95% confidence ellipse model. It gives more than 60% of misclassification; from this, we conclude that the Alogp and PSA (polar surface area) descriptors are not enough for this dataset to classify well-absorbed and poor-absorbed molecules. Calc prop tool was used to calculate Alogp and PSA. 95% confidence means that we get 95 times out of 100 molecules having good absorption inside the ellipse. So, it is called 95% confidence ellipse.

DTREG [20] software was used to generate single decision tree (Figure 2); this model also gives a lot of misclassification (more than 40%). Both absorption ellipse method and single decision tree methods show a lot of misclassification. Alogp and PSA descriptors are not enough to classify well-absorbed and poor compounds.

3.2. General Regression Neural Network Method (DTREG Software). Partial least square method and multiple linear regression present in CIMPL tool were employed to get good linear models; none of the models found was satisfactory. It became evident that the diversity of this data set and the number of data points possessing greater than 50% absorption values are largely responsible for producing poor quality linear models. Therefore, this data set became a good candidate for developing nonlinear neural network model.

3.3. Neural Network Analysis. 11-member reduced pool of descriptors was fed to the general regression neural network method for the purpose of developing a nonlinear model. The original regression data set was split randomly into a neural network training set of 159 compounds and a test set of 50 compounds using sphere exclusion method. The original 8-member external prediction set was used to validate any neural network models. The test set was used to monitor overtraining of the network, and the training set was used to actually train the network (Figures 3 and 4).

To decrease the possibility of chance effects influencing neural network training, the ratio of observations to total adjustable parameters should be at or above 2.0. A neural network consisting of 11 input neurons (descriptors), 3 hidden neurons, and 1 output neuron (target and percentage of bioavailability), thus producing an 11-3-1 architecture, was used since it produced the maximum number of adjustable parameters recommended for a dataset of this size. For this

11-3-1 architecture, the ratio of training set observations to adjustable parameters F was 67/33 or 2.03 (Table 1). Table 1 refers to statistics results of R^2, RMSE, MAE, and standard deviation for training set, test set, external prediction set.

4. Conclusion

11-descriptor general regression neural network model has been developed for the estimation of the percentage of oral bioavailability values for a data set of 217 drug and drug-like compounds. The training set rms error was 4.55% oral bioavailability units, and the test set rms error was 14.32% oral bioavailability units. Based on the rms errors of the training and test sets, it is clear that a link between the structure and the percentage of oral bioavailability does exist. However, the strength of that link is best measured by the quality of the external prediction set. With rms error of 19.0% HIA units and a good visual plot, the external prediction set ensures the quality of the model. Given the structural diversity and bias of the data set, this is a good first attempt at modeling oral bioavailability using QSPR methods. The optimization of individual properties along the absorption process must be integrated in a multiobjective scenario for studying oral bioavailability behavior in the early drug discovery and development [21].

The model can be used as a potential virtual screen or property estimator. With a larger data supply less biased toward the high end values of the percentage of oral bioavailability, a more successful and reviling model can be developed. This study illustrates the potential of using QSPR methods to aid in the drug development process.

Abbreviations

CIMPL: Chemo informatics platform tool
ADME/Tox: Absorption, distribution, metabolism, excretion, and toxicity
QSPR: Quantitative structure property relationship
HIA: Human intestinal absorption
PSA: Polar surface area
RMSE: Root mean square error
MAE: Mean absolute error
IV: Intravenous.

Acknowledgments

The authors thank Jubilant Biosys Company, Department of Computational Chemistry, Bangalore, for the support and infrastructure provided and also TOCE and the PES institute of technology, Bangalore, for their undiminished encouragement and valuable inputs in presenting the work.

References

[1] P. J. Eddershaw, A. P. Beresford, and M. K. Bayliss, "ADME/PK as part of a rational approach to drug discovery," *Drug Discovery Today*, vol. 5, no. 9, pp. 409–414, 2000.

[2] M. S. Alavijeh and A. M. Palmer, "The pivotal role of drug metabolism and pharmacokinetics in the discovery and development of new medicines," *IDrugs*, vol. 7, no. 8, pp. 755–763, 2004.

[3] S. A. Roberts, "Drug metabolism and pharmacokinetics in drug discovery," *Current Opinion in Drug Discovery & Development*, vol. 6, pp. 66–80, 2003.

[4] R. J. Riley, I. J. Martin, and A. E. Cooper, "The influence of DMPK as an integrated partner in modern drug discovery," *Current Drug Metabolism*, vol. 3, no. 5, pp. 527–550, 2002.

[5] J. Lin, D. C. Sahakian, S. M. de Morais, J. J. Xu, R. J. Polzer, and S. M. Winter, "The role of absorption, distribution, metabolism, excretion and toxicity in drug discovery," *Current Topics in Medicinal Chemistry*, vol. 3, no. 10, pp. 1125–1154, 2003.

[6] P. J. Eddershaw, A. P. Beresford, and M. K. Bayliss, "ADME/PK as part of a rational approach to drug discovery," *Drug Discovery Today*, vol. 5, no. 9, pp. 409–414, 2000.

[7] W. A. Ritschel, "The LADMER system: liberation, absorption,distribution, metabolism, elimination and response," in *Handbook of Basic Pharmacokinetics Including Clinical Applications.*, W. A. Ritschel, Ed., pp. 19–24, Drug Intelligence, Hamilton, Canada, 1992.

[8] D. P. Nicolau, "Pharmacokinetic and pharmacodynamic properties of meropenem," *Clinical Infectious Diseases*, vol. 47, supplement 1, no. 1, pp. S32–S40, 2008.

[9] W. Tong, D. R. Lowis, R. Perkins et al., "Evaluation of quantitative structure-activity relationship methods for large-scale prediction of chemicals binding to the estrogen receptor," *Journal of Chemical Information and Modeling*, vol. 38, no. 4, pp. 669–677, 1998.

[10] P. L. Toutain and A. Bousquet-Me'lou, "Bioavailability and its assessment," *Journal of Veterinary Pharmacology and Therapeutics*, vol. 27, pp. 455–466, 2004.

[11] A. Gobbi and M. L. Lee, "DISE: directed sphere exclusion," *Journal of Chemical Information and Computer Sciences*, vol. 43, no. 1, pp. 317–323, 2003.

[12] ChemBioBase, November 2012, http://www.jubilantbiosys.com/index.php/services/functional-discovery-services/discovery-informatics.

[13] *Suite: LigPrep, Version 2.5*, Schrödinger, LLC, New York, NY, USA, 2012.

[14] A. R. Katritzky, M. Karelson, and R. Petrukhin, "Comprehensive descriptors for structural and statistical," Tech. Rep. CODESSA PRO Project, 2002.

[15] D. M. Mudie, G. L. Amidon, and G. E. Amidon, "Physiological parameters for oral delivery and *in vitro*," *Molecular Pharmaceutics*, vol. 7, no. 5, pp. 1388–1405, 2010.

[16] M. D. Wessel, P. C. Jurs, J. W. Tolan, and S. M. Muskal, "Prediction of human intestinal absorption of drug compounds from molecular structure," *Journal of Chemical Information and Computer Sciences*, vol. 38, no. 4, pp. 726–735, 1998.

[17] C. A. Lipinski, F. Lombardo, B. W. Dominy, and P. J. Feeney, "Experimental and computational approaches to estimate solubility and permeability in drug discovery and development settings," *Advanced Drug Delivery Reviews*, vol. 23, pp. 3–25, 1995.

[18] W. J. Egan, K. M. Merz, and J. J. Baldwin, *Prediction of Drug Absorption Using Multivariate Statistics Center for Informatics & Drug Discovery and Pharmacopeia Laboratories*, Pharmacopeia, Princeton, NJ, USA, 2000.

[19] P. Artursson and J. Karlsson, "Correlation between oral drug absorption in humans and apparent drug permeability," *Biochemical and Biophysical Research Communications*, vol. 175, no. 3, pp. 880–885, 1991.

[20] CIMPL 1.0, November 2012, http://www.barissumengen.com/cimpl/.

[21] M. A. Cabrera-Perez, H. Pham-The, M. Bermejo, I. G. Alvarez, M. G. Alvarez, and T. M. Garrigues, "QSPR in oral bioavailability: specificity or integrality?" *Mini-Reviews in Medicinal Chemistry*, vol. 12, no. 6, pp. 534–550, 2012.

2′,6′-Dimethylphenylalanine: A Useful Aromatic Amino Acid Surrogate for Tyr or Phe Residue in Opioid Peptides

Yusuke Sasaki and Akihiro Ambo

Department of Biochemistry, Tohoku Pharmaceutical University, 4-4-1 Komatsushima, Aoba-ku, Sendai 981-8558, Japan

Correspondence should be addressed to Yusuke Sasaki, ysasaki@tohoku-pharm.ac.jp

Academic Editor: Yoshio Okada

Two aromatic amino acids, Tyr^1 and Phe^3 or Phe^4, are important structural elements in opioid peptides because they interact with opioid receptors. The usefulness of an artificial amino acid residue 2′,6′-dimethylphenylalanine (Dmp) was investigated as an aromatic amino acid surrogate for several opioid peptides, including enkephalin, dermorphin, deltorphin, endomorphin, dynorphin A, and nociceptin peptides. In most peptides, substitutions of Phe^3 by a Dmp residue produced analogs with improved receptor-binding affinity and selectivity, while the same substitution of Phe^4 induced markedly reduced receptor affinity and selectivity. Interestingly, replacement of Tyr^1 by Dmp produced analogs with unexpectedly high affinity or produced only a slight drop in receptor affinity and bioactivity for most peptides. Thus, Dmp is also a useful surrogate for the N-terminal Tyr residue in opioid peptides despite the lack of a phenolic hydroxyl group, which is considered necessary for opioid activity. The Dmp^1-substituted analogs are superior to 2′,6′-dimethyltyrosine $(Dmt)^1$-substituted analogs for high receptor selectivity since the latter generally have poor receptor selectivity. Thus, Dmp is very useful as an aromatic amino acid surrogate in opioid peptides and may be useful for developing other novel peptide mimetics with high receptor specificity.

1. Introduction

Three major types of opioid receptors, μ, δ, and κ, have been cloned and assigned to the superfamily of rhodopsin-like G-protein-coupled receptors [1–3]. The μ-receptors are involved in supraspinal analgesia, respiratory depression, euphoria, sedation, decreased gastrointestinal motility, and physical dependence [4]. The δ-receptors appear to affect cardiovascular function, contribute to analgesia, and cause changes in affective behavior [4]. The κ-receptors are responsible for spinal analgesia, miosis, a modest degree of sedation, and some respiratory depression [4]. *In vivo,* opioid peptides exert pharmacological actions *via* the opioid receptors. Enkephalins (Tyr-Gly-Gly-Phe-Leu/Met) [5] and endomorphins (Tyr-Pro-Phe/Trp-Phe-NH_2) [6] are endogenous ligands for the δ- and μ-opioid receptors, respectively. Dynorphin A (DYN: Tyr-Gly-Gly-Phe-Leu-Arg-Arg-Ile-Arg-Pro-Lys-Leu-Lys-Trp-Asp-Asn-Gln) is the endogenous ligand of the κ-opioid receptor [7]. Dermorphin (DM, Tyr-D-Ala-Phe-Gly-Tyr-Pro-Ser-NH_2) [8] and [D-Ala^2]deltorphin II (DT, Tyr-D-Ala-Phe-Glu-Val-Val-Gly-NH_2) [9] are naturally occurring opioid peptides isolated from South American frogs and possess high selectivity toward μ- and δ-opioid receptors, respectively.

A fourth cloned member of the opioid-receptor family is the opioid receptor-like 1 (ORL1) receptor that shares high sequence homology with traditional opioid receptors [10]. The heptadecapeptide nociceptin (NOC) [11] or orphanin FQ [12] (NOC: Phe-Gly-Gly-Phe-Thr-Gly-Ala-Arg-Lys-Ser-Ala-Arg-Lys-Leu-Ala-Asn-Gln) was discovered as an endogenous ligand for the ORL1 receptor. NOC and DYN have a similar structural feature including the N-terminal tetrapeptide sequence Phe-Gly-Gly-Phe in NOC and Tyr-Gly-Gly-Phe in DYN and the existence of basic residues, although with different distributions, at the C-terminal. Despite the sequence homology, NOC and DYN have different pharmacological profiles [13, 14]. NOC possesses hyperalgesia and allodynia activity when applied supraspinally at low doses, while spinal delivery of NOC causes analgesia at high doses [11, 13–15]. Two aromatic amino acids, Phe[1,4] at the N-terminal of NOC [14, 16, 17] and Tyr[1] and Phe[4] in DYN [18, 19], are important residues required for receptor binding and/or biological activity and are needed for discriminating between them. In particular, the presence of N-terminal Phe[1] is indispensable for NOC activity, and the C-terminal half of NOC may serve as a domain that prevents binding to opioid receptors [20].

In the field of opioid peptides, a number of synthetic analogs have been prepared based on structure-activity studies focusing on the aromatic amino acids Tyr[1] and either Phe[3] or Phe[4], which are important structural elements that interact with opioid receptors [21]. Among these, the most active analogs were those that substituted 2′,6′-dimethyltyrosine (Dmt) for Tyr[1], which vastly improved opioid receptor binding affinity [22–36]. Structure-activity relation studies of opioid peptides using Dmt revealed that introduction of Dmt[1] could improve receptor affinity and opioid potency. In addition, combination of Dmt with 1,2,3,4-tetraisoquinoline-3-carboxylic acid (Tic), Dmt-Tic pharmacophore, also produced potent δ-selective antagonists, including Dmt-Tic-OH [31, 32], N,N′-diMeDmt-Tic-OH [33], DIPP-NH_2 [34], and DIPP [Ψ] [35]. The 2,6-dimethylation of the aromatic moiety in Leu-enkephalin (ENK) imparted high enzymatic stability to the peptide [36]. These findings prompted a study to modify a Phe aromatic moiety at position 3 or 4 of opioid peptides through 2,6-dimethylation because no derivatives with phenyl ring-methylated Phe incorporated into opioid peptides have been reported, only other biologically active peptides have been prepared [37, 38]. The usefulness of incorporating the artificial aromatic amino acid, 2′,6′-dimethylphenylalanine (Dmp) (Figure 1) as an aromatic amino acid surrogate in opioid peptides to develop opioid ligands specific for opioid receptors was investigated. This paper includes all studies that involved substitution of the Dmp residue into opioid peptides reported in the last decade.

2. Preparation of 2′,6′-Dimethylphenylalanine (Dmp)

Initially, Dmp and D-Dmp were synthesized by the route illustrated in Scheme 1 [39]. Commercially available **1** reacted with sodium trimethyl stannane according to the method of Yamamoto et al. [40], followed by reaction with iodine according to the method of Ohno et al. [41]. A key intermediate, 2-iodo-*m*-xylene (**3**), reacted with methyl 2-acetamidoacrylate by Dygos' method [42], which led to **4**, followed by saponification to yield **5**. Catalytic hydrogenation of **5** yielded racemic Ac-Dmp (**6**). For optical resolution, **6** was converted to its dipeptide derivatives, Ac-DL-Dmp-Arg-OMe (**7**), which were easily separated into diastereoisomers by preparative HPLC. Acid hydrolysis and neutralization of each isomer yielded Dmp (**8**) and D-Dmp (**9**). The L and D configurations were determined using L-amino acid oxydase according to a method reported by Toth et al. [43].

In addition, Li et al. prepared Dmp using the asymmetric synthetic method of Dygos et al. [42], which was applied to the endomorphin analogs [44].

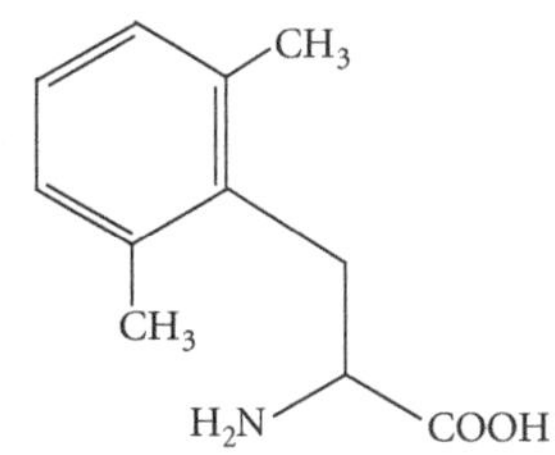

FIGURE 1: Structure of 2′,6′-dimethylphenylalanine (Dmp).

3. Dmp Replacement of Phenylalanine Residue at Position 3 or 4 in Opioid Peptides

The usefulness of Dmp was first investigated as a surrogate for Phe[4] in ENK analogs [39]. The receptor-binding affinities of synthetic analogs were determined using rat brain, as previously reported [45]. As shown in Table 1, replacement of Phe[4] in ENK by Dmp led to analog **10** with μ-receptor affinity comparable to ENK, but with approximately 12-fold reduction in δ-receptor affinity, resulting in a change of preferential receptor from δ to μ. The D-Dmp replacement of ENK (**11**) induced significant reduction in binding affinity for both receptors. Combined replacement of Dmp[4] and Dmt[1] produced **12** with markedly improved affinities for both receptors, 40- and 110-fold greater in affinity than **10** for the μ- and δ-receptors, respectively. However, **12** possessed 5-fold lower affinity than that of [Dmt[1]]ENK for both receptors, possibly due to slight changes in the active conformation by simultaneous dimethylation of two aromatic moieties. D-Dmp[4] replacement of **12** led to **13**, which retained μ affinity equivalent to that of ENK and modest δ affinity, demonstrating the effectiveness of Dmt[1] replacement for maintaining high receptor affinity. *In vitro* biological activity of ENK analogs was evaluated using isolated guinea pig ileum (GPI) and mouse vas deferens (MVD) tissue samples, as previously reported [46]. The GPI tissue contains predominantly μ-receptors, while MVD tissue contains δ-receptors [47]. As shown in Table 1, compound **10** possessed 8- and 30-fold lower activity compared to ENK in the GPI and MVD assays, respectively. In contrast, **11** was devoid of activity in both assays, as expected from the binding data. Analog **13** also lacked activity in both assays, even though this analog showed potent μ affinity and modest δ affinity. This analog turned out to be a potent μ-antagonist and a weak δ-antagonist. The pA_2 values of **13** were 6.90 against EM2 as a μ agonist in the GPI assay and 5.57 against DT as a δ agonist, in the MVD assay. The results of **13** are in line with observations that Dmt-D-Phe-NH_2 and its C-terminally extended analog are μ-receptor antagonists [48].

Effects of Dmp substitution for phenylalanine at position 3 or position 4 in EM2 were examined (Table 1) [49]. The EM2 possessed great affinity and selectivity for the μ-receptor. Interestingly, Dmp substitution for Phe[3] in EM2 (**14**) produced a compound with 10-fold greater affinity than that of EM2 for both the μ- and δ-receptors and still retained high μ-receptor selectivity comparable to that of EM2. The Dmp substitution of Phe in position 4 (**16**), however, resulted

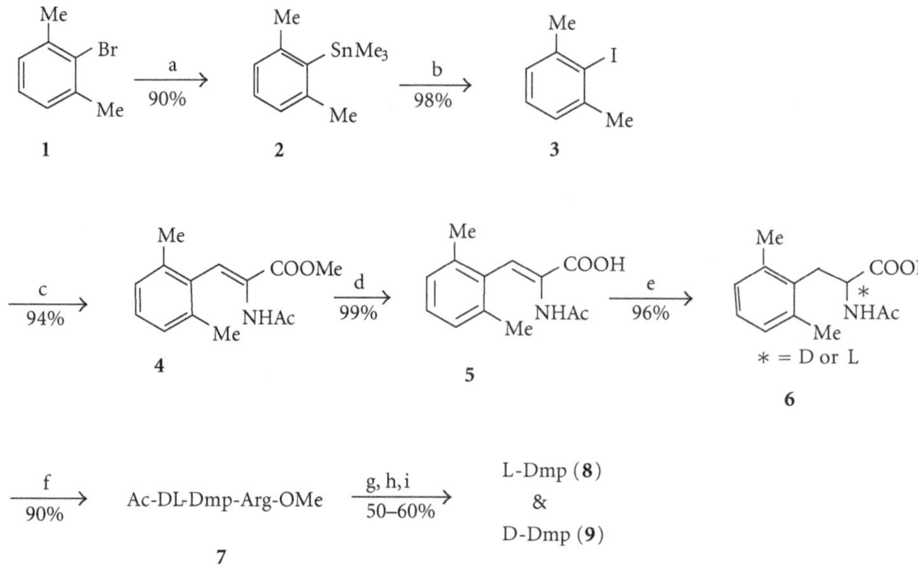

Scheme 1: Synthetic route to L-Dmp and D-Dmp, (a) Me_3SnNa, DME, ice-salt bath, 2 h; (b) I_2/THF, rt, 3 h; (c) methyl 2-acetamidoacrylate/$Pd(OAc)_2$/Et_3N/MeCN, $(2\text{-}MeC_6H_4)_3P$, reflux, 24 h; (d) 1 M NaOH/dioxane, rt, 2 h; (e) H_2 (4 kgf/cm^2)/10% Pd-C/AcOH, 70°C, 48 h; (f) HCl·Arg-OMe/Et_3N/DCC/HOBt/DMF, 0°C to rt, 5 h; (g) preparative HPLC; (h) concd HCl, reflux, 8 h; (i) pH 4–6/H_2O.

in a 23-fold decrease in μ affinity and a slight increase in δ affinity, resulting in a significant decrease in μ-receptor selectivity. The introduction of D-Dmp at either position 3 or 4 (**15** or **17**, resp.) resulted in a significant decrease in μ affinity and selectivity, which agreed with the results from D-Phe-replaced analogs [50]. Analog **15** retained moderate μ affinity with a Ki value of 2.4 nM, whereas **16** and **17** exhibited significantly decreased μ affinity. In the *in vitro* assay, **14** exhibited considerably greater GPI potency than EM2 as expected; however, this analog exhibited more potent MVD activity than that expected from δ-binding. This may be due to μ-receptors, which coexist in the MVD tissues, because the high MVD potency was strongly inhibited by the specific μ-receptor antagonist CTAP [51]. A similar trend was observed with other μ-receptor ligands [52, 53]. These results suggest that Dmp substitution of Phe^3 of EM2 promotes μ-receptor specificity and that Phe^3 is more amenable to Dmp or its D-isomer substitution compared to Phe^4.

Use of Dmp as a Phe surrogate in DM and DT heptapeptides was also examined [52]. Replacement of Phe by Dmp in the μ-specific ligand DM (**18**) induced a significant increase (170-fold) in μ affinity and only a modest increase in δ affinity, resulting in marked improvement of μ-receptor selectivity. The D-Dmp^3 replacement (**19**), however, resulted in marked decrease in both μ and δ affinities. Interestingly, the Dmp^3 replacement in δ-specific ligand DT produced **20** with a 22-fold increase in δ affinity and a 3-fold decrease in μ affinity, resulting in a 75-fold increase in δ-receptor selectivity with unprecedented δ-receptor selectivity (μ/δ = 1,045,714). The configurational inversion of Dmp in DT (**21**) was detrimental to δ-receptor selectivity. Results of the *in vitro* bioassay of these analogs showed that **18** exhibited a slight increase in GPI potency and a greater increase in MVD potency, while **19** showed marked decreases in both assays as expected from binding affinities. The discrepancy between the degree of increase (3-fold) in the GPI assay observed with **18** and the μ-binding data (170-fold) may be due to differences in μ-receptors in the brain and peripheral tissues. The Dmp^3-substituted DT analog **20** showed markedly increased MVD potency, resulting in a very high GPI/MVD ratio of 304,772. As expected, D-Dmp^3-substituted DT (**21**) possessed very low MVD potency. Analogs **18** and **20** are among the most potent and selective ligands for μ- and δ-opioid receptors, respectively, and therefore are candidates for investigations of opioid systems.

A dermorphin tetrapeptide analog, Tyr-D-Arg-Phe-βAla-NH_2 (YRFB), is a highly potent and selective ligand for the μ-opioid receptor [53]. The usefulness of Dmp replacement for Phe^3 in this tetrapeptide was examined [54]. Substituting Dmp for Phe^3 in YRFB (**22**) induced a 5-fold increase in μ-receptor affinity without significant change in δ-receptor affinity, as compared to the parent peptide. Results from the GPI assay using this compound coincided well with the binding data, but a slight increase in activity in the MVD assay was found. As shown in Table 2, low K_e values for the μ-antagonist CTAP and high K_e values for the δ antagonist N, N$(Me)_2$Dmt-Tic-OH [33], against **22** suggest that the GPI activity of these analogs occurred mainly via the μ-opioid receptor. Analog **22** was also tested for analgesic activity in the formalin test in mice and was compared to results for YRFB and morphine. As shown in Table 3, subcutaneous injection of this analog produced dose-dependent antinociceptive activity in mice in both the first and second phases. Its analgesic activity was approximately 40- and 70-fold more potent than that of morphine in the first and second phases, respectively. These results indicate that Dmp is effective as a Phe surrogate for improving functional activity and maintaining μ-selectivity

[52]. In contrast, [D-Dmp[3]] YRFB (**23**) exhibited μ-receptor affinity similar to the parent peptide, but it exhibited an order of magnitude lower GPI potency.

Next, the effect of Dmp replacement of N-terminal aromatic residues in DYN and NOC was compared [55]. Six analogs (**26–31**) containing Dmp in position 1 and/or 4 of DYN (1-13)-NH_2 and NOC (1-13)-NH_2 were synthesized and tested for their binding affinity to opioid receptors derived from rat (μ- and δ-receptors) or guinea pig (κ-receptor) brains and to membrane preparations derived from HEK293 cells expressing human ORL-1 receptor. Results are shown in Table 4. In a series of DYN(1-13)-NH_2 analogs, the parent peptide DYN(1-13)-NH_2 showed high affinity toward κ-, μ-, and δ-opioid receptors with κ-receptor selectivity, that is, an IC_{50} ratio of 1/15.6/40.1 and significantly low affinity toward the ORL1 receptor, similar to an observation of intact DYN [11, 20]. Dmp[4] replacement afforded **26**, which had greater κ-opioid receptor affinity than that of the parent peptide and significantly improved κ-receptor selectivity (IC_{50} ratios: $1(\kappa)/509(\mu)/21159(\delta)$ versus DYN(1-13)-NH_2, $1(\kappa)/15.6(\mu)/40.1(\delta)$). Compound **26** exhibited an order of magnitude decrease in affinity, indicating that the Dmp[4] modification in DYN peptides is detrimental to ORL1-receptor affinity, as was observed with NOC peptides. NOC (1-13)-NH_2 possessed high ORL1 receptor affinity and poor affinity for κ-, μ-, and δ-opioid receptors. The Dmp[4]-NOC analog (**29**) showed a 70-fold decrease in ORL1 affinity without significant changes in affinity toward the opioid receptors. These results indicate the critical importance of the Phe[4] residue for interactions with the ORL1 receptor. A Dmp residue at this position appears to influence the conformation of the NOC peptide by 2′,6′-dimethylation of the Dmp side chain aromatic moiety. This occurs because, according to the proposed model of the ORL1 receptor and its complex with NOC, the Phe[4] residue of NOC located at a hydrophobic pocket in a cavity formed by TM helices 3, 5, 6, and 7 and the Phe[4] side chain interact with Phe[220] of the ORL1 receptor through an edge-face interaction [56]. Two methyl groups on Dmp[4] may interfere with the receptor interaction due to a reduction in conformational flexibility and/or enhanced lipophilicity. To further examine the usefulness of Dmp-containing DYN peptides as κ-opioid receptor ligands, the *in vitro* bioactivity of DYN peptides was determined using the GPI assay (Table 5). Contrary to the high κ-opioid receptor-binding profile, **26** exhibited unexpectedly low GPI potency, which was one order of magnitude lower than the parent peptides. Low K_e values for the κ-receptor antagonist and high K_e values for the μ- and δ-antagonists of **26** suggest that the GPI activity of these analogs occurred mainly via the κ-opioid receptor.

4. Dmp Replacement of N-Terminal Tyr Residue in Opioid Peptides

The usefulness of Dmp[1] substitution for Tyr[1] in the δ-opioid receptor-selective ligands, ENK and DT, and the μ-opioid receptor-selective ligands, EM2 and YRFB, has been investigated [49, 53, 57]. Results of receptor-binding and *in vitro* assays are shown in Table 6. The replacement of Tyr[1] by Dmp in ENK led to **32**, which possessed similar receptor affinity and selectivity as ENK, whereas Dmt[1] replacement produced marked increases in both μ and δ affinities but did not increase receptor selectivity. Replacement of Phe[1] in ENK (**33**) decreased the affinity by 50- and 70-fold at the δ- and μ-receptors, respectively. Replacement of Tyr[1] by Dmp in DT (**34**) markedly decreased the binding affinity and selectivity toward the δ-receptor. Introduction of the inverse configuration at this position (**35**) markedly reduced δ affinity and selectivity. In contrast, [Dmt[1]]DT (**36**) possessed a 50-fold increase in δ affinity and a 1200-fold increase in μ affinity, resulting in substantial reduction in δ-receptor selectivity, which agrees with previous results [28]. [Phe[1]]DT (**37**) retained significant δ affinity and good δ selectivity, similar to the results for [Phe[1]] deltorphin I [58]. Analog **32** possessed lower MVD and GPI potency compared to ENK but showed significant MVD potency and GPI/MVD selectivity in agreement with the binding data. Analog **34** exhibited 2-fold greater potency for MVD and a greater GPI/MVD ratio compared to DT (selectivity ratio: 14,835 versus 9342) or [Dmt[1]]DT (**36**) (selectivity ratio: 14,835 versus 1700). The [Phe[1]]DT (**37**) exhibited a 20-fold decrease in MVD potency compared to DT but retained significant potency and MVD specificity. These results demonstrate that the Dmp[1] peptide is superior to the corresponding Dmt[1] peptide in receptor selectivity because the latter generally possesses poor receptor selectivity.

In EM2 analogs, replacement of Tyr[1] by Dmp led to **38**, which showed a 4-fold reduction, retaining significant μ affinity. Note that **38** retained high potency for μ-receptors despite the lack of a phenolic hydroxyl group at the N-terminal, which agreed with results for YRFB analogs [53]. The D-Dmp[1]- or Phe[1]-substituted analogs of EM2 (**39** and **40**, resp.) showed a significant decrease in GPI potency, which was expected from the binding affinity. In contrast, **39** and **40** were more than 100-fold less potent than EM2. Results from analog **38** supported the observation that a Dmp residue can mimic the N-terminal Tyr of opioid peptides [53].

Replacing Tyr with Dmp at position 1 in YRFB (**24**) produced greater μ affinity and considerably lower δ affinity compared to YRFB and improved μ-receptor selectivity by 15-fold. This compound, however, showed slightly lower GPI and MVD potency compared to YRFB. The D-Dmp substitution for Tyr[1] (**41**) markedly reduced affinities for both receptors and for *in vitro* biological activity, suggesting that the L-configuration at this position is crucial for receptor interactions. The dual substitutions of Dmp for the aromatic amino acids at positions 1 and 3 produced **25** with binding affinity and selectivity for the μ-receptor that were slightly improved relative to those of **22** or **24**. This analog also showed slightly greater GPI potency than YRFB. As shown in Table 2, the low K_e value found for the μ-receptor selective antagonist CTAP in the GPI assay demonstrated inhibition of the high activity of Dmp[1]-containing analogs (**24** and **25**) and suggests that the activity is mediated via μ-opioid receptors. The low K_e values for CTAP in the MVD assay

Table 1: Opioid receptor-binding affinity and biological activities of opioid peptide analogs containing Dmp at position 3/4.

Peptide	Receptor binding affinity, $Ki \pm$ S. E. (nM)		δ/μ	μ/δ	GPI (μ)	MVD (δ)	MVD/GPI	GPI/MVD
	μ^a	δ^b			IC_{50} (nM)	IC_{50} (nM)		
(1) Leu-enkephalin								
Tyr-Gly-Gly-Phe-Leu (ENK)	2.42 ± 0.93	1.43 ± 0.71	—	1.69	103 ± 30	22.2 ± 4.3	—	4.64
[Dmp^4]ENK **(10)**	1.25 ± 0.29	17.7 ± 4.2	—	0.07	808 ± 101	624 ± 103	—	1.29
[D-Dmp^4]ENK **(11)**	2505 ± 169	8924 ± 4098	—	0.28	>10,000	>10,000	—	—
[Dmt^1]ENK[c]	0.0068 ± 0.003	0.031 ± 0.011	—	0.22	0.55 ± 0.17	0.17 ± 0.02	—	3.24
[Dmt^1, Dmp^4]ENK **(12)**	0.030 ± 0.011	0.1 58 ± 0.034	—	0.19	2.00 ± 0.51	1.45 ± 0.26	—	1.38
[Dmt^1, D-Dmp^4]ENK **(13)**	5.61 ± 0.59	40.9 ± 11.5	—	0.14	>10,000[d]	>10,000[e]		
(2) Endomorphin-2								
Tyr-Pro-Phe-Phe-NH_2 (EM2)	0.557 ± 0.306	14,070 ± 3346	25260	—	10.5 ± 1.2	317 ± 65	30.2	—
[Dmp^3]EM2 **(14)**	0.0304 ± 0.0208	1063 ± 336	34967	—	0.378 ± 0.104	1.39 ± 0.17	3.68	—
[D-Dmp^3]EM2 **(15)**	2.4 ± 0.56	4169 ± 954	1737	—	30.4 ± 2.8	187 ± 30	6.15	—
[Dmp^4]EM2 **(16)**	13.2 ± 1.9	7624 ± 2571	578	—	196 ± 40	320 ± 55	1.63	—
[D-Dmp^4]EM2 **(17)**	106 ± 20	1765 ± 834	17	—	587 ± 119	2267 ± 603	3.86	—
(3) Demrophin/deltorphin								
Dermorphin (DM)	0.092 ± 0.024	192 ± 51	2087	—	3.74 ± 0.57	34.4 ± 4.8	9.2	—
[Dmp^3]DM **(18)**	0.00054 ± 0.00021	45.7 ± 11.8	84630	—	1.21 ± 0.23	4.62 ± 0.82	3.8	—
[D-Dmp^3]DM **(19)**	4.43 ± 1.85	3300 ± 702	745	—	44.4 ± 6.1	358 ± 45	8.1	—
Deltorphin 11 (DT)	314 ± 53	0.0226 ± 0.0077	—	13894	5437 ± 812	0.582 ± 0.029	—	9342
[Dmp^3]DT **(20)**	1098 ± 111	0.00105 ± 0.00043	—	1045714	6705 ± 992	0.022 ± 0.003	—	304772
[D-Dmp^3]DT **(21)**	1956 ± 177	111 ± 17	—	18	8214 ± 872	145 ± 15	—	56
(4) Dermorphin-relative short peptide								
Tyr-D-Arg-Phe-βAla-NH_2 (YRFB)	0.172 ± 0.025	482 ± 121	2802	—	5.31 ± 0.72	116 ± 18	21.8	—
[Dmp^3]YRFB **(22)**	0.0350 ± 0.0167	544 ± 143	15543	—	1.67 ± 0.24	27.9 ± 5.0	16.7	—
[D-Dmp^3]YRFB **(23)**	0.0618 ± 0.0109	>2823	>45679	—	19.8 ± 1.9	305 ± 53	15.4	—

[a]Versus [^{3}H]DAMGO. [b]Versus [^{3}H]DT. [c]Data cited from [39]. [d]Antagonism was shown with pA2 = 6.90 against EM2. [e]Antagonism was shown with pA2 = 5.57 against DT.

TABLE 2: K_e values of opioid receptor antagonists against Dmp-containing YRFB analogs in the GPI and MVD assays.

Peptides		K_e (nM)		
		GPI	MVD	
		CTAP	CTAP	N,N(Me)$_2$Dmt-Tic-OH
[Dmp^3]YRFB	(22)	22.7	11.4	>1000
[Dmp^1]YRFB	(24)	25.1	13.4	>1000
[Dmp^{1,3}]YRFB	(25)	21.3	10.2	>1000
YRFB		26.8	21.2	>1000
[Dmt^1]YRFB[b]		85.5	7.89	192
DT		NT[a]	>1000	0.64

[a]Not tested.

TABLE 3: Antinociceptive activities of Dmp-containing YRFB analogs after subcutaneous injection in the formalin test.

Peptides		ED$_{50}$ (95% C. L.)[a], nmol/kg			
		First phase		Second phase	
[Dmp^3]YRFB	(22)	98.6	(26.7–364)	113	(48.6–264)
[Dmp^1]YRFB	(24)	1946	(1026–3691)	1529	(1199–1950)
YRFB		628	(364–1280)	514	(378–700)
Morphine		3811	(2921–4973)	7319	(4058–13198)

[a]ED$_{50}$ values and 95% confidence limits.

indicate its inhibition of the analogs MVD activity, but the δ-receptor selective antagonist *N,N*(Me)$_2$Dmt-Tic-OH did not inhibit this activity. This result may be due to μ-receptor cooccurring with the δ-receptor in MVD tissue. Analogs **24** and **25** retained high μ-receptor affinity and potent GPI activity despite the lack of a phenolic hydroxyl group in the side chain of the N-terminal residue, which is considered crucial for binding and activating opioid receptors. However, some cyclic somatostatin- or DPDPE-based analogs possess high affinity for and/or potency toward the μ-receptor despite the absence of this group at the N-terminal residue [59–63]. Analogs **24** and **25** are examples of linear peptides lacking an N-terminal phenolic hydroxyl group but possess high opioid activity. The present results support reports of the interactions of cyclic compounds that indicate the Tyr hydroxyl moiety at the N-terminal residue of opioid peptides is not an absolute requirement for interaction with opioid receptors and signal transduction. Because replacing the Tyr^1 residue with Phe (**42**) drastically reduced μ-receptor affinity and GPI potency, effects of Dmp substitution on receptor interactions are attributable mainly to enhanced hydrophobicity and/or increased conformational stability of the side chain of the aromatic ring. The basic functional group of the D-Arg residue at position 2 may also be responsible for the potent receptor interaction because the affinity of **41** and **42** was very low, but significant affinity for the μ-receptor was retained. In addition, the Dmt^1-substituted YRFB exhibited great affinity for both the μ- and δ-receptors, which often resulted in low receptor selectivity. Such trends have also been observed with other Dmt^1-substituted opioid peptides [39, 64, 65]. In contrast, substitution of Dmp^1 for Tyr^1 improved μ-receptor selectivity exclusively, a result distinct from the effects of Dmt^{1}substitution. In the formalin test in mice, analog **24** also exhibited approximately 3-fold (first phase) and 5-fold (second phase) greater potency than that of morphine, but the potencies were approximately 3-fold less than those of YRFB. The analgesic potencies of these analogs correlated well with their GPI potencies.

In a series of DYN(1-13)-NH$_2$ analogs, Dmp^1 replacement afforded **27** with greater κ-opioid receptor affinity than that of the parent peptide; Dmp^1 replacement also significantly improved κ-receptor selectivity (IC$_{50}$ ratios: **27**, 1(κ)/293(μ)/180(δ) versus DYN(1-13)-NH$_2$, 1(κ)/15.6(μ)/40.1(δ)). These results support our recent finding that Dmp is an effective surrogate for the Tyr^1 residue in opioid peptides [49, 53, 57]. Analog **27**, however, exhibited low GPI potency two orders of magnitude less than DYN(1-13)-NH$_2$. Low K_e values for the κ-receptor antagonist nor-BNI suggests that its GPI activity occurred mainly via the κ-opioid receptor, similar to the observations for **26**. The discrepancy between κ-opioid receptor binding and GPI potency observed with **27** can be attributed to the lack of hydroxyl side chains on the N-terminal residue because the [Dmt^1]DYN peptide was as active as the parent peptide in a GPI assay [66]. Similar results have been reported for the Phe^1-DYN (1–11) peptide [67]. Unexpectedly, however, **27** possessed 3-fold greater affinity toward the ORL1 receptor, whereas **26** exhibited an order of magnitude decrease in ORL1 receptor affinity, indicating that Dmp^4 modification in DYN peptides is detrimental to ORL1-receptor affinity. Simultaneous Dmp replacements in positions 1 and 4 (**28**) resulted in a two order of magnitude decrease in κ-receptor affinity and dramatically reduced GPI potency with loss of receptor selectivity. These results indicate that the N-terminal phenolic hydroxyl group of the DYN peptide is not mandatory for κ-receptor binding but is critically important for receptor activation.

As shown in Table 4, Dmp^1-NOC peptide (**30**) possessed high ORL1 receptor affinity comparable to the parent peptide NOC (1-13)-NH$_2$. Interestingly, this analog exhibited improved affinity toward the three opioid receptors, with 5- and 16-fold improved affinity for the κ- and μ-receptors, respectively, perhaps due to the effect of Dmp^1, which can mimic Tyr^1 in some opioid peptides without a substantial decrease in receptor affinity [53, 58]. Dmp substitutions in positions 1 and 4 afforded **31** with a moderate decrease in affinity toward the ORL1 and opioid receptors, indicating that a Dmp residue in position 1 can compensate for the decrease caused by the Dmp^4 substitution.

Substitution of the peptide with an artificial amino acid often improves metabolic stability, which is useful when conducting *in vivo* and *in vitro* studies. As shown in Table 7, Dmp^1-substituted analogs **27** and **30** showed greater stability toward aminopeptidase M (AP-M) and rat brain synaptosomal enzymes compared to the parent peptides, which suggests the involvement of aminopeptidase(s) in the brain that breaks down these analogs as observed with intact NOC [68, 69] and DYN [70, 71]. However, the stability of the Dmp^4-substituted analogs (**26** and **29**) was similar to or somewhat less than that of the parent peptide toward rat brain enzymes. These results imply that a Dmp residue in

Table 4: Receptor-binding affinity of DYN analogs and NOC analogs containing Dmp for opioid receptors and ORLl receptor.

Peptides		$IC_{50}\pm$ SEM (nM)				
		ORL1 receptor	Opioid receptor			
		$[^3H]NOC^a$	$[^3H]$U-69593 $(\kappa)^b$	$[^3H]$DAMGO $(\mu)^c$	$[^3H]$DT $(\delta)^c$	$\kappa/\mu/\delta$
DYN(I-13)-NH_2		18.8 ± 3.01	0.162 ± 0.049	2.53 ± 0.38	6.49 ± 1.11	1/15.6/40.1
[Dmp^4]DYN (1-13)- NH_2	**(26)**	188 ± 18.2	0.044 ± 0.035	22.4 ± 10.2	931 ± 723	1/509/21159
[Dmp^1]DYN(1-13)-NH_2	**(27)**	6.60 ± 0.952	0056 ± 0026	16.4 ± 2.35	10.1 ± 6.02	1/293/180
[$Dmp^{1,4}$]DYN (1-13)- NH_2	**(28)**	51.5 ± 1.62	5.45 ± 1.65	251 ± 56.3	415 ± 185	1/46/76.1
NOC		0.151 ± 0.058	643 ± 218	1540 ± 601	>10000	—
NOC(1-13)-NH_2		0.743 ± 0.125	193 ± 54	319 ± 88	>10000	—
[Dmp^4]NOC(1-13)- NH_2	**(29)**	51.6 ± 12.9	299 ± 63	629 ± 433	>10000	—
[Dmp^1]NOC(1-13)-NH_2	**(30)**	0.814 ± 0.090	38.8 ± 16.7	25.0 ± 6.5	292 ± 61	—
[$Dmp^{1,4}$]NOC(1-13)- NH_2	**(31)**	21.3 ± 3.2	100 ± 29	56.8 ± 12.3	3407 ± 990	—

[a] Using cell membrane expressing human ORL1 receptor in Hek-293 cells. [b] Using guinea pig brain homogenate.
[c] Using rat brain homogenate.

Table 5: GPI assay and opioid receptor preference of DYN analogs.

Peptides		$IC_{50}\pm$ SEM (nM)	K_e (nM) value of receptor selective antagonist		
			nor-BNI (κ)	CTAP (μ)	N,N$(Me)_2$Dmt-Tic-OH (δ)
DYN(l-13)-NH_2		3.14 ± 1.13	1.1	99	98
[Dmp^4]DYN(I-13)-NH_2	**(26)**	32.2 ± 9.16	0.63	108	198
[Dmp^1]DYN(l-13)-NH_2	**(27)**	306 ± 68	10	115	> 1000
[$Dmp^{1,4}$]DYN(l-13)-NH_2	**(28)**	1341 ± 303	809	595	> 100

position 4 offers no additional metabolic stability for either peptide and that endopeptidases play a major role in brain metabolism. A doubly Dmp-replaced NOC analog **31** also possessed no additional stability, whereas its counterpart DYN analog **28** possessed improved metabolic stability. Comparison of the metabolism of Dmp-containing NOC and DYN analogs suggested that the NOC peptides generally are more susceptible to aminopeptidases and endopeptidases although other results in human blood have been reported [71].

5. Conclusions

The usefulness of the artificial amino acid residue Dmp was investigated as an aromatic amino acid surrogate for opioid peptides and related peptides, including ENK, DM, YRFB, DT, EM2, DYN, and NOC peptides. In most opioid peptides, substitution of Phe^3 by Dmp produced analogs with improved receptor-binding affinity and selectivity, for example, [Dmp^3]EM2 (**14**), [Dmp^3]DM (**18**), [Dmp^3]DT (**20**), and [Dmp^3]YRFB (**22**), while substitution by the D-enantiomer resulted in decreased receptor affinity and selectivity. A small analog [D-Dmp^3]YRFB (**23**) was the only exception because it possessed high μ affinity similar to the parent peptide. However, Dmp-substitution in position 4 produced analogs with markedly reduced receptor affinity and selectivity, for example, [Dmp^4]ENK (**10**), [Dmp^4]EM2 (**16**), and [Dmp^4]NOC (1-13)-NH_2 (**29**), while their D-Dmp^4-analogs were almost devoid of receptor affinity and opioid activity. [Dmp^4]DYN(1-13)-NH_2 (**26**) exceptionally possessed significantly improved receptor affinity for the κ-opioid receptor and outstanding κ-receptor selectivity. Interestingly, replacement of $Tyr^1$by Dmp residue produced analogs with equipotent or only slightly reduced receptor affinity and *in vitro* bioactivity, for example, [Dmp^1]EM2 (**38**), [Dmp^1]DT (**34**), [Dmp^1]YRFB (**24**), and [Dmp^1]DYN(1-13)-NH_2 (**27**). Thus, Dmp is also a useful surrogate for the N-terminal Tyr residue in opioid peptides despite the lack of a phenolic hydroxyl group, which has been considered to be indispensable for opioid activity. The Dmp^1-substituted analogs are superior to Dmt^1-substituted analogs in opioid receptor selectivity because the Dmt^1 analogs generally possess outstandingly high affinity to opioid receptors but poor receptor selectivity. These results demonstrate that Dmp is very useful as an aromatic amino acid surrogate in opioid peptides and may be applicable to other biologically active peptides for the development of novel peptide mimetics with high receptor specificity.

Abbreviations

Dmp: 2′,6′-Dimethylphenylalanine
Dmt: 2′,6′-Dimethyltyrosine
DAMGO: [D-Ala^2, $MePhe^4$, Gly-ol^5]enkephalin

Table 6: Opioid receptor-binding affinity and biological activities of opioid peptide analogs containing Dmp at position 1.

Peptide	Receptor binding affinity, $Ki \pm$ S. E. (nM)				GPI (μ)	MVD (δ)		
	μ[a]	δ[b]	δ/μ	μ/δ	IC_{50} (nM)	IC_{50} (nM)	MVD/GPI	GPI/MVD
Tyr-Gly-Gly-Phe-Leu (ENK)	2.42 ± 0.93	1.43 ± 0.71	—	1.69	103 ± 30	22.2 ± 4.3	—	4.64
[Dmp^{1}]ENK **(32)**	5.94 ± 1.45	1.86 ± 0.61	—	3.19	710 ± 69	66.6 ± 12.7	—	10.66
[Dmt^{1}]ENK[c]	0.0068 ± 0.003	0.031 ± 0.011	—	0.22	0.55 ± 0.17	0.17 ± 0.02	—	3.24
[Phe^{1}]ENK **(33)**	169 ± 12	73.7 ± 19.5	—	2.29	> 10000	> 10000	—	—
Deltorphin II (DT)	314 ± 53	0.0226 ± 0.0077	—	13894	5437 ± 812	0.582 ± 0.029	—	9342
[Dmp^{1}]DT **(34)**	156 ± 33	0.329 ± 0.077	—	475	4038 ± 1118	0.272 ± 0.054	—	14835
[D-Dmp^{1}]DT **(35)**	> 2178	1394 ± 495	—	—	> 10000	662 ± 147	—	—
[Dmt^{1}]DT **(36)**	0.261± 0.060	0.012 ± 0.002	—	21.8	88.4 ± 22.7	0.052 ± 0.007	—	1700
[Phe^{1}]DT **(37)**	> 2178	2.68 ± 1.22	—	—	> 10000	10.4 ± 1.5	—	—
Tyr-Pro-Phe-Phe-NH_2 (EM2)	0.557 ± 0.306	14,070 ± 3346	25260	—	10.5 ± 1.2	317 ± 65	30.2	—
[Dmp^{1}]EM2 **(38)**	2.48 ± 1.46	6762 ± 590	2727	—	76.9 ± 20.7	661± 316	8.61	—
[D-Dmp^{1}]EM2 **(39)**	40.4 ± 2.6	9714 ± 3820	241	—	1392 ± 221	2329 ± 943	167	—
[Dmt^{1}]EM2[d]	0.15 ± 0.04	—	—	—	0.07± 0.02	1.87 ± 0.61	26.7	—
[Phe^{1}]EM2 **(40)**	54.1± 23.4	18,851 ± 10,487	348	—	1073 ± 309	5,199 ± 2,584	4.85	—
Tyr-D-Arg-Phe-βAla-NH_2 (YRFB)	0.172 ± 0.025	482 ± 121	2802	—	531± 0.72	116 ± 18	21.8	—
[Dmp^{1}]YRFB **(24)**	0.0623 ± 0.0140	2572 ± 947	41284	—	9.88 ± 1.04	188 ± 52	19	—
[D-Dmp^{1}]YRFB **(41)**	7.62 ± 1.75	> 2823	> 370	—	320 ± 30	1474 ± 283	4.6	—
[Dmp^{1,3}]YRFB **(25)**	0.0216 ± 0.0062	1688 ± 458	78148	—	2.76 ± 0.56	501 ± 86	18.2	—
[Dmt^{1}]YRFB[c]	0.00205 ± 0.00069	1.13 ± 0.13	551	—	0.034 ± 0.065	0.398 ± 0.085	11.7	—
[Phe^{1}]YRFB **(42)**	7.17 ± 1.03	> 2823	> 393	—	633 ± 89	7143 ± 950	11.2	—

[a] Versus [^{3}H]DAMGO. [b] Versus [^{3}H]DT. [c] Data cited from [39]. [d] Data cited from [29].

TABLE 7: Comparison of stability of Dmp-containing peptides toward enzymatic degradation.

Peptides		Half-life time (min)[a]	
		Aminopeptidase M	Rat brain homogenate
DYN(I-13)-NH_2		15.5	435
[Dmp^4]DYN(l-13)-NH_2	**(26)**	NT[b]	315
[Dmp^1]DYN(1-13)-NH_2	**(27)**	> 30	577
[$Dmp^{1,4}$]DYN(1-13)-NH_2 **(28)**		NT[b]	770
NOC(1-13)-NH_2		12	41.5
[Dmp^4]NOC(1-13)-NH_2	**(29)**	NT[b]	33.6
[Dmp^1]NOC(1-13)-NH_2	**(30)**	28	60.3
[$Dmp^{1,4}$]NOC(l-13)-NH_2 **(31)**		NT[b]	27.1
Met-enkephalin		< 5	8.5

[a]Determined by HPLC. [b]Not tested.

ENK: Leu-enkephalin, Tyr-Gly-Gly-Phe-Leu
DM: Dermorphin Tyr-D-Ala-Phe-Gly-Tyr-Pro-Ser-NH_2
DT: [D-Ala^2]deltorphin II, Tyr-D-Ala-Phe-Glu-Val-Val-Gly-NH_2
YRFB: Tyr-D-Arg-Phe-βAla-NH_2
EM2: Endomorphin 2, Tyr-Pro-Phe-Phe-NH_2
CTAP: D-Phe-c[Cys-Tyr-D-Trp-Arg-Thr-Pen]-Thr-NH_2
NOC: Nociceptin/orphanin FQ
DYN: Dynorphin A
ORL1: Opioid receptor-like 1
AP-M: Aminopeptidase M
GPI: Guinea pig ileum
MVD: Mouse vas deferens.

References

[1] T. Reisine, "Neurotransmitter receptors V. Opiate receptors," *Neuropharmacology*, vol. 34, no. 5, pp. 463–472, 1995.

[2] B. N. Dhawan, F. Cesselin, R. Raghubir et al., "International Union of Pharmacology. XII. Classification of opioid receptors," *Pharmacological Reviews*, vol. 48, no. 4, pp. 567–592, 1996.

[3] P. A. Zaki, E. J. Bilsky, T. W. Vanderah, J. Lai, C. J. Evans, and F. Porreca, "Opioid receptor types and subtypes: the δ receptor as a model," *Annual Review of Pharmacology and Toxicology*, vol. 36, pp. 379–401, 1996.

[4] R. J. Bodnar, "Endogenous opiates and behavior: 2010," *Peptides*, vol. 32, no. 12, pp. 2522–2552, 2011.

[5] J. Hughes, T. W. Smith, and H. W. Kosterlitz, "Identification of two related pentapeptides from the brain with potent opiate agonist activity," *Nature*, vol. 258, no. 5536, pp. 577–579, 1975.

[6] J. E. Zadina, L. Hackler, L. J. Ge, and A. J. Kastin, "A potent and selective endogenous agonist for the μ-opiate receptor," *Nature*, vol. 386, no. 6624, pp. 499–502, 1997.

[7] A. Goldstein, *The Peptides: Analysis, Synthesis, Biology*, S. Udenfriend and J. Meienhofer, Eds., Academic Press, New York, NY, USA, 1984.

[8] P. C. Montecucchi, R. De Castiglione, S. Piani, L. Gozzini, and V. Erspamer, "Amino acid composition and sequence of dermorphin, a novel opiate-like peptide from the skin of Phyllomedusa sauvagei," *International Journal of Peptide and Protein Research*, vol. 17, no. 3, pp. 275–283, 1981.

[9] V. Erspamer, P. Melchiorri, G. Falconieri-Erspamer et al., "Deltrophins: a family of naturally occurring peptides with high affinity and selectivity for δ opioid binding sites," *Proceedings of the National Academy of Sciences of the United States of America*, vol. 86, no. 13, pp. 5188–5192, 1989.

[10] J. R. Bunzow, C. Saez, M. Mortrud et al., "Molecular cloning and tissue distribution of a putative member of the rat opioid receptor gene family that is not a μ, δ or κ opioid receptor type," *FEBS Letters*, vol. 347, no. 2-3, pp. 284–288, 1994.

[11] J. C. Meunier, C. Mollereau, L. Toll et al., "Isolation and structure of the endogenous agonist of opioid receptor-like ORL1 receptor," *Nature*, vol. 377, no. 6549, pp. 532–535, 1995.

[12] R. K. Reinscheid, H. P. Nothacker, A. Bourson et al., "Orphanin FQ: a neuropeptide that activates an opioidlike G protein-coupled receptor," *Science*, vol. 270, no. 5237, pp. 792–794, 1995.

[13] J.-C. Meunier, "Nociceptin/orphanin FQ and the opioid receptor-like ORL1 receptor," *European Journal of Pharmacology*, vol. 340, no. 1, pp. 1–15, 1997.

[14] R. K. Reinscheid, A. Ardati, F. J. Monsma, and O. Civelli, "Structure-activity relationship studies on the novel neuropeptide orphanin FQ," *Journal of Biological Chemistry*, vol. 271, no. 24, pp. 14163–14168, 1996.

[15] E. Okuda-Ashitaka, S. Tachibana, T. Houtani et al., "Identification and characterization of an endogenous ligand for opioid receptor homologue ROR-C: its involvement in allodynic response to innocuous stimulus," *Molecular Brain Research*, vol. 43, no. 1-2, pp. 96–104, 1996.

[16] C. T. Dooley and R. A. Houghten, "Orphanin FQ: receptor binding and analog structure activity relationships in rat brain," *Life Sciences*, vol. 59, no. 1, pp. PL23–PL29, 1996.

[17] R. Guerrini, G. CalO', A. Rizzi et al., "Address and message sequences for the nociceptin receptor: a structure- activity study of nociceptin-(1-13)-peptide amide," *Journal of Medicinal Chemistry*, vol. 40, no. 12, pp. 1789–1793, 1997.

[18] V. J. Hruby and R. S. Agnes, "Conformation-activity relationships of opioid peptides with selective activities at opioid receptors," *Biopolymers*, vol. 51, no. 6, pp. 391–410, 1999.

[19] Y. Shimohigashi, R. Hatano, T. Fujita et al., "Sensitivity of opioid receptor-like receptor ORL1 for chemical modification on nociceptin, a naturally occurring nociceptive peptide," *Journal of Biological Chemistry*, vol. 271, no. 39, pp. 23642–23645, 1996.

[20] R. K. Reinscheid, J. Higelin, R. A. Henningsen, F. J. Monsma, and O. Civelli, "Structures that delineate orphanin FQ and dynorphin A pharmacological selectivities," *Journal of Biological Chemistry*, vol. 273, no. 3, pp. 1490–1495, 1998.

[21] V. J. Hruby and C. A. Gehrig, "Recent developments in the design of receptor specific opioid peptides," *Medicinal Research Reviews*, vol. 9, no. 3, pp. 343–401, 1989.

[22] D. W. Hansen Jr, R. H. Mazur, and M. Clare, "The synthesis and SAR of orally active enkephalin analogs with modified N-terminal tyrosine residue," in *Peptides: Structure and Function*, C. M. Deber, V. J. Hruby, and K. D. Kopple, Eds., pp. 491–494, Pierce Chemical, Rockford, Ill, USA, 1985.

[23] N. S. Chandrakumar, P. K. Yonan, A. Stapelfeld et al., "Preparation and opioid activity of analogues of the analgesic

dipeptide 2,6-dimethyl-L-tyrosyl-N-(3-phenylpropyl)-D-alaninamide," *Journal of Medicinal Chemistry*, vol. 35, no. 2, pp. 223–233, 1992.

[24] D. W. Hansen Jr., A. Stapelfeld, M. A. Savage et al., "Systemic analgesic activity and δ-opioid selectivity in [2,6-dimethyl-Tyr[1], D-Pen[2], D-Pen[5]]enkephalin," *Journal of Medicinal Chemistry*, vol. 35, no. 4, pp. 684–687, 1992.

[25] B. S. Pitzele, R. W. Hamilton, K. D. Kudla et al., "Enkephalin analogs as systemically active antinociceptive agents: O- and N-alkylated derivatives of the dipeptide amide L-2,6-dimethyltyrosyl-N-(3-phenylpropyl)-D-alaninamide," *Journal of Medicinal Chemistry*, vol. 37, no. 7, pp. 888–896, 1994.

[26] P. W. Schiller, G. Weltrowska, R. Schmidt et al., "Four different types of opioid peptides with mixed mu agonist/delta antagonist properties," *Analgesia*, vol. 1, pp. 703–706, 1995.

[27] S. Salvadori, M. Attila, G. Balboni et al., "Delta opioidmimetic antagonists: prototypes for designing a new generation of ultraselective opioid peptides.," *Molecular Medicine*, vol. 1, no. 6, pp. 678–689, 1995.

[28] R. Guerrini, A. Capasso, L. Sorrentino et al., "Opioid receptor selectivity alteration by single residue replacement: synthesis and activity profile of [Dmt[1]]deltorphin B," *European Journal of Pharmacology*, vol. 302, no. 1–3, pp. 37–42, 1996.

[29] Y. Okada, Y. Fujita, T. Motoyama et al., "Structural studies of [2′,6′-dimethyl-L-tyrosine[1]]endomorphin-2 analogues: enhanced activity and cis orientation of the Dmt-Pro amide bond," *Bioorganic and Medicinal Chemistry*, vol. 11, no. 9, pp. 1983–1994, 2003.

[30] C. Wang, I. J. McFadyen, J. R. Traynor, and H. I. Mosberg, "Design of a high affinity peptidomimetic opioid agonist from peptide pharmacophore models," *Bioorganic and Medicinal Chemistry Letters*, vol. 8, no. 19, pp. 2685–2688, 1998.

[31] A. Capasso, R. Guerrini, G. Balboni et al., "Dmt-Tic-OH a highly selective and potent delta-opioid dipeptide receptor antagonist after systemic administration in the mouse.," *Life sciences*, vol. 59, no. 8, pp. PL93–PL98, 1996.

[32] S. D. Bryant, S. Salvadori, P. S. Cooper, and L. H. Lazarus, "New δ-opioid antagonists as pharmacological probes," *Trends in Pharmacological Sciences*, vol. 19, pp. 42–46, 1999.

[33] S. Salvadori, R. Guerrini, G. Balboni et al., "Further studies on the Dmt-Tic pharmacophore: hydrophobic substituents at the C-terminus endow δ antagonists to manifest μ agonism or μ antagonism," *Journal of Medicinal Chemistry*, vol. 42, no. 24, pp. 5010–5019, 1999.

[34] P. W. Schiller, R. Schmidt, G. Weltrowska et al., "Conformationally constrained opioid peptide analogs with novel activity profiles," *Letters in Peptide Science*, vol. 5, no. 2-3, pp. 209–214, 1998.

[35] P. W. Schiller, M. E. Fundytus, L. Merovitz et al., "The opioid μ agonist/δ antagonist DIPP-NH_2[Ψ] produces a potent analgesic effect, no physical dependence, and less tolerance than morphine in rats," *Journal of Medicinal Chemistry*, vol. 42, no. 18, pp. 3520–3526, 1999.

[36] Y. Sasaki, T. Suto, A. Ambo, H. Ouchi, and Y. Yamamoto, "Biological properties of opioid peptides replacing Tyr at position 1 by 2,6-dimethyl-Tyr," *Chemical and Pharmaceutical Bulletin*, vol. 47, no. 10, pp. 1506–1509, 1999.

[37] P. Majer, J. Slaninova, and M. Lebl, "Synthesis of methylated phenylalanines via hydrogenolysis of corresponding 1,2,3,4-tetrahydroisoquinoline-3-carboxylic acids. Synthesis and biological activity of oxytocin analogs with methylated phenylalanines in position 2," *International Journal of Peptide and Protein Research*, vol. 43, no. 1, pp. 62–68, 1994.

[38] S. Sagan, H. Josien, P. Karoyan, A. Brunissen, G. Chassaing, and S. Lavielle, "Tachykinin NK-1 receptor probed with constrained analogues of substance P," *Bioorganic and Medicinal Chemistry*, vol. 4, no. 12, pp. 2167–2178, 1996.

[39] Y. Sasaki, M. Hirabuki, A. Ambo, H. Ouchi, and Y. Yamamoto, "Enkephalin analogues with 2′,6′-dimethylphenylalanine replacing phenylalanine in position 4," *Bioorganic and Medicinal Chemistry Letters*, vol. 11, no. 3, pp. 327–329, 2001.

[40] Y. Yamamoto, H. Ouchi, T. Tanaka, and Y. Morita, "Direct introduction of acyl and ethoxycarbonyl groups into pyrimidine ring through the trimethylstannyl derivatives," *Heterocycles*, vol. 41, no. 6, pp. 1275–1290, 1995.

[41] A. Ohno, A. Tsutsumi, N. Yamazaki, M. Okamura, Y. Mikata, and M. Fujii, "NAD(P)$^+$-NAD(P)H Models. 87. Nonsteric Stereochemistry Controlled by a Carbonyl Dipole," *Bulletin of the Chemical Society of Japan*, vol. 69, no. 6, pp. 1679–1685, 1996.

[42] J. H. Dygos, E. E. Yonan, M. G. Scaros et al., "A convenient asymmetric synthesis of the unnatural amino acid 2,6-dimethyl-L-tyrosine," *Synthesis*, no. 8, pp. 741–743, 1992.

[43] G. Toth, M. Lebl, and V. J. Hruby, "Chiral TLC separation of phenylalanine and tyrosine derivatives," *Journal of Chromatography*, vol. 504, pp. 450–455, 1990.

[44] T. Li, K. Shiotani, A. Miyazaki et al., "Bifunctional [2′,6′-dimethyl-L-tyrosine[1]] endomorphin-2 analogues substituted at position 3 with alkylated phenylalanine derivatives yield potent mixed μ-agonist/δ-antagonist and dual μ-agonist/δ-agonist opioid ligands," *Journal of Medicinal Chemistry*, vol. 50, no. 12, pp. 2753–2766, 2007.

[45] Y. Sasaki, A. Ambo, and K. Suzuki, "[D-Ala[2]]deltorphin II analogs with high affinity and selectivity for delta-opioid receptor," *Biochemical and Biophysical Research Communications*, vol. 180, no. 2, pp. 822–827, 1991.

[46] Y. Sasaki, A. Ambo, K. Midorikawa, and K. Suzuki, "Solid-phase synthesis and opioid activities of [D-Ala[2]]deltorphin II analogs," *Chemical and Pharmaceutical Bulletin*, vol. 41, no. 8, pp. 1391–1394, 1993.

[47] F. M. Leslie, "Methods used for the study of opioid receptors," *Pharmacological Reviews*, vol. 39, no. 3, pp. 197–249, 1987.

[48] A. Capasso, P. Amodeo, G. Balboni et al., "Design of μ selective opioid dipeptide antagonists," *FEBS Letters*, vol. 417, no. 1, pp. 141–144, 1997.

[49] Y. Sasaki, A. Sasaki, H. Niizuma, H. Goto, and A. Ambo, "Endomorphin 2 analogues containing Dmp residue as an aromatic amino acid surrogate with high μ-opioid receptor affinity and selectivity," *Bioorganic and Medicinal Chemistry*, vol. 11, no. 5, pp. 675–678, 2003.

[50] Y. Okada, A. Fukumizu, M. Takahashi et al., "Synthesis of stereoisomeric analogues of endomorphin-2, H-Tyr-Pro-Phe-Phe-NH_2, and examination of their opioid receptor binding activities and solution conformation," *Biochemical and Biophysical Research Communications*, vol. 276, no. 1, pp. 7–11, 2000.

[51] J. T. Pelton, W. Kazmierski, K. Gulya, H. I. Yamamura, and V. J. Hruby, "Design and synthesis of conformationally constrained somatostatin analogues with high potency and specificity for μ opioid receptors," *Journal of Medicinal Chemistry*, vol. 29, no. 11, pp. 2370–2375, 1986.

[52] A. Ambo, H. Murase, H. Niizuma, H. Ouchi, Y. Yamamoto, and Y. Sasaki, "Dermorphin and deltorphin heptapeptide analogues: replacement of Phe residue by Dmp greatly improves opioid receptor affinity and selectivity," *Bioorganic and Medicinal Chemistry Letters*, vol. 12, no. 6, pp. 879–881, 2002.

[53] Y. Sasaki, A. Ambo, and K. Suzuki, "Studies on analgesic oligopeptides. VII. Solid phase synthesis and biological properties of Tyr-D-Arg-Phe-βAla-NH_2 and its fluorinated aromatic amino acid derivatives," *Chemical and Pharmaceutical Bulletin*, vol. 39, no. 9, pp. 2316–2318, 1991.

[54] A. Ambo, H. Niizuma, A. Sasaki, H. Kohara, and Y. Sasaki, "Dermorphin tetrapeptide analogues with 2′,6′-dimethylphenylalanine (Dmp) substituted for aromatic amino acids have high μ opioid receptor binding and biological activities," *Bioorganic and Medicinal Chemistry Letters*, vol. 13, no. 7, pp. 1269–1272, 2003.

[55] Y. Sasaki, S. Kawano, H. Kohara, H. Watanabe, and A. Ambo, "ORL1 and opioid receptor preferences of nociceptin and dynorphin A analogues with Dmp substituted for N-terminal aromatic residues," *Bioorganic and Medicinal Chemistry*, vol. 14, no. 7, pp. 2433–2437, 2006.

[56] C. M. Topham, L. Mouledous, G. Poda, B. Maigret, and J. C. Meunier, "Molecular modelling of the ORL1 receptor and its complex with nociceptin," *Protein Engineering*, vol. 11, no. 12, pp. 1163–1179, 1998.

[57] Y. Sasaki, A. Sasaki, T. Ariizumi et al., "2′,6′-dimethylphenylalanine (Dmp) can mimic the N-terminal Tyr in opioid peptides," *Biological and Pharmaceutical Bulletin*, vol. 27, no. 2, pp. 244–247, 2004.

[58] R. Schmidt, D. Menard, C. Mrestani-Klaus, N. N. Chung, C. Lemieux, and P. W. Schiller, "Structural modifications of the N-terminal tetrapeptide segment of [D-Ala^2]Deltorphin I: effects on opioid receptor affinities and activities in vitro and on antinociceptive potency," *Peptides*, vol. 18, no. 10, pp. 1615–1621, 1997.

[59] J. T. Pelton, W. Kazmierski, K. Gulya, H. I. Yamamura, and V. J. Hruby, "Design and synthesis of conformationally constrained somatostatin analogues with high potency and specificity for μ opioid receptors," *Journal of Medicinal Chemistry*, vol. 29, no. 11, pp. 2370–2375, 1986.

[60] K. Gulya, J. T. Pelton, V. J. Hruby, and H. I. Yamamura, "Cyclic somatostatin octapeptide analogues with high affinity and selectivity toward mu opioid receptors," *Life Sciences*, vol. 38, no. 24, pp. 2221–2229, 1986.

[61] H. I. Mosberg, J. C. Ho, and K. Sobczyk-Kojiro, "A high affinity, Mu-opioid receptor-selective enkephalin analogue lacking an N-terminal tyrosine," *Bioorganic and Medicinal Chemistry Letters*, vol. 8, no. 19, pp. 2681–2684, 1998.

[62] I. J. McFadyen, K. Sobczyk-Kojiro, M. J. Schaefer et al., "Tetrapeptide derivatives of [D-Pen^2, D-Pen^5]-enkephalin (DPDPE) lacking an N-terminal tyrosine residue are agonists at the μ-opioid receptor," *Journal of Pharmacology and Experimental Therapeutics*, vol. 295, no. 3, pp. 960–966, 2000.

[63] J. E. Burden, P. Davis, F. Porreca, and A. F. Spatola, "Synthesis and biological activities of position one and three transposed analogs of the opioid peptide YKFA," *Bioorganic and Medicinal Chemistry Letters*, vol. 9, no. 24, pp. 3441–3446, 1999.

[64] N. S. Chandrakumar, A. Stapelfeld, P. M. Beardsley et al., "Analogs of the δ opioid receptor selective cyclic peptide [2-D-penicillamine,5-D-penicillamine]-enkephalin: 2′,6′-Dimethyltyrosine and Gly^3-Phe^4 amide bond isostere substitutions," *Journal of Medicinal Chemistry*, vol. 35, no. 16, pp. 2928–2938, 1992.

[65] L. H. Lazarus, S. D. Bryant, P. S. Cooper, and S. Salvadori, "What peptides these deltorphins be," *Progress in Neurobiology*, vol. 57, no. 4, pp. 377–420, 1999.

[66] Y. Lu, T. M. D. Nguyen, G. Weltrowska et al., "[2′,6′-dimethyltyrosine]dynorphin A(1–11)-NH_2 analogues lacking an N-terminal amino group: potent and selective κ opioid antagonists," *Journal of Medicinal Chemistry*, vol. 44, no. 19, pp. 3048–3053, 2001.

[67] A. M. Kawasaki, R. J. Knapp, A. Walton et al., "Syntheses, opioid binding affinities, and potencies of dynorphin A analogues substituted in positions 1, 6, 7, 8 and 10," *International Journal of Peptide and Protein Research*, vol. 42, no. 5, pp. 411–419, 1993.

[68] J. Yu, B. T. Chait, L. Toll, and M. J. Kreek, "Nociception in vitro biotransformation in human blood," *Peptides*, vol. 17, no. 5, pp. 873–876, 1996.

[69] J. L. Montiel, F. Cornille, B. P. Roques, and F. Noble, "Nociceptin/orphanin FQ metabolism: role of aminopeptidase and endopeptidase 24.15," *Journal of Neurochemistry*, vol. 68, no. 1, pp. 354–361, 1997.

[70] J.-P. Meyer, T. J. Gillespie, V. J. Hruby, and T. P. Davis, "In vitro stability of some reduced peptide bond pseudopeptide analogs of dynorphin A," *Peptides*, vol. 16, no. 7, pp. 1215–1219, 1995.

[71] J. Z. Chou, B. T. Chait, R. Wang, and M. J. Kreek, "Differential biotransformation of dynorphin A(1-17) and dynorphin A(1-13) peptides in human blood, ex vivo," *Peptides*, vol. 17, no. 6, pp. 983–990, 1996.

18

AutoGPA: An Automated 3D-QSAR Method Based on Pharmacophore Alignment and Grid Potential Analysis

Naoyuki Asakawa,[1] Seiichi Kobayashi,[1] Junichi Goto,[1] and Noriaki Hirayama[2]

[1] *Science and Technology Systems Division, Computational Science Department, Ryoka Systems Inc., 1-28-38 Shinkawa, Chuo-ku, Tokyo 104-0033, Japan*
[2] *Basic Medical Science and Molecular Medicine, Tokai University School of Medicine, 143 Shimokasuya, Isehara, Kanagawa 259-1143, Japan*

Correspondence should be addressed to Noriaki Hirayama, hirayama@is.icc.u-tokai.ac.jp

Academic Editor: Armando Rossello

3D-QSAR approach has been widely applied and proven to be useful in the case where no reliable crystal structure of the complex between a biologically active molecule and the receptor is available. At the same time, however, it also has highlighted the sensitivity of this approach. The main requirement of the traditional 3D-QSAR method is that molecules should be correctly overlaid in what is assumed to be the bioactive conformation. Identifying an active conformation of a flexible molecule is technically difficult. It has been a bottleneck in the application of the 3D-QSAR method. We have developed a 3D-QSAR software named AutoGPA especially based on an automatic pharmacophore alignment method in order to overcome this problem which has discouraged general medicinal chemists from applying the 3D-QSAR methods to their "real-world" problems. Applications of AutoGPA to three inhibitor-receptor systems have demonstrated that without any prior information about the three-dimensional structure of the bioactive conformations AutoGPA can automatically generate reliable 3D-QSAR models. In this paper, the concept of AutoGPA and the application results will be described.

1. Introduction

There are two major types of *in silico* drug discovery techniques: structure-based and ligand-based techniques. Quantitative structure-activity relationship (QSAR) approach only based on biological activities and chemical structures of a series of molecules with the modest biological activities is one of the ligand-based techniques. The QSAR approach explicitly considering three-dimensional shape of molecules is called 3D-QSAR. The CoMFA method proposed by Cramer et al. [1] is one of the 3D-QSAR approaches which has been widely applied and proven that the 3D-QSAR approach is better than the traditional QSAR one. The CoMFA method is based on the idea that biological activity can be analyzed by relating the shape-dependent steric and electrostatic field of molecules to their biological activity.

The results of a 3D-QSAR depend on a number of factors, each of which must be carefully considered. One of the most important considerations involves the selection of biologically active conformations and their alignment prior to the analysis. This may be relatively straightforward when one is working with a congeneric series of compounds that all have some key structural features that can be overlaid. For example, the original CoMFA paper [1] examined a series of steroid molecules which can be overlaid easily using the rigid steroid nucleus. In most cases, however, molecules of interest for drug discovery have flexible structural features and overlaying them objectively is not easy.

Expansion of the possibility of the 3D-QSAR is highly important to substantially promote many drug discovery projects where obtaining reliable X-ray structures of complexes between the active molecule and the relevant receptor is technically difficult. We have developed an automated 3D-QSAR method named AutoGPA in order to solve the above-mentioned difficult problem practically and make the 3D-QSAR easier to use by ordinary medicinal chemists.

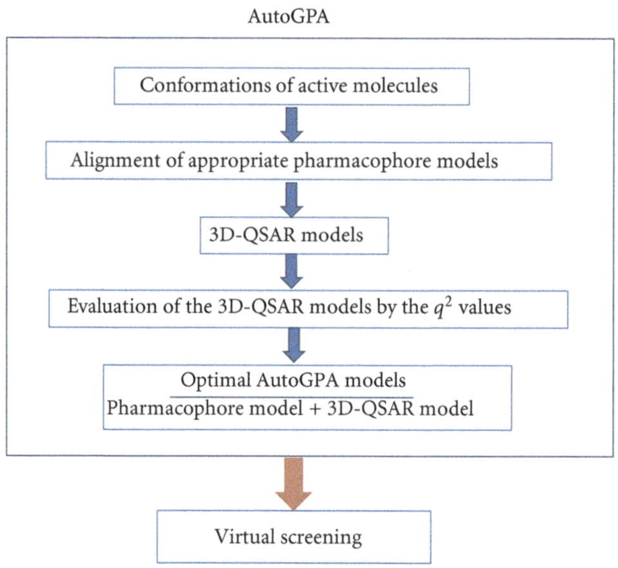

Figure 1: The process of developing AutoGPA models.

A pharmacophore is defined as an ensemble of steric and electronic features that is necessary to ensure the optimal supramolecular interactions with a specific biological target and to trigger (or block) its biological response. Since biologically active molecules for the same active site should share the common interactions at the site, their active conformations should possess common three-dimensional arrangements of pharmacophores. It is hoped that the geometries of pharmacophore features common to many of the actives will contain information related to the important interactions between the bound conformations of the actives and the receptors. Therefore, it is naturally expected that the selection of the active conformations and overlaying them can be undertaken objectively by searching the arrangement of pharmacophore features that induce good overlay of the most active molecules. In AutoGPA, pharmacophores in a set of molecules with biological activities are fully exploited to find the conformations closely related to their biological activities and overlay them.

Applications of AutoGPA to three inhibitor-receptor systems have demonstrated that AutoGPA can automatically generate reliable 3D-QSAR models from the 2D chemical structures and the biological activities of sets of the inhibitors.

2. Method

The software AutoGPA was coded by Scientific Vector Language implemented in Molecular Operating Environment (MOE) [2]. The process of AutoGPA is illustrated in Figure 1.

2.1. Pharmacophore-Based Alignments of Molecules. A function named conformation import implemented in MOE was applied to generate possible conformations with low strain energy for each molecule. The molecular mechanics setting using MMFF94x [3] force field with generalized Born solvation model [4] was applied. Typical pharmacophore features such as hydrogen bond acceptor, hydrogen bond donor, hydrophobic area, and positively and negatively ionized areas are assigned to each conformation. The three-dimensional arrangements of the pharmacophore features are compared for a set of biologically active molecules and the common 3D arrangements of pharmacophore features (hereafter pharmacophore queries) are extracted. The pharmacophore queries are used for selection of the conformations and their alignment prior to the 3D-QSAR modeling. The function named pharmacophore elucidation implemented in MOE can exhaustively search for all pharmacophore queries that induce good overlay of most of the active molecules and distinguish actives from inactives. This function is suitable for our purpose and is implemented in AutoGPA. Pharmacophore elucidation normally gives multiple possible alignments. Appropriate pharmacophores induce good alignment of the active molecules. Alignments are scored according to atomic overlap [5].

2.2. Building of 3D-QSAR Models. The algorithm of the comparative molecular field analysis (CoMFA) [1] is employed to develop 3D-QSAR models. Hydrogen atoms are added to all conformers of the molecules and the partial charges for all atoms are evaluated using the MMFF94x force field. A regular three-dimensional grid with a 2.0 Å separation surrounding all of the molecules is created. A "probe atom" is placed at each intersection grid point. An sp^3 carbon atom with charge +1.0 is used as the probe atom. Molecular fields around each molecule are evaluated by calculating the electrostatic (Coulombic, with a $1/r$ dielectric) and steric (van der Waals 6–12) interaction energies between the molecule and a series of probe atoms placed at each grid point. The potential energy in kcal/mol is assigned to each grid point and a set of the grids is designated as a grid potential model in this paper. Wherever the probe atom experiences a steric repulsion greater than +30 kcal/mol, the relevant grid point is discarded from consideration. Other grid points named scoring grids are used to develop 3D-QSAR models.

The partial-least-squares analysis is used to derive the 3D-QSAR models. The optimal number of components up to ten is identified by leave-one-out cross-validation. Each alignment yields the corresponding grid potential model and the cross-validated $r^2(q^2)$. The set of pharmacophore model and the grid potential model that gives the best q^2 value is saved as an AutoGPA model for each alignment.

3. Results and Discussion

PDK1 is a promising target for developing anticancer drugs. Islam et al. have reported the inhibitory activities against PDK1 of 70 indolinone-based molecules [6, 7]. Using the chemical and biological data, Abdul-Hameed et al. have developed good CoMFA models [8]. They used the conformation of an inhibitor bound to PDK1 in the crystal structure (PDB ID: 2PE2) [7] of the complex between the inhibitor and PDK1 as a template for alignment of molecular structures. It means that they restricted the bioactive conformation from the beginning. In "real-world" situations, however, high-quality crystal structure of the complex between an active molecule and the target protein is not always available and identification of the bioactive conformation is usually extremely difficult. Practically this difficulty has severely limited the applicability of CoMFA and the objectivity of the results obtained. On the contrary, AutoGPA models can be obtained without prior knowledge of the three-dimensional molecular structures of the bioactive conformations. The greatest merit of AutoGPA lies in the fact that reasonably reliable 3D-QSAR models can be automatically obtained from a set of molecules with a suitable range of biological activities. It is interesting to compare the 3D-QSAR models obtained by CoMFA and AutoGPA.

The dataset was separated into a training set for which AutoGPA models are derived and a test set that will prove the external predictivity of the resulting models. For comparison with the results by Abdul-Hameed et al. by CoMFA, we adopted the same training and test sets as they used. Representative molecules in the training and test sets are given in Figure 2 together with their pIC_{50} values. The molecules in the test set are asterisked. All molecules share a common chemical skeleton. Ten different AutoGPA models were generated and the results are summarized in Table 1. The statistics of the CoMFA model is also shown for comparison. In this table, various parameters that can assess the appropriateness of the models are given. The statistical parameters of the best model in terms of the cross-validated correlation coefficient (q^2) suggest that this model is reasonable and should have a good predictive ability. The q^2 and r^2 values are significantly higher and the mean squared error is significantly lower than those obtained for the CoMFA model. It is noteworthy that the AutoGPA model can also suggest the pharmacophores which are essential for ligand binding at the binding site of the expected target protein. The best AutoGPA model indicates that there are three aromatic or π ring centers, and a hydrogen donor. Such information about pharmacophore features at the binding site is helpful to grasp the chemical characteristics of bioactive molecules. In Figure 3(a), the alignment of the conformations of the molecules in the training set is shown together with the pharmacophore features. Three aromatic or π ring centers agree nicely to the corresponding three rings in the conformations. The satisfactory overlay clearly indicates that AutoGPA can successfully identify the bioactive conformation and the molecular alignment objectively in this case. In Figure 3(b), the molecular fields obtained by AutoGPA are depicted by steric and electrostatic contour maps. The conformation of molecule **35** is also shown in Figure 3(b). The combination of pharmacophore features and the grid potentials gives us an image of the characteristics of the binding site. In Figure 3(c), the molecular interactions observed in the crystal structure of the complex between molecule 35 and PDK1 [6] are illustrated. The pharmacophore features and the grid potentials generally correspond to the experimentally observed interactions. The projected hydrogen donor feature (Don2) shown in Figure 3(a) obviously corresponds to Ser160. The red polyhedra located over the urea group accept the approach of Lys111 and Thr222. The significant green polyhedron corresponds to the open space observed in the crystal structure of the complex.

Table 1: Statistics of ten AutoGPA models based on the training set.

Model	Overlap	PH4	NOC	Grids	MSE	r^2	q^2
1	46.91	RRRd	6	28	0.092	0.923	0.760
2	46.45	RRHa	7	16	0.141	0.882	0.731
3	46.44	RRHd	4	28	0.214	0.821	0.724
4	46.83	RRda	4	22	0.174	0.854	0.695
5	46.91	RRRa	4	17	0.218	0.818	0.675
6	46.84	RRHd	3	21	0.252	0.789	0.661
7	46.27	RHda	4	25	0.201	0.832	0.660
8	46.49	RRda	3	19	0.233	0.805	0.650
9	46.92	RRRd	3	26	0.249	0.791	0.626
10	46.41	RHda	5	19	0.192	0.839	0.612
CoMFA*	—	—	5	—	0.354	0.907	0.737

Overlap: atomic overlapping score in pharmacophore-based alignment.
PH4: pharmacophore feature labels; R: aromatic or π-ring center,
H: hydrophobic, d: projected donor, a: projected acceptor.
NOC: number of components.
Grids: number of grid points for QSAR model.
MSE: mean squared error.
r^2: correlation coefficient.
q^2: predictive coefficient in leave-one-out cross-validation.
*The CoMFA model obtained by Abdul-Hameed et al. [8].

The best AutoGPA model with a q^2 value of 0.760 was further validated by use of an external test set of 14 molecules. Since the biological activities of all the conformations which satisfy both the pharmacophore and grid potential models are evaluated in AutoGPA, multiple predicted values are assigned for each molecule. In this study, the maximum value is adopted as the predicted activity for each molecule. The correlations between observed and predicted pIC_{50} values are depicted in Figure 4. The best AutoGPA model

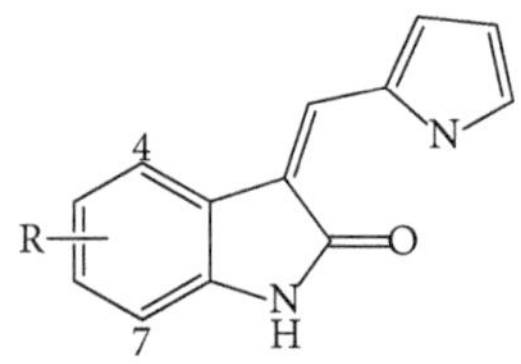

Number	R	pIC_{50}
1	H	5.74
2	4-Me	6.29
3	5-Me	5.57
4	7-Me	4.68
5	4-OH	6.00
6	5-OH	6.47
7	6-OH	6.05
8	7-OH	4.58

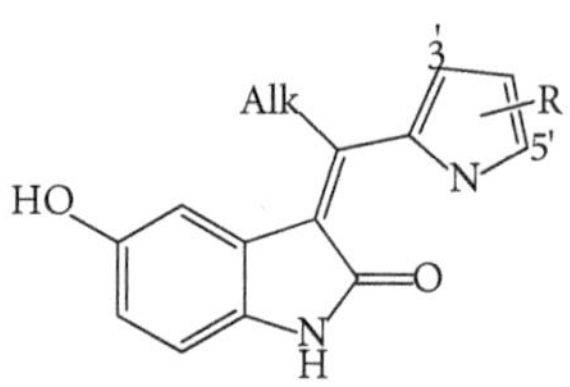

Number	Alk	R	pIC_{50}
9	Me	H	7.10
10*	H	3′-Me	6.55
11	H	4′-Me	6.55
12	H	5′-Me	6.17
13	H	5′-Et	5.64
14	H	3′, 5′-Me	5.96

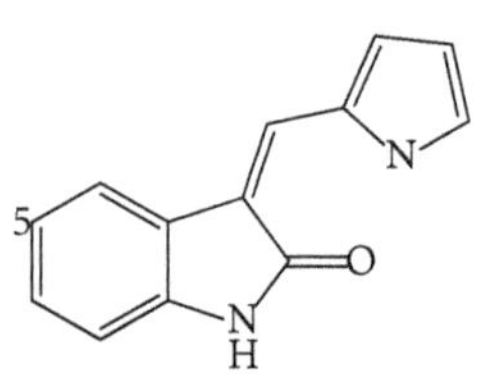

Number	R	pIC_{50}
15	OMe	6.24
16	SO_2NH_2	6.54
17	CO_2Me	6.17
18	CO_2H	5.59
19	$CONH_2$	6.92
20	5-terazole	6.70
21*	NH_2	6.24
22*	CN	6.00
23	CH_2	6.00
24	$NHSO_2Me$	6.28
25*	NHCOMe	7.26
26	$NHCONH_2$	7.74

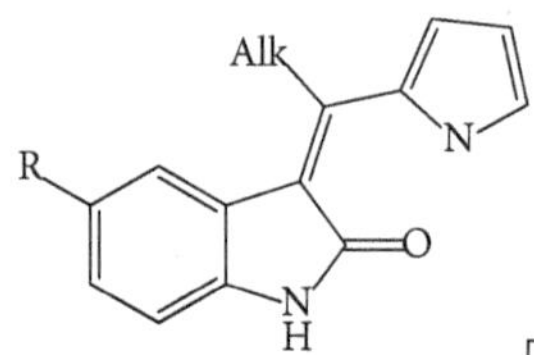

Number	R	Alk	pIC_{50}
27	SO_2NH_2	Me	7.17
28*	SO_2NH_2	Et	7.85
29	SO_2NH_2	Ph	7.54
30	SO_2NH_2	CO_2Et	7.47
31	SO_2NH_2	CO_2H	5.38
32	SO_2NH_2	$CONH_2$	5.41
33	SO_2NH_2	CONHEt	5.21
34	SO_2NH_2	$CONEt_2$	5.27
35	$NHCONH_2$	Me	8.30
36	$NHCONH_2$	Et	8.52
37	$NHCONH_2$	Ph-3-NH_2	8.05
38*	$NHCONH_2$	4-pyridine	8.00

FIGURE 2: Continued.

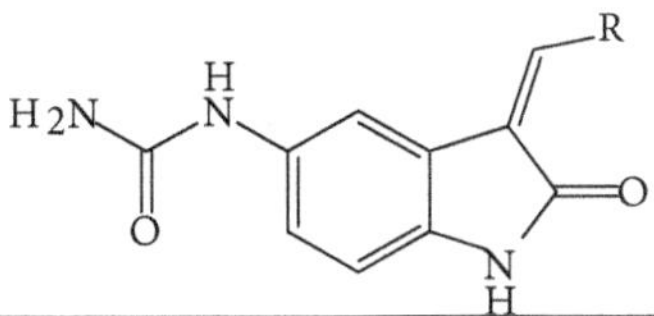

Number	R	pIC50
39	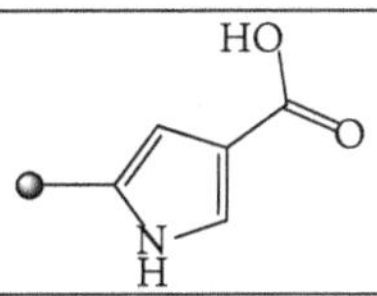	8.05
40	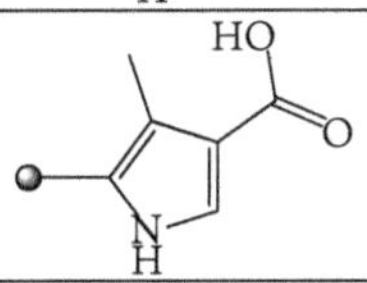	8.05
41	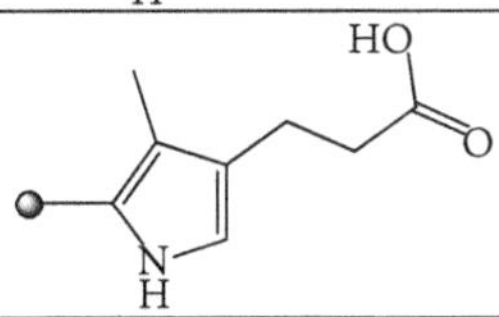	7.41
42	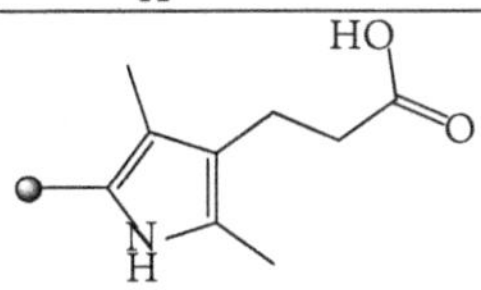	7.24
43	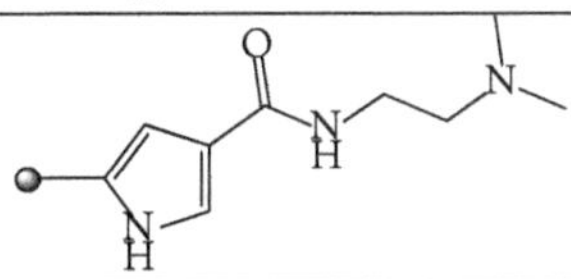	7.57
44	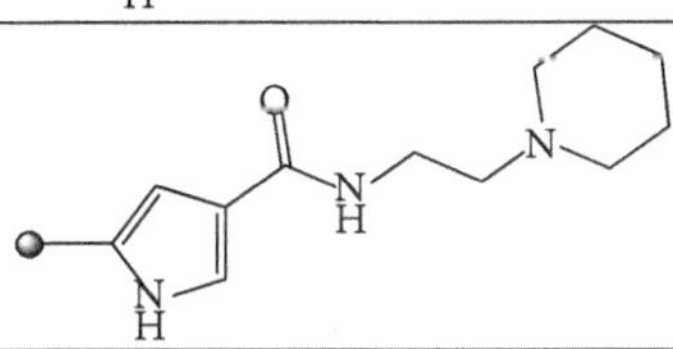	7.96
45	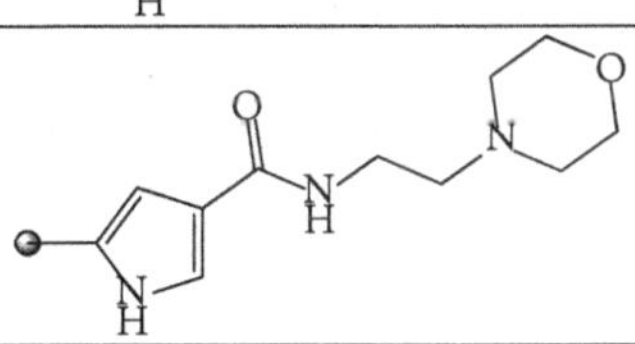	7.54
46*	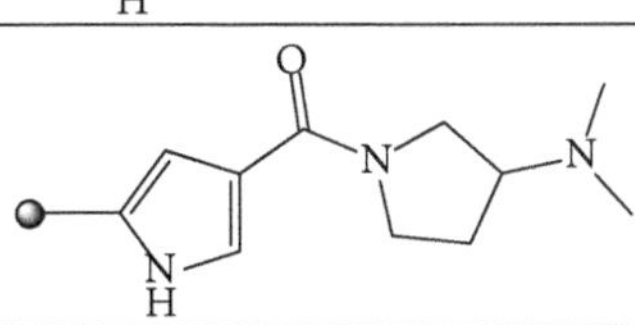	7.51
47	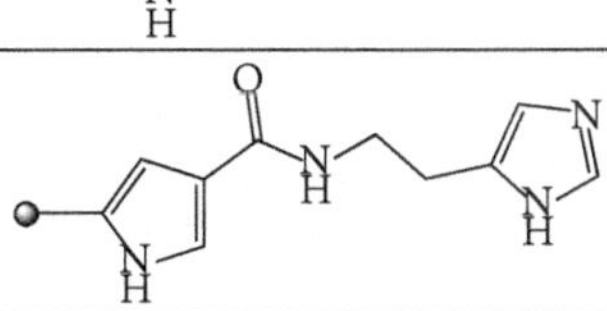	7.85
48	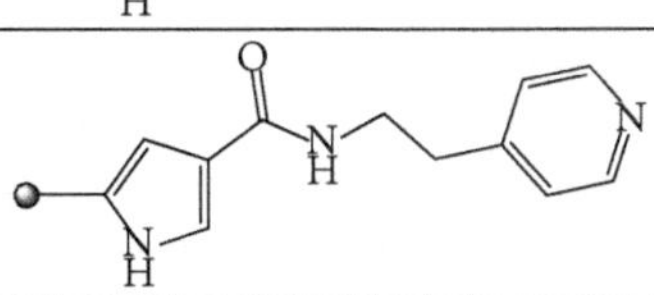	7.77

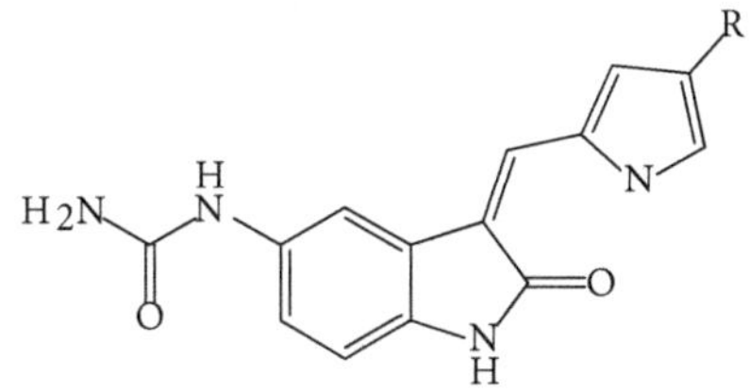

Number	R	pIC50
49	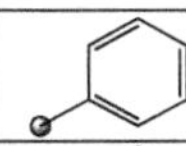	7.68
50	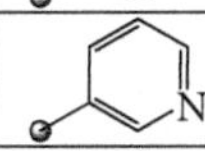	8.10
51	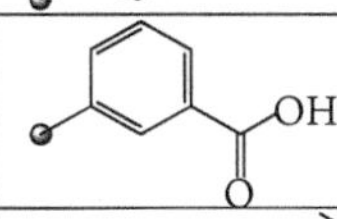	7.35
52	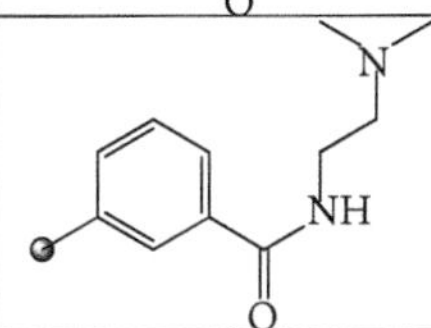	8.40
53	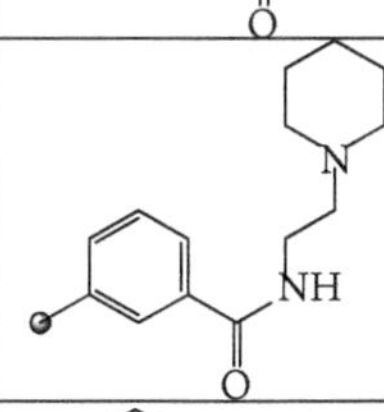	8.40
54*	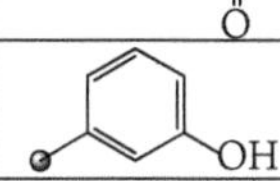	8.22
55	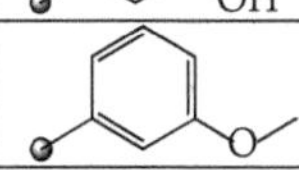	7.72
56*	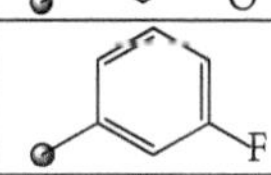	7.29
57	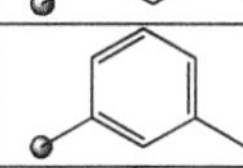	7.43
58	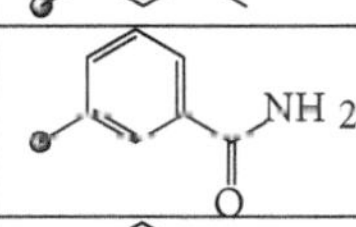	8.3
59*	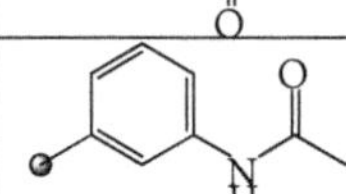	8.10
60	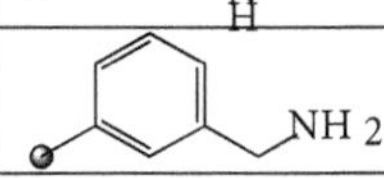	8.52

Figure 2: Continued.

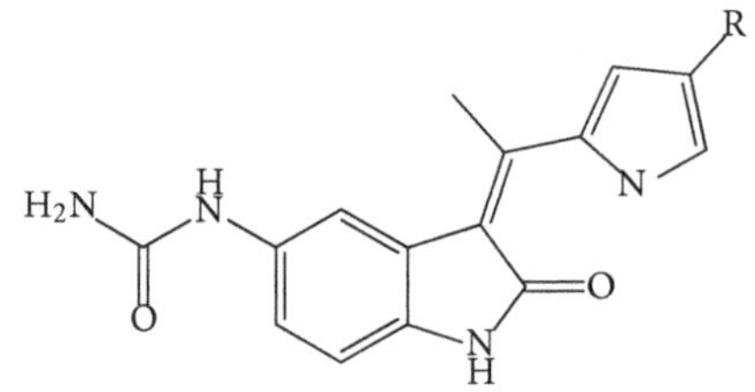

Number	R	pIC_{50}
61	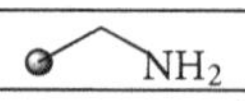	8.00
62*	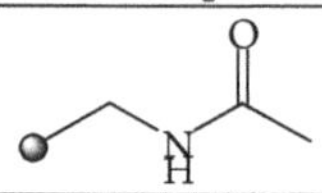	8.10
63	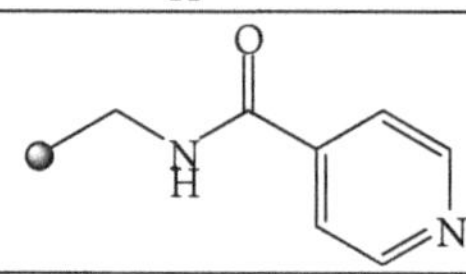	8.30
64	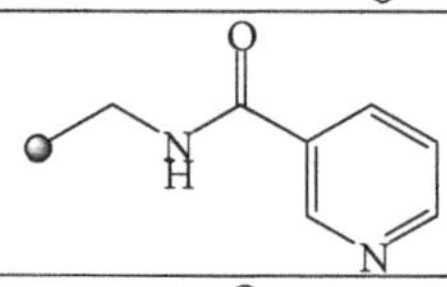	8.00
65	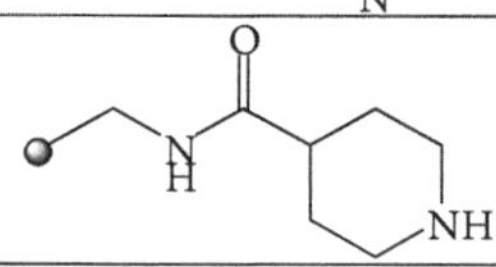	8.22
66	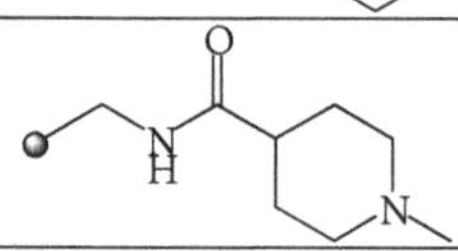	8.40
67*	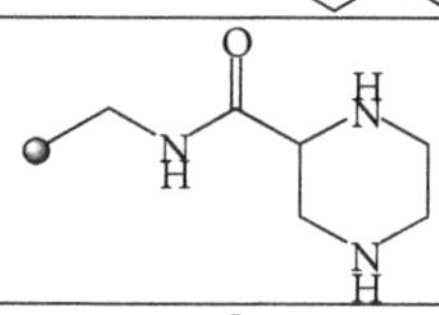	8.30
68	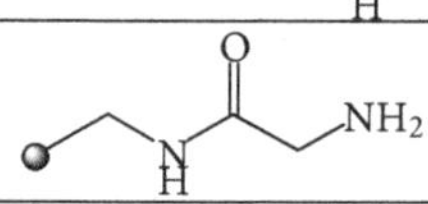	8.30
69	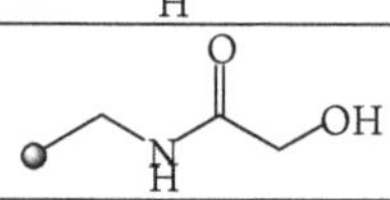	8.00
70	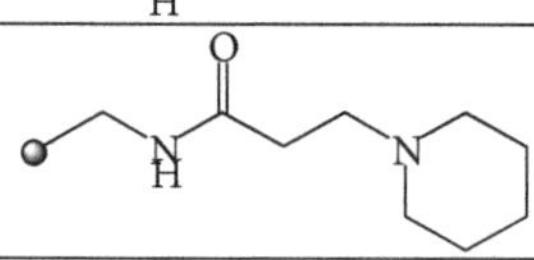	8.52

FIGURE 2: Chemical structures and the pIC50 values of PDK1 inhibitors used for validation of AutoGPA. The molecules used for a test set are asterisked, and other molecules are used for a training set.

predicts adequately the biological activities of molecules not included in the training set. The r_{pred^2} value of 0.811 attained is almost identical to that of 0.812 obtained by the CoMFA model. Despite the fact that no prior knowledge about the three-dimensional structure of the active conformation is employed to develop the 3D-QSAR model, the predictive ability of AutoGPA is highly satisfactory.

Additional two 3D-QSAR studies were undertaken to assess the performance of AutoGPA. The results will be concisely described below. The first study is on inhibitors of epidermal growth factor receptor (EGFR). Pasha et al. [9] developed 3D-QSAR model by CoMFA and CoMSIA. In this study, the X-ray structure of EGFR was used to obtain best-fit docking-based 3D-QSAR model. 46 and 12 molecules were used for training and test sets, respectively. The docked geometry-based CoMFA and CoMSIA models gave q^2 of 0.66 and 0.59, respectively, for the training set. For the test set, r^2 were 0.72 and 0.63, respectively. Although the AutoGPA model has given q^2 of 0.564 that is lower than those obtained by CoMFA and CoMSIA, r^2 of 0.768 is better than

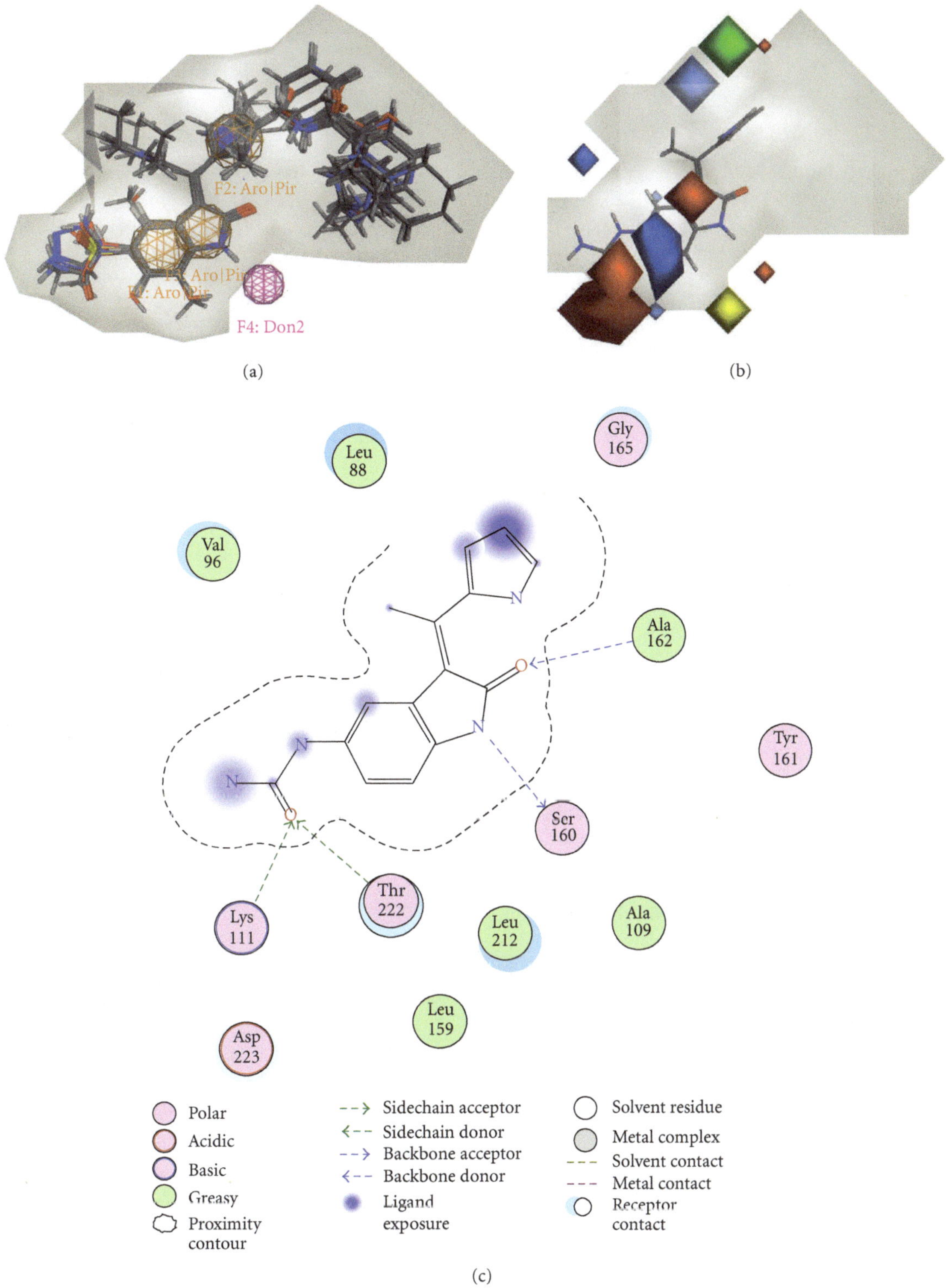

Figure 3: The best AutoGPA model obtained from the training set of PDK1 inhibitors. (a) The pharmacophore query of the best model is shown together with 56 molecules in the training set. (b) AutoGPA steric and electrostatic contours field plot. The position and the structure of molecule35 are superposed. Green and yellow contours indicate regions where bulky groups increase and decrease activity, respectively. Blue and red contours indicate regions where positive and negative electrostatic groups increase activity, respectively. (c) The interactions between the molecule 35 and PDK1 observed in the crystal structure (PDB code: 2PE1).

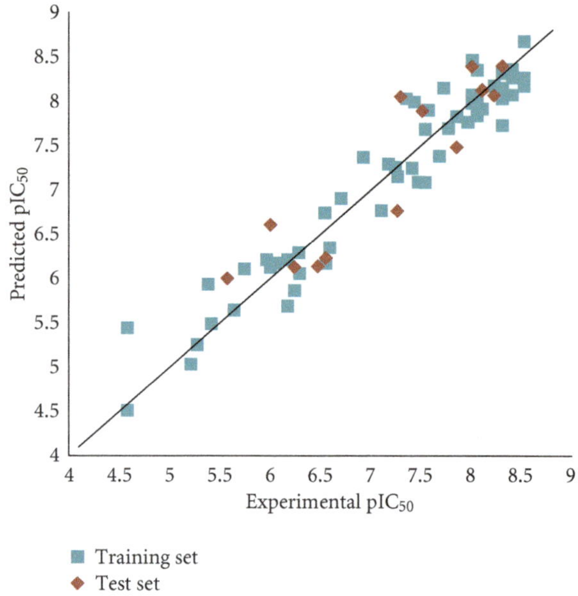

FIGURE 4: Plots of the experimental versus predicted pIC50 values using the training set of 56 molecules and the test set of 14 molecules.

those obtained by CoMFA and COMSIA. Judging from the fact that AutoGPA used no three-dimensional information of the receptor in the analysis, the result indicates that AutoGPA gave a reasonably good 3D-QSAR model. The second study is on inhibitors of human dihydrofolate reductase (DHFR). Dixon et al. [10] introduced the software named PHASE and developed a 3D-QSAR model for a set of human DHFR inhibitors. Since PHASE is based on identification of common pharmacophore among active molecules and does not require experimentally determined three-dimensional structures of both ligands and receptors, it is conceptually similar to AutoGPA. They used training and test sets which comprise of 20 and 57 molecules, respectively. The q^2 value for the training set and the r^2 value for the test set are 0.932 and 0.492, respectively. The corresponding values obtained by AutoGPA are 0.910 and 0.500, respectively. The correlation coefficients obtained by PHASE and AutoGPA are only marginally different. However, the numbers of molecules ranked high are significantly different. Six molecules of the top 10 active molecules were predicted in the top 10 by AutoGPA, but only three active molecules were predicted in the top 10 by PHASE. In addition, the best molecule with the experimental pIC50 value of 8.57 was ranked 24 and 42 by AutoGPA and PHASE, respectively. These results obviously indicate that AutoGPA is relatively superior to PHASE in the prediction ability.

In addition to PHASE, various ligand-based approaches employing pharmacophoric hypotheses have been reported in the literatures [11, 12]. Since in these approaches several different computer programs are used in combination, rather complicated tasks are required and relatively deep knowledge and experience on 3D QSAR are essential. On the contrary, the great merit of AutoGPA is that researchers, even novices on 3D QSAR, can obtain the comparable 3D QSAR only from a dataset of compounds with the chemical structures and the biological activities by executing a practically single job.

4. Conclusions

The present study has clearly shown that AutoGPA can develop reliable 3D-QSAR models without prior knowledge about the three-dimensional structure of bioactive conformations. The results obtained for three inhibitor-receptor systems have demonstrated that AutoGPA can substantially resolve the problem of objective identification of bioactive conformations and their alignments that has severely limited the application of 3D-QSAR method such as CoMFA so far. Since prerequisite of AutoGPA is only a set of 2D structures of inhibitors with their biological activities, the analysis can be undertaken almost without human intervention. It is highly expected that AutoGPA can be practically applicable to various "real-world" drug discovery projects.

Acknowledgment

This work was partly supported by Grant-in-Aid for Scientific Research on Innovative Areas (22133012) from MEXT (Ministry of Education, Culture, Sports, Science and Technology) for N. Hirayama.

References

[1] R. D. Cramer, D. E. Patterson, and J. D. Bunce, "Comparative molecular field analysis (CoMFA). 1. Effect of shape on binding of steroids to carrier proteins," *Journal of the American Chemical Society*, vol. 110, no. 18, pp. 5959–5967, 1988.

[2] "Molecular Operating Environment (MOE)," Chemical Computing Group Inc., 2010.

[3] T. A. Halgren, "MMFF VI. MMFF94s option for energy minimization studies," *Journal of Computational Chemistry*, vol. 20, no. 7, pp. 720–729, 1999.

[4] M. Wojciechowski and B. Lesyng, "Generalized Born model: analysis, refinement, and applications to proteins," *Journal of Physical Chemistry B*, vol. 108, no. 47, pp. 18368–18376, 2004.

[5] P. Labute, C. Williams, M. Feher, E. Sourial, and J. M. Schmidt, "Flexible alignment of small molecules," *Journal of Medicinal Chemistry*, vol. 44, no. 10, pp. 1483–1490, 2001.

[6] I. Islam, J. Bryant, Y.-L. Chou et al., "Indolinone based phosphoinositide-dependent kinase-1 (PDK1) inhibitors. Part 1: design, synthesis and biological activity," *Bioorganic and Medicinal Chemistry Letters*, vol. 17, no. 14, pp. 3814–3818, 2007.

[7] I. Islam, G. Brown, J. Bryant et al., "Indolinone based phosphoinositide-dependent kinase-1 (PDK1) inhibitors. Part 2: optimization of BX-517," *Bioorganic and Medicinal Chemistry Letters*, vol. 17, no. 14, pp. 3819–3825, 2007.

[8] M. D. M. Abdul-Hameed, A. Hamza, J. Liu, and C. G. Zhan, "Combined 3D-QSAR modeling and molecular docking study on indolinone derivatives as inhibitors of 3-phosphoinositide-dependent protein kinase-1," *Journal of Chemical Information and Modeling*, vol. 48, no. 9, pp. 1760–1772, 2008.

[9] F. A Pasha, M. Muddassar, A. K. Srivastava, and S. J. Cho, "In silico QSAR studies of anilinoquinolines as EGFR inhibitors,"

Journal of Molecular Modeling, vol. 16, no. 2, pp. 263–277, 2010.

[10] S. L. Dixon, A. M. Smondyrev, E. H. Knoll, S. N. Rao, D. E. Shaw, and R. A. Friesner, "PHASE: a new engine for pharmacophore perception, 3D QSAR model development, and 3D database screening: 1. Methodology and preliminary results," *Journal of Computer-Aided Molecular Design*, vol. 20, no. 10-11, pp. 647–671, 2006.

[11] F. Manetti, S. Schenone, F. Bondavalli et al., "Synthesis and 3D QSAR of new pyrazolo[3,4-b]pyridines: potent and selective inhibitors of A1 adenosine receptors," *Journal of Medicinal Chemistry*, vol. 48, no. 23, pp. 7172–7185, 2005.

[12] A. Tafi, C. Bernardini, M. Botta et al., "Pharmacophore based receptor modeling: the case of adenosine A3 receptor antagonists. An approach to the optimization of protein models," *Journal of Medicinal Chemistry*, vol. 49, no. 14, pp. 4085–4097, 2006.

19

Synthesis, Characterization, and Biological Evaluation of Some New Functionalized Terphenyl Derivatives

Seranthimata Samshuddin,[1] Badiadka Narayana,[1] Balladka Kunhanna Sarojini,[2] Divya N. Shetty,[1] and Nalilu Suchetha Kumari[3]

[1] *Department of Studies in Chemistry, Mangalore University, Mangalagangotri, Karnataka, Mangalore 574 199, India*
[2] *Research Department of Chemistry, P.A. College of Engineering, Nadupadavu, Karnataka, Mangalore 574153, India*
[3] *Department of Biochemistry, K. S. Hegde Medical Academy, Deralakatte 574162, India*

Correspondence should be addressed to Badiadka Narayana, nbadiadka@yahoo.co.uk

Academic Editor: Armando Rossello

New functionalized terphenyl derivatives incorporating various heterocyclic rings are prepared by using 4,4″-difluoro-5′-hydroxy-1,1′:3′,1″-terphenyl-4′-carbohydrazide as a key intermediate derived from 4,4′-difluoro chalcone, a versatile synthone. All the derivatives are characterized by ^{1}H NMR, IR, and mass spectral data. All the synthesized products are screened for their *in vitro* antimicrobial and antioxidant properties. The majority of the tested compounds exhibited significant antioxidant activity and some of them showed good antimicrobial activity.

1. Introduction

Terphenyl is a common structural motif found in various natural products, largely isolated from microbes and mushrooms [1]. In recent years, it has been reported that some terphenyls exhibit considerable biological activities, for example, potent anticoagulants, immunosuppressants, antithrombotic, neuroprotective, specific 5-lipoxygenase inhibitory, and cytotoxic activities [2]. Because of their promising biological activities and important properties, terphenyls have been produced increasing research interest. Therefore, synthesis of terphenyl derivatives has been a fascinating area in organic field [3].

Recently we have reported a new and simple method for the preparation of ester derivative of terphenyl starting from 4,4′-difluoro chalcone [4]. The esters can be easily converted into the corresponding carbohydrazide by the reaction with hydrazine hydrate. Hydrazides and related compounds have been described as useful precursors for the assembly of various heterocyclic rings. A large number of hydrazides are reported to be of biological interest [5, 6], while oxadiazole derivatives and thiosemicarbazides have been reported to possess antibacterial [7, 8], antifungal [9, 10], and other biological activities. Furthermore, a number of substituted thiazolines and thiazolidinones are found to exhibit appreciable antimicrobial activities [11–15].

In the course of investigation, we have found that 4,4″-difluoro-5′-hydroxy-1,1′:3′,1″-terphenyl-4′-carbohydrazide is an excellent building block for the synthesis of numerous heterocyclic systems. In view of the pharmacological importance of terphenyls and in continuation of our work on synthesis of various derivatives of 4,4′-difluoro chalcone [16–23], it was decided to prepare new functionalized terphenyl derivatives by using 4,4″-difluoro-5′-hydroxy-1,1′:3′,1″-terphenyl-4′-carbohydrazide as the key intermediate and to study their biological activities.

2. Chemistry

A novel terphenyl derivative, ethyl 4,4″-difluoro-5′-hydroxy-1,1′:3′,1″-terphenyl-4′-carboxylate **3**, was synthesized by the oxidative aromatization of cyclohexenone derivative of 4,4′-difluoro chalcone using chloramine-T [4]. The ester **3** was converted into the corresponding carbohydrazide **4** in high yield by the reaction with hydrazine hydrate (Scheme 1). In the IR spectrum of carbohydrazide **4** broad stretching

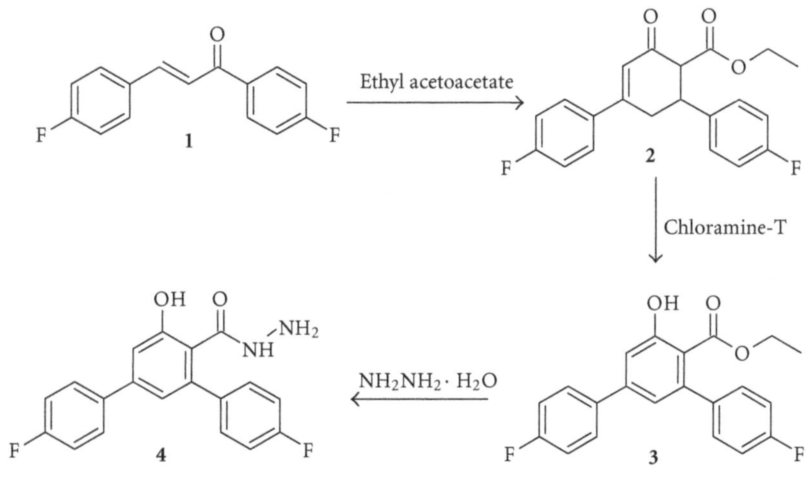

Scheme 1

band at 3290 cm^{-1} was due to NH while strong stretching band at 1672 cm^{-1} was attributed to carbonyl. ^1H NMR spectrum showed two singlets at δ 9.19 ppm and δ 4.18 ppm which were accounted for NH_2 and NH protons. The mass spectrum of **4** showed a molecular ion peak at m/z 341.0 which confirmed its molecular weight.

The carbohydrazide is a versatile key intermediate for the synthesis of several heterocyclic systems such as pyrazoles, thiazoles, 1,3,4-oxadiazoles, and 1,2,4-triazoles [24, 25]. Hence, the carbohydrazide **4** was treated with various reagents to afford pyrazole and 1,3,4-oxadiazole derivatives (Scheme 2). Condensation reaction of carbohydrazide **4** with dicarbonyl compounds, namely, acetylacetone and ethyl acetoacetate, in the presence of catalytic amount of triethylamine afforded the corresponding substituted pyrazole derivatives **5** and **6**, respectively. The IR spectrum of pyrazole **5** showed the presence of vibrations at 2926 cm^{-1} due to methyl group. The ^1H NMR spectrum of **5** showed two singlets at δ 2.29 and 2.70 ppm due to three each protons of two methyl groups at C-3 and C-5 carbon of pyrazole ring, respectively, and hence confirming the presence of 3,5-dimethyl-1*H*-pyrazole moiety. Similarly, in the IR spectrum of pyrazole **6**, vibrations at 1654 cm^{-1} indicates the presence of carbonyl group of the pyrazolone nuclei. Singlets at δ 2.84 and 2.03 ppm in the ^1H NMR spectrum were observed for the methyl and methylene protons of the pyrazolone moiety.

Cyclization to the oxadiazoles was carried out by treating the carbohydrazide **4** with triethyl orthoformate/toluene, cyanogen bromide/ethanol and with aromatic acids/phosphorous oxychloride. The IR spectra of oxadiazoles **7** and **9** showed the absence of vibrations due to NH and NH_2 groups. Their ^1H NMR spectra also indicated the absence of NH protons, thus confirmed the cyclization reaction. In the IR spectrum of oxadiazole **8**, a stretching vibration observed at 3385 cm^{-1} which indicates the presence of amino group. A singlet at δ 6.93 ppm in the ^1H NMR spectrum of **8** confirmed the presence of NH_2 group and hence proved that ring closure starting from carbohydrazide **4** resulted in the formation of 5-amino-1,3,4-oxadiazole moiety.

An arylidine hydrazone incorporated into the parent terphenyl was also synthesized. Thus, condensation of carbohydrazide **4** with aromatic aldehydes in absolute ethanol in presence of few drops of concentrated hydrochloric acid afforded the corresponding Schiff bases **10a–c**. The Schiff bases **10a–c** thus obtained were cyclized into 1,3,4-oxadiazole derivatives **11a–c** by refluxing them with acetic anhydride (Scheme 3). IR spectra of **10a–c**, showed an NH stretching band at around 3329 cm^{-1} and a carbonyl absorption band near 1647 cm^{-1}. The ^1H NMR spectrum of **10c**, as an example, showed a singlet signal at δ 8.15 ppm corresponding to CH of the benzylidene group and another singlet signal at δ 11.71 ppm due to NH proton, hence confirming the formation of Schiff base hydrazone. The IR spectra of compounds **11a–c** had different characteristics since they showed no N–H stretching bands, but carbonyl stretching near 1770 cm^{-1} which could be attributed to the carbonyl of acetyl group. ^1H NMR spectra of **11a–c** showed a singlet in the spectra in the range δ 6.41–6.52 ppm indicating CH resonance of the oxadiazoline ring which is in accordance with the reported values found in the literature [26].

The treatment of hydrazides with carbon disulfide in the presence of alkali hydroxide was a general method leading to the formation of 1,3,4-oxadiazole-2-thione derivatives [27]. As the starting material, carbohydrazide **4** was used for the synthesis of 1,3,4-oxadiazole-2-thione derivative **12** (Scheme 4). The IR spectrum of compound **12** showed a sharp absorption band at 3224 cm^{-1} representing NH group. Its ^1H NMR spectrum displayed a signal at δ 14.53 ppm due to the NH proton, while no signal representing a hydrazide structure appeared. The 4-amino-1,2,4-triazole-3-thione derivative **13** was synthesized starting from compound **12** by treating it with hydrazine hydrate. The IR spectrum of compound **13** showed two sharp absorption bands at 3373 cm^{-1} and 3259 cm^{-1} representing –NH and –NH_2 groups. The singlet signal integrating two protons seen at δ 5.45 ppm in the ^1H NMR spectrum of compound **13** was attributed to the –NH_2 group. Furthermore, an additional

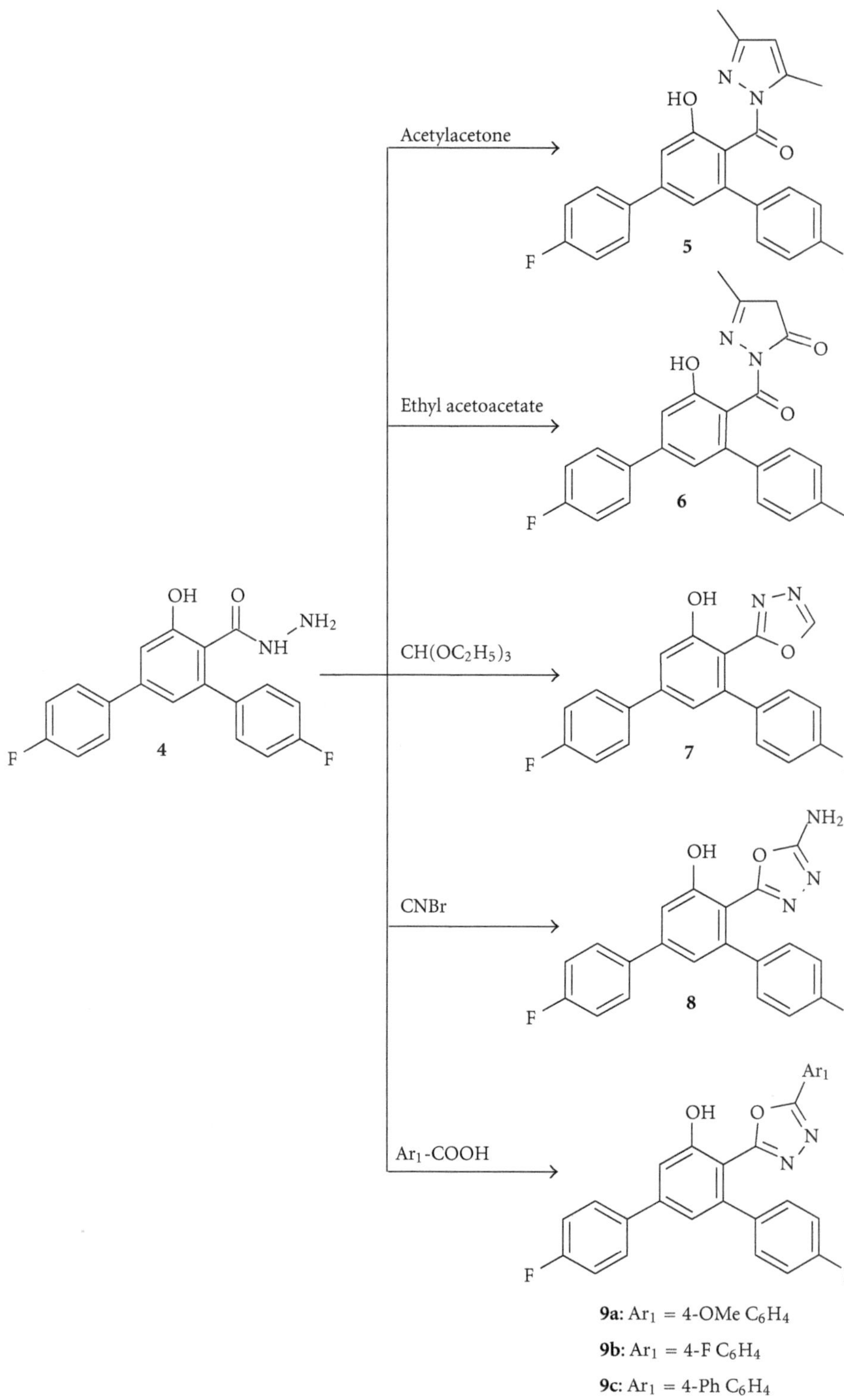

Scheme 2

singlet was observed due to the –NH group at δ 13.54 ppm integrating one proton.

Mannich reaction of 1,3,4-oxadiazole-2-thione derivative **12** with secondary amines and formaldehyde in absolute ethanol at room temperature yielded the desired compounds **14a–c**. When compared with the compound **12**, the IR spectra of Mannich bases **14a–c** showed the absence of NH group. Their ^{1}H NMR spectra also supported the proposed structure as there was no signal corresponding to NH proton. Further, 1,3,4-oxadiazole-2-thione derivative **12** was

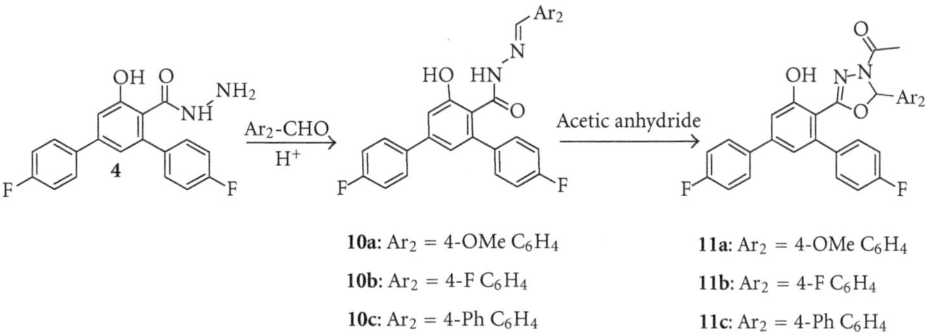

Scheme 3

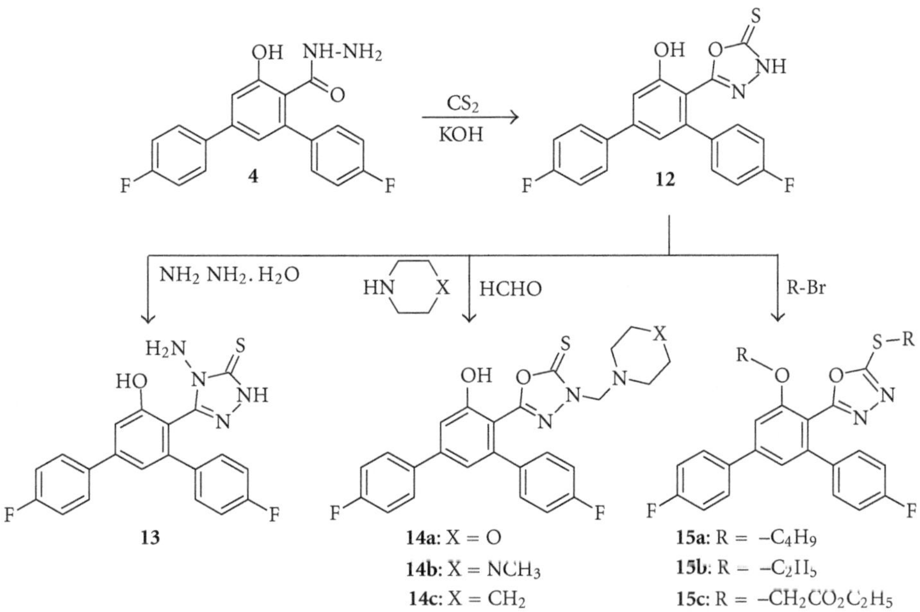

Scheme 4

alkylated using different alkyl halides. Compound **12** can exist in thione or thiol tautomeric forms. However, it has been established that under the action of alkyl halides *S*-substitution took place instead of *N*-substitution. The mass spectrum of **15c** showed a molecular ion peak at *m/z* 555.1 which reveals that the compound not only *S*-alkylated but also *O*-alkylated, because of the presence of phenolic group.

Furthermore, the intermediate thiosemicarbazide derivative **16** representing a versatile building block for the synthesis of new heterocycles incorporating thiazole, imidazole, and triazole nucleus was synthesized by heating the carbohydrazide **4** with potassium thiocyanate in the presence of hydrochloric acid [28, 29] (Scheme 5). The ^{1}H NMR spectrum of **16** exhibited singlet signals at δ 6.96 ppm and 8.19 ppm due to NH_2 protons as one of the proton involved in intramolecular hydrogen bonding with carbonyl group; two protons were unequal. In addition to that, it showed two singlets at δ 9.57 and 10.45 ppm due to two NH protons. The IR spectrum of **16** exhibited two sharp absorption bands at 3464 cm^{-1} and 3329 cm^{-1} representing –NH and –NH_2 groups. The thiazole derivatives **17a–c** was synthesized by the treatment of thiosemicarbazide derivative **16** with different phenacyl bromides in boiling ethanol. These reactions were assumed to proceed via *S*-alkylation followed by dehydration. The IR spectra of thiazoles **17a–c** showed absorption band at 3331 cm^{-1} representing –NH groups. Its ^{1}H NMR spectra exhibited two singlets in the region δ 9.4–10.5 ppm for hydrazide (NH) protons, and hence confirming the formation of thiazole moiety.

Compound **16** was cyclized via oxidative cyclization in basic medium (10% NaOH), under reflux condition with subsequent acidification to give 1,2,4-triazole-3-thiol derivative **18**. Further, reflux of compound **16** with monochloroacetic acid and anhydrous sodium acetate in glacial acetic acid afforded a product identified as 2-imino-4-oxo-1,3-thiazolidine derivative **19** (Scheme 5). The structure of the synthesized products **18** and **19** was established on the basis of spectral data. For example, in compound **18**, the absence of absorption band of carbonyl group in IR spectrum and lack of ^{1}H NMR resonances observed with NH and

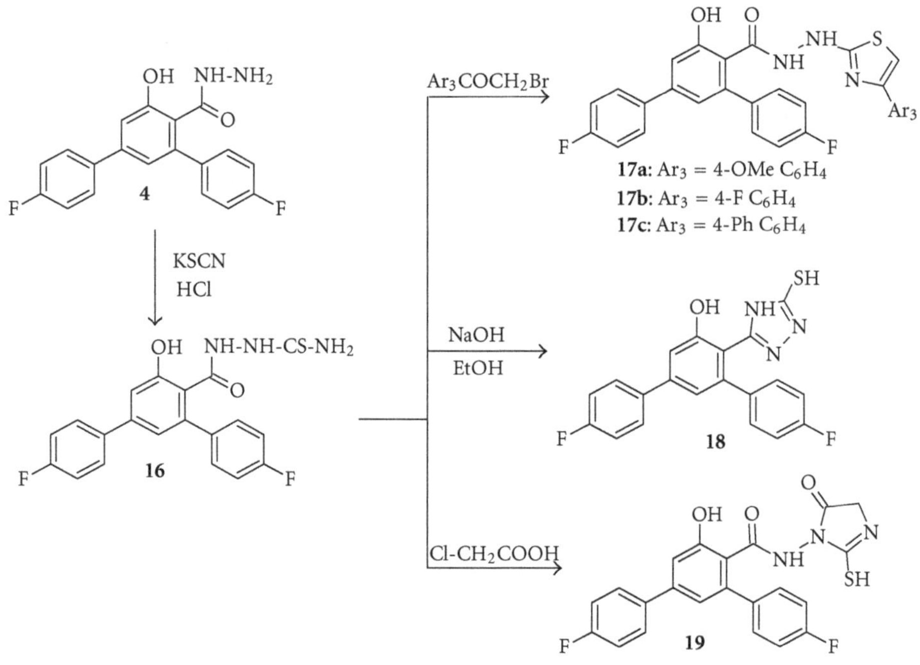

Scheme 5

NH_2 functions proved the intramolecular cyclization of compound **16** resulted in the formation of 1,2,4-triazole ring. Similarly, IR spectrum of thiazolidine derivative **19** showed an absorption band at $1724\,cm^{-1}$ representing carbonyl group of imidazolidinone moiety. A singlet at δ 3.88 ppm present in the ^1H NMR spectrum of **19** indicated the two protons of methylene group of imidazolidinone moiety, in addition to a singlet at δ 4.10 ppm due to SH proton and another singlet at δ 10.59 ppm due to that NH proton confirmed the presence of imidazolidinone moiety.

3. Biological Evaluation

3.1. Antibacterial Studies. The newly synthesized compounds were screened for their antibacterial activity against *Escherichia coli* (ATTC-25922), *Staphylococcus aureus* (ATTC-25923), *Pseudomonas aeruginosa* (ATCC-27853), and *Klebsiella pneumoniae* (recultured) bacterial strains by serial plate dilution method [30, 31]. Ciprofloxacin was used as standard antibiotics for antibacterial activity.

Compound **12** was found to be the most active derivative against *Escherichia coli, Pseudomonas aeruginosa*, and *Klebsiella pneumoniae* at MIC value of 6.25 μg/mL among the tested compounds and showed the same potency with ciprofloxacin. The compounds **3**, **4**, **5**, **6**, **16**, **17a**, **17b,** and **17c** were found to be active against all the tested microorganisms at MIC value of 12.5 μg/mL. The high activity of **12** was due to the presence of 1,3,4-oxadiazole-2-thione moiety containing free NH group. Also the derivatives containing hydrazide moiety (**4**, **16**, **17a**, **17b**, **17c**), ester group (**3**), and pyrazole moiety (**5** & **6**) were helped to inhibit the growth of the microorganisms.

3.2. Antifungal Studies. Newly prepared compounds were screened for their antifungal activity against *Aspergillus flavus* (NCIM No. 524), *Aspergillus fumigates* (NCIM No. 902), *Penicillium (S. aureus)* (recultured), and *Trichophyton mentagrophytes* (recultured) in DMSO by serial plate dilution method [32, 33]. Activity of each compound was compared with itraconazole as standard.

Compounds **6**, **8**, and **10b** were found to be the most active derivatives against all the tested microorganisms at an MIC value of 6.25 μg/mL among the tested compounds, and showed the same potency with itraconozole. Although compounds **9a**, **9c**, **10a**, **11a**, **11b**, **11c**, **17a**, and **19** showed activity having MIC value of 12.5 μg/mL against selected microorganisms, they exhibited lower potencies than the compared standard drug. Hence, it could be concluded that different functionality present in the molecules contributed to the antifungal activity in different manner comparable with that of standard drug tested.

3.3. DPPH Radical Scavenging Assay. A rapid, simple, and inexpensive method to measure antioxidant capacity of substances involves the use of the free radical, 2,2-Diphenyl-1-picrylhydrazyl (DPPH). DPPH is widely used to test the ability of compounds to act as free radical scavengers or hydrogen donors. Antioxidants tested on DPPH were also found extremely effective in cell systems. This simple test further provides information on the ability of a compound to donate electrons during antioxidant action [34]. The radical scavenging mechanism is based on the transfer of acidic H-atom from the compound to DPPH radical to form DPPH-H.

Among the tested compounds, many compounds showed significant radical scavenging capacity due to the presence of phenolic hydrogen in the parent molecule. Compounds **12** and **13** showed good radical scavenging capacity while compounds **9a**, **9b**, **14a**, **14b**, **15c**, and **16** exhibited moderate radical scavenging capacity with concentration of 1 mg/mL in comparison with the standard glutathione. The good radical scavenging capacity of compounds **12** and **13** was due to the presence of acidic protons in the oxadiazole moiety. The variation exhibited in DPPH scavenging capacity could be attributed to the effect of different moiety present in the compounds.

3.4. Reducing Power Assay. Reducing power is associated with antioxidant activity and may serve as a significant reflection of the antioxidant activity [35]. Compounds with reducing power indicate that they are electron donors and can reduce the oxidized intermediates of lipid peroxidation processes, so that they can act as primary and secondary antioxidants [36]. Substances, which have reduction potential, react with potassium ferricyanide (Fe^{3+}) to form potassium ferrocyanide (Fe^{2+}), which then reacts with ferric chloride to form ferrous complex that has an absorption maximum at 700 nm. Increased absorbance of the reaction mixture indicates the increased reducing power.

Reducing power assay is expressed in effective concentration (mg/mL) equivalent to 0.5 absorbance glutathione. Among the tested compounds, many compounds showed significant reducing power capacity. Compounds **4**, **12**, **13**, and **16** showed good reducing power capacity while compounds **5**, **8**, **10b**, **17a**, **17b**, **17c**, **18**, and **19** exhibited moderate reducing power capacity in comparison with the standard Glutathione. The good reducing power capacity of these compounds was due to the presence of free NH group in the molecules.

4. Conclusions

New functionalized terphenyl derivatives incorporating various heterocyclic rings are prepared by using 4,4″-difluoro-5′-hydroxy-1,1′:3′,1″-terphenyl-4′-carbohydrazide as a key intermediate derived from 4,4′-difluoro chalcone as a versatile synthone. All these derivatives are characterized by spectral data. All the synthesized products are screened for their *in vitro* antimicrobial and antioxidant properties. The majority of the tested compounds exhibited significant antimicrobial activity and some of them showed good DPPH scavenging activity. Hence this study has widened the scope of developing easy method to synthesize fluorinated functionalized terphenyl derivatives as promising antimicrobial and antioxidant agents.

5. Experimental

5.1. Chemistry. Melting points were taken in open capillary tubes and are uncorrected. The purity of the compounds was confirmed by thin layer chromatography using Merck silica gel 60 F_{254} coated aluminium plates. IR spectra were recorded on Shimadzu-FTIR Infrared spectrometer in KBr (ν_{max} in cm^{-1}). ^{1}HNMR (400 MHz) spectra were recorded on a Bruker AMX 400 spectrometer, with 5 mm PABBO BB-1H TUBES with TMS as internal standard. LCMS were obtained using Agilent 1200 series LC and Micromass zQ spectrometer. Elemental analysis was carried out by using VARIO EL-III (Elementar Analysensysteme GmBH).

5.2. General Procedures for the Synthesis

Synthesis of Ethyl 4,4″-Difluoro-5′-hydroxy-1,1′:3′,1″-terphenyl-4′-carboxylate (3). Ethyl 4,4″-difluoro-5′-hydroxy-1,1′:3′,1″-terphenyl-4′-carboxylate (**3**) was synthesized by the oxidative aromatization of ethyl 4,6-bis(4-fluorophenyl)-2-oxocyclohex-3-ene-1-carboxylate (**2**) using chloramine-T, which in turn was prepared by the condensation of ethyl acetoacetate to the 4,4′-difluoro chalcone (**1**) according to the methods described in our previous work [4].

Synthesis of 4,4″-Difluoro-5′-hydroxy-1,1′:3′,1″-terphenyl-4′-carbohydrazide (4). A mixture of ester **3** (17.7 g, 0.05 mol) and hydrazine hydrate (10 mL, 99%) in absolute ethanol (60 mL) was heated under reflux for 16 h and cooled at room temperature. The resultant solid was filtered and recrystallized from ethanol to give colorless crystals. Yield 86%; m.p 193–196°C. IR (KBr, ν cm^{-1}): 3290 (NH), 1672 (C=O), 1219 (C–F). ^{1}H NMR (400 MHz, DMSO-d_6, δ ppm): 4.18 (s, 2H, NH_2), 6.99–7.69 (m, 10H, Ar–H), 9.19 (s, 1H, NH), 9.87 (s, 1H, OH). LCMS: m/z 341.0 (M^++1). C H N Analysis; calculated for $C_{19}H_{14}F_2N_2O_2$: C, 67.05; H, 4.15; N, 8.23. Found: C, 67.02; H, 4.16; N, 8.20.

Synthesis of (4,4″-Difluoro-5′-hydroxy-1,1′:3′,1″-terphenyl-4′-yl)(3,5-dimethyl-1H-pyrazol-1-yl)methanone (5). To a mixture of carbohydrazide **4** (3.4 g, 0.01 mol) and acetylacetone (1.0 g, 0.01 mol) in ethanol (20 mL) triethylamine (1 mL) was added, and the mixture was refluxed for 10 h. The precipitated solid was filtered off and recrystallized from ethanol. Yield 52%; m.p 166–168°C. IR (KBr, ν cm^{-1}): 3327 (OH), 2926 (C–H), 1604 (C=N), 1215 (C–F). ^{1}H NMR (400 MHz, DMSO-d_6, δ ppm): 2.29 (s, 3H, CH_3), 2.70 (s, 3H, CH_3), 6.28 (d, 1H, CH), 6.95–7.71 (m, 10H, Ar–H), 9.70 (d, 1H, OH). LCMS: m/z 407.0 (M^++1). C H N Analysis; calculated for $C_{24}H_{20}F_2N_2O_2$: C, 71.28; H, 4.49; N, 6.93. Found: C, 71.24; H, 4.47; N, 6.91.

Synthesis of 2-[(4,4″-Difluoro-5′-hydroxy-1,1′:3′,1″-terphenyl-4′-yl)carbonyl]-5-methyl-2,4-dihydro-3H-pyrazol-3-one (6). To a mixture of carbohydrazide **4** (3.4 g, 0.01 mol) and ethyl acetoacetate (1.3 g, 0.01 mol) in ethanol (20 mL) triethylamine (1 mL) was added, and the mixture was refluxed for 12 h. The precipitated solid was filtered off and recrystallized from ethanol. Yield 48%; m.p 96–98°C. IR (KBr, ν cm^{-1}): 3055 (OH), 2985 (C-H), 1654 (C=O), 1608 (C=N), 1211 (C–F). ^{1}H NMR (400 MHz, DMSO-d_6, δ ppm): 2.03 (s, 2H, CH_2), 2.84 (s, 3H, CH_3), 7.04–7.72 (m, 10H, Ar–H), 10.23 (s, 1H, OH). LCMS: m/z 407.1 (M^++1). C H N Analysis; calculated for $C_{23}H_{16}F_2N_2O_3$: C, 67.98; H, 3.97; N, 6.89. Found: C, 67.94; H, 3.97; N, 6.84.

Synthesis of 4,4''-Difluoro-4'-(1,3,4-oxadiazol-2-yl)-1,1':3',1''-terphenyl-5'-ol (7). A mixture of carbohydrazide **4** (3.4 g, 0.01 mol) and triethyl orthoformate (1 mL) in toluene (60 mL) was heated under reflux for 10 h. The excess solvent was evaporated and the reaction mass was cooled to room temperature. The resultant solid was filtered and recrystallized from ethanol. Yield 58%; m.p 193–196°C. IR (KBr, $\nu\,cm^{-1}$): 3149 (OH), 1620 (C=N), 1232 (C–F). ^1H NMR (400 MHz, DMSO-d_6, δ ppm 7.12–7.79 (m, 10H, Ar–H), 9.23 (s, 1H, CH), 10.57 (s, 1H, OH). LCMS: *m/z* 351.0 (M^++1). C H N Analysis; calculated for $C_{20}H_{12}F_2N_2O_2$: C, 68.57; H, 3.45; N, 8.00. Found: C, 68.54; H, 3.44; N, 7.98.

*Synthesis of 4'-(5-Amino-1,3,4-oxadiazol-2-yl)-4,4''-difluoro-1,1':3',1''-terphenyl-5'-ol (**8**).* A mixture of carbohydrazide **4** (3.4 g, 0.01 mol) and cyanogen bromide (1.57 g, 0.015 mol) in absolute ethanol (20 mL) was refluxed for 5 h. The reaction mass was cooled to room temperature and neutralized with saturated sodium bicarbonate solution. The precipitated solid was filtered off and recrystallized from ethanol. Yield 61%; m.p 248–250°C. IR (KBr, $\nu\,cm^{-1}$): 3385 (NH_2), 3116 (OH), 1600 (C=N), 1217 (C–F). ^1H NMR (400 MHz, DMSO-d_6, δ ppm): 6.93 (s, 2H, NH_2), 7.13–7.77 (m, 10H, Ar–H), 10.44 (s, 1H, OH). LCMS: *m/z* 366.0 (M^++1). C H N Analysis; calculated for $C_{20}H_{13}F_2N_3O_2$: C, 65.75; H, 3.59; N, 11.50. Found: C, 65.73; H, 3.57; N, 11.47.

*Synthesis of 4,4''-Difluoro-4'-(5-aryl-1,3,4-oxadiazol-2-yl)-1,1':3',1''-terphenyl-5'-ol (**9a–c**).* A mixture of carbohydrazide **4** (3.4 g, 0.01 mol) and aromatic carboxylic acid (0.015 mol) in phosphorus oxychloride (10 mL) was refluxed for 8 h. The excess phosphorus oxychloride was evaporated and the cooled reaction mass was then poured into ice cold water with stirring. The separated solid was filtered and washed with saturated sodium bicarbonate solution and then with water. The product was recrystallized from ethanol-dimethyl formamide mixture.

*4,4''-Difluoro-4'-[5-(4-methoxyphenyl)-1,3,4-oxadiazol-2-yl]-1,1':3',1''-terphenyl-5'-ol (**9a**).* Yield 61%; m.p 198–201°C. IR (KBr, $\nu\,cm^{-1}$): 3042 (OH), 2956 (CH), 1604 (C=N), 1220 (C–F). ^1H NMR (400 MHz, DMSO-d_6, δ ppm): 3.64 (s, 3H, CH_3), 6.95–7.73 (m, 14H, Ar–H), 10.32 (s, 1H, OH). LCMS: *m/z* 457.1 (M^++1). C H N Analysis; calculated for $C_{27}H_{18}F_2N_2O_3$: C, 71.05; H, 3.97; N, 6.14. Found: C, 71.03; H, 3.98; N, 6.12.

*4,4''-Difluoro-4'-[5-(4-fluorophenyl)-1,3,4-oxadiazol-2-yl]-1,1':3',1''-terphenyl-5'-ol (**9b**).* Yield 66%; m.p 190–193°C. IR (KBr, $\nu\,cm^{-1}$): 3032 (OH), 1604 (C=N), 1231 (C–F). ^1H NMR (400 MHz, DMSO-d_6, δ ppm): 6.98–7.70 (m, 14H, Ar–H), 9.85 (s, 1H, OH). LCMS: *m/z* 445.1 (M^++1). C H N Analysis; calculated for $C_{26}H_{15}F_3N_2O_2$: C, 70.27; H, 3.40; N, 6.30. Found: C, 70.25; H, 3.42; N, 6.28.

*4,4''-Difluoro-4'-[5-(biphenyl-4-yl)-1,3,4-oxadiazol-2-yl]-1,1':3',1''-terphenyl-5'-ol (**9c**).* Yield 68%; m.p 210–213°C. IR (KBr, $\nu\,cm^{-1}$): 3039 (OH), 1606 (C=N), 1226 (C–F). ^1H NMR (400 MHz, DMSO-d_6, δ ppm): 7.13–8.11 (m, 19H, Ar–H), 10.13 (s, 1H, OH). LCMS: *m/z* 503.1 (M^++1). C H N Analysis; calculated for $C_{32}H_{20}F_2N_2O_2$: C, 76.48; H, 4.01; N, 5.57. Found: C, 76.44; H, 4.02; N, 5.55.

*Synthesis of Schiff Base Derivatives (**10a–c**).* A mixture of carbohydrazide **4** (3.4 g, 0.01 mol) and aromatic aldehyde (0.01 mol) in absolute ethanol (25 mL) was heated under reflux for 4 h in presence of concentrated HCl (0.5 mL). The reaction mixture was then poured into ice cold water and filtered. The pure product was recrystallized from ethanol-dimethyl formamide mixture.

*4,4''-Difluoro-N'-[(Z)-(4-methoxyphenyl)methylidene]-5'-hydroxy-1,1':3',1''-terphenyl-4'-carbohydrazide (**10a**).* Yield 84%; m.p 177–180°C. IR (KBr, $\nu\,cm^{-1}$): 3336 (NH), 3056 (OH), 1637 (C=O), 1606 (C=N), 1232 (C–F). ^1H NMR (400 MHz, DMSO-d_6, δ ppm): 3.21 (s, 3H, OCH_3), 6.93–7.81 (m, 14H, Ar–H), 8.23 (s, 1H, CH), 10.32 (s, 1H, OH), 11.48 (s, 1H, NH). LCMS: *m/z* 459.1 (M^+). C H N Analysis; calculated for $C_{27}H_{20}F_2N_2O_3$: C, 70.73; H, 4.40; N, 6.11. Found: C, 70.71; H, 4.39; N, 6.08.

*4,4''-Difluoro-N'-[(Z)-(4-fluorophenyl)methylidene]-5'-hydroxy-1,1':3',1''-terphenyl-4'-carbohydrazide (**10b**).* Yield 87%; m.p 108–110°C. IR (KBr, $\nu\,cm^{-1}$): 3354 (NH), 3046 (OH), 1640 (C=O), 1600 (C=N), 1223 (C–F). ^1H NMR (400 MHz, DMSO-d_6, δ ppm): 6.91–7.78 (m, 14H, Ar–H), 7.93 (s, 1H, CH), 9.85 (s, 1H, OH), 10.91 (s, 1H, NH). LCMS: *m/z* 447.1 (M^+). C H N Analysis; calculated for $C_{26}H_{17}F_3N_2O_2$: C, 69.95; H, 3.84; N, 6.28. Found: C, 69.94; H, 3.86; N, 6.24.

*4,4''-Difluoro-N'-[(Z)-(biphenyl-4-yl)methylidene]-5'-hydroxy-1,1':3',1''-terphenyl-4'-carbohydrazide (**10c**).* Yield 82%; m.p 226–228°C. IR (KBr, $\nu\,cm^{-1}$): 3329 (NH), 3059 (OH), 1647 (C=O), 1604 (C=N), 1217 (C–F). ^1H NMR (400 MHz, DMSO-d_6, δ ppm): 7.07–7.97 (m, 19H, Ar–H), 8.15 (s, 1H, CH), 10.15 (s, 1H, OH), 11.71 (s, 1H, NH). LCMS: *m/z* 505.1 (M^++1). C H N Analysis; calculated for $C_{32}H_{22}F_2N_2O_2$: C, 76.18; H, 4.40; N, 5.55. Found: C, 76.14; H, 4.38; N, 5.52.

*Acetylation of Schiff Base Derivatives (**11a–c**).* A mixture of Schiff base (0.01 mol) and acetic anhydride (10 mL) was dissolved in ethanol (25 mL) and the reaction mixture was refluxed for 6 h. The reaction mixture was then concentrated and allowed to cool. The solid product obtained was filtered, washed with water, and recrystallized using ethanol-dimethyl formamide mixture.

*1-[5-(4,4''-Difluoro-5'-hydroxy-1,1':3',1''-terphenyl-4'-yl)-2-(4-methoxyphenyl)-1,3,4-oxadiazol-3(2H)-yl]ethanone (**11a**).* Yield 42%; m.p 124–126°C. IR (KBr, $\nu\,cm^{-1}$): 3059 (OH), 2933, 2837 (CH), 1770 (C=O), 1608 (C=N), 1224 (C–F). ^1H NMR (400 MHz, DMSO-d_6, δ ppm): 1.84 (s, 3H, $COCH_3$), 3.45 (s, 3H, OCH_3), 6.41 (s, 1H, CH), 6.74–7.85 (m, 14H, Ar–H), 9.98 (s, 1H, OH). LCMS: *m/z* 501.1

(M^++1). C H N Analysis; calculated for $C_{29}H_{22}F_2N_2O_4$: C, 69.59; H, 4.43; N, 5.60. Found: C, 69.57; H, 4.44; N, 5.58.

1-[5-(4,4''-Difluoro-5'-hydroxy-1,1':3',1''-terphenyl-4'-yl)-2-(4-fluorophenyl)-1,3,4-oxadiazol-3(2H)-yl]ethanone ***(11b)***. Yield 45%; m.p 142–145°C. IR (KBr, ν cm^{-1}): 3043 (OH), 2889 (CH), 1762 (C=O), 1604 (C=N), 1232 (C–F). ^{1}H NMR (400 MHz, DMSO-d$_6$, δ ppm): 1.72 (s, 3H, $COCH_3$), 6.52 (s, 1H, CH), 6.82–7.91 (m, 14H, Ar–H), 9.85 (s, 1H, OH). LCMS: *m/z* 489.1 (M^+). C H N Analysis; calculated for $C_{28}H_{19}F_3N_2O_3$: C, 68.85; H, 3.92; N, 5.74. Found: C, 68.83; H, 3.90; N, 5.73.

1-[2-(Biphenyl-4-yl)-5-(4,4''-difluoro-5'-hydroxy-1,1':3',1''-terphenyl-4'-yl)-1,3,4-oxadiazol-3(2H)-yl]ethanone ***(11c)***. Yield 50%; m.p 133–126°C. IR (KBr, ν cm^{-1}): 3032 (OH), 2915 (CH), 1742 (C=O), 1611 (C=N), 1215 (C–F). ^{1}H NMR (400 MHz, DMSO-d$_6$, δ ppm): 1.94 (s, 3H, $COCH_3$), 6.48 (s, 1H, CH), 7.03–7.99 (m, 19H, Ar–H), 10.15 (s, 1H, OH). LCMS: *m/z* 547.1 (M^+). C H N Analysis; calculated for $C_{34}H_{24}F_2N_2O_3$: C, 74.71; H, 4.43; N, 5.13. Found: C, 74.72; H, 4.42; N, 5.10.

Synthesis of 5-(4,4''-Difluoro-5'-hydroxy-1,1':3',1''-terphenyl-4'-yl)-1,3,4-oxadiazole-2(3H)-thione ***(12)***. To a solution of carbohydrazide **4** (3.4 g, 0.01 mol) in ethanol (10 mL) a solution of carbon disulphide (1.14 g, 0.015 mol) in water (3 mL) and potassium hydroxide (1 g) was added, and the mixture was refluxed until the evolution of H_2S ceased (7-8 h). The reaction mixture was allowed to cool and then acidified with dilute hydrochloric acid. The solid obtained was collected by filtration, washed with least amount of water, and recrystallized from ethanol-water (1 : 1). Yield 62%; m.p 121–124°C. IR (KBr, ν cm^{-1}): 3224 (NH), 3124 (OH), 1604(C=N), 1228 (C–F). ^{1}H NMR (400 MHz, DMSO-d$_6$, δ ppm): 7.17–7.78 (m, 10H, Ar–H), 10.75 (s, 1H, OH), 14.53 (s, 1H, NH). LCMS: *m/z* 383.0 (M^++1). C H N Analysis; calculated for $C_{20}H_{12}F_2N_2O_2S$: C, 62.82; H, 3.16; N, 7.33. Found: C, 62.80; H, 3.17; N, 7.30.

Synthesis of 4-Amino-5-(4,4''-difluoro-5'-hydroxy-1,1':3',1''-terphenyl-4'-yl)-2,4-dihydro-3H-1,2,4-triazole-3-thione ***(13)***. A mixture of the oxadiazole **12** (3.82 g, 0.01 mol) and hydrazine hydrate (5 mL, 95%) in water (20 mL) was refluxed with stirring for 4 h, diluted with cold water (200 mL), acidified by the drop wise addition of concentrated hydrochloric acid, and filtered off. The solid formed was washed with the least amount of cold water and recrystallized from ethanol-dimethyl formamide mixture. Yield 58%; m.p 240–242°C. IR (KBr, ν cm^{-1}): 3373 (NH_2), 3259 (NH), 3163 (OH), 1616 (C=N), 1219 (C–F). ^{1}H NMR (400 MHz, DMSO-d$_6$, δ ppm): 5.45 (s, 2H, NH_2), 7.11–7.76 (m, 10H, Ar–H), 10.43 (s, 1H, OH), 13.54 (s, 1H, NH). LCMS: *m/z* 397.0 (M^++1). C H N Analysis; calculated for $C_{20}H_{14}F_2N_4OS$: C, 60.60; H, 3.56; N, 14.13. Found: C, 60.56; H, 3.57; N, 14.10.

Synthesis of Mannich Bases ***(14a–c)***. To a solution of oxadiazole **12** (3.82 g, 0.01 mole) in ethanol (5 mL), formaldehyde (0.5 mL, 37%) and secondary amine (0.01 mol) were added. The reaction mixture was stirred overnight. After cooling, the precipitate was filtered and recrystallized from ethanol.

5-(4,4''-Difluoro-5'-hydroxy-1,1':3',1''-terphenyl-4'-yl)-3-(morpholin-4-ylmethyl)-1,3,4-oxadiazole-2(3H)-thione ***(14a)***. Yield 72%; m.p 200–202°C. IR (KBr, ν cm^{-1}): 3157 (OH), 2937, 2858 (CH), 1620 (C=N), 1236 (C–F). ^{1}H NMR (400 MHz, DMSO-d$_6$, δ ppm): 2.35 (t, 4H, N-CH_2 of morpholine), 3.45 (t, 4H, O–CH_2 of morpholine), 4.90 (s, 2H, N–CH_2–N), 7.19–7.79 (m, 10H, Ar–H), 10.83 (s, 1H, OH). C H N Analysis; calculated for $C_{25}H_{21}F_2N_3O_3S$: C, 62.36; H, 4.40; N, 8.73. Found: C, 62.34; H, 4.41; N, 8.71.

5-(4,4''-Difluoro-5'-hydroxy-1,1':3',1''-terphenyl-4'-yl)-3-[(4-methylpiperazin-1-yl)methyl]-1,3,4-oxadiazole-2(3H)-thione ***(14b)***. Yield 68%; m.p 191–194°C. IR (KBr, ν cm^{-1}): 3163 (OH), 2862 (CH), 1616 (C=N), 1219 (C–F). ^{1}H NMR (400 MHz, DMSO-d$_6$, δ ppm): 2.29 (s, 3H, N–CH_3), 2.45 (t, 2H, CH_2), 2.95 (t, 4H, CH_2), 5.07 (s, 2H, N–CH_2–N), 7.17–7.89 (m, 10H, Ar–H), 10.72 (s, 1H, OH). C H N Analysis; calculated for $C_{26}H_{24}F_2N_4O_2S$: C, 63.14; H, 4.89; N, 11.33. Found: C, 63.12; H, 4.87; N, 11.30.

5-(4,4''-Difluoro-5'-hydroxy-1,1':3',1''-terphenyl-4'-yl)-3-(piperidin-1-ylmethyl)-1,3,4-oxadiazole-2(3H)-thione ***(14c)***. Yield 61%; m.p 98–100°C. IR (KBr, ν cm^{-1}): 3173 (OH), 2912 (CH), 1608 (C=N), 1222 (C–F). ^{1}H NMR (400 MHz, DMSO-d$_6$, δ ppm): 1.55–1.61 (m, 6H, CH_2), 2.82 (t, 4H, N–CH_2), 5.23 (s, 2H, N–CH_2–N), 7.11–7.76 (m, 10H, Ar–H), 10.63 (s, 1H, OH). C H N Analysis; calculated for $C_{26}H_{23}F_2N_3O_2S$: C, 65.12; H, 4.83; N, 8.76. Found: C, 65.10; H, 4.84; N, 8.74.

Alkylation of 1,3,4-Oxadiazole-2-thione Derivative ***(15a–c)***. Oxadiazole **12** (3.82 g, 0.01 mol) was dissolved in ethanol (20 mL) and 10% aqueous sodium hydroxide solution (3.5 mL) was added. Alkyl halide (0.02 mol) was added and heated to reflux for 4 h. The reaction mixture was then poured into ice cold water, filtered, and crystallized from ethanol.

5-(Butylsulfanyl)-2-(5'-butoxy-4,4''-difluoro-1,1':3',1''-terphenyl-4'-yl)-1,3,4-oxadiazole ***(15a)***. Yield 63%; m.p 270–272°C. IR (KBr, ν cm^{-1}): 2964 (CH), 1560(C=N), 1170 (C–F). LCMS: *m/z* 494.3 (M^+). C H N Analysis; calculated for $C_{28}H_{28}F_2N_2O_2S$: C, 67.99; H, 5.71; N, 5.66. Found: C, 67.96; H, 5.69; N, 5.63.

2-(4,4''-Difluoro-5'-ethoxy-1,1':3',1''-terphenyl-4'-yl)- 5-(Ethylsulfanyl)-1,3,4-oxadiazole ***(15b)***. Yield 58%; m.p 222–224°C. IR (KBr, ν cm^{-1}): 2972 (CH), 1572(C=N), 1220 (C–F). LCMS: *m/z* 438.4 (M^+). C H N Analysis; calculated for $C_{24}H_{20}F_2N_2O_2S$: C, 65.74; H, 4.60; N, 6.39. Found: C, 65.71; H, 4.57; N, 6.35.

Ethyl ({5-[4,4''-difluoro-5'-(2-ethoxy-2-oxoethoxy)-1,1':3',1''-terphenyl-4'-yl]-1,3,4-oxadiazol-2-yl}sulfanyl)acetate ***(15c)****.* Yield 72%; m.p 103–105°C. IR (KBr, ν cm^{-1}): 2954 (CH), 1724 (C=O ester), 1220 (C–F). ^{1}H NMR (400 MHz, DMSO-d$_6$, δ ppm): 1.14 (m, 6H, 2 × CH_3), 3.34 (s, 2H, SCH_2), 4.09 (m, 4H, 2 × CH_2), 5.05 (s, 2H, OCH_2), 7.16–7.91 (m, 10H, Ar–H). LCMS: *m/z* 555.1 (M^++1). C H N Analysis; calculated for $C_{28}H_{24}F_2N_2O_6S$: C, 60.64; H, 4.36; N, 5.05. Found: C, 60.61; H, 4.333; N, 5.01.

Synthesis of 2-[(4,4''-Difluoro-5'-hydroxy-1,1':3',1''-terphenyl-4'-yl)carbonyl]hydrazine carbothioamide ***(16)****.* A mixture of carbohydrazide **4** (3.4 g, 0.01 mol), potassium thiocyanate (0.015 mol), and hydrochloric acid (5 mL) in water (25 mL) was refluxed for 4 h with stirring. The resulting solid that separated was washed with hot water and recrystallized from 75% ethanol. Yield 83%; m.p 200–202°C. IR (KBr, ν cm^{-1}): 3464, 3329 (NH), 3236 (OH), 1672 (C=O), 1608 (C=N), 1226 (C–F). ^{1}H NMR (400 MHz, DMSO-d$_6$, δ ppm): 6.96, 8.19 (two singlets, 2H, NH_2), 7.09–7.71 (m, 10H, Ar–H), 9.57 (s, 1H, NH), 10.45 (s, 1H, NH), 10.75 (s, 1H, OH). LCMS: *m/z* 400.0 (M^++1). C H N Analysis; calculated for $C_{20}H_{15}F_2N_3O_2S$: C, 60.14; H, 3.79; N, 10.52. Found: C, 60.11; H, 3.79; N, 10.50.

Synthesis of 1,3-Thiazol Derivatives ***(17a–c)****.* An equimolar mixture of thiosemicarbazide derivative **16** (0.399 g, 0.001 mol) and phenacyl bromide derivative (0.001 mol) in 20 mL of absolute ethanol was heated under reflux for 4 h. The product that separated after cooling was collected and recrystallized from ethanol.

4,4''-Difluoro-5'-hydroxy-N'-[4-(4-methoxyphenyl)-1,3-thiazol-2-yl]-1,1':3',1''-terphenyl-4'-carbohydrazide ***(17a)****.* Yield 64%; m.p 213–215°C. IR (KBr, ν cm^{-1}): 3331(NH), 3196 (OH), 1654 (C=O), 1606 (C=N), 1228 (C–F). ^{1}H NMR (400 MHz, DMSO-d$_6$, δ ppm): 3.75 (s, 3H, OCH_3), 6.91–7.73 (m, 14H, Ar–H), 9.47 (s, 1H, NH), 10.23 (s, 1H, NH), 10.37 (s, 1H, OH). LCMS: *m/z* 530.1 (M^++1). C H N Analysis; calculated for $C_{29}H_{21}F_2N_3O_3S$: C, 65.77; H, 4.00; N, 7.93. Found: C, 65.74; H, 3.96; N, 7.90.

4,4''-Difluoro-5'-hydroxy-N'-[4-(4-fluorophenyl)-1,3-thiazol-2-yl]-1,1':3',1''-terphenyl-4'-carbohydrazide ***(17b)****.* Yield 67%; m.p 200–202°C. IR (KBr, ν cm^{-1}): 3329 (NH), 3236 (OH), 1662 (C=O), 1608 (C=N), 1226 (C–F). ^{1}H NMR (400 MHz, DMSO-d$_6$, δ ppm): 6.81–7.67 (m, 14H, Ar–H), 9.39 (s, 1H, NH), 10.17 (s, 1H, NH), 10.41 (s, 1H, OH). LCMS: *m/z* 518.2 (M^++1). C H N Analysis; calculated for $C_{28}H_{18}F_3N_3O_2S$: C, 64.98; H, 3.51; N, 8.12. Found: C, 64.96; H, 3.50; N, 8.08.

4,4''-Difluoro-5'-hydroxy-N'-[4-(biphenyl-4-yl)-1,3-thiazol-2-yl]-1,1':3',1''-terphenyl-4'-carbohydrazide ***(17c)****.* Yield 60%; m.p 169–172°C. IR (KBr, ν cm^{-1}): 3324 (NH), 3214 (OH), 1652 (C=O), 1601 (C=N), 1221 (C–F). ^{1}H NMR (400 MHz, DMSO-d$_6$, δ ppm): 6.69–8.14 (m, 19H, Ar–H), 9.41 (s, 1H, NH), 10.29 (s, 1H, NH), 10.74 (s, 1H, OH). LCMS: *m/z* 576.3 (M^++1). C H N Analysis; calculated for $C_{34}H_{23}F_2N_3O_2S$: C, 70.94; H, 4.03; N, 7.30. Found: C, 70.91; H, 4.01; N, 7.27.

Synthesis of 4,4''-Difluoro-4'-(5-sulfanyl-4H-1,2,4-triazol-3-yl)-1,1':3',1''-terphenyl-5'-ol ***(18)****.* A suspension of the thiosemicarbazide derivative **16** (0.399 g, 0.001 mol) in sodium hydroxide solution (10 mL, 10%) was heated under reflux for 10 h. The reaction mixture was allowed to cool and then adjusted to pH 6 with 10% hydrochloric acid. The precipitate that formed was filtered off, washed with water, dried, and recrystallized from ethanol. Yield 60%; m.p 189–192°C. IR (KBr, ν cm^{-1}): 3433, 3329 (NH), 3236 (OH), 2372 (SH), 1608 (C=N), 1228 (C–F). ^{1}H NMR (400 MHz, DMSO-d$_6$, δ ppm): 4.07 (s, 1H, SH), 7.05–7.72 (m, 10H, Ar–H), 10.23 (s, 1H, OH), 13.54 (s, 1H, NH). LCMS: *m/z* 382.0 (M^++1). C H N Analysis; calculated for $C_{20}H_{13}F_2N_3OS$: C, 62.98; H, 3.44; N, 11.02. Found: C, 62.94; H, 3.42; N, 10.99.

Synthesis of 4,4''-Difluoro-5'-hydroxy-N-(5-oxo-2-sulfanyl-4,5-dihydro-1H-imidazol-1-yl)-1,1':3',1''-terphenyl-4'-carboxamide ***(19)****.* A mixture of thiosemicarbazide derivative **16** (0.399 g, 0.001 mol), monochloroacetic acid (0.001 mol), and anhydrous sodium acetate (0.002 mol) in glacial acetic acid (30 mL) was refluxed for 8 h. After cooling, the reaction mixture was poured on crushed ice. The resulting solid that separated was washed with hot water and recrystallized from ethanol. Yield 65%; m.p 262–264°C. IR (KBr, ν cm^{-1}): 3327 (NH), 3132 (OH), 1724, 1635 (C=O), 1598 (C=N), 1222 (C–F). ^{1}H NMR (400 MHz, DMSO-d$_6$, δ ppm): 3.88 (s, 2H, CH_2), 4.10 (s, 1H, SH), 6.94–7.78 (m, 10H, Ar–H), 10.59 (s, 1H, NH), 10.60 (s, 1H, OH). LCMS: *m/z* 440.0 (M^++1). C H N Analysis; calculated for $C_{22}H_{15}F_2N_3O_3S$: C, 60.13; H, 3.44; N, 9.56. Found: C, 60.10; H, 3.42; N, 9.52.

5.3. Biological Evaluation

5.3.1. Antibacterial Studies. The newly synthesized compounds were screened for their antibacterial activity against *Escherichia coli* (ATTC-25922), *Staphylococcus aureus* (ATTC-25923), *Pseudomonas aeruginosa* (ATCC-27853), and *Klebsiella pneumoniae* (recultured) bacterial strains by serial plate dilution method. Serial dilutions of the drug in Mueller-Hinton broth were taken in tubes and their pH was adjusted to 5.0 using phosphate buffer. A standardized suspension of the test bacterium was inoculated and incubated for 16–18 h at 37°C. The minimum inhibitory concentration (MIC) was noted by seeing the lowest concentration of the drug at which there was no visible growth. A number of antimicrobial discs were placed on the agar for the sole purpose of producing zones of inhibition in the bacterial lawn. About 20 mL of agar media was poured into each petri dish. Excess of suspension was decanted and plates were dried by placing in an incubator at 37°C for an hour. Using a punch, wells were made on these seeded agar plates and minimum inhibitory concentrations of the test compounds in dimethylsulfoxide (DMSO) were added into each labeled well. A control was also prepared for the

Table 1: Antibacterial activity of the synthesized compounds: MIC in μg/mL (Zone of inhibition in mm).

Compound	*Staphylococcus aureus* (ATTC-25923)	*Escherichia coli* (ATTC-25922)	*Pseudomonas Aeruginosa* (ATTC-27853)	*Klebsiella pneumonia* (recultured)
3	12.5 (6)	12.5 (6)	12.5 (6)	12.5 (6)
4	12.5 (6)	12.5 (8)	12.5 (6)	12.5 (6)
5	12.5 (6)	12.5 (6)	12.5 (6)	12.5 (8)
6	12.5 (6)	12.5 (6)	12.5 (6)	12.5 (6)
7	R	100 (2)	100 (4)	100 (5)
8	R	R	R	R
9a	R	R	R	R
9b	R	R	R	R
9c	R	R	R	R
10a	R	R	R	R
10b	R	R	R	R
10c	R	R	R	R
11a	R	R	R	R
11b	R	R	R	R
11c	R	R	R	R
12	100 (4)	6.25 (6)	6.25 (6)	6.25 (6)
13	R	R	R	R
14a	R	R	R	R
14b	R	R	R	R
14c	R	R	R	R
15a	R	R	R	R
15b	R	R	R	R
15c	R	R	R	R
16	12.5 (6)	12.5 (6)	12.5 (6)	12.5 (6)
17a	12.5 (6)	12.5 (6)	12.5 (6)	12.5 (6)
17b	R	12.5 (6)	12.5 (6)	12.5 (6)
17c	12.5 (6)	12.5 (6)	12.5 (6)	12.5 (6)
18	R	R	R	R
19	R	R	R	R
Ciprofloxacin	6.25 (20)	6.25 (17)	6.25 (19)	6.25 (18)

Note: R indicates bacterial strains are resistant to the compounds >100 μg/mL.

plates in the same way using solvent DMSO. The petri dishes were prepared in triplicate and maintained at 37°C for 3-4 days. Antibacterial activity was determined by measuring the diameter of inhibition zone. Activity of each compound was compared with ciprofloxacin as standard. Zone of inhibition was determined for newly synthesized compounds and the results are summarized in Table 1.

5.3.2. Antifungal Studies. Newly prepared compounds were screened for their antifungal activity against *Aspergillus flavus* (NCIM No. 524), *Aspergillus fumigates* (NCIM No. 902), *Penicillium* (*S. aureus*) (recultured), and *Trichophyton mentagrophytes* (recultured) in DMSO by serial plate dilution method. Sabourands agar media were prepared by dissolving peptone (1 g), D-glucose (4 g) and agar (2 g) in distilled water (100 mL) and adjusting the pH to 5.7. Normal saline was used to make a suspension of spore of fungal strains for lawning. A loopful of particular fungal strain was transferred to 3 mL saline to get a suspension of corresponding species. Twenty milliliters of agar media was poured into each Petri dish. Excess of suspension were decanted and plates were dried by placing in incubator at 37°C for 1 h. Using a punch, wells were made on these seeded agar plates minimum inhibitory concentrations of the test compounds in DMSO were added into each labeled well. A control was also prepared for the plates in the same way using solvent DMSO. The Petri dishes were prepared in triplicate and maintained at 37°C for 3-4 days. Antifungal activity was determined by measuring the diameter of inhibition zone. Activity of each compound was compared with itraconozole as standard. Zones of inhibition were determined and the results are summarized in Table 2.

5.3.3. DPPH Radical Scavenging Assay. The DPPH assay was based on the reported method [37]. Briefly, 1 mM solution of DPPH in ethanol was prepared, and this solution (1 mL) was added to sample solutions 1 mg/mL of DMSO. The mixture was shaken vigorously and allowed to stand at room temperature for 20 min. Then the absorbance was measured at 517 nm in a spectrophotometer. Lower absorbance of the

TABLE 2: Antifungal activity of the synthesized compounds: MIC in μg/mL (Zone of inhibition in mm).

Compound	*Penicillium marneffei* (recultured)	*Trichophyton mentagrophytes* (recultured)	*Aspergillus flavus* (NCIM No. 524)	*Aspergillus fumigates* (NCIM No. 902)
3	R	R	R	R
4	R	R	R	R
5	R	R	R	R
6	6.25 (8)	6.25 (8)	6.25 (8)	6.25 (8)
7	R	R	R	R
8	6.25 (8)	6.25 (8)	6.25 (8)	6.25 (8)
9a	100 (5)	12.5 (6)	12.5 (6)	R
9b	R	R	R	R
9c	100 (2)	12.5 (6)	12.5 (6)	R
10a	12.5 (6)	R	R	R
10b	6.25 (8)	6.25 (8)	6.25 (8)	6.25 (8)
10c	R	R	R	R
11a	100 (4)	12.5 (6)	12.5 (6)	R
11b	12.5 (6)	12.5 (6)	12.5 (6)	R
11c	100 (3)	12.5 (6)	12.5 (6)	R
12	R	R	R	R
13	R	R	R	R
14a	R	R	R	R
14b	R	R	R	R
14c	R	R	R	R
15a	R	R	R	R
15b	R	R	R	R
15c	R	R	R	R
16	R	R	R	R
17a	12.5 (6)	12.5 (6)	12.5 (6)	12.5 (6)
17b	R	R	R	R
17c	12.5 (6)	R	R	R
18	R	R	R	R
19	R	12.5 (8)	12.5 (6)	12.5 (8)
Itraconozole	6.25 (17)	6.25 (19)	6.25 (18)	6.25 (18)

Note: R indicates fungal strains are resistant to the compounds >100 μg/mL.

reaction mixture indicated higher free radical scavenging activity. The capability to scavenge the DPPH radical was calculated using the following equation:

$$\text{DPPH scavenging effect (\%)} = \left(A_0 - \frac{A_1}{A_0}\right) \times 100, \quad (1)$$

where A_0 is the absorbance of the control reaction and A_1 is the absorbance in the presence of the samples or standards. Each sample was assayed at 1 mg/mL and all experiments were carried out in triplicate and the % RSC is shown in Table 3.

5.3.4. Reducing Power Assay. The reducing power of the synthesized compounds was determined according to the method of Oyaizu [38]. Different concentrations of the samples (100–1000 μg/mL) in DMSO (1 mL) were mixed with phosphate buffer (2.5 mL, 0.2 M, pH = 6.6) and potassium ferricyanide (2.5 mL, 1% solution). The mixture was incubated at 50°C for 20 min. after which 10% trichloroacetic acid (2.5 mL) was added to the mixture, which was then centrifuged for 10 min. The upper layer of solution (2.5 mL) was mixed with distilled water (2.5 mL) and $FeCl_3$ (0.5 mL, 0.1%) and then the absorbance at 700 nm was measured using a spectrophotometer. Higher absorbance of the reaction mixture indicated greater reducing power. All experiments were carried out in triplicate and the reducing power assay was represented by effective concentration (mg/mL) equivalent to 0.5 absorbance glutathione (Table 3).

Acknowledgments

The authors are thankful to the Director of IISC, Bangalore, for NMR data. B. Narayana thanks the UGC for financial assistance through BSR one time Grant for the purchase

TABLE 3: Antioxidant activity of synthesized compounds.

Compounds	% DPPH scavenging	Reducing power assay
3	8.46 ± 4.49	18.18 ± 0.46
4	48.40 ± 1.12	0.78 ± 0.01
5	35.44 ± 10.47	6.39 ± 1.52
6	28.30 ± 1.12	10.32 ± 0.45
7	10.05 ± 2.73	11.90 ± 0.01
8	29.09 ± 1.42	6.15 ± 1.25
9a	69.54 ± 4.35	14.43 ± 3.96
9b	54.94 ± 3.14	10.45 ± 0.92
9c	42.32 ± 5.35	12.11 ± 2.14
10a	7.66 ± 1.85	10.22 ± 0.59
10b	15.34 ± 2.23	6.99 ± 1.16
10c	19.04 ± 5.97	8.23 ± 2.22
11a	36.81 ± 0.64	63.49 ± 11.22
11b	32.80 ± 0.01	10.82 ± 3.31
11c	47.27 ± 3.86	21.47 ± 16.77
12	72.22 ± 2.62	1.49 ± 0.11
13	73.63 ± 9.00	1.65 ± 0.19
14a	61.36 ± 9.64	7.82 ± 1.46
14b	57.26 ± 6.42	20.51 ± 8.21
14c	49.54 ± 3.07	8.77 ± 1.18
15a	1.01 ± 0.01	26.06 ± 4.72
15b	19.99 ± 3.29	8.25 ± 0.96
15c	68.63 ± 9.64	7.67 ± 2.23
16	51.58 ± 4.11	1.12 ± 0.07
17a	16.43 ± 2.54	6.18 ± 1.87
17b	49.09 ± 4.99	4.29 ± 0.54
17c	27.50 ± 4.49	6.07 ± 1.66
18	22.22 ± 7.48	2.16 ± 0.25
19	15.40 ± 2.91	5.92 ± 1.44
Glutathione	92.09 ± 1.09	0.595 ± 0.01

of chemicals. S. Samshuddin thanks UGC for providing financial help for the research work through UGC Research Fellowship in Sciences for Meritorious Students scheme.

References

[1] M. Gill and W. Steglich, "Pigments of fungi (*Macromycetes*)," *Fortschritte der Chemie Organischer Naturstoffe*, vol. 51, pp. 1–317, 1987.

[2] J. K. Liu, "Natural terphenyls: developments since 1877," *Chemical Reviews*, vol. 106, no. 6, pp. 2209–2223, 2006.

[3] D. Astrue, *Modern Arene Chemistry*, Wiley-VCH, Weinheim, Germany, 2002.

[4] S. Samshuddin, B. Narayana, and B. K. Sarojini, "Ethyl 4, 4′′-difluoro-5′-hydroxy-1, 1′:3′, 1′′-terphenyl-4′-carboxylate," *Molbank*, vol. 2011, no. 4, article M745, 2011.

[5] W. Wenner, "Antitubercular agents. Derivatives of pyridine-carboxylic acid hydrazides," *Journal of Organic Chemistry*, vol. 18, no. 10, pp. 1333–1337, 1953.

[6] S. S. Parmar and R. Kumar, "Substituted quinazolone hydrazides as possible antituberculous agents," *Journal of Medicinal Chemistry*, vol. 11, no. 3, pp. 635–636, 1968.

[7] R. P. Bhamaria, R. A. Bellare, and C. V. Deliwala, "In intro effect of 1-acyl-4-alkyl-(oraryl)-thiosemicarbazides 1-(5-chlorosalicylidine)-4-alkyl-(oraryl)-thiosemicarbazones and some hydrazones of 5-chlorosalicylaldehyde against pathogenic bacteria including mycobacterium tuberculosis (H37Rv)," *Indian Journal of Experimental Biology*, vol. 6, no. 1, pp. 62–63, 1968.

[8] E. H. Abdel-Al, M. I. Al-Ashamawi, and B. Abdel Fattah, "Synthesis and antimicrobial testing of certain oxadiazoline and triazole derivatives," *Pharmazie*, vol. 38, no. 12, pp. 833–835, 1983.

[9] A. K. Sen Gupta, M. Garg, and U. Chandra, "Synthesis of some new Mannich bases derived from substituted benzimidazole, benzoxazol-2-one, benzoxazol-2-thione, oxadiazol-2-thiones and their biological activities," *Journal of the Indian Chemical Society*, vol. 56, no. 12, pp. 1230–1232, 1979.

[10] A. K. Mansour, M. M. Eid, and N. S. A. M. Khalil, "Synthesis and reactions of some new heterocyclic carbohydrazides and related compounds as potential anticancer agents," *Molecules*, vol. 8, no. 10, pp. 744–755, 2003.

[11] M. M. Dutta, B. N. Goswami, and J. C. S. Kataky, "Studies on biologically active heterocycles. Part I. Synthesis and antifungal activity of some new aroyl hydrazones and 2,5-disubstituted-1,3,4-oxadiazoles," *Journal of Heterocyclic Chemistry*, vol. 23, no. 3, pp. 793–795, 1986.

[12] S. C. Sharma, "Synthesis of new fungicides. 2-(4′-Arylthiazolyl-2′-imino)-3-aryl-4-thiazolidones," *Bulletin of the Chemical Society of Japan*, vol. 40, no. 10, pp. 2422–2424, 1967.

[13] V. N. Choubey and H. Singh, "Synthesis of some new fungicides," *Bulletin of the Chemical Society of Japan*, vol. 43, no. 7, pp. 2233–2236, 1970.

[14] W. O. Foye and P. Tovivich, "*N*-glucopyranosyl-5-aralkylidenerhodanines: synthesis and antibacterial and antiviral activities," *Journal of Pharmaceutical Sciences*, vol. 66, no. 11, pp. 1607–1611, 1977.

[15] E. B. Akerblom, "Synthesis and structure- activity relationships of a series of antibacterially active 5-(5-nitro-2-furfurylidene)thiazolones, 5 (5 nitro 2 furylpropenylidene)thiazolones, and 6-(5-nitro- 2-furyl)-4*H*-1,3-thiazinones," *Journal of Medicinal Chemistry*, vol. 17, no. 6, pp. 609–615, 1974.

[16] S. Samshuddin, B. Narayana, D. N. Shetty, and R. Raghavendra, "An efficient synthesis of 2,4,6-triaryl pyridines and their biological evaluation," *Der Pharma Chemica*, vol. 3, no. 3, pp. 232–240, 2011.

[17] S. Samshuddin, B. Narayana, B. K. Sarojini et al., "Antimicrobial, analgesic, DPPH scavenging activities and molecular docking study of some 1,3,5-triaryl-2-pyrazolines," *Medicinal Chemistry Research*, vol. 21, no. 8, pp. 2012–2022, 2011.

[18] S. Samshuddin, B. Narayana, D. N. Shetty, R. Srinivasan, and B. K. Sarojini, "(2*E*)-3-[4-(1*H*-Benzimidazol-2-ylmethoxy)-3-methoxyphenyl]-1-(4, 4′′-difluoro-5′-methoxy-1, 1′:3′, 1′′-terphenyl-4′-yl)prop-2-en-1-one," *Molbank*, vol. 2012, no. 3, article M764, 2012.

[19] S. Samshuddin, B. Narayana, B. K. Sarojini, R. Srinivasan, Vinayachandra, and K. R. Chandrashekar, "Synthesis, characterization and biological evaluation of some pyrazoles derived from α,β-dibromo 4,4′-difluoro chalcone," *Der Pharma Chemica*, vol. 4, no. 2, pp. 587–592, 2012.

[20] S. Samshuddin, B. Narayana, Z. Baktir, M. Akkurt, and H. S. Yathirajan, "Synthesis, characterization and crystal structure of 1-[3,5-bis(4-fluorophenyl)-4,5-dihydro-1H-pyrazol-1-yl]propan-1-one," *Der Pharma Chemica*, vol. 3, no. 6, pp. 487–493, 2011.

[21] S. Samshuddin, R. J. Butcher, M. Akkurt, B. Narayana, H. S. Yathirajan, and B. K. Sarojini, "1,3-Bis(4-fluorophenyl)-N,N'-(propane-1,3-diylidene)dihydroxylamine," *Acta Crystallographica Section E*, vol. 67, no. 8, pp. o1954–o1955, 2011.

[22] H. -K. Fun, M. Hemamalini, S. Samshuddin, B. Narayana, and H. S. Yathirajan, "1-(4, 4''-Difluoro-5'-methoxy-1, 1':3', 1''-terphenyl-4'-yl) ethanone," *Acta Crystallographica Section E*, vol. 68, no. 1, p. o163, 2010.

[23] H. K. Fun, M. Hemamalini, S. Samshuddin, B. Narayana, and B. K. Sarojini, "(E)-1-(4, 4''-Difluoro-5'-methoxy-1, 1':3', 1''-terphenyl-4'-yl)-3-(6-methoxynaphthalen-2-yl) prop-2-en-1-one," *Acta Crystallographica Section E*, vol. 67, no. 12, pp. o3327–o3328, 2011.

[24] B. Narayana, K. K. V. Raj, B. V. Ashalatha, and N. S. Kumari, "Synthesis of some new 2-(6-methoxy-2-naphthyl)-5-aryl-1,3,4-oxadiazoles as possible non-steroidal anti-inflammatory and analgesic agents," *Archiv der Pharmazie*, vol. 338, no. 8, pp. 373–377, 2005.

[25] B. Narayana, B. V. Ashalatha, K. K. Vijaya Raj, and B. K. Sarojini, "Synthesis and studies on antimicrobial, anti-inflammatory and antiproliferative activities of heterocycles derived from 4-/5-/6-/7-nitro/ 5fluoro/chloro/bromoindole-2-carbohydrazides," *Indian Journal of Chemistry*, vol. 48, no. 12, pp. 1794–1805, 2009.

[26] S. Rollas, N. Gulerman, and H. Erdeniz, "Synthesis and antimicrobial activity of some new hydrazones of 4-fluorobenzoic acid hydrazide and 3-acetyl-2,5-disubstituted-1,3,4-oxadiazolines," *IL Farmaco*, vol. 57, no. 2, pp. 171–174, 2002.

[27] A. N. Mayekar, H. S. Yathirajan, B. Narayana, B. K. Sarojini, and N. S. Kumari, "Synthesis and antimicrobial studies on new substituted 1, 3, 4-oxadiazole derivatives bearing 6-bromonaphthalene moiety," *International Journal of Chemistry*, vol. 2, no. 1, pp. 38–54, 2010.

[28] K. K. Vijaya Raj, B. Narayana, B. V. Ashalatha, N. Suchetha Kumari, and B. K. Sarojini, "Synthesis of some bioactive 2-bromo-5-methoxy-*N*'-[4-(aryl)-1,3-thiazol-2-yl]benzohydrazide derivatives," *European Journal of Medicinal Chemistry*, vol. 42, no. 3, pp. 425–429, 2007.

[29] B. V. Ashalatha, B. Narayana, and N. S. Kumari, "Synthesis and biological activity of novel 4-/5-/6-/7-nitro-*N*'-(4-aryl-1,3-thiazol-2-yl)1*H*-indole-2-carbohydrazide derivatives," *Phosphorus, Sulfur and Silicon and the Related Elements*, vol. 181, no. 12, pp. 2785–2795, 2006.

[30] A. L. Barry, "Procedure for testing antimicrobial agents in agar media," in *Antibiotics in Laboratory Medicine*, V. L. Corian, Ed., Williams and Wilkins, Baltimore, Md, USA, 1991.

[31] D. James, M. Lowry, M. J. Jaqua, and S. T. Selepak, "Detailed methodology and implementation of a semi automated serial dilution micro technique for antimicrobial susceptibility testing," *Applied Microbiology*, vol. 20, no. 1, pp. 46–53, 1970.

[32] C. H. Fenlon and M. H. Cynamon, "Comparative *in vitro* activities of ciprofloxacin and other 4-quinolones against mycobacterium tuberculosis and mycobacterium intracellulare," *Antimicrobial Agents and Chemotherapy*, vol. 29, no. 3, pp. 386–388, 1986.

[33] B. A. Arthington-Skaggs, M. Motley, D. W. Warnock, and C. J. Morrison, "Comparative evaluation of PASCO and national committee for clinical laboratory standards M27-A broth microdilution methods for antifungal drug susceptibility testing of yeasts," *Journal of Clinical Microbiology*, vol. 38, no. 6, pp. 2254–2260, 2000.

[34] A. K. Tiwari, "Antioxidants: new-generation therapeutic base for treatment of polygenic disorders," *Current Science*, vol. 86, no. 8, pp. 1092–1102, 2004.

[35] M. Oktay, I. Gulcin, and O. I. Kufrevioglu, "Determination of *in vitro* antioxidant activity of fennel (*Foeniculum vulgare*) seed extracts," *LWT—Food Science and Technology*, vol. 36, no. 2, pp. 263–271, 2003.

[36] S. Chanda and R. Dave, "*In vitro* models for antioxidant activity evaluation and some medicinal plants possessing antioxidant properties: an overview," *African Journal of Microbiology Research*, vol. 3, no. 13, pp. 981–996, 2009.

[37] M. S. Blois, "Antioxidant determinations by the use of a stable free radical," *Nature*, vol. 181, no. 4617, pp. 1199–1200, 1958.

[38] M. Oyaizu, "Studies on products of the browning reaction. Antioxidative activities of browning reaction products prepared from glucosamine," *Japanese Journal of Nutrition*, vol. 44, no. 6, pp. 307–315, 1986.

20

A DFT and Semiempirical Model-Based Study of Opioid Receptor Affinity and Selectivity in a Group of Molecules with a Morphine Structural Core

Tamara Bruna-Larenas and Juan S. Gómez-Jeria

Department of Chemistry, Faculty of Sciences, University of Chile, P.O. Box 653, Santiago, Chile

Correspondence should be addressed to Juan S. Gómez-Jeria, facien03@uchile.cl

Academic Editor: O. Bruno

We report the results of a search for model-based relationships between mu, delta, and kappa opioid receptor binding affinity and molecular structure for a group of molecules having in common a morphine structural core. The wave functions and local reactivity indices were obtained at the ZINDO/1 and B3LYP/6-31G** levels of theory for comparison. New developments in the expression for the drug-receptor interaction energy expression allowed several local atomic reactivity indices to be included, such as local electronic chemical potential, local hardness, and local electrophilicity. These indices, together with a new proposal for the ordering of the independent variables, were incorporated in the statistical study. We found and discussed several statistically significant relationships for mu, delta, and kappa opioid receptor binding affinity at both levels of theory. Some of the new local reactivity indices incorporated in the theory appear in several equations for the first time in the history of model-based equations. Interaction pharmacophores were generated for mu, delta, and kappa receptors. We discuss possible differences regulating binding and selectivity in opioid receptor subtypes. This study, contrarily to the statistically backed ones, is able to provide a microscopic insight of the mechanisms involved in the binding process.

1. Introduction

Molecular recognition processes control a huge number of aspects of life on Earth. The ability of molecules to recognize a certain pattern of atom distribution and not another is central to catalysis, drug effects, chemical reactivity, and so forth. Concerning the recognition by a drug of one or more receptors, this is a phenomenon that still needs to be fully understood to design new agonists or antagonists for a given receptor type. The central problem of the drug-receptor interaction is the following: how can a certain molecule be recognized by two or more receptors and display different affinities for them? Among the molecules having this interesting property we may cite dopaminergic, serotoninergic, and opioid compounds. In the following we shall focus on the latter.

Regarding opioids there is abundant evidence for the existence of four major classes of receptors in the central nervous system (CNS), designated as μ, δ, κ, and nociceptin, as well as subtypes within the first three classes (we employed capital letters to avoid confusions because similar small Greek letters are used to design reactivity indices used below). Each receptor type has a distinct selectivity profile and a unique distribution within the CNS. They are activated both by endogenously produced opioid peptides and by exogenously administered opiate compounds, some of which are not only among the most effective analgesics known but also highly addictive and abused (they are not the only ones to be activated by exogenous compounds; serotoninergic 5-HT_2 receptors are activated by hallucinogens like LSD; e.g.).

The great medical importance of these and similar molecular systems requires research on their quantitative structure-activity relationships (QSAR) in order to improve our knowledge about how receptor binding, selectivity, and pharmacological effects are achieved. The set of molecular features which are necessary for binding are summarized in the "interaction pharmacophore" and those necessary for pharmacological effects (agonism, antagonism, etc.) in the "agonist pharmacophore," "antagonist pharmacophore", and so forth. In the case of opiates there are several lines of research dealing with the synthesis and pharmacological evaluation of derivatives of endogenous opioid peptides (enkephalins, which are δ receptor ligands, β-endorphin, which binds to all opioid receptors, and dynorphins, which exert their effects primarily through the κ-opioid receptor). Another line of research does the same but with rigid opiates derived from morphine, pethidine, naltrexone, and other exogenous molecules. Finally, there have been some efforts to combine experimental results with crystallographic data and quantum chemical calculations into several pharmacophore models.

Regarding the last line of research, Gorin and Marshall defined a model for the opiate receptor by using a computer-based molecular display, and X-ray crystallographic input data. The model can explain the stereochemistry of the way in which the morphine, morphinan, and oripavine classes of compounds interact with the receptor. The minimal structural unit of the enkephalins demonstrated to be pharmacologically active, Tyr-lyGly-Phe, was also fitted to this model by using a systematic search of conformational space. This model for the analgesic pharmacophore utilizes the previously recognized requirement of the phenolic ring and tertiary amine of morphine. To be consistent with the stereospecific activity of the morphine, morphinan, and oripavine classes of compounds, it was also proposed that atoms C5 and C6 of the C ring of morphine are an additional requirement [1]. Burt et al. have identified molecular features and types of receptor interactions that modulate agonist/antagonist potencies in diverse classes of opiates. Using the fused-ring opiates as a template for the interaction of opiates at the receptor site, they have developed hypotheses that can explain pure agonist, pure antagonist, and mixed agonist/antagonist activity not only for fused-ring opiates, but also for peptide opiates and flexible opiates such as the 4-phenylpiperidines and 3-phenylpiperidines [2]. Fournie-Zaluski et al. suggested that μ receptors bind preferentially to highly hydrophobic compounds with compact structures while δ receptors exhibit a stronger affinity for larger peptides with hydrophilic components. For this they used a mixture of experimentally measured IC_{50} values and opioid activities [3]. Loew et al. examined the conformational behavior of four tetrapeptide enkephalin analogues (Tyr-Gly-Gly- Phe-OH, Tyr-Gly-Gly-Phe-NH_2, Tyr-D-Ala-Gly-Phe-NH_2, and Tyr-D-Ala-Gly- (NMe)Phe-NH_2) to identify conformations that are active and inactive at the opiate analgesic receptor. Thus, on the basis of conformational data, the Tyr-Gly-Gly-Phe-NH_2 analogue is predicted to have very weak opiate activity [4]. Lavecchia et al. modeled the κ-opioid receptor-agonists interactions using pharmacophore-based and docking simulations [5]. Their data provide additional evidence that δ-opioid agonists and antagonists interact within the same ligand-binding domain in opioid receptors and that hydrophobic substituents at the C-terminus of the Dmt-Tic pharmacophore augment μ-opioid receptor affinity. Thus, Dmt-Tic (Dmt = 2′,6′-dimethyl-L-tyrosine, Tic = 1,2,3,4-tetrahydroisoquinoline-3-carboxylic acid) with hydrophobic C-terminal substituents enhanced μ affinity to provide δ antagonists with dual receptor affinities and bifunctional activity. Bonner et al. prepared topographically constrained analogues of the highly μ-opioid-receptor-selective antagonist CTAP (H-D-Phe-c [Cys-Tyr-D-Trp-Arg-Thr-Pen]-Thr-NH_2) [6]. Molecular modeling based on 2D NMR revealed that low energy conformers of peptides with similar biological activities had similar aromatic pharmacophore orientations and interaromatic distances. Peptides that exhibit μ antagonism have interaromatic distances of 7.0–7.9 Å and have their amino terminal aromatic moiety pointing in a direction opposite to the orientation of the amino terminus. Peptides with δ opioid activity displayed an interaromatic distance of <7 Å and had their amino terminal aromatic moiety pointing in the same direction as the amino terminus. Bernard et al. extended the conformationally sampled pharmacophore (CSP) method to peptide ligands using replica exchange molecular dynamics simulation for conformational sampling [7]. The developed 2D CSP indicates that the spatial relationship of the basic nitrogen and the hydrophobic moiety in the δ opioid ligands differentiates activity. Such an overlap is expected because all the ligands bind to the same receptor and support a model where both classes of ligands interact with the δ receptor via the same binding mode. However, there exist high-probability regions that are primarily sampled by agonists versus antagonists and vice versa for both the peptide and nonpeptide ligands. Kuz'mina et al. presented in a series of papers a general model of the opiate pharmacophore [8–10]. Bernard et al. extended the CSP method to obtain quantitative models of δ opioid ligand efficacy and affinity [11]. The models obtained for a structurally diverse set of peptide and nonpeptide δ opioid ligands offer good predictions with $R2$ values > 0.9, and the predicted efficacy for a set of test compounds was consistent with the experimental values. Later Shim et al. applied the CSP method to develop a predictive model of the efficacy of μ-opioid receptor ligands [12]. Their model predicts (1) that interactions of ligands with the B site, as with the 19-alkyl substituents of oripavines, modulate the extent of agonism; (2) that agonists with long N-substituents, as with fentanyl and N-phenethylnormorphine, can bind in an orientation such that the N substituent interacts with the B site that also allows the basic N-receptor Asp interaction essential for agonism; (3) that the μ agonist herkinorin, that lacks a basic nitrogen, binds to the receptor in a manner similar to the traditional opioids via interactions mediated by water or an ion.

On the experimental side, biological studies of endogenous opioid peptides, as well as synthetic analogues, have led to the hypothesis that the tetrapeptide sequence from Tyr^1 to Phe^4 is an important requirement for activity. It was proposed that the N-terminal tetrapeptide sequence of endogenous peptides carries the "message", which is responsible for mediating the opioid effect. The C-terminal segments of these peptides, which differ in length and physical-chemical character, play an "address" role in conferring selectivity for different opioid receptor types. Lipkowski et al. synthesized analogues of leucine-enkephalin and dynorphin(1–8) in which the N-terminal dipeptide message sequence has been replaced by oxymorphone or naltrexone (these molecules are called Alkaloid-Peptide Hybrids) [13]. Their results suggest that the selectivity for different opioid receptor types depends on a balance between the affinities of the message and address components. In cases where these components have comparable receptor affinities, the address can significantly shift selectivity by increasing affinity for one receptor type while reducing affinity for other types. When the message component has high affinity for a particular receptor type, the modulatory role of the address is expressed mainly by reducing the affinity of the ligand for other opioid receptor types. Portoghese et al. also synthesized bivalent ligands consisting of oxymorphamine and [D-Glu^2]enkephalin pharmacophores linked through a spacer attached to the 6-amino group of the former and D-Glu of the latter to investigate the possible coexistence of μ and δ recognition sites in the same opioid receptor complex [14]. They suggest that the results are consistent with the simultaneous occupation of μ and δ by a single bivalent ligand, but they are also in harmony with the interaction of the bivalent ligands with an opioid receptor and an accessory binding site. In another paper Portoghese et al. investigated whether one or two pharmacophores are required for the κ opioid receptor selectivity of the bivalent opioid antagonist norbinaltorphimine [15]. They suggested that the κ selectivity of this kind of molecules is derived from the portions of the second halves of these molecules in that they mimic key address components of dynorphin at κ opioid receptors. Larson et al. also examined the effect of structural modifications on the affinity of norBNI analogues for wild-type and mutant κ and μ opioid receptors expressed in COS-7 cells [16]. It is suggested that the antagonist pharmacophore is bound within this highly conserved region of the κ or mutant μ receptor and that an anionic residue at the top of transmembrane helix 6 provides additional binding affinity. Lazarus et al. prepared analogues of the Dmt-Tic pharmacophore to test the hypothesis that a spacer and a third aromatic center in opioid peptides are required to convert a δ-antagonist into ligands with δ-agonist or with mixed δ-antagonist/μ-agonist properties [17, 18]. These data confirm that the distance between the Dmt-Tic pharmacophore and a third aromatic nucleus is an important criterion in converting Dmt-Tic from a highly potent δ-antagonist into a potent δ-agonist or into ligands with mixed δ- and μ-opioid properties. Grundt et al. start from the fact that the trans-(3,4)-dimethyl-4-(3-hydroxyphenyl)piperidines are a unique class of opioid antagonists that have recently provided selective antagonists for μ-opioid receptors and κ-opioid receptors. Molecular modeling indicated a strong structural similarity between the parent of this series and 2-amino-1,1-dimethyl-7-hydroxytetralin. Introduction of a methoxy group in the 3-position increased potency at μ and κ receptors, suggesting that this aminotetralin skeleton can be utilized as a new scaffold for the design of selective opioid receptor antagonists [19]. Balboni et al. developed a series of 17 analogues on the basis of the general formula H-Dmt-Tic-NH-*CH(R)-R′ (* denotes chirality; R charged, neutral, or aromatic functional group; R′ = –OH or –NH_2) [20]. Thus, these C-terminally extended analogues indicated that an amino acid residue containing a single charge, amino or guanidino functionality, or aromatic group, substantially altered the δ-opioid receptor activity profile (selectivity and antagonism) of the Dmt-Tic pharmacophore, which suggests that the C-terminal constituent plays a major role in determining opioid receptor activity as an address domain. Daniels et al. synthesized and evaluated bivalent ligands (KDAN series) containing δ-antagonist (naltrindole) and κ_1-agonist (ICI-199,441) pharmacophores [21]. The data suggested that KDAN-18 bridges phenotypic δ_2- and κ_1- receptors. They presented a conceptual model to explain the organizational differences between the opioid receptors that give rise to the phenotypes (δ_1, δ_2, κ_1, κ_2). Peng et al. synthesized a series of homo- and heterodimeric ligands containing κ agonist and μ agonist/antagonist pharmacophores joined by a linker chain of varying lengths. They were evaluated in vitro for their binding affinity at μ, δ, and κ opioid receptors. The functional activities of these compounds were measured in the [35S]-GTPγS binding assay [22]. The data suggest that the stereochemistry of the pharmacophores, the N-substituents of the pharmacophore, ester linkages, and the spacer length were crucial factors for optimal interactions of such ligands at opioid receptor binding sites. Agnes et al. designed a single peptide which can interact with δ and μ opioid receptors as agonists and with CCK receptors as antagonists [23]. These results provide evidence supporting the concept that opioid and CCK receptors have overlapping pharmacophores required for binding affinity and biological activity and that design of overlapping pharmacophores of two peptides into a single peptide is a valid approach. Balboni et al. analyzed the substitution of Gly with side-chain-protected or unprotected Lys in lead compounds containing the opioid pharmacophore Dmt-Tic [H-Dmt-Tic-Gly-NH-CH_2-Ph, μ agonist/δ antagonist; H-Dmt-Tic-Gly-NH-Ph, μ agonist/δ agonist; and H-Dmt-Tic-NH-CH_2-Bid, δ agonist (where Bid = 1H-benzimidazol-2-yl)] obtaining a new series of compounds endowed with distinct pharmacological activities [24]. The presence of a Lys linker provides new lead compounds in the formation of opioid peptidomimetics containing the Dmt-Tic pharmacophore with distinct agonist and/or antagonist properties. The change of biological activity, receptor binding, and selectivity upon changes in the molecular structure have been analyzed in several recent publications [25–28].

The abovementioned experimental work suggests very interesting and challenging hypotheses about opiate receptor binding and selectivity that need to be explained at the microscopic level. In our Laboratory, and using a model-based method, we have addressed this question through the analysis of the different modes of binding of molecules to μ, δ, and κ opiate receptors in order to generate a binding pharmacophore [29–31].

As, on the one hand, the model-based method has been perfected by the formal addition of new terms having definite chemical meaning and, on the other hand, we have written new computer codes to extract new and useful information from standard quantum chemical packages, we now have newer tools that are able to detect the microscopic factors regulating affinity and selectivity. This paper uses quantum chemical methods in an attempt to advance the knowledge of the relationship between electronic structure and the binding to μ, δ, and κ receptors in a series of high affinity opioid receptor ligands, whereby the phenolic OH group of nalbuphine, naltrexone methiodide, 6-desoxonaltrexone, hydromorphone, and naltrindole was replaced by a carboxamido group and the furan ring was opened to the corresponding 4-OH derivatives to the receptor.

2. Methods, Models, and Calculations

It was during the second half of the 1950 decade when the first applications of Molecular Orbital (MO) theory to the study of questions of pharmacological interest began to appear in the scientific literature [32–35]. Quantum chemistry provided the Linear Combination of Atomic Orbital approximation (LCAO) [36] and the work of Fukui et al. about reactivity and weak molecular interactions [37–39]. At that time progress was slow due to the lack of a reasonable theoretical framework within which to work and to the complexity and long calculation time of the electronic structures. This last barrier was overcome with the availability of faster and faster computers (calculations lasting about 24 hours or more in the 1960s are done today in a matter of seconds or minutes). This, together with the new all-valence electron semiempirical [40], and the *ab initio* Hartree-Fock (HF) [41] and Density Functional (DFT) [42] methods finally placed biologically important substances in the realm of feasible MO studies.

If compounds in a congeneric or a homologous series of compounds differ only by one substituent, it is not unreasonable to assume that the difference in their binding to a receptor site is due solely to differing properties of the substituents. Since there is no *a priori* or *a posteriori* knowledge of the particular atoms primarily responsible for the activity of a drug, we will require in principle that each atom of the drug interact with each atom of the drug receptor site. In such a case, if the model works, that is, can represent the data, it may give new insights into the mechanism of binding. We must note that the assumption implicit in this treatment is that receptor structure and mechanism of action remain constant while only the binding (affinity) varies throughout the series. In 1967 Klopman and Hudson published a general perturbation treatment of chemical reactivity, not restricted to π-conjugated molecules, in which allowance is made for ionic interactions [43–45]. As this theory represents the interaction energy in terms of atom-atom interactions it was only a question of time before the first papers applying it to the study of the activity of biological molecules appeared [46–48]. Given that, as far as we know, the last paper not belonging to our group, and using the model we are using here, was published in 1979 [49], we shall present in the following a detailed description of this model. We are doing so to distinguish it very clearly from the statistics-backed methodologies.

Let us consider the state of thermodynamic equilibrium, and a 1 : 1 stoichiometry in the formation of the drug-receptor complex:

$$D_i + R \rightleftharpoons D_iR, \tag{1}$$

where D_i is the drug, R is the receptor, and D_iR is the drug-receptor complex. According to statistical thermodynamics the equilibrium constant, K_i, is expressed as [50]

$$K_i = \frac{Q_{D_iR}}{Q_{D_i}Q_R}\exp\left(-\frac{\Delta\varepsilon_0^i}{kT}\right), \tag{2}$$

where $\Delta\varepsilon_0^i$ is the difference between the ground-state energy of D_iR and the energies of the ground states of D_i and R:

$$\Delta\varepsilon_0^i = \varepsilon_{D_iR} - \left(\varepsilon_{D_i} + \varepsilon_R\right) \tag{3}$$

and the Q's are the total partition functions (PF) measured from the ground state (in solution). T and k are the temperature and the Boltzmann constant, respectively. First of all, if we consider that for almost all polyatomic molecules the Boltzmann factors of the excited electronic states are negligible compared to those of the ground state, we may consider only the electronic ground state in the PF. Second, we shall consider that the rotational and vibrational motions are independent and uncoupled and that at body temperature, the vibrational PFs have a value close to 1 [51]. Third, we shall employ the classical expression for the rotational PF together with the assumption that the rotational PFs of the receptor and the drug receptor are similar (this requires that the receptor molecule be much greater than the drug molecule). In logarithmic form, (2) transforms into [52]

$$\log K_i = a + bM_{D_i} + c\log\left[\left(\frac{\sigma_{D_i}}{\mathrm{ABC}}\right)^{1/2}\right] + d\Delta\varepsilon_i, \tag{4}$$

where a, b, c, and d are constants, M is the drug's mass, σ its symmetry number, and ABC the product of the drug's moment of inertia about the three principal axes of rotation.

The interaction energy, $\Delta\varepsilon_i$, cannot be determined directly, either due to the size of the receptor or to the lack of knowledge of its molecular structure. Nevertheless, as we are dealing with a weak drug-receptor interaction, we can employ Perturbation Theory in the Klopman-Hudson form to evaluate $\Delta\varepsilon_i$. According to this method, the change in electron energy, ΔE, associated with the interaction of atoms i and j is

$$\Delta E = \sum_p \left[\frac{Q_iQ_j}{R_{ij}} + \left(\frac{1}{2}\right) \times \left(\beta_{ij}^2\right)\sum_m\sum_n \frac{D_{mi}D_{n'j}}{\left(E_m - E_{n'}\right)}\left(\frac{1}{2}\right) - \left(\beta_{ij}^2\right)\sum_m\sum_{\acute{n}} \frac{D_{m'i}D_{nj}}{\left(E_{m'} - E_n\right)}\right], \tag{5}$$

where Q_i is the net charge of atom i, D_{mi} is the orbital charge of atom i in the MO m, β_{ij} is the resonance integral, and E_m

$(E_{m'})$ is the energy of the m-th (m'-th) occupied (virtual) MO of the drug, with n and n' standing for the receptor. The value of β_{ij} is kept independent of the kind of AO because the drug-receptor complex does not involve covalent bonds. The summation on p is over all pairs of interacting atoms of the drug and the receptor.

The first term of the right side of (5) represents the electrostatic interaction between two atoms having net charges Q_i and Q_j. The second and third terms introduce the interactions between the occupied and empty MOs of the drug and those of the receptor.

As the MO energies of the receptor are not known, in the first applications of this method these values were replaced by constants [48, 49]. We followed a different approach. Noting that $1/(E_m - E_{n'})$ and similar terms can be written in the form $1/(1-x)$, we may expand them as a convergent infinite series. After a little algebra we obtain [53]

$$\begin{aligned}\Delta E = a &+ \sum_i \left[e_i Q_i + f_i S_i^E + s_i S_i^N\right] \\ &+ \sum_i \sum_m \left[h_i(m) F_i(m) + j_i(m) S_i^E(m)\right] \\ &+ \sum_i \sum_{m'} \left[r_i(m') F_i(m') + t_i(m') S_i^N(m')\right] + \Phi,\end{aligned} \quad (6)$$

where a, e, f, g, h, j, r, and t are constants and the summation on i is now only over the drug's atoms. S_i^E and S_i^N are, respectively, the total atomic electrophilic and nucleophilic superdelocalizabilities of Fukui et al. [54]. $F_{i,m}$ is the Fukui index of atom i in occupied (empty) MO m (m'). The total atomic electrophilic superdelocalizability (ESD) of atom i is defined as

$$S_i^E = \sum_m \frac{F_{i,m}}{E_m} = \sum_m S_i^E(m), \quad (7)$$

where the summation on m runs only over the occupied MOs. $S_i^E(m)$ is called the orbital electrophilic superdelocalizability of atom i at MO m.

The total atomic nucleophilic superdelocalizability (NSD) of atom i is defined as

$$S_i^N = \sum_{m'} \frac{F_{i,m'}}{E_{m'}} = \sum_{m'} S_i^N(m'), \quad (8)$$

where the summation on m' runs only over the empty MOs. $S_i^N(m')$ is called the orbital nucleophilic superdelocalizability of atom i at MO m'. Φ stands for the remaining series expansion terms. The summation on p (i.e., the atoms involved in the interaction) is hidden for the sake of clarity. S_i^E is associated with the total electron-donating capacity of atom i and S_i^N with its total electron-accepting capacity. These indices are very useful to compare the reactivity of a similar atomic position through a series of molecules because they incorporate the eigenvalue spectrum which is usually different in each molecular system. The orbital components, $S_i^E(m)$ and $S_i^N(m')$, become important when fine aspects of the drug-receptor interaction are needed for a more complete explanation.

The most important feature of (6) *is that it includes terms belonging only to the drug molecule.* During the year 2011 we analyzed several terms composing Φ and, after long and tedious algebra, we found that we may associate some of them with a set of local atomic reactivity indices recently proposed by one of us [55]. These indices are defined within the Hartree-Fock LCAO-MO framework as follow.

The local atomic electronic chemical potential of atom i, μi:

$$\mu_i = \frac{E_{\rm oc}^* - E_{\rm em}^*}{2}, \quad (9)$$

where $E_{\rm oc}^*$ is the upper occupied MO with a nonzero Fukui index and $E_{\rm em}^*$ is the lowest empty MO with a non-zero Fukui index.

The local atomic hardness of atom i, η_i:

$$\eta_i = E_{\rm em}^* - E_{\rm oc}^*. \quad (10)$$

The local atomic softness of atom i, ς_i, defined as the inverse of the local atomic hardness.

The local electrophilic index of atom i, ω_i:

$$\omega_i = \frac{\mu_i^2}{2\eta_i}. \quad (11)$$

The maximal amount of electronic charge that an electrophile may accept, $Q_i^{\max}$:

$$Q_i^{\max} = \frac{-\mu_i}{\eta}. \quad (12)$$

What is the relationship between these local atomic indices and the global indices obtained within Density Functional Theory? [56–70]. Let us analyze a couple of examples. DFT defines the global (molecular) electronic chemical potential, μ, and the global hardness, η, as (Koopman's theorem was used)

$$\mu = \frac{E_{\rm HOMO} + E_{\rm LUMO}}{2}, \quad (13)$$

$$\eta = E_{\rm LUMO} - E_{\rm HOMO}. \quad (14)$$

Here, $E_{\rm HOMO}$ and $E_{\rm LUMO}$ are, respectively, the eigenvalues of the higher occupied and lowest unoccupied MOs of the molecule. DFT chemical potential measures the escaping tendency of electrons from a system, so electrons flow from regions with higher chemical potential to areas with lower chemical potential until μ becomes uniform throughout [70]. Global hardness is just the HOMO-LUMO gap and can be interpreted as the resistance of the chemical potential to change in the number of electrons.

In the case of the local atomic electronic chemical potential μ_i (9), $E_{\rm occ}^*$ is the eigenvalue of the highest occupied MO that has a non-zero electron population on atom i. As Molecular Orbitals of large molecules are not localized over all the system, the value of μ_i will not be the same for all the atoms. The same definition holds for the local atomic hardness, the local atomic softness, the local electrophilic index, and the maximal amount of electronic charge that an electrophile may accept (10)–(12).

For this reason these local atomic indices are the local atomic analogues of similar global reactivity indices currently used in today's quantum chemistry [56–70] and can be interpreted in a similar way. The general meaning of these local atomic indices is one μ_i which is a measure of the tendency of an atom to gain or lose electrons; a large negative value indicates a good electron acceptor atom while a small negative value implies a good electron donor atom. The local atomic hardness can be interpreted as the resistance of an atom to exchange electrons with the environment. The local electrophilic index is associated with the electrophilic power and includes the tendency of the electrophile atom to receive extra electronic charge together with its resistance to exchange charge with the medium. This index can be viewed as a measure of the electrophilicity power because it is an analogous of the classical electrostatics power, $V2/R$, and μ and η serve the purpose of potential (V) and resistance (R), respectively [70].

The insertion of (6) and (9) to (12) into (4) leads to the following final equation:

$$\begin{aligned}\log K_i = {} & a + bM_{D_i} + c\log\left[\left(\frac{\sigma_{D_i}}{\mathrm{ABC}}\right)^{1/2}\right] \\ & + \sum_j \left[e_j Q_j + f_j S_j^E + s_j S_j^N\right] \\ & + \sum_j \sum_m \left[h_j(m) F_j(m) + x_j(m) S_j^E(m)\right] \\ & + \sum_j \sum_{m'} \left[r_j(m') F_j(m') + t_j(m') S_j^N(m')\right] \\ & + \sum_j \left[g_j \mu_j + k_j \eta_j + o_j \omega_j + z_j \varsigma_j + w_j Q_j^{\max}\right].\end{aligned} \qquad (15)$$

Then, for N ($i = 1, N$) molecules we have a set of simultaneous equations 15. In principle, this system of simultaneous equations holds for the atoms j of the molecule directly perturbed by their interaction with the receptor. Combined with the usual multiple-regression techniques, these equations can be usefully applied to estimate the relative variation of log K_i in the family of molecules analyzed. Also, they can be used to determine which atoms are directly concerned in the formation of the drug-receptor complex. Here statistical analysis is used, *not to see whether there is a structure-activity relationship, but to find the best one.*

On the other hand, the moment of inertia term was analyzed and it was possible to show that it can be expressed in a first approximation as [71–73]

$$\log\left[(\mathrm{ABC})^{-1/2}\right] = \sum_t \sum_t m_{i,t} R_{i,t}^2 = \sum_t O_t, \qquad (16)$$

where the summation over t is over the different substituents of the molecule, and $m_{i,t}$ is the mass of the ith atom belonging to the t-th substituent, $R_{i,t}$ being its distance to the atom to which the substituent is attached. This approximation allows us to transform a molecular property into a sum of substituent properties. As the physical interpretation of these terms it was proposed that they represent the fraction of molecules attaining the proper orientation to interact with the receptor. We have called them Orientation Parameters.

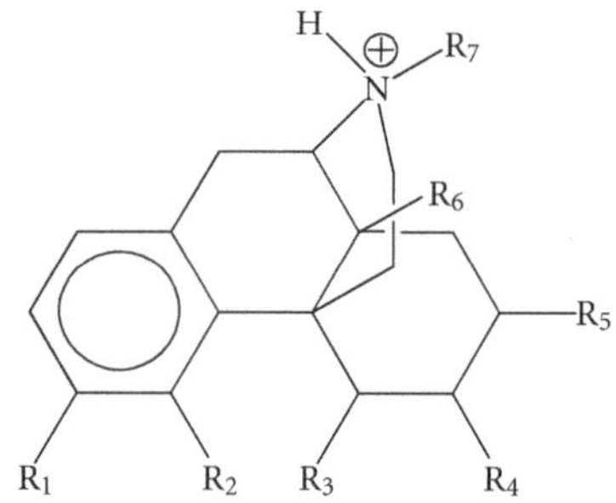

FIGURE 1: General formula of the molecules employed in this study.

The success of this method was appreciated when it was applied to a great variety of systems: molecules interacting with serotoninergic receptors [74–78], opiates interacting with opioid receptors [29–31], carbamate insecticides [72], molecules interacting with dopaminergic receptors [79], kynurenic acid derivatives interacting with Gly/NMDA sites [80], cannabinoid derivatives interacting with the CB1 and CB2 receptors [81, 82], the inhibition of wild-type and drug-resistant HTV-1 reverse transcriptase by some thiazolidenebenzenesulfonamide derivatives [83], and the interaction of some N,N-dialkyl-2-phenylindol-3-ylglyoxylamide derivatives with the peripheral benzodiazepine receptor [84]. Two results stress the goodness of this model. The first one was the successful prediction of the hallucinogenic activity of (±)-1-(2,5-dimethoxy-4-nitrophenyl)-2-aminopropane and the approximate dose humans could take [78]. The second one was when we could not obtain a statistically significant equation for the dopaminergic D_2 binding affinity of some apomorphines. We were able to show that the experimental results reported were wrong [79]. Recent results (accepted for publication) were obtained for relationships between accumulation data and molecular structure in a group of pollutant molecules (polychlorinated dibenzo-p-dioxins, polychlorinated dibenzofurans, and polychlorinated biphenyls) in Gold Rush, Black Beauty, and Patty Green zucchini subspecies. Because this accumulation process seems not to be a molecule-site equilibrium but a partition between two phases, and because there are early data showing that the logarithm of the octanol-water partition coefficient (a representation of lipophilicity) can be represented by the atomic indices used here [85–88], we conjecture that the method presented above can give better results than the Hansch or Hammett approaches.

The molecules selected for this study are shown in Figure 1 and Table 1, together with their corresponding experimental opioid receptor binding affinities [89].

Our experience has shown that many times one or more of the components of (15) may be constant or negligible, but there is no *a priori* way to say with certainty which of these components can be eliminated in a specific study. On

TABLE 1: Molecules and their experimental opioid receptor binding affinities[a,b].

Molecule	R_1	R_2	R_3	R_4	R_5	R_6	R_7	Exp. log μ	Exp. Log δ	Exp. log κ
1	$-CONH_2$	–OH	–H	=O	–H	–OH	$-CH_2$–c–Pr	−1.28	0.41	−0.63
2	$-CONH_2$	–O–		=O	–H	–OH	$-CH_2$–c–Pr	−0.14	2.74	1.04
3	–OH	–O–		=O	–H	–OH	$-CH_2$–c–Pr	−0.95	1.77	−0.72
4	$-CONH_2$	–OH	–H	α-OH	–H	–OH	$-CH_2$–c–Bu	−0.28	1.89	0.95
5	$-CONH_2$	OH	–H	–OH	–H	–OH	$-CH_2$–c–Bu	0.57	2.17	−0.33
6	–OH	–O–		–OH	–H	–OH	$-CH_2$–c–Bu	0.2	2.76	0.47
7	$-CONH_2$	–OH	–H	β-OH	–H	–OH	$-CH_2$–c–Bu	−1.14	0.59	−0.46
8	$-CONH_2$	–OH	–H	–H	–H	–OH	$-CH_2$–c–Pr	−0.79	0.62	−0.53
9	$-CONH_2$	–O–		–H	–H	–OH	$-CH_2$–c–Pr	0.39	2.79	1.2
10	–OH	–O–		–H	–H	–OH	$-CH_2$–c–Pr	−0.61	1.89	−0.61
11	$-CONH_2$	–OH	–H	=O	–H	–H	$-CH_3$	−0.52	0.91	0.36
12	$-CONH_2$	–O–		=O	–H	–H	$-CH_3$	0.07	2.41	1.27
13	–OH	–O–		=O	–H	–H	$-CH_3$	−0.55	1.57	0.44
14	$-CONH_2$	–OH	–H	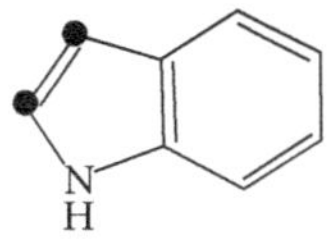		–H	$-CH_2$–c–Pr	−0.29	−1.6	−0.09
15	$-CONH_2$	–O–		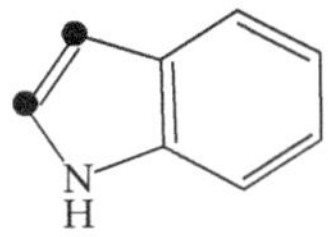		–OH	$-CH_2$–c–Pr	1.66	−0.52	1.67
16	–OH	–O–		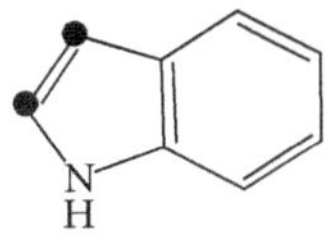		–OH	$-CH_2$–c–Pr	0.72	−0.85	0.46

[a]From [89]. [b]In molecules 15–17 the carbon atoms marked with black dots belong to the common skeleton.

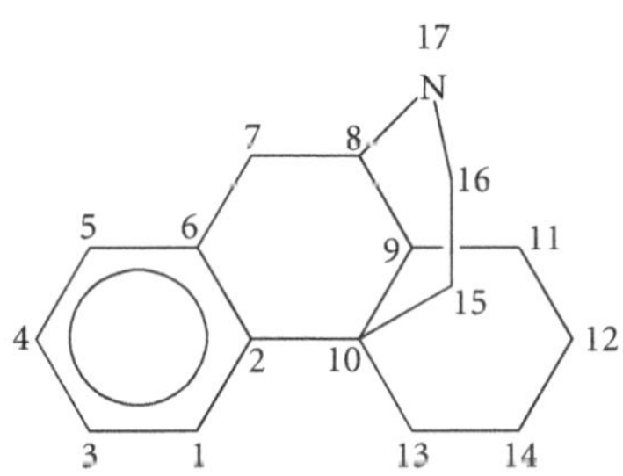

FIGURE 2: Common skeleton with atom numbering.

the other hand, regarding the drug atoms interacting with the receptor, the *common skeleton hypothesis* has given good results. This hypothesis states that there is a certain group of atoms, common to all molecules analyzed (called the common skeleton), that accounts for almost all the binding to the receptor. The action of the substituents consists in modifying the electronic structure of this skeleton (and influencing the correct alignment of the drug through the orientational parameters). For the case studied here the common skeleton is shown in Figure 2.

The obtention of the numerical values for the reactivity indices can be achieved at the semiempirical, HF or DFT levels of the theory. In our previous studies we have employed the semiempirical CNDO/2 and ZINDO/1 methods, *ab initio* HF and DFT methodologies. CNDO/2, HF, and DFT methods presented problems for the calculation of S^N (see (8)); CNDO/2, because the empty MO eigenvalues it provides are highly dependent on the conformation [90]. HF, DFT, and CNDO/2 methodologies provide empty MOs with negative and positive energy eigenvalues producing algebraic zeros around the Fermi Level and leading to bad results for the total NSD. The ZINDO/1 method is designed in such a way that the empty MOs eigenvalues are always positive for neutral molecules, thus avoiding this algebraic problem. For the sake of comparison we carried out the calculations at the semiempirical ZINDO/1 and DFT B3LYP/6-31G(d,p) levels. The molecules were analyzed in their protonated form. Full geometry optimization was carried out (with OPLS for ZINDO/1 and B3LYP/6-31G(d,p) for DFT). ZINDO/1 calculations were performed with the Hyperchem software [91] and DFT ones with the Gaussian package [92]. With software written in our laboratory all the necessary information was extracted from the above commercial software and the local atomic reactivity indices were calculated. For the calculation of the local atomic reactivity indices all electron populations lesser or equal to 0.01 e were considered as zero. In the case of the DFT calculations, negative electron populations arising from Mulliken Population Analysis were corrected according to a recently proposed method [93]. We need to stress that,

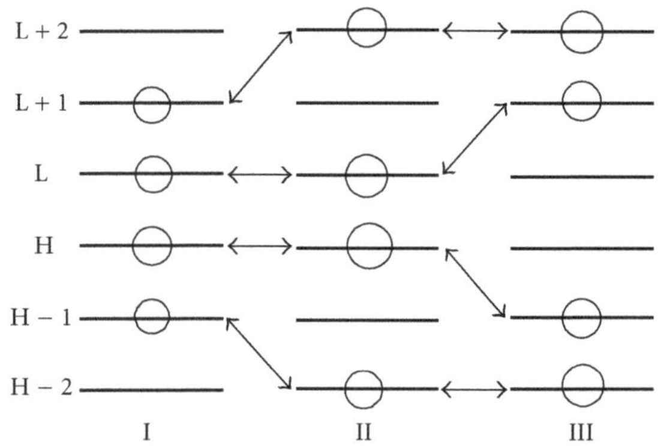

FIGURE 3: Atom *i* of molecules I, II, and III. Circles depict those MOs in which atom *i* has nonzero electron populations.

as we have shown during the last few years, the correct choice of more "primitive" methods (Extended Hückel Theory) can be very helpful in interpreting experimental results [[94–96] and references therein]. Linear multiple regression analysis was carried out with Statistica software [97]. The dependent variable is log IC_{50} and the independent variables are the set of local atomic reactivity indices plus the orientational parameters of the substituents R_1–R_8 (see Figure 1).

A last word about variable ordering: Figure 3 shows an equivalent atom for molecules I, II, and III, each one with three "occupied" (H, H − 1, H − 2) and three "empty" (L, L + 1, L + 2) MOs. The circle in one MO means that the corresponding Fukui index of atom i has a non-zero value for this particular MO (e.g., in molecule I MOs H − 1, H, L, and L + 1). As the drug-receptor interaction is achieved through the existing (i.e., with nonzero electron populations) MOs of atom *i*, the data to be entered in the regression matrix must be modified as shown in Figure 4. The same holds for the orbital superdelocalizabilities. This must be done for all atoms entering in the multiple regression analysis.

3. Results

3.1. μ Receptor Binding Affinity

3.1.1. ZINDO/1 Results. The best equation obtained was

$$\log \mu = 7.18(\pm 0.52) + 0.20(\pm 0.05)S_3^N(\text{LUMO}+1) - 16.22(\pm 1.13)\eta_6 + 1.85(\pm 0.37)Q_{13} + 0.81(\pm 0.16)S_{13}^N(\text{LUMO}+2), \quad (17)$$

with $n = 16$, $R = 0.98$, $R^2 = 0.97$, $R^2_{\text{adj}} = 0.96$, $F(4,11) = 89.10$ ($P < 0.000001$), outliers $> 2S = 0$, and SD $= 0.16$. Here S_3^N (LUMO + 1) is the contribution of atom 3 to the second empty MO with non-zero electronic population, η_6 is the hardness of atom 6, Q_{13} is the net charge of atom 13, and S_{13}^N (LUMO + 2), its contribution to the third empty MO with non-zero electronic population. Tables 2 and 3 show, respectively, the beta coefficients, the results of the *t*-test for significance of coefficients, and the matrix of squared correlation coefficients for the variables appearing in (17).

Table 4 shows the experimental and calculated values for $\log(\mu)$. Figure 5 displays calculated versus observed values.

3.1.2. B3LYP/6-31G(d,p) Results. For $\log(\mu)$ receptor binding affinity the best equation obtained is

$$\log \mu = -15.91(\pm 0.99) + 5.50(\pm 0.41)Q_{15}^{\max} + 49.75(\pm 4.71)\eta_5 - 29.43(\pm 2.92)F_{11}(\text{LUMO}) + 10.11(\pm 2.18)\omega_{13}, \quad (18)$$

with $n = 16$, $R = 0.99$, $R^2 = 0.98$, $R^2_{\text{adj}} = 0.97$, $F(4,11) = 113.62$ ($P < 0.00001$), outliers $> 2S = 0$, and SD $= 0.14$. Here $Q_{15}^{\max}$ is the maximal amount of electronic charge that atom 15 may accept, η_5 is the hardness of atom 5, F_{11} (LUMO) is the Fukui index of atom 11 at the first empty MO with nonzero electronic population and ω_{13} is the electrophilic index of atom 13. Tables 5 and 6 show, respectively, the beta coefficients, the results of the *t*-test for significance of coefficients and the matrix of squared correlation coefficients for the variables appearing in (18). Table 4 shows the experimental and calculated values for $\log(\mu)$. Figure 6 displays calculated versus observed values.

3.2. δ Receptor Binding Affinity

3.2.1. ZINDO/1 Results. When the best statistical equation was obtained, it was observed that in the case of molecule 6 the corresponding standard residual fell outside the ±2 sigma limit (2.18). When a new linear multiple regression analysis was carried out without molecule 6, the standard residual for molecule 10 fell outside the ±2 sigma limit (2.30). As our interest is to compare results for the whole set, no attempt was made to get equations for smaller sets. We consider that the variation of the actual ZINDO/1 variables is not able to account for the variation of $\log(\delta)$.

3.2.2. B3LYP/6-31G(d,p) Results. For the δ receptor binding affinity the best equation obtained is

$$\log \delta = 56.07(\pm 3.18) + 7.21(\pm 0.41)S_{12}^E - 6.82(\pm 1.48)F_{13}(\text{HOMO}) - 0.19(\pm 0.004) \times S_8^N(\text{LUMO}+1) - 0.78(\pm 0.21)S_{10}^N(\text{LUMO}+1), \quad (19)$$

with $n = 16$, $R = 0.99$, $R^2 = 0.98$, $R^2_{\text{adj}} = 0.97$, $F(4,11) = 107.48$ ($P < 0.000001$), outliers $> 2S = 0$ and SD $= 0.25$. Here S_{12}^E is the total atomic electrophilic superdelocalizability of atom 12, F_{13} (HOMO) is the Fukui index of atom 13 at the highest occupied MO with non-zero electronic population, S_8^N (LUMO + 1) is the orbital nucleophilic superdelocalizability of atom 8 at the second empty MO with non-zero electronic population, and S_{10}^N (LUMO + 1) is the orbital nucleophilic superdelocalizability of atom 10 at the second empty MO with non-zero electronic population. Tables 7 and 8 show, respectively, the beta coefficients, the

$$\left\{\begin{array}{lll} \text{I} & \text{II} & \text{III} \\ F_i(\text{L}+1)\ldots & F_i(\text{L}+1)\ \ldots & F_i(\text{L}+1) \\ F_i(\text{L})\ \cdots & F_i(\text{L})\ \cdots & F_i(\text{L}) \\ F_i(\text{H})\ \cdots & F_i(\text{H})\ \cdots & F_i(\text{H}) \\ F_i(\text{H}-1)\cdots & F_i(\text{H}-1)\cdots & F_i(\text{H}-1) \end{array}\right\} \longrightarrow \left\{\begin{array}{lll} \text{I} & \text{II} & \text{III} \\ F_i(\text{L}+1)\ldots & F_i(\text{L}+2)\ \ldots & F_i(\text{L}+2) \\ F_i(\text{L})\ \cdots & F_i(\text{L})\ \cdots & F_i(\text{L}+1) \\ F_i(\text{H})\ \ldots & F_i(\text{H})\ \cdots & F_i(\text{H}-1) \\ F_i(\text{H}-1)\cdots & F_i(\text{H}-2)\cdots & F_i(\text{H}-2) \end{array}\right\}$$

FIGURE 4: Left side: part of the original matrix data for atom i built from Figure 3. Right side: part of the final matrix data for atom i containing only nonzero values.

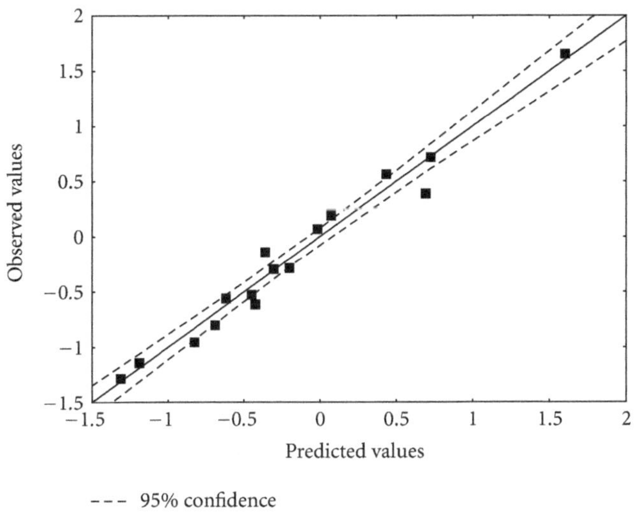

FIGURE 5: Observed and ZINDO/1 calculated values (17) of log(μ). Dashed lines denote the 95% confidence interval.

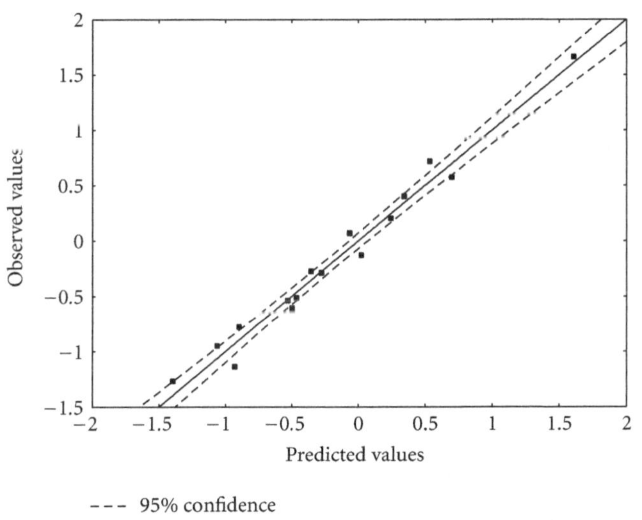

FIGURE 6: Observed and DFT calculated values (18) of log(μ). Dashed lines denote the 95% confidence interval.

results of the t-test for significance of coefficients, and the matrix of squared correlation coefficients for the variables appearing in (19). Table 9 shows the experimental and calculated values for log(δ). Figure 7 displays calculated versus observed values.

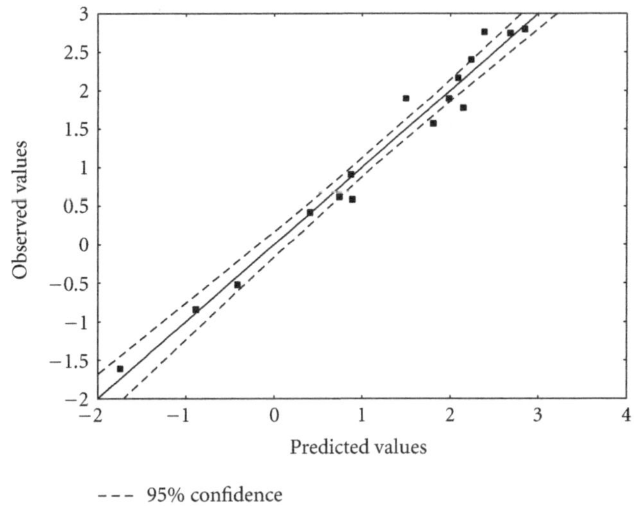

FIGURE 7: Observed and DFT calculated values (19) of log(δ). Dashed lines denote the 95% confidence interval.

3.3. κ Receptor Binding Affinity Results

3.3.1. ZINDO/1 Results. The best statistical equation obtained is

$$\begin{aligned} \log\kappa = & -8.33(\pm1.36) + 0.71(\pm0.18)S_7^N(\text{LUMO}+1) \\ & + 0.60(\pm0.14)S_7^N + 41.66(\pm9.10) \\ & \times F_{16}(\text{HOMO}-1) + 1.36(\pm0.45) \\ & \times S_{11}^N(\text{LUMO}+1) + 11.56(\pm4.02) \\ & \times S_{11}^E(\text{HOMO}), \end{aligned} \tag{20}$$

with $n = 16$, $R = 0.96$, $R^2 = 0.92$, $R^2_{\text{adj}} = 0.89$, $F(5,10) = 24.20$ ($P < 0.00003$), outliers $> 2S = 0$, and SD $= 0.27$. Here S_7^N (LUMO + 1) is the orbital nucleophilic superdelocalizability of atom 7 at the second empty MO with non-zero electronic population, S_7^N is the total atomic nucleophilic superdelocalizability of atom 7, F_{16} (HOMO − 1) is the Fukui index of atom 16 at the second occupied MO with non-zero electronic population, S_{11}^N (LUMO + 1) is the orbital nucleophilic superdelocalizability of atom 11 at the second empty MO with non-zero electronic population, and S_{11}^E (HOMO) is the orbital electrophilic superdelocalizability of atom 11 at the highest occupied MO with non-zero electronic population. Tables 10 and 11 show, respectively,

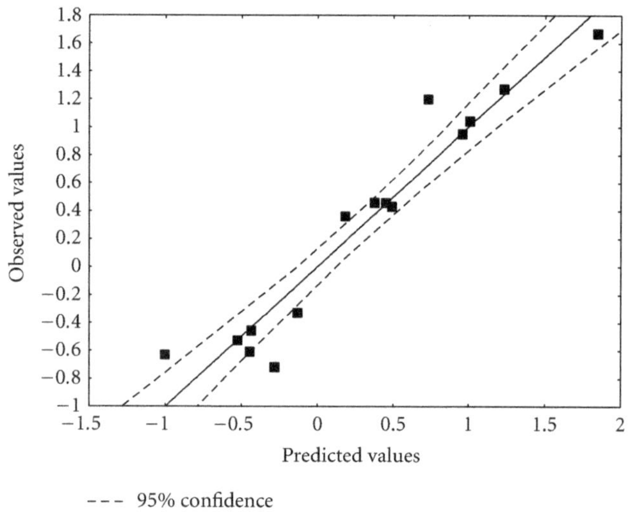

Figure 8: Observed versus ZINDO/1 calculated values (20) of $\log(\kappa)$. Dashed lines denote the 95% confidence interval.

Table 2: Beta coefficients and t-test for significance of coefficients in (17).

Variable	Beta coefficient	t	P
S_3^N(LUMO + 1)	0.24	3.93	<0.0023
η_6	−0.84	−14.39	<0.000001
Q_{13}	0.32	4.94	<0.0004
S_{13}^N(LUMO + 2)	0.31	5.16	<0.0003

Table 3: Squared correlation coefficients for the variables appearing in (17).

	S_{13}^N(LUMO + 1)	η_6	Q_{13}	S_{13}^N(LUMO + 2)
S_3^N(LUMO + 1)	1.0			
η_6	0.14	1.0		
Q_{13}	0.005	0.026	1.0	
S_{13}^N(LUMO + 2)	0.036	0.002	0.19	1.0

the beta coefficients, the results of the t-test for significance of coefficients and the matrix of squared correlation coefficients for the variables appearing in (20). Table 12 shows the experimental and calculated values for $\log(\kappa)$. Figure 8 displays calculated versus observed values.

3.3.2. B3LYP/6-31G(d,p) Results. For the κ receptor binding affinity the best equation obtained is:

$$\log \kappa = -3.22(\pm 0.71) + 27.44(\pm 6.61)\omega_{15} + 0.008(\pm 0.001)S_{11}^N - 1.46(\pm 0.32) \times S_1^E(\text{HOMO}-1) - 16.55(\pm 4.56)F_{11}(\text{LUMO}) \quad (21)$$

with $n = 16$, $R = 0.96$, $R^2 = 0.93$, $R^2_{adj} = 0.91$, $F(4,11) = 36.78$ ($P < 0.00001$), outliers $> 2S = 0$ and SD $= 0.24$. Here ω_{15} is the is the electrophilic index of atom 15, S_{11}^N is the total atomic nucleophilic superdelocalizability of atom 11, S_1^E(HOMO − 1) is the orbital electrophilic superdelocalizability of atom 1 at the second highest occupied MO with non-zero electronic population and F_{11} (LUMO) is the Fukui index of atom 11 at the first empty MO with non-zero electronic population. Tables 13 and 14 show, respectively,

Table 4: Observed and calculated log(mu) values.

Molecule	Exp log (μ)[a]	Calculated log (μ)[b]ZINDO/1	Calculated log (μ)[c]DFT
1	−1.28	−1.32	−1.39
2	−0.14	−0.37	0.03
3	−0.95	−0.83	−1.06
4	−0.28	−0.21	−0.35
5	0.57	0.43	0.70
6	0.2	0.07	0.25
7	−1.14	−1.19	−0.92
8	−0.79	−0.70	−0.89
9	0.39	0.69	0.35
10	−0.61	−0.43	−0.49
11	−0.52	−0.45	−0.46
12	0.07	−0.02	−0.06
13	−0.55	−0.63	−0.52
14	−0.29	−0.31	−0.28
15	1.66	1.60	1.61
16	0.72	0.72	0.54

[a]Reference [89]. [b]With (17). [c]With (18).

Table 5: Beta coefficients and t-test for significance of coefficients in (18).

Variable	Beta coefficient	t	P
Q_{15}^{max}	0.73	13.19	<0.000001
η_5	0.51	10.56	<0.000001
F_{11}(LUMO)	−0.54	−10.06	<0.000001
ω_{13}	0.29	4.65	<0.0007

Table 6: Squared correlation coefficients for the variables appearing in (18).

	η_5	F_{11}(LUMO)	ω_{13}	Q_{15}^{max}
η_5	1.0			
F_{11}(LUMO)	0.004	1.0		
ω_{13}	0.05	0.20	1.0	
Q_{15}^{max}	0.0006	0.02	0.25	1.0

Table 7: Beta coefficients and t-test for significance of coefficients in (19).

Variable	Beta coefficient	t	P
S_{12}^E	0.90	17.53	<0.0000001
F_{13}(HOMO)	−0.24	−4.63	<0.0007
S_8^N(LUMO + 1)	−0.29	−5.70	<0.0001
S_{10}^N(LUMO + 1)	−0.18	−3.66	<0.0038

Table 8: Squared correlation coefficients for the variables appearing in (19).

	F_8(HOMO − 2)	F_{10}(HOMO − 2)	S_{11}^E	μ_{12}
F_8(HOMO − 2)	1.0			
F_{10}(HOMO−2)	0.04	1.0		
S_{11}^E	0.06	0.03	1.0	
μ_{12}	0.02	0.04	0.06	1.0

Table 9: Observed and calculated log(delta) values.

Molecule	Exp log(delta)[a]	Calculated log(delta)[b] DFT
1	0.41	0.40
2	2.74	2.68
3	1.77	2.16
4	1.89	1.50
5	2.17	2.09
6	2.76	2.39
7	0.59	0.89
8	0.62	0.74
9	2.79	2.85
10	1.89	1.98
11	0.91	0.88
12	2.41	2.24
13	1.57	1.81
14	−1.6	−1.75
15	−0.52	−0.41
16	−0.85	−0.89

[a] Reference [89]. [b] With (19).

Table 10: Beta coefficients and t-test for significance of coefficients in (20).

Variable	Beta coefficient	t	P
S_7^N(LUMO + 1)	0.45	3.96	<0.0027
S_7^N	0.65	5.52	<0.0003
F_{16}(HOMO − 1)	0.45	4.55	<0.0011
S_{11}^N(LUMO + 1)	0.29	3.01	<0.0130
S_{11}^N(HOMO)	0.29	2.87	<0.0166

the beta coefficients, the results of the t-test for significance of coefficients, and the matrix of squared correlation coefficients for the variables appearing in (21). Table 12 shows the experimental and calculated values for log(κ). Figure 9 displays the calculated versus observed values.

4. Discussion

4.1. μ Receptor Results. The associated indices of (17) (ZINDO/1 results) and (18) (DFT results) show that both equations are statistically significant. Nevertheless, the results of the t-test for significance of the coefficients indicate that the DFT results are better. We note that DFT calculations provide a better representation of the electronic structure of molecules. Our previous studies were done at the CNDO/2 and ZINDO/1 levels of theory [29–31]. We cannot compare

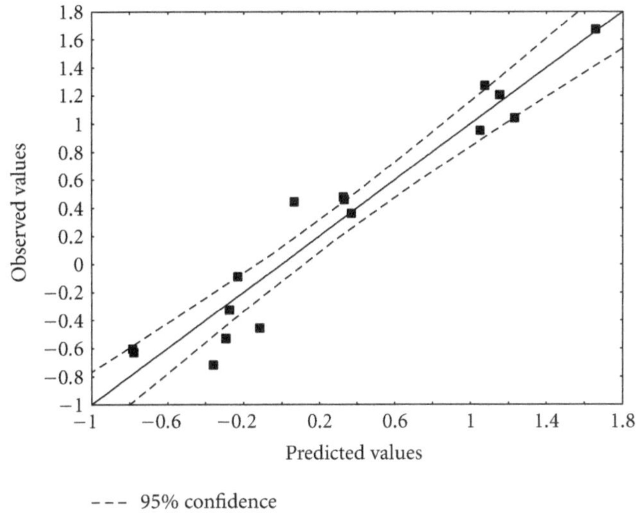

Figure 9: Observed versus DFT calculated values (21) of log(κ). Dashed lines denote the 95% confidence interval.

them with none of the results obtained here because the variable ordering, in the sense of Figures 3 and 4, was not carried out. We shall therefore only discuss the ZINDO/1 results for μ binding as an example of analysis.

The ZINDO/1 results show that the variation of log(μ) is associated with the variation of several local atomic reactivity indices located at atoms 3, 6, and 13 of the common skeleton (see Figure 2). Table 2 shows that the most important variable is the hardness (η) of atom 6. As η is positive, optimal binding is therefore associated with a high value of η_6, indicating that atom 6 resists exchanging electrons with the environment (the receptor) and that it is a bad electrophile. At this moment it is not clear to us what the exact nature is of the region of the receptor near this atom. The appearance of S_3^N (LUMO + 1) indicates that atom 3 faces an electron donor center of the μ receptor, probably a MO with π character. The appearance of the two variables for atom 13 is challenging. Table 3 shows that they are not correlated and the beta values of Table 2 indicate that their relative effects on log(μ) are similar. On one hand S_{13}^N (LUMO + 2) suggests that atom 13 receives charge from an occupied π MO of the receptor. On the other, the optimal value of Q_{13} should be negative. A possibility is that atom 13 binds the μ receptor in such a way that one side faces the receptor MO while another side interacts with a positively charged site. Figure 10 summarizes these ideas.

In the case of the DFT results (18) we find that the variation of the affinity is associated with the hardness of atom 5, with an interaction of an empty MO in atom 11 with an electron donor site in the receptor, with the electrophilicity of atom 13, and with the maximal amount of electron charge that atom 15 may accept as an electrophile (see Figure 2 for atom numbering). The most important index is the maximal amount of electron charge that atom 15 may accept (see Table 5). The value for the hardness of atom 5 should be low, implying that this atom should be a good electrophile. Atom 15 needs to have a low capacity

Table 11: Squared correlation coefficients for the variables appearing in (20).

	S_7^N	$S_7^N(\text{LUMO}+1)$	$S_{11}^E(\text{HOMO})$	$S_{11}^N(\text{LUMO}+1)$	$F_{16}(\text{HOMO}-1)$
S_7^N	1.0				
$S_7^N(\text{LUMO}+1)$	0.25	1.0			
$S_{11}^E(\text{HOMO})$	0.03	0.06	1.0		
$S_{11}^N(\text{LUMO}+1)$	0.0004	0.02	0.05	1.0	
$F_{16}(\text{HOMO}-1)$	0.16	0.04	0.008	0.04	1.0

Table 12: Observed and calculated log(κ) values.

Molecule	Exp log (κ)[a]	Calculated log (κ)[b] ZINDO/1	Calculated log (κ)[c] DFT
1	−0.63	−1.00	−0.78
2	1.04	1.01	1.23
3	−0.72	−0.28	−0.36
4	0.95	0.96	1.06
5	−0.33	−0.13	−0.27
6	0.47	0.45	0.33
7	−0.46	−0.43	−0.11
8	−0.53	−0.52	−0.29
9	1.2	0.73	1.16
10	−0.61	−0.44	−0.78
11	0.36	0.19	0.37
12	1.27	1.23	1.08
13	0.44	0.48	0.07
14	−0.09	0.03	−0.23
15	1.67	1.85	1.66
16	0.46	0.38	0.34

[a]Reference [89]. [b]With (20). [c]With (21).

to accept charge, which is expected for a CH_2 carbon atom. Atom 11 interacts through one of its empty MOs with an electron donor center of the receptor. Atom 13 should have low electrophilicity. As this index includes the tendency of atom 13 to receive extra electric charge together with its resistance to exchange charge with the medium we may say, in a first approach, that in this atom the resistance to exchange charge with the medium is higher than its tendency to receive charge. All these facts are summarized in Figure 11. It is interesting to note that μ binding seems to be charge and orbital controlled [43–45].

4.2. δ Receptor Results. The associated indices of (19) (DFT results) indicate that this equation is statistically significant. The most important index is the total atomic electrophilic superdelocalizability of atom 12 (see Table 7). We may see that atoms 8 and 10 act as electron acceptors through the second empty MO with non-zero (virtual) electron populations. Atoms 12 and 13 act as electron donors. Contrary to the DFT results for μ binding, δ binding involves Molecular Orbitals other than the HOMO and LUMO. Perhaps this is one of the mechanisms regulating not only affinity, but also selectivity. The proposed δ interaction pharmacophore is shown in Figure 12.

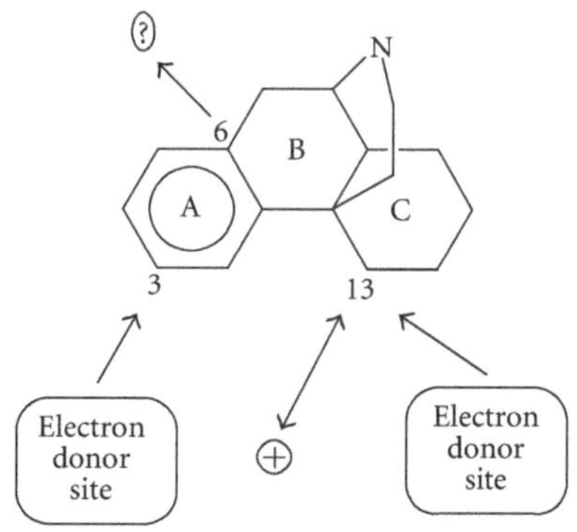

Figure 10: μ interaction pharmacophore from the ZINDO/1 results.

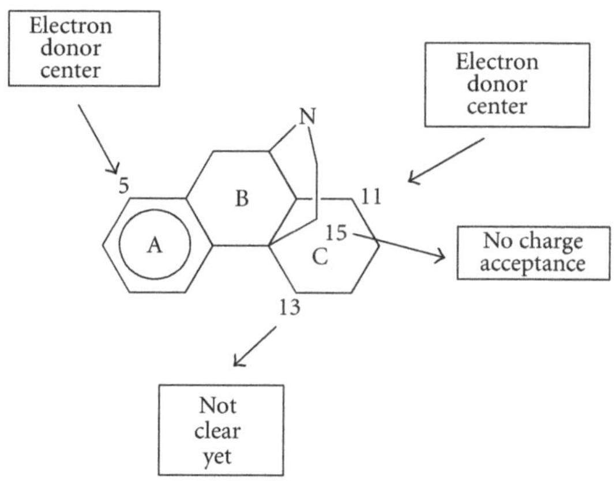

Figure 11: μ interaction pharmacophore from the DFT results.

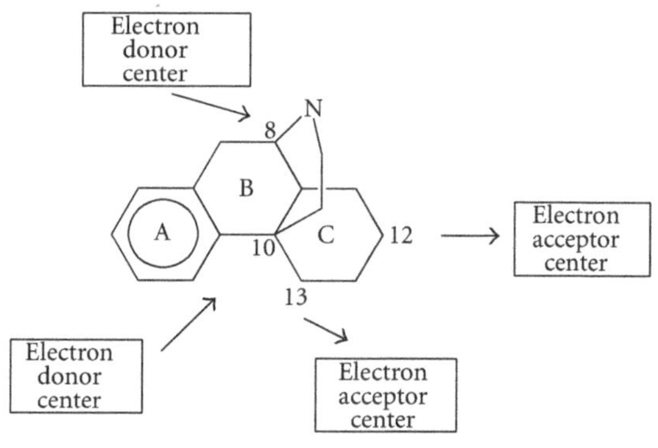

Figure 12: δ interaction pharmacophore from the DFT results.

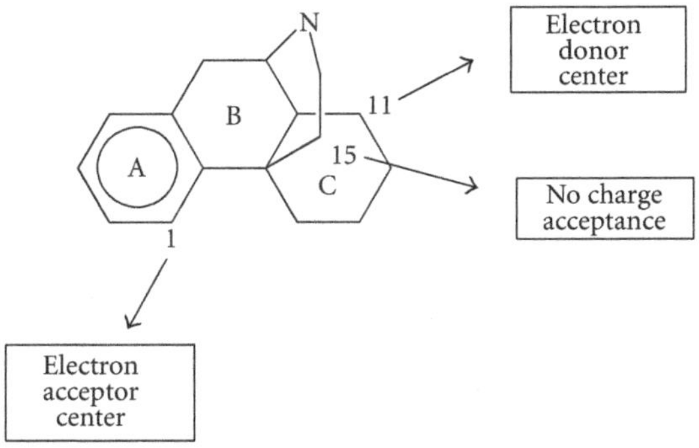

FIGURE 13: κ interaction pharmacophore from the DFT results.

TABLE 13: Beta coefficients and t-test for significance of coefficients in (21).

Variable	Beta coefficient	t	P
ω_{15}	0.44	4.15	<0.0008
S_{11}^{N}	0.51	6.34	<0.002
S_{1}^{E}(HOMO − 1)	−0.47	−4.51	<0.00006
F_{11}(LUMO)	−0.30	3.63	<0.004

TABLE 14: Squared correlation coefficients for the variables appearing in (21).

	S_{1}^{E}(HOMO − 1)	S_{11}^{N}	F_{11}(LUMO)	ω_{15}
S_{1}^{E}(HOMO − 1)	1.0			
S_{11}^{N}	0.008	1.0		
F_{11}(LUMO)	0.02	0.0004	1.0	
ω_{15}	0.41	0.004	0.05	1.0

4.3. κ Receptor Results. The associated indices of (21) (DFT results) indicate that this equation is statistically significant. Here, ω_{15}, S_{11}^{N} and S_{1}^{E} (HOMO − 1) have the same importance (see Table 13). Atom 15 should have low electrophilicity implying, as in the case of μ binding, that in this atom the resistance to exchange charge with the medium is higher than its tendency to receive charge. S_{11}^{N} and F_{11} (LUMO) indicate that atom 11 transfers charge to an electron-deficient center in the receptor. These two variables are not correlated (see Table 14). Atom 1 also transfers charge to an electron-deficient center. Figure 13 summarizes these ideas.

The appearance of local reactivity indices belonging to atom 15 is interesting because they can give an account of the difference in receptor affinity between (+) and (−) isomers. This fact needs to be supported by more studies like this one.

Figures 10–13 are planar representations of tridimensional interaction pharmacophores. What must be done to improve our knowledge? We need to apply this method to more experimental data. In this sense, as the IC_{50} values must be measured under almost the same experimental conditions, we are limited to papers reporting a large number of experimental results. IC_{50} values coming from different sources cannot be merged into one set [98]. Results with more extended sets of molecules may help to identify the exact nature of the receptor sites with which opiate molecules interact. Another line of research is to test this model for agonist and antagonist properties to see if it is possible to build the corresponding pharmacophores. The study of enkephalin analogues seems to be possible because there are several articles reporting binding data and pharmacological profiles for large sets of molecules.

The main conclusions of this work are as follows 1. some of the new local atomic reactivity indices of the extended model appear in the resulting statistical equations; 2. the interaction pharmacophores are different for μ, δ, and κ receptors and this difference may account for selectivity; 3. μ and κ interaction pharmacophores seem to share common features.

Aknowledgments

Prof. Dr. Bruce K. Cassels is gratefully acknowledged for very helpful comments. This work received financial help from AOG.

References

[1] F. A. Gorin and G. R. Marshall, "Proposal for the biologically active conformation of opiates and enkephalin," *Proceedings of the National Academy of Sciences of the United States of America*, vol. 74, no. 11, pp. 5179–5183, 1977.

[2] S. K. Burt, G. H. Loew, and G. M. Hashimoto, "Quantum chemical studies of molecular features and receptor interactions that modulate opiate agonist and antagonist activity," *Annals of the New York Academy of Sciences*, vol. 367, pp. 219–239, 1981.

[3] M. C. Fournie-Zaluski, G. Gacel, and B. Maigret, "Structural requirements for specific recognition of μ or δ opiate receptors," *Molecular Pharmacology*, vol. 20, no. 3, pp. 484–491, 1981.

[4] G. Loew, G. Hashimoto, and L. Williamson, "Conformational-energy studies of tetrapeptide opiates. Candidate active and inactive conformations," *Molecular Pharmacology*, vol. 22, no. 3, pp. 667–677, 1982.

[5] A. Lavecchia, G. Greco, E. Novellino, F. Vittorio, and G. Ronsisvalle, "Modeling of κ-opioid receptor/agonists interactions using pharmacophore-based and docking simulations," *Journal of Medicinal Chemistry*, vol. 43, no. 11, pp. 2124–2134, 2000.

[6] G. G. Bonner, P. Davis, D. Stropova et al., "Opiate aromatic pharmacophore structure-activity relationships in CTAP analogues determined by topographical bias, two-dimensional NMR, and biological activity assays," *Journal of Medicinal Chemistry*, vol. 43, no. 4, pp. 569–580, 2000.

[7] D. Bernard, A. Coop, and A. D. MacKerell Jr., "Conformationally sampled pharmacophore for peptidic δ opioid ligands," *Journal of Medicinal Chemistry*, vol. 48, no. 24, pp. 7773–7780, 2005.

[8] N. E. Kuz'mina, E. S. Osipova, V. S. Kuz'min, and V. B. Sitnikov, "A general model of the opiate pharmacophore 1. Regions of the opiate pharmacophore responsible for nonselective affinity for the opiate receptor," *Russian Chemical Bulletin*, vol. 55, no. 9, pp. 1523–1529, 2006.

[9] N. E. Kuz'mina, E. S. Osipova, V. S. Kuz'min, and V. B. Sitnikov, "General model of the opiate pharmacophore : 22. Regions of the opiate pharmacophore defining agonistic

properties of the opiate receptor ligands," *Russian Chemical Bulletin*, vol. 57, no. 6, pp. 1277–1284, 2008.

[10] N. E. Kuz'mina, E. S. Osipova, V. S. Kuz'min, and V. B. Sitnikov, "General model of the opiate pharmacophore : 33. Regions of the opiate pharmacophore determining antagonistic properties of the opiate receptor ligands," *Russian Chemical Bulletin*, vol. 57, no. 6, pp. 1285–1298, 2008.

[11] D. Bernard, A. Coop, and A. D. MacKerell, "Quantitative conformationally sampled pharmacophore for δ opioid ligands: reevaluation of hydrophobic moieties essential for biological activity," *Journal of Medicinal Chemistry*, vol. 50, no. 8, pp. 1799–1809, 2007.

[12] J. Shim, A. Coop, and A. D. MacKerell, "Consensus 3D model of μ-opioid receptor ligand efficacy based on a quantitative conformationally sampled pharmacophore," *Journal of Physical Chemistry B*, vol. 115, no. 22, pp. 7487–7496, 2011.

[13] A. W. Lipkowski, S. W. Tam, and P. S. Portoghese, "Peptides as receptor selectivity modulators of opiate pharmacophores," *Journal of Medicinal Chemistry*, vol. 29, no. 7, pp. 1222–1225, 1986.

[14] P. S. Portoghese, D. L. Larson, G. Ronsisvalle et al., "Hybrid bivalent ligands with opiate and enkephalin pharmacophores," *Journal of Medicinal Chemistry*, vol. 30, no. 11, pp. 1991–1994, 1987.

[15] P. S. Portoghese, H. Nagase, and A. E. Takemori, "Only one pharmacophore is required for the κ opioid antagonist selectivity of norbinaltorphimine," *Journal of Medicinal Chemistry*, vol. 31, no. 7, pp. 1344–1347, 1988.

[16] D. L. Larson, R. M. Jones, S. A. Hjorth, T. W. Schwartz, and P. S. Portoghese, "Binding of norbinaltorphimine (norBNI) congeners to wild-type and mutant mu and kappa opioid receptors: molecular recognition loci for the pharmacophore and address components of kappa antagonists," *Journal of Medicinal Chemistry*, vol. 43, no. 8, pp. 1573–1576, 2000.

[17] G. Balboni, R. Guerrini, S. Salvadori et al., "Evaluation of the Dmt-Tic pharmacophore: conversion of a potent δ-opioid receptor antagonist into a potent δ agonist and ligands with mixed properties," *Journal of Medicinal Chemistry*, vol. 45, no. 3, pp. 713–720, 2002.

[18] G. Balboni, S. Salvadori, R. Guerrini et al., "Potent δ-opioid receptor agonists containing the Dmt-Tic pharmacophore," *Journal of Medicinal Chemistry*, vol. 45, no. 25, pp. 5556–5563, 2002.

[19] P. Grundt, I. A. Williams, J. W. Lewis, and S. M. Husbands, "Identification of a new scaffold for opioid receptor antagonism based on the 2-amino-1,1-dimethyl-7-hydroxytetralin pharmacophore," *Journal of Medicinal Chemistry*, vol. 47, no. 21, pp. 5069–5075, 2004.

[20] G. Balboni, S. Salvadori, R. Guerrini et al., "Direct influence of C-terminally substituted amino acids in the Dmt-Tic pharmacophore on δ-opioid receptor selectivity and antagonism," *Journal of Medicinal Chemistry*, vol. 47, no. 16, pp. 4066–4071, 2004.

[21] D. J. Daniels, A. Kulkarni, Z. Xie, R. G. Bhushan, and P. S. Portoghese, "A bivalent ligand (KDAN-18) containing δ-antagonist and κ-agonist pharmacophores bridges δ_2 and κ_1 opioid receptor phenotype," *Journal of Medicinal Chemistry*, vol. 48, no. 6, pp. 1713–1716, 2005.

[22] X. Peng, B. I. Knapp, J. M. Bidlack, and J. L. Neumeyer, "Synthesis and preliminary in vitro investigation of bivalent ligands containing homo- and heterodimeric pharmacophores at μ, δ, and κ opioid receptors," *Journal of Medicinal Chemistry*, vol. 49, no. 1, pp. 256–262, 2006.

[23] R. S. Agnes, Y. S. Lee, P. Davis et al., "Structure-activity relationships of bifunctional peptides based on overlapping pharmacophores at opioid and cholecystokinin receptors," *Journal of Medicinal Chemistry*, vol. 49, no. 10, pp. 2868–2875, 2006.

[24] G. Balboni, V. Onnis, C. Congiu et al., "Effect of lysine at C-terminus of the Dmt-Tic opioid pharmacophore," *Journal of Medicinal Chemistry*, vol. 49, no. 18, pp. 5610–5617, 2006.

[25] G. Balboni, S. Fiorini, A. Baldisserotto et al., "Further studies on lead compounds containing the opioid pharmacophore Dmt-Tic," *Journal of Medicinal Chemistry*, vol. 51, no. 16, pp. 5109–5117, 2008.

[26] D. Kracht, E. Rack, D. Schepmann, R. Fröhlich, and B. Wünsch, "Stereoselective synthesis and structure-affinity relationships of bicyclic κ receptor agonists," *Organic and Biomolecular Chemistry*, vol. 8, no. 1, pp. 212–225, 2010.

[27] G. Balboni, S. Salvadori, C. Trapella et al., "Evolution of the bifunctional lead μ agonist/δ Antagonist containing the 2′,6′-dimethyl- l -tyrosine-1,2,3,4- tetrahydroisoquinoline-3-carboxylic acid (Dmt-Tic) opioid pharmacophore," *ACS Chemical Neuroscience*, vol. 1, no. 2, pp. 155–164, 2010.

[28] Y. Tang, J. Yang, M. M. Lunzer, M. D. Powers, and P. S. Portoghese, "A κ opioid pharmacophore becomes a spinally selective κ-δ Agonist when modified with a basic extender arm," *ACS Medicinal Chemistry Letters*, vol. 2, no. 1, pp. 7–10, 2011.

[29] J. S. Gomez-Jeria and P. Sotomayor, "Quantum chemical study of electronic structure and receptor binding in opiates," *Journal of Molecular Structure: THEOCHEM*, vol. 166, no. C, pp. 493–498, 1988.

[30] J. S. Gómez-Jeria, L. Lagos-Arancibia, and E. Sobarzo-Sánchez, "Theoretical study of the opioid receptor selectivity of some 7-arylidenenaltrexones," *Boletin de la Sociedad Chilena de Quimica*, vol. 48, no. 1, pp. 61–66, 2003.

[31] J. S. Gómez-Jeria, L. A. Gerli-Candia, and S. M. Hurtado, "A structure-affinity study of the opioid binding of some 3-substituted morphinans," *Journal of the Chilean Chemical Society*, vol. 49, no. 4, pp. 307–312, 2004.

[32] K. Fukui, C. Nagata, and T. Yonezawa, "Electronic structure and auxin activity of benzoic acid derivatives," *Journal of the American Chemical Society*, vol. 80, no. 9, pp. 2267–2270, 1958.

[33] O. Chalvet, R. Daudel, and C. Moser, "A note on the interaction of carcinogenic molecules with cellular protein," *Cancer Research*, vol. 18, no. 9, pp. 1033–1037, 1958.

[34] O. Chalvet and R. Daudel, "Contribution à l'étude du mechanism d'action des hydrocarbures cancérogènes," *Comptes Rendus de l'Académie des Sciences*, vol. 242, pp. 416–418, 1956.

[35] A. Pullman and B. Pullman, *Érisation par les Substances Chimiques et Structure Moléculaire*, Masson et Cie, Paris, France, 1955.

[36] C. C. J. Roothaan, "New developments in molecular orbital theory," *Reviews of Modern Physics*, vol. 23, no. 2, pp. 69–89, 1951.

[37] K. Fukui, T. Yonezawa, and H. Shingu, "A molecular orbital theory of reactivity in aromatic hydrocarbons," *The Journal of Chemical Physics*, vol. 20, no. 4, pp. 722–725, 1952.

[38] K. Fukui, T. Yonezawa, C. Nagata, and H. Shingu, "Molecular orbital theory of orientation in aromatic, heteroaromatic, and other conjugated molecules," *The Journal of Chemical Physics*, vol. 22, no. 8, pp. 1433–1442, 1954.

[39] K. Fukui, T. Yonezawa, and C. Nagata, "Interrelations of quantum-mechanical quantities concerning chemical reactivity of conjugated molecules," *The Journal of Chemical Physics*, vol. 26, no. 4, pp. 831–841, 1957.

[40] J. Sadlej, *Semiempirical Methods in Quantum Chemistry*, Ellis Horwood, New York, NY, USA, 1984.

[41] W. J. Hehre, L. Radom, P. von Schleyer, and J. Pople, *Ab Initio Molecular Orbital Theory*, Wiley Interscience, New York, NY, USA, 1986.

[42] R. G. Parr and Y. Weitao, *Density-Functional Theory of Atoms and Molecules*, Oxford University Press, New York, NY, USA, 1994.

[43] G. Klopman and R. F. Hudson, "Polyelectronic perturbation treatment of chemical reactivity," *Theoretica Chimica Acta*, vol. 8, no. 2, pp. 165–174, 1967.

[44] R. F. Hudson and G. Klopman, "A general perturbation treatment of chemical reactivity," *Tetrahedron Letters*, vol. 8, no. 12, pp. 1103–1108, 1967.

[45] G. Klopman, "Chemical reactivity and the concept of charge- and frontier-controlled reactions," *Journal of the American Chemical Society*, vol. 90, no. 2, pp. 223–234, 1968.

[46] A. Cammarata, "Some electronic factors in drug-receptor interactions," *Journal of Medicinal Chemistry*, vol. 11, no. 6, pp. 1111–1115, 1968.

[47] A. J. Wohl, "Electronic molecular pharmacology: the benzothiadiazine antihypertensive agents. II. Multiple regression analyses relating biological potency and electronic structure," *Molecular Pharmacology*, vol. 6, no. 3, pp. 195–205, 1970.

[48] F. Peradejordi, A. N. Martin, and A. Cammarata, "Quantum chemical approach to structure-activity relationships of tetracycline antibiotics," *Journal of Pharmaceutical Sciences*, vol. 60, no. 4, pp. 576–582, 1971.

[49] F. Tomás and J. M. Aulló, "Monoamine oxidase inhibition by β-carbolines: a quantum chemical approach," *Journal of Pharmaceutical Sciences*, vol. 68, no. 6, pp. 772–776, 1979.

[50] O. K. Rice, *Statistical Mechanics, Thermodynamics and Kinetics*, W. H. Freeman, San Francisco, Calif, USA, 1967.

[51] G. Herzberg, *Spectra and Molecular Structure. II. Infrared and Raman Spectra of Diatomic Molecules*, Van Nostrand, New York, NY, USA, 1945.

[52] J. S. Gómez-Jeria, "On some problems in quantum pharmacology. I. The partition functions," *International Journal of Quantum Chemistry*, vol. 23, pp. 1969–1972, 1983.

[53] J. S. Gómez-Jeria, "Modeling the drug-receptor interaction in quantum pharmacology," in *Physics, Chemistry and Biology*, J. Maruani, Ed., vol. IV, pp. 215–231, Kluwer Academic Publishers, Dordrecht, The Netherlands, 1989.

[54] K. Fukui, T. Yonezawa, and C. Nagata, "Theory of substitution in conjugated molecules," *Bulletin of the Chemical Society of Japan*, vol. 27, no. 7, pp. 423–427, 1954.

[55] J. S. Gómez-Jeria, *New reactivity indices in discrete and extended systems [Ph.D. thesis]*, Universidad Andrés Bello, Santiago, Chile, 2008.

[56] R. G. Parr and R. G. Pearson, "Absolute hardness: companion parameter to absolute electronegativity," *Journal of the American Chemical Society*, vol. 105, no. 26, pp. 7512–7516, 1983.

[57] R. G. Pearson, "Chemical hardness and bond dissociation energies," *Journal of the American Chemical Society*, vol. 110, no. 23, pp. 7684–7690, 1988.

[58] R. G. Pearson, "Absolute electronegativity and hardness: application to inorganic chemistry," *Inorganic Chemistry*, vol. 27, no. 4, pp. 734–740, 1988.

[59] Z. Zhou and R. G. Parr, "New measures of aromaticity: absolute hardness and relative hardness," *Journal of the American Chemical Society*, vol. 111, no. 19, pp. 7371–7379, 1989.

[60] R. G. Pearson, "Absolute electronegativity and hardness: applications to organic chemistry," *Journal of Organic Chemistry*, vol. 54, no. 6, pp. 1423–1430, 1989.

[61] R. G. Pearson, "Chemical hardness and the electronic chemical potential," *Inorganica Chimica Acta*, vol. 198–200, pp. 781–786, 1992.

[62] R. G. Pearson, "The principle of maximum hardness," *Accounts of Chemical Research*, vol. 26, no. 5, pp. 250–255, 1993.

[63] R. G. Parr, L. V. Szentpály, and S. Liu, "Electrophilicity index," *Journal of the American Chemical Society*, vol. 121, no. 9, pp. 1922–1924, 1999.

[64] R. K. Roy, V. Usha, J. Paulovič, and K. Hirao, "Are the local electrophilicity descriptors reliable indicators of global electrophilicity trends?" *Journal of Physical Chemistry A*, vol. 109, no. 20, pp. 4601–4606, 2005.

[65] R. G. Parr and W. Yang, "Density-functional theory of the electronic structure of molecules," *Annual Review of Physical Chemistry*, vol. 46, no. 1, pp. 701–728, 1995.

[66] H. Chermette, "Chemical reactivity indexes in density functional theory," *Journal of Computational Chemistry*, vol. 20, no. 1, pp. 129–154, 1999.

[67] P. Geerlings, F. De Proft, and W. Langenaeker, "Conceptual density functional theory," *Chemical Reviews*, vol. 103, no. 5, pp. 1793–1874, 2003.

[68] P. W. Ayers, J. S. M. Anderson, and L. J. Bartolotti, "Perturbative perspectives on the chemical reaction prediction problem," *International Journal of Quantum Chemistry*, vol. 101, no. 5, pp. 520–534, 2005.

[69] J. L. Gázquez, "Perspectives on the density functional theory of chemical reactivity," *Journal of the Mexican Chemical Society*, vol. 52, no. 1, pp. 3–10, 2008.

[70] S. B. Liu, "Conceptual density functional theory and some recent developments," *Acta Physico-Chimica Sinica*, vol. 25, no. 3, pp. 590–600, 2009.

[71] M. Ojeda-Vergara, *Chemistry [M.S. thesis]*, University of Chile, Faculty of Sciences, 1991.

[72] J. S. Gomez-Jeria, M. Ojeda-Vergara, and C. Donoso-Espinoza, "Quantum-chemical structure-activity relationships in carbamate insecticides," *Molecular Engineering*, vol. 5, no. 4, pp. 391–401, 1995.

[73] J. S. Gómez-Jeria and M. Ojeda-Vergara, "Parametrization of the orientational effects in the drug-receptor interaction," *Journal of the Chilean Chemical Society*, vol. 48, no. 4, pp. 119–124, 2003.

[74] J. S. Gomez-Jeria and D. R. Morales-Lagos, "Quantum chemical approach to the relationship between molecular structure and serotonin receptor binding affinity," *Journal of Pharmaceutical Sciences*, vol. 73, no. 12, pp. 1725–1728, 1984.

[75] J. S. Gómez-Jeria and D. Morales-Lagos, "The mode of binding of phenylalkylamines to the Serotonergic Receptor," in *QSAR in Design of Bioactive Drugs*, J. R. Prous, Ed., pp. 145–173, J. R. Prous International Publishers, Barcelona, Spain, 1984.

[76] J. S. Gómez-Jeria, D. Morales-Lagos, J. I. Rodriguez-Gatica, and J. C. Saavedra-Aguilar, "The relation between electronic structure and pA2 in a series of 5-substituted tryptamines," *International Journal of Quantum Chemistry*, vol. 28, pp. 421–428, 1985.

[77] J. S. Gomez-Jeria, D. Morales-Lagos, B. K. Cassels, and J. C. Saavedra-Aguilar, "Electronic structure and serotonin receptor binding affinity of 7-substituted tryptamines QSAR

of 7-substituted tryptamines," *Quantitative Structure-Activity Relationships*, vol. 5, no. 4, pp. 153–157, 1986.

[78] J. S. Gomez-Jeria, B. K. Cassels, and J. C. Saavedra-Aguilar, "A quantum-chemical and experimental study of the hallucinogen (±)-1-(2,5-dimethoxy-4-nitrophenyl)-2-aminopropane (DON)," *European Journal of Medicinal Chemistry*, vol. 22, no. 5, pp. 433–437, 1987.

[79] J. S. Gómez-Jeria and M. Ojeda-Vergara, "Electrostatic medium effects and formal quantum structure-activity relationships in apomorphines interacting with D_1 and D_2 dopamine receptors," *International Journal of Quantum Chemistry*, vol. 61, no. 6, pp. 997–1002, 1997.

[80] J. S. Gómez-Jeria and L. Lagos-Arancibia, "Quantum-chemical structure-affinity studies on kynurenic acid derivatives as Gly/NMDA receptor ligands," *International Journal of Quantum Chemistry*, vol. 71, no. 6, pp. 505–511, 1999.

[81] J. S. Gómez-Jeria, F. Soto-Morales, J. Rivas, and A. Sotomayor, "A theoretical structure-affinity relationship study of some cannabinoid derivatives," *Journal of the Chilean Chemical Society*, vol. 53, no. 1, pp. 1382–1388, 2008.

[82] J. S. Gómez-Jeria, F. Soto-Morales, and G. Larenas-Gutierrez, "A zindo/1 study of the cannabinoid-mediated inhibition of adenylyl cyclase," *Iranian International Journal of Science*, vol. 4, pp. 151–164, 2003.

[83] F. Soto-Morales and J. S. Gómez-Jeria, "A theoretical study of the inhibition of wild-type and drug-resistant HTV-1 reverse transcriptase by some thiazolidenebenzenesulfonamide derivatives," *Journal of the Chilean Chemical Society*, vol. 52, no. 3, pp. 1214–1219, 2007.

[84] J. S. Gómez-Jeria, "A DFT study of the relationships between electronic structure and peripheral benzodiazepine receptor affinity in a group of N,N-dialkyl-2- phenylind0l-3-ylglyoxylamides," *Journal of the Chilean Chemical Society*, vol. 55, no. 3, pp. 381–384, 2010.

[85] B. Duperray, "Corrélations quantitatives entre activité pharmacologique et paramètres physicochimiques," *Chimie Thérapeutique*, no. 4, pp. 305–334, 1971.

[86] A. Cammarata, "Some electronic factors in drug-receptor interactions," *Journal of Medicinal Chemistry*, vol. 11, no. 6, pp. 1111–1115, 1968.

[87] A. Cammarata, R. C. Allen, J. K. Seydel, and E. Wempe, "Cautions regarding the physical interpretation of statistically based structure-activity relationships," *Journal of Pharmaceutical Sciences*, vol. 59, no. 10, pp. 1496–1499, 1970.

[88] A. Cammarata, "Electronic representation of the lipophilic parameter, π," *Journal of Medicinal Chemistry*, vol. 14, no. 4, pp. 269–274, 1971.

[89] M. P. Wentland, R. Lou, Q. Lu et al., "Syntheses of novel high affinity ligands for opioid receptors," *Bioorganic and Medicinal Chemistry Letters*, vol. 19, no. 8, pp. 2289–2294, 2009.

[90] J. S. Gómez-Jeria, "Calculation of the nucleophilic superdelocalizability by the CNDO/2 method," *Journal of Pharmaceutical Sciences*, vol. 71, no. 12, pp. 1423–1424, 1982.

[91] Hypercube, 1115 NW 4th Street, Gainesville, Fl. 32601, USA.

[92] M. J. Frisch, G. W. Trucks, H. B. Schlegel et al., *Gaussian 98, Revision A.11.3*, Gaussian, Pittsburgh, Pa, USA, 2002.

[93] J. S. Gómez-Jeria, "An empirical way to correct some drawbacks of mulliken population analysis," *Journal of the Chilean Chemical Society*, vol. 54, no. 4, pp. 482–485, 2009.

[94] J. S. Gómez-Jeria, "Minimal molecular models for the study of nanostructures," *Journal of Computational and Theoretical Nanoscience*, vol. 6, no. 6, pp. 1361–1369, 2009.

[95] A. E. Aliaga, H. Ahumada, K. Sepúlveda et al., "Surface-enhanced Raman scattering, molecular dynamics and molecular orbital studies of the MRKDV immunostimulant peptide," *The Journal of Physical Chemistry C*, vol. 115, no. 10, pp. 3982–3989, 2011.

[96] C. Garrido, A. E. Aliaga, J. S. Gómez-Jeria, J. J. Cárcamo, E. Clavijo, and M. M. Campos-Vallette, "Interaction of the C-terminal peptide from pigeon cytochrome C with silver nanoparticles. A Raman, SERS and theoretical study," *Vibration Scopy*, vol. 61, pp. 94–98, 2012.

[97] StatSoft, *Statistica, v. 8. 0*, 2007.

[98] J. M. Boeynaems and E. Jacques Dumont, *Outlines of Receptor Theory*, Elsevier Science, 1980.

Permissions

The contributors of this book come from diverse backgrounds, making this book a truly international effort. This book will bring forth new frontiers with its revolutionizing research information and detailed analysis of the nascent developments around the world.

We would like to thank all the contributing authors for lending their expertise to make the book truly unique. They have played a crucial role in the development of this book. Without their invaluable contributions this book wouldn't have been possible. They have made vital efforts to compile up to date information on the varied aspects of this subject to make this book a valuable addition to the collection of many professionals and students.

This book was conceptualized with the vision of imparting up-to-date information and advanced data in this field. To ensure the same, a matchless editorial board was set up. Every individual on the board went through rigorous rounds of assessment to prove their worth. After which they invested a large part of their time researching and compiling the most relevant data for our readers. Conferences and sessions were held from time to time between the editorial board and the contributing authors to present the data in the most comprehensible form. The editorial team has worked tirelessly to provide valuable and valid information to help people across the globe.

Every chapter published in this book has been scrutinized by our experts. Their significance has been extensively debated. The topics covered herein carry significant findings which will fuel the growth of the discipline. They may even be implemented as practical applications or may be referred to as a beginning point for another development. Chapters in this book were first published by Hindawi Publishing Corporation; hereby published with permission under the Creative Commons Attribution License or equivalent.

The editorial board has been involved in producing this book since its inception. They have spent rigorous hours researching and exploring the diverse topics which have resulted in the successful publishing of this book. They have passed on their knowledge of decades through this book. To expedite this challenging task, the publisher supported the team at every step. A small team of assistant editors was also appointed to further simplify the editing procedure and attain best results for the readers.

Our editorial team has been hand-picked from every corner of the world. Their multi-ethnicity adds dynamic inputs to the discussions which result in innovative outcomes. These outcomes are then further discussed with the researchers and contributors who give their valuable feedback and opinion regarding the same. The feedback is then collaborated with the researches and they are edited in a comprehensive manner to aid the understanding of the subject.

Apart from the editorial board, the designing team has also invested a significant amount of their time in understanding the subject and creating the most relevant covers. They scrutinized every image to scout for the most suitable representation of the subject and create an appropriate cover for the book.

The publishing team has been involved in this book since its early stages. They were actively engaged in every process, be it collecting the data, connecting with the contributors or procuring relevant information. The team has been an ardent support to the editorial, designing and production team. Their endless efforts to recruit the best for this project, has resulted in the accomplishment of this book. They are a veteran in the field of academics and their pool of knowledge is as vast as their experience in printing. Their expertise and guidance has proved useful at every step. Their uncompromising quality standards have made this book an exceptional effort. Their encouragement from time to time has been an inspiration for everyone.

The publisher and the editorial board hope that this book will prove to be a valuable piece of knowledge for researchers, students, practitioners and scholars across the globe.

List of Contributors

Isaac Asusheyi Bello, George Iloegbulam Ndukwe and Joseph Olorunju Amupitan
Department of Chemistry, Ahmadu Bello University, Zaria 810001, Nigeria

Rachael Gbekele Ayo
Division of Agricultural Colleges, College of Agriculture, Ahmadu Bello University, Zaria 810001, Nigeria

Francis Oluwole Shode
School of Chemistry and Biochemistry, University of Zululand, Empangeni 3880, South Africa

Md. Irshad
Department of Biosciences, Jamia Millia Islamia, New Delhi 110025, India
School of Chemical Sciences, Central University of Gujarat, Gujarat, Gandhinagar 382030, India
Department of Medical Education, College of Medicine, King Saud University, Riyadh 11321, Saudi Arabia

Md. Zafaryab and M. Moshahid A. Rizvi
Department of Biosciences, Jamia Millia Islamia, New Delhi 110025, India

Man Singh
School of Chemical Sciences, Central University of Gujarat, Gujarat, Gandhinagar 382030, India

Neha Pandit and Birendra Shrivastava
Department of Pharmaceutical Chemistry, School of Pharmaceutical Sciences, Jaipur National University, Jagatpura-Jaipur, Rajasthan 302025, India

Rajeev K. Singla
Sadbhavna College of Management & Technology, Jalaldiwal, Ludhiana-Barnala State Highway-13, Raikot (Ludhiana), Punjab, India

Marina Roussaki, Anna-Maria Kypreou and Anastasia Detsi
Laboratory of Organic Chemistry, School of Chemical Engineering, National Technical University of Athens, Heroon Polytechniou 9, Zografou Campus, 15780 Athens, Greece

Sofia Costa Lima
Parasite Disease Group, IBMC-Institute for Molecular and Cell Biology, University of Porto, Rua do Campo Alegre, 823, 4150-180 Porto, Portugal

Panagiotis Kefalas
Department of Food Quality and Chemistry of Natural Products, Mediterranean Agronomic Institute of Chania, International Centre for Advanced Mediterranean Agronomic Studies, 73100 Chania, Crete, Greece

Anjna Bhatewara, Srinivasa Rao Jetti, Tanuja Kadre, Pradeep Paliwal and Shubha Jain
Laboratory of Heterocycles, School of Studies in Chemistry and Biochemistry, Vikram University, Ujjain, Madhya Pradesh 456010, India

Anabela Cordeiro da Silva
Parasite Disease Group, IBMC-Institute for Molecular and Cell Biology, University of Porto, Rua do Campo Alegre, 823, 4150-180 Porto, Portugal
Department of Biological Sciences, Faculty of Pharmacy, University of Porto, Rua de Jorge Viterbo Ferreiera, 228, 4050-313 Porto, Portugal

Thomas L. Shaak
Department of Integrative Life Sciences, Virginia Commonwealth University, Richmond, VA 23298, USA
Virginia Commonwealth University Reanimation Engineering Science Center (VCURES), Virginia Commonwealth University, Richmond, VA 23298, USA
Department of Biochemistry, Virginia Commonwealth University, 1101 E. Marshall Street, Sanger Hall, Room 2-004, Richmond, VA 23298, USA

Dayanjan S. Wijesinghe
Hunter Holmes McGuire VA Medical Center, Richmond, VA 23249, USA
Department of Biochemistry and Molecular Biology, Virginia Commonwealth University, Richmond, VA 23298, USA

Charles E. Chalfant
Hunter Holmes McGuire VA Medical Center, Richmond, VA 23249, USA
Department of Biochemistry and Molecular Biology, Virginia Commonwealth University, Richmond, VA 23298, USA
Massey Cancer Center, Richmond, VA 23298, USA

Kevin R. Ward
Michigan Critical Injury and Illness Research Center, Department of Emergency Medicine, University of Michigan, Ann Arbor, MI 48109, USA

Roger M. Loria
Virginia Commonwealth University Reanimation Engineering Science Center (VCURES), Virginia Commonwealth University, Richmond, VA 23298, USA
Department of Microbiology, Immunology, Pathology and Emergency Medicine, Virginia Commonwealth University, Richmond, VA 23298, USA

Robert F. Diegelmann
Virginia Commonwealth University Reanimation Engineering Science Center (VCURES), Virginia Commonwealth University, Richmond, VA 23298, USA
Department of Biochemistry and Molecular Biology, Virginia Commonwealth University, Richmond, VA 23298, USA

Helmut Schmidhammer and Mariana Spetea
Department of Pharmaceutical Chemistry, Institute of Pharmacy and Center for Molecular Biosciences Innsbruck (CMBI), University of Innsbruck, Innrain 52a, 6020 Innsbruck, Austria

Surendra N. Pandeya and Neha Rajput
Department of Pharmaceutical Sciences, Saroj Institute of Technology and Management, Sultanpur Road, Lucknow 226002, India

Korany A. Ali
Applied Organic Chemistry Department, National Research Centre, Dokki, Cairo 12622, Egypt

Mokhles M. Abd-Elzaher
Inorganic Chemistry Department, National Research Centre, Dokki, Cairo 12622, Egypt

Khaled Mahmoud
Department of Pharmacognosy, National Research Centre, Dokki, Cairo 12622, Egypt

Utpal Bora
Computational Biology Laboratory, Department of Biotechnology, Indian Institute of Technology Guwahati, Assam, Guwahati 781039, India
Biotech Hub, Centre for the Environment, Indian Institute of Technology Guwahati, Assam, Guwahati 781039, India

Anil Kumar
Computational Biology Laboratory, Department of Biotechnology, Indian Institute of Technology Guwahati, Assam, Guwahati 781039, India

Mario D. Aceto, Louis S. Harris, S. Stevens Negus, Matthew L. Banks and Larry D. Hughes
Department of Pharmacology and Toxicology, School of Medicine, Virginia Commonwealth University, Richmond, VA 23298, USA

Eyup Akgun and Philip S. Portoghese
Department of Medicinal Chemistry, College of Pharmacy, University of Minnesota, Minneapolis, MN 55455, USA

Bhawna Sharma, Sunil Prajapati and Upendra Kumar Sharma
Institute of Pharmacy, Bundelkhand University, Jhansi 284128, India

Amita Verma
Department of Pharmaceutical Sciences, Faculty of Health Sciences, Sam Higginbottom Institute of Agriculture, Technology and Sciences-Deemed University, Allahabad 211007, India

Olayinka O. Ajani and Johnbull O. Echeme
Department of Chemistry, Covenant University, Canaanland, P.M.B. 1023, Ogun State, Ota, Nigeria

Oluwole B. Familoni
Department of Chemistry, University of Lagos, Lagos State, Akoka 100001, Nigeria

Feipeng Wu
New Functional Polymeric Material Group, Technical Institute of Physics and Chemistry, Chinese Academy of Sciences (CAS), Beijing 100190, China

Zheng Sujiang
Test Center of Antimicrobial Materials, Technical Institute of Physics and Chemistry, Chinese Academy of Sciences (CAS), Beijing 100190, China

Stephane L. Raeppel, Frederic Gaudette and Arkadii Vaisburg
Department of Medicinal Chemistry, Methyl Gene Inc., 7150 Rue Frederick-Banting, Suite 200, Montreal, QC, Canada H4S 2A1

Hannah Nguyen
Department of Cell Biology and Pharmacology, Methyl Gene Inc., 7150 Rue Frederick-Banting, Suite 200, Montreal, QC, Canada H4S 2A1

Normand Beaulieu, Christiane Maroun and Jeffrey M. Besterman
Department of Cell Biology and Pharmacology, Methyl Gene Inc., 7150 Rue Frederick-Banting, Suite 200, Montreal, QC, Canada

James Wang
Department of PK/Analytical Chemistry, Methyl Gene Inc., 7150 Rue Frederick-Banting, Suite 200, Montreal, QC, Canada

Paramjit Kaur, Vikas Sharma and Vipin Kumar
Institute of Pharmaceutical Sciences, Kurukshetra University, Haryana, Kurukshetra 136119, India

Deepu Bakasta
Department of Biotechnology, PES Institute of Technology, Bangalore 560068, India

M. G. Shambhu
Department of Biotechnology, The Oxford College of Engineering, Bangalore 560068, India

Yusuke Sasaki and Akihiro Ambo
Department of Biochemistry, Tohoku Pharmaceutical University, 4-4-1 Komatsushima, Aoba-ku, Sendai 981-8558, Japan

Naoyuki Asakawa, Seiichi Kobayashi and Junichi Goto
Science and Technology Systems Division, Computational Science Department, Ryoka Systems Inc., 1-28-38 Shinkawa, Chuo-ku, Tokyo 104-0033, Japan

Noriaki Hirayama
Basic Medical Science and Molecular Medicine, Tokai University School of Medicine, 143 Shimokasuya, Isehara, Kanagawa 259-1143, Japan

Seranthimata Samshuddin, Badiadka Narayana and Divya N. Shetty
Department of Studies in Chemistry, Mangalore University, Mangalagangotri, Karnataka, Mangalore 574 199, India

Balladka Kunhanna Sarojini
Research Department of Chemistry, P.A. College of Engineering, Nadupadavu, Karnataka, Mangalore 574153, India

Nalilu Suchetha Kumari
Department of Biochemistry, K. S. Hegde Medical Academy, Deralakatte 574162, India

Tamara Bruna-Larenas and Juan S. Gomez-Jeria
Department of Chemistry, Faculty of Sciences, University of Chile, P.O. Box 653, Santiago, Chile